RURAL ECONOMY,

IN ITS RELATIONS WITH

CHEMISTRY, PHYSICS, AND METEOROLOGY;

OR,

CHEMISTRY APPLIED TO AGRICULTURE.

BY

J. B. BOUSSINGAULT,

MEMBER OF THE INSTITUTE OF FRANCE, ETC., ETC.

TRANSLATED, WITH AN INTRODUCTION AND NOTES,

BY

GEORGE LAW, AGRICULTURIST.

NEW YORK:

D. APPLETON & CO., 200 BROADWAY.

PHILADELPHIA:

GEO. S. APPLETON, 164 CHESNUT-ST.

MDCCC L.

THE

AUTHOR'S PREFACE.

IN the work now published, I present the results of the inquiries in which I have been engaged for many years, and which were undertaken in the hope of throwing light upon various points of practical agriculture. My first idea was, to confine myself to the mere re-impression of the several papers which I had communicated from time to time to different periodicals, with the addition of a quantity of inedited matter which I had by me. But upon more mature consideration, I was led to believe that I should be doing a useful thing did I fill up the numerous gaps which must necessarily occur between papers published isolatedly, at dates more or less remote from one another, and treating of the most dissimilar subjects. I have thus been led, in addition to my own observations, to give those of numerous writers on almost every branch of agricultural science, being only careful to confine myself in each instance to the most authentic practical conclusions; for it is certain, that practical data have the most direct interest for rural economy, both in itself and in its bearings upon the general science of political economy.

I have a further reason for the plan which I have adopted, which I am the less disposed to pass by in silence, inasmuch as it may plead in excuse with those who might be disposed to criticize the tone, perhaps somewhat too didactic, of my work.

I was invited, in conjunction with several other professors attached to a great educational institution, to give a course embracing my views upon the science of agriculture. I consented to this, and prepared my lectures; but motives, to which I was entirely a stranger, having prevented the project from being carried out, I made up my mind to publish, not the lectures such as I should have delivered them, but the documents which would have formed the basis of my teaching.

The first part of this work treats in succession of the physical and chemical phenomena of vegetation ; of the composition of vegetables and their immediate principles ; of fermentation ; and of soils. The second comprises a summary of all that has yet been done on the subject of manures, organic and mineral ; a discussion of the subject of rotations ; general views of the maintenance and economy of live-stock ; finally, some considerations on meteorology and climate, and on the relations between organized beings and the atmosphere.

I have endeavored, therefore, to give a summary view of all the questions of rural economy that admit of scientific treatment. It may be found, perhaps, that the number of these questions is still extremely small. Nevertheless, in regarding the multitude of inquiries that have been instituted within a very few years only, in viewing especially the ever-increasing interest attached to researches bearing upon practical agriculture, we are bound to anticipate progress, and to hope for conclusions important as regards science, profitable to practice, and useful to humanity.

EDITOR'S INTRODUCTION

THE following work is submitted to the agricultural public in the fullest confidence that it stands in need of no recommendatory strictures on the part of those who have undertaken to present it in its present form to the English agriculturist. In the person of its distinguished author the man of science is happily associated with the practical farmer—the accomplished naturalist, the profound chemist and natural philosopher. The friend and fellow-laborer of Arago, Biot, Dumas, and all the leading minds of his age and country—M. Boussingault's title to consideration is recognised wherever letters and civilization have extended their influence.

Surely the collected and carefully recorded experience of such a man, experience by which the conclusions of the *member of the Institute* have been tested and weighed by the results of *the farmer of Bechelbronn*, must have value in the estimation of every educated mind, and cannot fail to be especially welcome to that class of readers who are professionally engaged in the practical application of that noble science which his labors have contributed to illustrate and advance.

When the following pages were confided to the editor, it was the impression both of the publisher and himself, that in the course of the work many points would necessarily arise demanding elucidation, others calculated to provoke controversy or challenge investigation, and others again which could be rendered available or instructive to the British agriculturist only by means of copious explanation, showing with what modification and under what circumstances the views advanced might be applicable to the art as exercised in the climate and soil of this country. But the minute and analytical perusal indispensable in the operation of investing the Author's thoughts and expressions with an English garb, has demonstrated the fallacy of this impression, and if possible has augmented the admiration of the untiring patience, the vast experience, and the astute, circumstantial, and elaborate accuracy of the accomplished Author, in whose researches the reader will find the profoundest sagacity, combined with a childlike simplicity which communicates to his work a charm not necessarily inherent in the subject.

This is not intended to imply an unqualified approval of the illustrious philosopher's manner of dealing with his own facts and observations; still less of his style of writing, which is often wandering and diffuse, and which, in order to render it presentable to the English reader, has required much compression and retrenchment. Still, however, instead of having, as was expected, to pause at each step of the Author's progress, and dissert upon his views upon this or that particular branch of his subject, the observations of the commentator must of necessity be restricted in a great degree to an indication of such parts of the work as in his judgment are the most valuable and instructive, together with such incidental objections as appear to be of sufficient importance to require stating at length.

The chemical portion of the work is of inestimable value and conducted with consummate skill and knowledge; and with a minuteness and accuracy perfectly unexampled. At the same time the results of the writer's researches, as well as the means and processes by which these results were obtained, are displayed with such absolute perspicuity as to be intelligible and instructive to every agricultural inquirer, however superficial his previous acquaintance may be with the details of chemical science.

Nothing from the pen of the Editor could throw additional light upon the Author's brilliant and most interesting elucidation of vegetable physiology: his exposition is at once masterly and complete, and contains much that is both valuable and new. And even when the novelty of the facts which he adduces, or the originality of the inferences deduced and unfolded may admit of question, they are still deserving of the most respectful attention from the new and striking lights in which he places them, and presents them to the agricultural reader, and the clear and convincing way in which he demonstrates their inter-dependency and their most intimate connection with many of the most important pecuniary and professional interests of the cultivator. Every intelligent farmer will find his account not merely in a repeated perusal of this portion of the work, but in regarding it as a text-book and manual to be kept by him for permanent reference and consultation.

The arrangement of the subject, naturally and judiciously adopted by the writer, presents the consideration of soils as the first topic for the observations of the agricultural commentator; but on this head the distinguished author is so thoroughly explanatory and judicious, that nothing is left for the Editor but to approve, to acquiesce, and to recommend him with admiring confidence to the patient consideration and study of the intelligent inquirer.

At page 237 the subject of manures is taken up, and discussed with characteristic minuteness through many succeeding pages.

It may perhaps be objected, that the various theories respecting the origin, nature, efficacy, and relative nature of the different manures in use, as well as the various modes of their production, concoction, and application, which M. Boussingault has here collated and elucidated, contain nothing new; that they have, in fact, under one form or other, been long familiar to practical men; but without impugning the justness of this opinion, the Editor has long been convinced that the subject has received, generally, far less care and attention than it so eminently deserves; and, in short, that it is much neglected by many who are accounted not merely intelligent, but scientific agriculturists; and while admitting that much valuable information has been frequently given to the agricultural world by the repeated experiments of several enterprising individuals both in Scotland and England, he still most urgently recommends a careful study of this part of the work, which will probably lead the reader to the conclusion that the methods and practice recommended by the Author are, upon the whole, those best worthy of adoption. In page 260 will be found some very urgent warnings against what he (M. Boussingault) regards as the prevalent and pernicious custom of turning dung-heaps "frequently." If, however, by the term "frequently," a course not exceeding three complete turnings of the heap be comprehended, the Editor can by no means coincide with this opinion; a long experience having convinced him that there are many circumstances under which the Author's recommendations would be found not merely over-cautious, but positively injurious. For drill crops, for instance, when it happens that the farm-dung is somewhat rough, which must generally be the case towards the close of every season, when the animal dejections are scanty and the great bulk of the already ripened manure has been carried out upon the land, and the fresh additions have not had the advantage of being compounded with matter already concocted, an extra turning is very advantageous.

Every farmer will, of course, turn his heap once, for the purpose of thoroughly mixing the various ingredients and different qualities of manure which it contains; the extra turning, even admitting that it may to a certain extent promote the over-decomposition of the manure, and dissipate the ammoniacal principles which it is important to preserve, is not attended with so great a loss in this respect as that which is inevitable from keeping open the drill by the application of coarse dung, which cannot fail to be attended with a most serious loss of the more volatile principles, sometimes even laying the manure quite bare, and in the case of turnips, materially obstructing the operation of sowing.

Our Author brings forward the authority of several eminent inquirers in support of his own favorite view of the use of fresh or unfermented manure; but however plausible their theories may appear, and however just may be their views in the abstract, there are many intermitting circumstances connected with the general economy of a farm, which must govern and determine their adoption, and in which the practical cultivator must be guided by his own judgment alone.

To the Author's 6th chapter the reader may be advantageously referred, as containing a very full and valuable description and discussion, under the head of mineral manures, of the different varieties of the class usually denominated stimulants, and concluding with a brief but lucid and interesting account of WATER, considered as an agent of vegetation, and of its importance for manuring purposes.

The composition and preparation of liquid manures, as well as the various means of procuring and preserving them, will be found to have engaged much of the Author's attention; and he justly points to the rapidity of their ameliorating action as a peculiar excellence, not otherwise attainable; at the same time admitting that in the great majority of cases, the great and unavoidable expense attending their application, however moderate may be the prime-cost of the material, has always operated as an insuperable obstacle to their general adoption. In the justice of this vital objection, most practical agriculturists who have used them to any extent, will readily concur; and it will not be uninteresting to the reader to learn that there is reason to believe that it will henceforth be effectually obviated by the use of a very simple and convenient apparatus, devised by Mr. Smith of Deanston, a zealous and able friend of agriculture, who at the Highland Society's meeting at Glasgow in autumn last, explained the details of his contrivance; and the Editor has reason to suppose that the particulars will be given in a report of the proceedings of the meeting, in the forthcoming January publication of the Highland Society's Transactions.

The Editor is anxious to direct especial attention to the Author's 7th chapter, wherein he treats of the organic and inorganic manures, and of crops—of the elements of manures and of crops with their relations *inter se*, &c.—a section of the work which presents, in synopsis, a more copious and complete body of new, interesting, and important facts, of a nature more valuable to the practical farmer, than has ever been collected in any previous treatise on agricultural science. The great mass of this invaluable information is condensed, as it were, for practical reference, and displayed in copious and elaborate tabular *data*—a form which, though not attractive, has enabled the writer to comprise within succinct and manageable limits, a quantity of instruction which, in a more discursive shape, must have distended the work to double its actual size. The tables adverted to, present not merely the results of multifarious experiments in illustration of the important subject of rotation-cropping, but also these results as especially affected by the application of the various manures to which the several experimenters had recourse. The rotations reported may appear strange and curious, and sometimes, perhaps, even amusing to the farmers of England and Scotland; but not more so, in all probability, than those which are followed in many parts of our Island would appear to the cultivators of that part of Europe where our Author's agricultural speculations have been carried on, and where the bulk of his analyses have been obtained: indeed, *locality* and climate, and their inseparable concomitants, will in every country be found to prescribe and control the sorts of crops which may be rendered the most subservient to the permanent advantage both pecuniary and economical of the husbandman. Thus, with regard to the Author's more didactic observations and positive directions on the subject of rotations, there is no reason to doubt that, in relation to the soil, climate, and geographical position of the east of France, where his experimental course of rotations has been conducted, they are highly judicious, and have not been prescribed and required without mature consideration. Moreover, they are marked, like the deductions and inferences upon which they are founded, by his unusual acumen, patience, and sagacity; but in their application to the more circumscribed range of culture to which the agriculturist is limited in the ruder and more fickle climates of north and of south Britain, the practice of the cultivator must be governed mainly by his own judgment and experience in the circumstances by which he finds himself surrounded.

The interesting and ample instruction conveyed in the observations of this acute and profound observer upon the food and alimentary treatment of cattle of every species, accompanied as they are by minute details of the results obtained in the shape of organic and inorganic elements, cannot be too urgently recommended to the attentive consideration of every one interested in that important branch of rural economy to which they more particularly relate.

The Author's strictures comprehend the economy of the domestic animals with the exception of sheep, a subject from which he professedly abstains, for the very sufficient reason, that in his opinion, his opportunities of obtaining accurate information thereupon have not been sufficiently ample to enable him to discuss it with confidence and advantage. His theory in favor of the superior fattening quality of hay and the grasses in general above that which is found in tubers and roots, (though apparently supported by his usual convincing appeal to experiment,) will be received with considerable allowance by the practical farmer.

We have many instances, in the present day, of theories ably, plausibly, nay even satisfactorily established, which are nevertheless met by opposite results in practice; and the hesitation which the Editor ventures to intimate upon the particular point in question, will, he doubts not, be readily concurred in by many experienced feeders. It will be generally admitted that the boiled or steamed potato possesses a much higher nutritive value than belongs to it when in the raw state. In the former case, however, it requires to be mixed with some of the other roots which are not characterized by the same property, such as beet, turnips, &c.; the Swede, (Rutabaga,) or any of the harder sorts are best adapted for this purpose, and form a complete counteractive to the dangerous constipating tendency of the boiled potato when given alone.

There are many different substitutes or equivalents in the shape of mashes, containing leguminous ingredients which are admitted to be fully as nutritious as the potato, still there are circumstances connected with *market value* which render it a most valuable resource in farm alimentation. The popular notion that (when used as the feed of horses) the boiled or steamed potato is what is vulgarly called "soft meat," tending to unfit them for active work, is daily losing ground; for not only is it rapidly getting into more general use among the farmers of England and Scotland, but even postmasters are adopting it for horses employed in road work.

The meteorological section of the volume will be found no less instructive to the agriculturist than fascinating to the general reader;

no equally complete and extensive body of new and interesting facts has ever before been presented in a collected form to the agricultural world.

It will be observed that the capital, the all-important subject of Draining, as the great master-engine of agricultural improvement, is merely touched upon by our Author in a cursory way; should this incite a feeling of disappointment, it must be borne in mind that he has accomplished all, and more than all, that he proposed to himself, which was not to write a complete work on practical tillage, but rather, as his title implies, on "Rural Economy," *i. e.*, the economic production and application of the produce of the soil under the guidance of chemistry.

Among the faults of execution for which the Translator ventures to solicit the agricultural reader's indulgence, is the occasional adoption of terms which are rather French than English. Many of these words are, in the original, not merely technical, but local and provincial, and are not inserted in any of the dictionaries. Moreover, in the description of certain processes and operations, the Author has occasionally employed terms for which there is no English equivalent; and the Translator had frequently no other choice than that of either leaving the sense of the passage obscure and defective, or, on the other hand, of adopting the barbarisms in question, which not only deform the English of the construction, but cannot fail to be offensive to the taste and professional prepossessions of the agricultural reader.

With reference to the weights and measures made use of in the original, it may be proper to state, that (against strong temptation to let them stand as in the French, merely adding a table of equivalents) they have, at the instance of the Publisher, been reduced into their corresponding quantities in the English standard. *Grammes*, in the more delicate experiments, have been reduced into grains troy, assuming the gramme as equal to 15.438 grains; in less delicate experiments, grammes have been converted into pennyweights (dwts.) and ounces troy. *Kilogrammes* are given in lbs. avoirdupois; and where the quantity was large, they are often brought into tons, cwts., qrs., &c., taking the French kilogramme at 2.2 lbs. avoirdupois. The *Litre*, or present French measure of liquids, has been reduced into pints, calculating the French measure at 1.76 pints English imperial measure. The *Hectolitre* employed in measuring grain, is rendered into bushels, estimating it at 22 gallons English dry measure. The old French *Quintal* is also sometimes employed: this measure of weight has been either reduced to its proper corresponding quantity, 1 cwt. 3 qrs. 24 lbs. English, or where odd numbers might be disregarded, it has been called 2 cwts. The *Are*, or French superficial measure of quantity, has been calculated throughout at 120 square yards English: the *Hectare* at 2.4 acres English.

The labor of reducing these measures into their English equivalents has been immense; and errors, in spite of the best care which could be exerted, have doubtless in various instances crept into the reductions. Slight discrepancies between aggregate sums and their component quantities will also be apparent here and there, an inexactness which arises from the number of decimal places not having always been carried out far enough.

Our Author often quotes English agricultural writers, whose weights, &c., he has always been at the pains to reduce into their corresponding French equivalents. Not having at all times the works referred to at command, the Editor was compelled to bring back the French weight or measure into the corresponding English one by calculation. Thus from not knowing the precise equivalents adopted by M. Boussingault, some trivial discrepancy between the computed and the original weights, &c., may have resulted; but as the quantities that have been treated in this way are especially important as *relative*, scarcely ever as *absolute* quantities, the error where it occurs can be of no real consequence. *Metres*, centimetres, and millimetres have been reduced into English feet and inches, assuming the metre as equal to 39.370 inches. Finally, and to conclude our list of reductions, (would that it had been shorter!) the degrees of the centigrade thermometer have been brought into degrees of the only scale in familiar use among us, viz. Fahrenheit's.

In the translation the Editor has endeavored (not always with perfect success) to be as little technical as possible, with a view to the convenience of the general reader. In a very few places he has even ventured slightly to condense the style of the original in order to keep the volume within moderate dimensions, occasionally throwing the information contained in a table into the text or narrative; and where the Author appeared to him to be forgetting the rural economist in the mere chemist, as where for example he describes the special modes of preparing and purifying indigo, &c. he has made bold to retrench details, and give the results or conclusions only. All analyses bearing on the practical subject, whether it was the soil that produced, the crop that was grown, or the animal which fed on that crop, have been scrupulously retained. In conclusion, the reader is earnestly recommended to read an admirable little work, the joint production of Messrs. Dumas and Boussingault, entitled in the original, "Essai de Statique Chimique des Etres organisés," which has been presented in a clear English translation, under the title of, "An Essay on the Chemical and Physiological Balance of Organic Nature," and may be regarded as a most valuable introductory aid to the perfect comprehension of Boussingault's Philosophy of Agriculture, and as a key to the more scientific and technical portions of the work now submitted to the public.

CONTENTS.

CHAPTER I.

CHAPTER II.

CHAPTER III.

CHAPTER IV.

CHAPTER V.

CHAPTER VI.

CHAPTER VII.

CHAPTER VIII.

CHAPTER IX.

CHAPTER X.

RURAL ECONOMY.

CHAPTER I.

PHYSICAL PHENOMENA OF VEGETATION.—VEGETABLE PHYSIOLOGY.

THE operations of agriculture having for their object the production of plants which are either essential as food, or useful in the arts and industrial processes of man, it is well to begin with a summary view of the principal organs of which vegetables are composed; and by the instrumentality of which, under certain influences which we shall seek to appreciate, all the phenomena of their existence are manifested.

Plants fixed in the soil by their roots, live in the atmosphere by the concurrence of their green parts under the combined actions of light, heat, and moisture. We shall by and by ascertain at the cost of what elements, and under what conditions, their growth and complete development are accomplished.

The seed, which is the final result of vegetable life, and of which the aim is the reproduction and multiplication of the species, should first receive our attention. The seed is, if we may so speak, the starting point of all husbandry; it is with very few exceptions the first point on which the industry of the farmer exerts itself.

Nature, to ensure the preservation of seeds, has had recourse to infinite care and foresight, which are in some measure an assurance of their importance. The seed is often placed in the middle of an abundant fleshy pulp, which serves to afford it nourishment or manure at the time of its future development. Sometimes, as in leguminous plants, it is lodged between thick and tough membranes, or is covered with hard but flexible scales, as in the gramineous plants; or again it is enveloped in a woody substance of extreme hardness, as in stone fruits.

Nature does not show herself less provident in furnishing means for scattering seeds, and propagating vegetable species at great distances. There are, indeed, seeds which, furnished with light silky plumes or wings, flutter in the air, and are transported afar by the winds. Others, by means of a viscous, hard, impermeable envelope, float on rivers, and descend their courses without suffering the slightest change, or losing their germinating power. There are seeds again of a sufficiently coherent texture to resist the digestive action of the stomachs of animals that feed on the fruits which contain them, and which are consequently often found deposited at great distances from the plant which produced them; they are thus frequently dropped to germinate and flourish at the tops of the steepest mountains. By these admirable provisions of nature, then, the air, the water, and even animals themselves become the vehicles by which the migration of various vegetable species over the surface of the globe is effected.

We distinguish in seeds the kernel, and the integument which covers or encloses it. In the kernel, the embryo exists, which, as its name indicates, is destined to reproduce the plant of which the seed is the issue. The embryo is formed of several essential parts: —1st. of the radicle; 2d. of the gemmule, plumule, or rudiment of the stem, which by its extension engenders the organs that are to vegetate above the ground; 3d. of cotyledons, which form the greatest portion of the kernel, and which are destined to support the plant during the first periods of its existence.

In most cases, the cotyledons are formed of two lobes which separate during the act of germination. The plumule presents itself under the form of a little white point which penetrates into the interior of both cotyledons. The radicle is of a slightly conical shape, and is first seen when it projects externally from the seed.

The seeds of gramineous plants do not separate into two parts at the commencement of their independent existence. They are, in fact, seeds which have but a single cotyledon. As plants which spring from seeds of one or of more cotyledons present capital differences in their organization at large, and mode of development, botanists have established two grand divisions among them—into monocotyledonous vegetables, and dicotyledonous or polycotyledonous vegetables.

When the seed is gathered in its state of perfect maturity it is completely inert, its vital functions are wholly suspended, and it may be kept often for a very long time without being made to grow.

The length of time during which seeds may be kept, however, varies extremely, according to the species. There are plants, for instance, the seeds of which preserve for an indefinite period their germinative power; there are others, on the contrary, which lose it very speedily.

From various observations which appear to deserve every confidence, the seeds—

Of Tobacco have germinated after having been kept for	10	years.	
" Stramonium	25	"	(Duhamel.)
" the Sensitive plant	60	"	
" Wheat	100	"	(Pliny.)
" Wheat	10	"	(Duhamel.)
" Melons	41	"	(Friewald.)
" Cucumbers	17	"	(Roger Galen.)
" Haricots	33	"	
" Idem	100	"	(Gerardin.)
" Rape	17	"	(Lefèbure.)
" Rye	140	"	(Home.)

The seeds of the coffee plant are perhaps those which lose the property of germinating most speedily; planters are well aware that they must be sown almost immediately after they are taken from the bush. Oleaginous seeds are generally preserved with great difficulty; so also are those of rubiaceous plants, of the laurels, myrtles, &c.*

In practical agriculture there is always much advantage, and additional security, in sowing the most recent seed, even of kinds which are known to be the longest lived. It frequently happens, even after a very short time, that a certain proportion of these seeds die: they have, perhaps, not been gathered under circumstances favorable to their complete preservation. It is, therefore, only when he is compelled to do so, that the farmer trusts wheat to the ground which has been gathered in former years; and experience has proved that in using such seed, it is necessary to increase very considerably the quantity sown.

The inactivity of the seed ceases so soon as it is brought into contact with water and the air under the influence of a sufficient temperature. Sown in moist earth, a seed absorbs water, and swells considerably; the pellicle which covers it becomes distended, and ends by bursting; the radicle and the plumule appear, and become more and more distinct; the root penetrates the ground; the plumule by and by grows into a stalk which gets greener and greener, increases rapidly, and augments the number of its leaves, so that the young plant acquires vigor every day. At a certain period, flowers appear; and these are succeeded by fruit, the final term of which is the maturity of the seed. The phenomena of vegetation then cease. The whole of the organs of annual plants now wither and die; the work of reproduction, of multiplication, is accomplished. Thus begins and ends the existence of the plants which are the usual subjects of our husbandry.

With regard to biennial plants and trees, which possess more than this ephemeral existence, things pass differently. The plant vegetates so long as the temperature of the atmosphere and moisture of the soil are favorable to it: during the cold season the leaves fall, and the growth is suspended; but the plant revives anew on the return of spring. Those vegetables, the stem of which is generally ligneous, and whose roots penetrate deeply into the ground, have a power of resisting cold, and brave the rigors of the winter. In these latitudes, the renewal of the vegetation of trees in the spring presents an obvious analogy to the process of germination: the evolution of the buds represents this process very closely; and the phenomena at large, which we observe in annual plants, are for the major part reproduced:—there is increase of size in the stem and root, sprouting of leaves, inflorescence, ripening of fruits, production of seeds, and then suspension of function.

In the tropics, where the temperature is nearly uniform throughout the year, vegetation goes on without interruption; it only varies in its vigor, and this is determined by the abundance or the paucity

* Decandolle, Physiology, page 622

of rains and dews. The leaves which have concurred in the production of the fruit, and in perfecting the seed, fall as it were exhausted; but they are soon replaced, and their fall is only perceived by their presence on the surface of the ground.

The perfect plant, therefore, whether it be studied among annuals, or among trees that endure for a century, has analogous organs, destined to fulfil the same functions, to conduce to the same end—the reproduction of the seed. These organs, which we shall study in succession, are, 1st. The roots; 2d. The stem; 3d. The leaves; 4th. The appendages of the fructification.

When we follow the progress of a seed set in a proper soil, we observe that from their very first appearance the roots seek or tend towards the interior of the earth; the plumule, or young stem, on the contrary, takes a direction diametrically opposite; it grows vertically and seeks the air.

The lateral shoots in herbaceous plants, and the young branches of shrubs, form various angles with the principal stem or trunk. The first tendency of the branches is to rise vertically; but as they gain length and weight, they bend more or less downward, yielding to the power of gravitation. Mr. Knight showed, by a series of ingenious experiments, that the direction taken by the roots and branches is mainly due to this force.

This able observer arranged a wheel of wood in such a way that he could make it turn with different velocities in planes variously inclined to the horizon. The wheel, which was kept in motion by a stream of water, could be made to revolve vertically or horizontally at will.

A number of beans were planted upon the circumference of the wheel, in circumstances known to be indispensable to their germination and growth. By giving the wheel a sufficient velocity, it was easy to make the centrifugal force greater than the centripetal force. In Mr. Knight's apparatus, this happened when the wheel in the vertical plane performed one hundred and fifty revolutions in a minute. The whole of the radicles were then seen to turn their suckers beyond the circumference in lines which were prolongations of the radii of the wheel, and their growth took place in planes perpendicular to its axis.

The stems took a completely opposite direction, and after a time their summits attained the centre or axis of the wheel.

In causing the wheel to revolve in a horizontal plane, the same effects were still observed, when the rapidity of rotation was sufficient to annul the action of terrestrial gravitation. But when the motion was so far diminished, as merely to modify or to lessen the force of attraction, without entirely annulling it, the plant took a course comprised in a plane which formed a certain angle with the circumference of the wheel. With a certain velocity, the roots were inclined 10° below the horizontal plane in which the wheel moved, and the stems then formed an angle of the same magnitude above the same plane. The angle of deviation formed in this position of the wheel was always smaller in proportion as the rapidity of rotation was greater.

Now, since gravitation influences the position which vegetables present, as these beautiful experiments of Mr. Knight demonstrate, a practical conclusion which seems to follow from the fact is this, that the number of plants which may be placed upon a certain extent of soil, does not depend solely on the extent of surface; and that the power of production of a field which is very much sloped, does not exceed its horizontal projection. With regard to creeping plants, and with reference to meadows, it is clear that this principle is not rigorously exact: but in so far as plants with isolated stems are concerned, many agricultural philosophers, and among the number Davy,* have admitted it as perfectly indisputable. This opinion, as M. Corrard† has judiciously observed, is founded on the geometrical principle, which in itself is perfectly true, that an inclined plane cannot be cut by a greater number of vertical perpendiculars of a determinate thickness, than the horizontal plane which serves it for a base. Thus, says Corrard, as buildings which rest on an inclined plane are raised perpendicularly to the horizon, it has been concluded that an inclined plane can hold no larger an extent of building than would the horizontal plane which it covers; so that inclinations of surface do not actually add to the extent of towns. It is further a matter of absolute certainty, that as rain falls vertically, the quantity of water collected upon the eaves of a house is precisely the same as that which would be gauged in the same place upon a horizontal surface, equal to that of the building. But we should very much deceive ourselves, adds Corrard, if upon the same principle we inferred that on the surface of an inclined plane we could not have a tree more than upon the much smaller horizontal plane which serves as its base.

For although plants grow perpendicularly to the horizon, and may in this respect be considered as so many vertical perpendiculars or laminæ, still, from circumstances which are peculiar to them, we cannot here apply with propriety the geometrical principle in question. Because, to make the application exact, it were necessary to suppose that plants required no space around them to thrive, and that the whole surface of the ground might be covered with their stems without any space being left between them, and without this proximity interfering with their growth and vigor.

But such a supposition is impossible, inasmuch as it is absolutely necessary that plants should have a certain amount of space, both in the ground and in the atmosphere, in which to extend their roots and stretch forth their branches. Supposing, therefore, the inclined plane to be of considerably greater extent than the horizontal plane which supports it, it will necessarily afford to a larger number of plants, the spaces which their roots require for their growth and nourishment. In other words, upon the inclined surface there will be a larger quantity of vegetable earth, and more of the nutritious juices favorable to vegetation; and for these reasons the space which

* Agricultural Chemistry, vol. i.

† B. Corrard, Verhandel. von der Maatsch. te Haarlem, vol. xv. p. 308.

must always exist between plants may be less than on the horizontal plane. Consequently, all the conditions necessary to fertility being assumed as equal, the inclined plane will be capable of supporting a larger number of vegetables having vertical stems than the horizontal plane.

The organization of different parts of plants, so worthy in all respects of exercising the sagacity of physiologists, need not be made a subject of minute research in this place. Generalities suffice in our agricultural science. This organization, however complex it is in appearance, is probably much more simple than is usually believed; we might perchance find the proof of this simplicity in the readiness with which organs, the most dissimilar in their external forms and so different in their functions, undergo modification and transformation one into another, it might almost be said at the will of the observer. Thus tubers, those fleshy amylaceous bodies, which accumulate on the subterranean stems of certain vegetables, such as the potato, give birth to a plant which differs in nothing from that which would arise from the seed of the same vegetable. Certain leaves,—those of the orange, of the ficus elastica, &c., will do the same. Woody stems, branches severed from the tree and planted in the ground, produce roots and become independent plants. If the branches of certain shrubs be buried, and their roots be exposed to the air, these last are soon seen covered with buds and leaves; while the buried branches acquire a fibrous capillary structure, and in no great length of time they both present the appearance and exercise the functions of roots. This singular mutation readily succeeds with the willow, and it was upon this plant that the English vegetable physiologist, Woodward, effected it for the first time.*

The intimate structure of the roots, trunk, and branches, present considerable similarity. Divided perpendicularly to their longitudinal axis, three different zones, so dissimilar that it is impossible to confound them, are discovered in the different concentric layers which make up their mass; these are the BARK, the WOOD, and the PITH. A more careful examination shows that each of these zones may be further subdivided.

The exterior of the bark is covered by an extremely thin, nearly transparent and porous pellicle, formed by an assemblage of little adherent scales; this is the cuticle, or epidermis, which encloses the entire vegetable. As it is extensible within certain narrow limits only, it gives way and cracks in proportion as the body of the tree increases in size. The pores of the epidermis are minute openings or mouths which communicate with the exterior by an oval orifice, surrounded by a kind of contractile margin. It has been remarked that moisture tends to close these pores, and that drought and the action of solar light tend on the contrary to make them open. The chemical nature of the cuticle which covers the bark appears to indicate that it is destined to defend the plant against the too direct

* Davy's Agricultural Chemistry

action of external influences. In certain trees, the cuticle is covered with wax or resin. The most remarkable example of this kind, which can be quoted, is that of the wax-tree (ceroxilon andicola) which grows abundantly upon the slopes of the Andes. This tree, (a palm,) which attains a height of between 130 and 164 English feet, is covered over the whole surface of its trunk with a mixture of wax and resin.* In gramineous plants, the epidermis is almost entirely formed of silica. The bark of the birch-tree is covered with a pellicle of an unctuous nature, capable under the agency of nitric acid of yielding a peculiar suberic (the) acid.†

After the epidermis, in going from the circumference towards the centre, a layer of cellular tissue appears, which is designated by many physiologists under the name of the herbaceous envelope. In the cork oak, the cork represents the tissue by which the *liber* or true bark is covered, an organ formed of a vascular tissue, which with care can be separated into numerous very thin flakes or layers, which have been aptly compared to the leaves of a book.

The origin of the *liber*, or bark, is found in the most central part of the trunk; it is the result of the exudation of the woody parts, as Duhamel, with the same wonderful sagacity which characterizes all his works, has proved. Having cut away a portion of the bark of a tree in full vigor, and taken care to preserve the incision from contact with the air, he perceived that from the surface of the wood laid bare, and the edges of the bark adhering to it, a viscous matter exudes, which accumulates, acquires consistency, and ends by becoming cellular, thus regenerating the liber which had been taken away. Grew called this viscous secretion *cambium*, a title which it still retains. It is now generally admitted that cambium proceeds from the descending sap.

The liber is a very important organ in vegetables; we know for instance that it is necessary for the success of a graft that its liber penetrate or be penetrated by that of the tree on which it is grafted.

The woody layers are situated under the liber. Those which are at the greatest distance from the axis of the trunk, although they present the fibrous structure, and the principal characteristics of the woody tissue, still differ from it in being less hard and less tenacious; this zone, which at the first glance is easily distinguished from the wood properly so called, is the *alburnum*, the soft or false wood. Its fibres are much looser, and its color paler than that of the wood beneath it, the difference of shade being particularly apparent in the dye and deeply colored woods.

The alburnum becomes harder and tougher with age, and passes into the woody tissue, the *duramen* or hard wood, properly so called. The wood begins where the alburnum terminates, and reaches to the centre, to the *pith* or medullary canal.

In dicotyledonous trees, a certain quantity of wood is formed

* Boussingault, sur le Palmier à cire. Annales de Chimie et de Physique, 2e série, t. 59, p. 19.

† From the observations of M. Chevreul.

during vegetation at the expense of the alburnum; while on the opposite side towards the bark, the alburnum increases in about an equal proportion: so that in our climates, the alburnum grows each year from a new concentric layer; but in tropical countries, where the dicotyledonous trees vegetate without interruption, the annual concentric layers are scarcely perceptible. To prove the conversion of alburnum into woody tissue, Duhamel inserted a metallic wire into it in several places. At the end of a few years he found that the wire had become engaged in the proper woody layers.

The most central zone of the trunk or stem is traversed by the medullary canal or sheath, which is usually filled with the pith, a diaphanous spongy matter, consisting almost entirely of cellular tissue.

The pith sends ramifications towards the external parts of the trunk. Its use is not exactly determined; and notwithstanding the high purposes ascribed to it by some physiologists, we have many reasons for believing that its functions are not of great importance. Experiment proves, in fact, that the pith may be removed from young trees without killing them, without even stopping their growth. One of the least unquestionable offices assigned to the pith, is that of its being a reservoir for moisture with which it supplies the plant in times of drought, and when the ground does not furnish a sufficient quantity of water.

The internal structure and progressive development of the stem of monocotyledonous plants differ essentially from those which we have just been describing in connection with dicotyledonous plants.

If a perpendicular transverse section of the trunk of a palm-tree be examined, the same arrangement of zones which is observed in the dicotyledonous plants of our climates will not be perceived. The regions of the outer bark, of the liber or true bark, of the alburnum, and of the wood, forming so many concentric circles round a canal which is their common centre, are no longer distinguishable. The trunk of the palm-tree presents a more homogeneous appearance. The pith is disposed through the whole substance of the stem, and the woody tissue, presenting a fibrous texture disposed longitudinally, is found intimately mixed, or felted, as it were, with the medullary substance. The bark, if there be any, is always very indistinct; sometimes reduced to a simple epidermis, it is with difficulty distinguished from other parts of the trunk. In the beginning, and when it first appears above the ground, a palm-tree puts forth a system of leaves, the adhering extremities of which are attached in the same plane, and usually surround the neck of the root. At the second shoot, a system similar to the preceding one appears, which throws off the outside leaves, and interrupts their power of vegetating. These leaves wither, bend towards the earth and fall off, leaving a projecting circular ring on the stem, the only vestige of their existence. The same phenomenon takes place periodically. In the centre of the crown of leaves or branches, which terminates the palm-tree plant, a bud appears which is at first small and blanched;* but soon displays the most vigorous powers of vegetation. Its growth, inflorescence, and progress towards maturity are indicated by the decay and fall of the leaves which had hitherto protected it. The age of a palm-tree, or rather the number of times that it has fructified, or become crowned with fresh leaves, is calculated by the number of woody circles which are found marked on the stem. Its power of lasting seems to have no other limits than the resistance which the base offers to the weight it supports. In these colossal trees, a sensible diminution in the diameter of the stem is often perceptible towards the top, and in most of the species it is a fact equally well proved, that the fruit decreases in quantity when they have attained a certain epoch of their existence. In the cocoa-nut tree (*lodicea cocus nucifera*) the period of this decrease shows itself at about the age of thirty years, although this tree continues to bear for nearly a century.†

The leaves, the forms of which are so various, present however the greatest analogy in their organization: the green membranous substance of which they are almost entirely composed, is an extension of the parenchyma; the envelope which covers them answers to the epidermis.

It is in the leaves that the sap is subjected to the action of the atmosphere; it is there concentrated and peculiarly modified. According to the position of leaves upon the plant, their under sides, or those turned towards the ground, are distinguished from their upper sides which meet the light from above.

The upper side of the leaf is covered with a thick and frequently shining epidermis; this epidermis is sometimes endued with a substance rich in silicious matter, as in rushes. In the Steppes of South America I observed a tree, called Chapparal, the leaves of which are so highly silicious, that they are used for polishing metals. Generally speaking, the covering of the upper surface of leaves is a matter which is something of the nature of wax or resin. The epidermis which covers the lower surface is formed in most cases of a very thin, rough membrane, full of cavities and frequently covered with hairs or down.

The appearance and position of the leaves are not the same during the day and night. In the dark, simple leaves incline to fold up; in compound leaves, as in those of the acacia and sensitive plant, the same thing is still more marked; the effect can even be produced at will. If during the day a sensitive plant is placed in a dark room, the leaves immediately close; on lighting the room even with candles, they open again as if under the influence of the solar light;‡ Linnæus, who first paid attention to this class of phenomena, admits that plants in the absence of light experience a sort of sleep.

* This bud, in certain species of palm-trees, is sought after as food, and is often spoken of as the *cabbage* of the palm-tree.

† Information communicated by Mr. Codazzi. The trunk of certain species of palm-trees shows an enlargement towards the middle of its height, as in the *palma barrigona* of Choco.

‡ Observation of M. de Candolle.

The flower is the forerunner of the fruit, the fruit is the medium in the heart of which the seed is developed. The organs which constitute the flower are the calyx and the corolla, destined to support, nourish, and protect the pistil and the stamina, which are the essential parts; the calyx is a green membrane which surrounds the corolla, and in certain flowers replaces it.

The corolla is monopetalous or polypetalous according as it is composed of one or of several pieces The stamens occupy the interior of the corolla; they are terminated by summits of a vascular texture; these are the anthers; the powder which covers and sticks slightly to them is designated under the name of pollen.

The pistil placed in the middle of the flower is composed of the ovary, the style, and the stigma.

The ovary encloses the germ, the embryo of the seed; but this embryo is only developed by the action of the pollen. The style is in some sort the tubular prolongation of the ovary; it supports the stigma, which is the glandular part that receives the fecundating influence of the pollen.

From what has now been said, the pistil may be considered as the female organ of the flower, the stamens as the male organs.

Many flowers combine the organs of the two sexes. These flowers are hermaphrodites; those which only contain one organ, are called unisexual. Both male and female flowers are produced together on certain plants; in others, the flowers are all only of one sex, male or female. Polygamous plants are those which show a union of male and female flowers, or which have hermaphrodite flowers on the same stem.

In some flowers, the sexual organs at the period of fecundation acquire the property of motion, so as to facilitate this grand act. The stamens, for example, are seen in certain plants to approach the stigma, to deposite their pollen on it, and then to withdraw. It occasionally happens again that stamens, which are at first naturally in a position inclined with reference to the pistil, become suddenly straightened in such a way as to cast their pollen on the female organ, after which they resume their original position. In certain flowers a very considerable evolution of caloric has been perceived on the approach of the period of fecundation. In some arums, for example, the temperature has been observed to rise to 40° and even 50° cent. (104° to 122° Fahr.) It is probable that this phenomenon is quite general, and that it only varies in point of intensity.

Fecundation accomplished, the office of the flower is at an end. It collapses, withers, and dies. But the impregnated ovary enlarges by degrees, until it has attained maturity, when it presents two distinct parts, which by their union compose the fruit: these parts are the pericarp, and the seed—the husk or shell and the grain. The pericarp always surrounds the seed; but it sometimes happens that it is so thin and delicate that it blends with the seed.

The germination of seeds, the evolution of new plants, is only accomplished under certain physical conditions which demand our consideration.

We have already said incidentally, that in order that a seed may germinate, it must be in contact with moisture, have communication with the air, and be under the influence of a certain temperature. The same conditions continue to be indispensable after the seed has sprung, and the plant has been organized; and in addition the access of light is now imperative.

Roots seek in the soil the moisture which is requisite to vivify the whole vegetable. These organs are terminated by hair-like fibres of extreme delicacy, and having spongioles at their extremities: it is by these spongioles that absorption is effected. The following experiment is sufficient to prove that this is the case: let such a plant as a turnip be placed with the hair-like extremities of its root plunged in water, and the plant will continue to live, although almost the whole body of the root is in the air; let things be now so arranged that the hair-like filaments of the root are not in the water, but that the bulb or body of the plant is so: the leaves will soon droop and wither.

The force which brings into play the suction power of the roots, resides in almost every part of the plant: thus a root deprived of its spongioles, a stem, a branch, and a leaf, exert this suction power when plunged in water. But the absorption effected in this way has a limit, and we soon discover the necessity of making fresh sections of the extremities, which have no power of renovation like the filaments furnished with spongioles, which terminate a root.

We are still ignorant of the cause which produces the ascent of liquids in vegetables, and which carries them to the remotest leaves, in spite as it were of the laws of hydrostatics. We readily conceive how the spongioles of the roots, surrounded by earth abundantly charged with moisture, should imbibe by the simple effect of porosity. We can also understand how, after having been modified by the spongioles, the water and the principles contained in it should be transformed into sap; but the porosity of the extremities of the roots, and the chemical modification effected by the spongioles upon the fluid imbibed, give no kind of explanation of the rapid ascent of the sap. The force which occasions this rise is very considerable, as was demonstrated by Dr. Stephen Hales in a series of ingenious experiments more than a century ago.

Hales adapted a tube bent at a right angle and filled with water, to the extremity of the root of a pear-tree, the point of which had been cut off; the extremity of the tube opposite to that which was connected with the root dipped into a bath of mercury. In a few minutes a portion of the water contained in the tube was absorbed, and the mercury rose above the surface of the bath to the extent of eight inches. In the beginning of April, Hales cut off a vine stem at the distance of thirty-three inches from the ground. The stem had no lateral branches, and its cut surface, which was nearly circular, had a diameter of $\frac{7}{8}$ths of an inch. To this section, he adapted a reversed syphon: and things being so disposed, he poured in a quantity of mercury, which after a time, and from the effect of the pressure exerted by the sap as it escaped, rose in one of the arms of the syphon, and remained stationary at the height of thirty-eight inches above its original level. This column of mercury, it is obvious, represents a pressure very much greater than that of our atmosphere.

The ascent of the sap in trees takes place by the woody layers. It is easy to obtain conviction of this by making a plant absorb a watery solution of cochineal. By making sections in the stem at different heights, we can readily trace the colored liquid in its progress; it is undoubtedly the course which the natural sap would have taken. We see no indication of the coloring matter in the pith nor in the bark, the woody tissue alone is colored, sometimes entirely, but more generally in its younger parts only. The dyeing which results from this injection of the wood is in lines, and parallel with the axis of the trunk, like the woody fibres themselves; but in some cases the sap may deviate from the rectilinear course. Hales showed this by the following experiment: upon a tree he made four notches, one above the other; each notch occupied one quarter of the trunk and reached to its centre. In this way the whole of the woody fibres were cut through at different heights, so that to continue its ascent the sap must necessarily experience a series of lateral deviation, which in fact took place.

The ascending sap of vegetables, as it has hitherto been procured for examination, is an extremely watery fluid, holding in solution very small quantities of divers saline and organic substances. Having attained the leaves, the sap there undergoes modification, and becomes concentrated by losing water. It at the same time experiences, through the agency of the atmospheric air, under the influence of light, a great modification in its constitution. Thus elaborated, the sap takes a descending course; following the *liber*, it retrogrades towards the soil, and therefore performs a kind of circulation in its passage through the plant. The descending course of the sap is demonstrated by throwing a ligature round the trunk of a tree; after a time there is formed, above the part that is tied, an enlargement which is owing to the accumulation of the principles of the sap; but below it the tree experiences no increase. The descending course of the elaborated sap is no effect of simple gravity; because, if the ligature be thrown around a pendent branch, the enlargement still forms between the ligature and the free extremity of the branch. The descending sap passing through the cortical layers must necessarily contribute to their formation; and it is almost certain, as appears from the capital experiment of Duhamel, that it is the cambium which is changed into liber, and so concurs in the growth of trees. The concentration of the ascending sap, which occurs during its passage through the leaves, by the simple effect of evaporation, is the phenomenon which is spoken of under the name of the exhalation of plants: this exhalation of plants, it is easily understood, is favored by a high temperature, dryness, and motion of the air. In favorable circumstances, the water escapes in the state of vapor. Hales compared the watery exhalation of plants to the perspiration of animals, and made many experiments to ascertain the quantity of watery vapor which they usually throw off.

Hales planted a sun-flower in an air-tight vessel, the top of which was sealed hermetically by a leaden cover. This cover was pierced by two holes: one for the passage of the stem of the plant, the other for the introduction of the water necessary to its growth. For a fortnight the apparatus was regularly weighed, and our ingenious experimenter found that the green parts of the sun-flower threw off on an average about twenty ounces of water in twelve hours of the day. The evaporation was always increased during dry and warm weather; moist air lessened it; during the night season, the evaporation was sometimes no more than three ounces, and it occasionally happened that it was nil.

Vegetable life appears to be intimately connected with the phenomenon of evaporation. From the inquiries which I have myself undertaken on this subject, so well deserving the attention of observers, it would appear that a plant grows only when it transpires, and that in hindering this transpiration, we in fact arrest vegetation.

We now associate with the phenomenon of exhalation the source or accumulation of certain substances which are met with in considerable quantity in the organization of plants, although scarcely a trace of them can be detected in the water with which they are supplied. The water evaporating, leaves these substances behind; and as the mass of liquid imbibed by the roots and exhaled by the green parts is very considerable, it is easy to conceive how they should finally come to be present in rather large quantity.

A portion of the water which a plant in full vigor absorbs, must necessarily enter into its constitution; the water exhaled by the leaves, therefore, cannot equal the whole of that which has been absorbed by the roots. Sennebier endeavored to ascertain the relation which exists between the absorption and the exhalation, and he found in the particular case which he observed, that about $\frac{1}{3}$ of the water absorbed was fixed, and became a constituent part of the vegetable.

---

§ II.—CHEMICAL PHENOMENA OF VEGETATION.

The chemical phenomena of vegetation are accomplished by the concurrence of the elements of the atmosphere, of water, and of certain organic substances which exist as constituents of the soil.

The action of the atmosphere upon plants presents two phases perfectly distinct from one another; germination, and vegetation properly so called, which last comprises the development, the growth, and the multiplication of species.

GERMINATION.

We have ascertained that a seed, considered with reference to its organization, consists, 1st. of an embryo which includes the germs of the root and of the stem; and 2d. of the cotyledon, or cotyledons. Considered with reference to their chemical compositions, seeds exhibit a certain similarity of constitution. They contain: 1st. starch and gum; 2d. a highly azotized matter analogous to the caseum of milk and animal albumen; this is the matter which is commonly and very improperly designated under the name of gluten, and of vegetable albumen; 3d. a fatty or oily matter, rich in carbon and hydrogen. Seeds contain either fixed oils, such as hemp-seed, rape-seed, &c., or volatile oils, as aniseed, cummin-seed, &c. The different principles which are associated in the seeds vary considerably in their relative proportions: they also vary slightly in their nature. One seed, that of the colewort, for example, will contain more than forty per cent. of its weight of oily matter, while another, such as wheat, will only contain a few hundredths. Oats may contain ten or twelve per cent. of caseum or gluten; in certain varieties of wheat, analysis indicates a much larger quantity. The proportions of starch, gum, sugar, or mucilage do not vary less. It almost always happens that these different substances are found associated in the same seed; sometimes one predominates and the others only enter in very small proportion.

After burning, the ashes of seeds are always found composed of phosphates, sulphates, and alkaline and earthy chlorides. These ashes also contain silica, and certain carbonates produced by the destruction of salts formed by organic acids.

If some seeds, sufficiently moistened, are placed under a bell-glass containing atmospheric air confined over quicksilver, all the signs of germination will soon be perceived. In the course of a few days, provided the temperature has been sufficiently high, germination will have made a certain progress. Supposing that the temperature of the bell-glass has not varied, and that the atmospheric pressure remains the same, we generally find that the air, in which germination has been proceeding, has not changed its original volume; but it has been modified in its composition: a notable quantity of carbonic acid has been formed, and a portion of oxygen has disappeared. The volume of carbonic acid produced, represents for the most part the volume of oxygen which has disappeared. Now we know that carbon being burnt in a certain volume of oxygen gas, produces sensibly an equal volume of carbonic acid gas. It was the knowledge of this fact that induced M. de Saussure to say, that in germination, carbonic acid is produced by the combustion of a portion of the carbon which enters into the composition of the seed.

Germination and the appearance of carbonic acid, (which is always its consequence,) take place as readily in pure oxygen gas, as in atmospheric air; but if placed in an atmosphere deprived of oxygen, seeds cease to germinate. Consequently, germination is out of the question in azote, in hydrogen, or in carbonic acid, however favorable they may be in reference to humidity and temperature. Some formation of carbonic acid is indeed to be observed under such circumstances, but then this gas is the result of the decomposition and putrid fermentation of the seed. It is therefore by means of the oxygen which it contains, that atmospheric air concurs in the germination of seeds.

Rollo was the first who ascertained the production of carbonic acid, during the germination of seeds in an atmosphere of oxygen; but it was M. Theodore de Saussure, who by delicate eudiometrical experiments, demonstrated the phenomena in all their nicety, by proving that the oxygen consumed was replaced by a corresponding volume of carbonic acid.*

There are some seeds, for instance, peas, and the seeds of aquatic plants, which have the property of germinating under water. Some observers have, from this fact, come to the premature conclusion that atmospheric air, and consequently oxygen, were by no means necessary to germination. Saussure has explained this anomaly by referring to the constant presence of air in a state of solution in water. In fact, having placed some seeds of the polygonum amphibium under water, deprived of its air by long boiling, Saussure proved that germination could not take place.†

Under like circumstances, the quantity of carbonic acid generated in a given time, is by so much greater, the larger the quantity of oxygen in the atmosphere which immediately surrounds the germinating seed. Carbonic acid gas is, of all the gases which have been tried, that which is most unfavorable to germination; and one way of hastening the process is to place under the receivers which cover the seed, some substance capable of absorbing it as fast as it is formed—quick-lime, for example. By this arrangement the radicular increase is sensibly accelerated.‡

The quantity of oxygen gas necessary to germination, is not the same in reference to all seeds; lettuce, the french-bean, and the field-bean require about $\frac{1}{100}$th part of their respective weights; while $\frac{1}{10}$th less is sufficient for wheat, barley, purslane, &c. Saussure moreover came to the conclusion that the carbonic acid generated by these different seeds in germinating is proportioned to their mass, and altogether independent of their number.§

Inasmuch as seeds during germination yield carbonic acid to the atmosphere, it is quite obvious that they must lose some part of their original weight. And this they do in fact; but the loss experienced by seeds which have germinated is always greater than that which would have resulted from the destruction of carbon that takes place. Saussure attributed this excess of loss to the volatilization of a portion of the water which entered into the composition of the seed.‖ According to Saussure, therefore, the phenomena of germination resolve themselves into the diminution of carbon and of the elements of water. It is, nevertheless, doubtful whether the chemical actions

* Saussure, Recherches chimiques sur la Végétation, p. 10.
† Idem, p. 3. ‡ Idem, p. 26. § Idem, p. 13. ‖ Idem, p. 20.

are so simple as this; we know, for example, that M. Becquerel considered the organic acid which appears during germination as acetic acid, whereas it is much more likely that it should be the lactic acid. There is certainty of the formation of an acid during germination; to prove its development it is sufficient to make a few moist seeds sprout on blue litmus paper, which speedily acquires the permanent red tint indicating the presence of an acid.

The volume of the air in which seeds germinate is not absolutely invariable. On examining, with renewed attention, the action of germinating seeds on a limited volume of air, M. de Saussure ascertained that certain seeds have the property of diminishing the bulk of this atmosphere, while others perceptibly augment it. It must be admitted, therefore, that during germination, the volume of carbonic acid produced is now greater, now less, than the volume of oxygen gas that is consumed. The nature of the results obtained appears, however, to vary in regard to the same class according to the stage of the germination.

Elementary analysis appeared to me the most satisfactory means of investigating the subject of germination. I shall here recapitulate a few attempts that have been made in this direction, less however with a view to the final settlement of the question, than to point out the general method of procedure to those who would enter farther upon this interesting portion of physiology. The experiments I allude to were made upon the seed of trefoil and on wheat.

The seed, on being dried at a heat of 110° cent. (230° Fahr.,) lost 0.120 of water. Duly moistened, it was placed to sprout on a porcelain plate. As soon as the radicle had attained a length of from $\frac{1}{40}$th to $\frac{1}{20}$th of an inch, each seed was placed in a stove, the temerature of which was sufficiently high to check the growth immediately. The complete desiccation was then terminated over an oil bath at a temperature of 110° cent. (230° Fahr.)

The seed put to germinate weighed 2.474 grammes, (38.193 grains troy;) perfectly dry, its weight would have been 2.405 grms. (37.128 grains troy.) When germinated, the seed, also quite dry, weighed 2.241 grms. (34.596 grains troy.)

Analysis gives us the composition of

	THE SEED BEFORE GERMINATION.	THE SEED AFTER GERMINATION.
Carbon	51.5	50.8
Hydrogen	6.0	6.3
Azote	7.2	8.0
Oxygen	36.0	34.2
	100.0	100.0

RESULTS OF EXPERIMENT.

	Grains troy.		Carbon.	Hydrogen.	Oxygen.	Azote.
Seed placed to germinate	37.128	containing	18.865	2.223	13.369	2.670
Seed after germination	34.596	"	17.815	2.176	11.840	2.763
Difference	— 2.532	"	— 1.050	— .047	— 1.529	+ .093

The total loss then during germination was 0.164 grm., (2.531 grs.) while the loss due to the carbon, only amounts to 0.068 grm. (1.049

grs. :) the analysis shows besides that in this particular case, the excess of the loss in the present case over and above that which is ascribed to the carbon, is not altogether due to the elements of water, inasmuch as it is partly ascribable to carbonic oxide : for

	1.049 grs.	of carbon,
	1.404 "	of oxygen,
Represent	2.453 "	of oxide of carbon.

Supposing this to be so, and the first period of the germination of the trefoil to have been conducted in a close vessel, the volume of atmospheric air would have been increased ; because 1 volume of carbonic oxide+½ volume of oxygen=1 volume of carbonic acid gas. It is consequently evident that for each volume of carbonic oxide produced from the seed, there is one half of this volume added to the total volume of the atmosphere.

It will not, perhaps, be useless to advert to the circumstance that the increase of volume, which in the experiment I have just related must have amounted to about twenty-five cubic inches, would certainly have passed undetected, if the experiment had been conducted in a close vessel. For inasmuch as several quarts of atmospheric air must have been used to place 38.193 grs. of seed in conditions favorable for germination, it may readily be imagined that the increase of volume must have been too small a fraction of the total mass of air to be appreciated with any certainty.

GERMINATION OF WHEAT.

The wheat employed, on being dried, lost 0.652 grain of moisture. Thirty-one grains were arranged for germination, which process was suspended immediately after the appearance of the radicles. The young stalks were hardly visible. The germinated grain looked slightly shrivelled : on being crushed, after having been dried, it scarcely differed in appearance from ordinary wheat reduced to powder, a considerable quantity of starch being still recognisable.

The wheat, before germinating, taken as dry, and free from ashes, weighed 2.439 grms., or 37.653 grs. troy.

The seed when germinated and gathered, under the same condition, weighed 2.365 grms., or 36.510 grs. troy.

Elementary analysis gives for the composition of—

	WHEAT NOT GERMINATED.	GERMINATED WHEAT.
Carbon	46.6	47.0
Hydrogen	5.8	5.9
Azote	3.45	3.7
Oxygen	44.15	43.4
	100.0	100.0

RESULTS OF EXPERIMENT.

	Grains troy.		Carbon.	Hydrogen.	Oxygen.	Azote.
Wheat placed to germinate	37.653	containing	17.47	2.176	16.56	1.281
Wheat when germinated	36.510	"	17.15	2.145	15.83	1.343
Difference	—1.143	"	—0.032	—0.031	—0.073	+0.062

0.324 of a grain of carbon +0.432 of a grain of oxygen represent 0.756 of a grain of carbonic oxide; 0.030 of a grain of hydrogen would require 0.247 of a grain of oxygen to form water. Now, the oxygen remaining, abstraction made of that which enters into the formation of the carbonic oxide is 0.282 of a grain.

In the first period of its germination, therefore wheat, like trefoil seed, experiences a loss which may in great part be referred to elimination of the carbonic oxide. The chemical composition of these two kinds of seed at more advanced periods of their germination, no longer presents relations so simple. We easily discover that carbon continues to be eliminated; but the loss no longer corresponds with that which the oxygen of the seed ought to suffer, in order that the total loss should be represented by a definite compound of carbon. The phenomenon, in fact, becomes extremely complex; and we can even perceive that it must be so, when we reflect that in proportion as the green parts are evolved, a new chemical action is set up entirely different from that which takes place in the earliest periods of the germination: the green matter of vegetables having, as we shall find, the singular faculty of decomposing carbonic acid gas, and assimilating its carbon under the agency of light.

This action of the green matter begins to be manifested long before the first phases of germination have entirely ceased; so that during a certain time two opposite forces are at work simultaneously. One of these, as we have seen, tends to discharge carbon from the seed; the other tends to accumulate this element within it. So long as the first of these forces predominates, the seed loses carbon; but with the appearance of the green matter the young plant recovers a portion of this principle; finally, when by the progress of the vegetation, the second force surpasses the first in energy, the plant grows, increases, and advances to maturity.

The presence of light is indispensable to the manifestation of the chemical force by which the green parts of plants appropriate the gaseous elements of the atmosphere. Germination, on the contrary, may take place in absolute darkness; and it would be curious to inquire into the issues of vegetation begun and ended under such circumstances, in which the organs produced by the seed would have no power to fix any of the principles of the atmosphere to repair the loss of carbon which the seed suffers. It is evident that this loss of carbon must have a limit, which is probably that of germination.

CONTINUED GERMINATION OF PEAS.

Ten peas, weighing together 2.237 grms. or 34.534 grs. troy, taken as quite dry, were put to germinate in a dusky room, the temperature of which was maintained between 12° and 17° cent. (54° and 63° Fahr.) The experiment, begun the 5th of May, was ended on the 1st of July.

The germinated peas, when dried, weighed 1.075 grm. or 16.595 grs. troy.

Composition of the peas :

	BEFORE GERMINATION.	AFTER GERMINATION.
Carbon	46.5	44.0
Hydrogen	6.1	6.0
Azote	4.2	6.7
Oxygen	40.1	36.9
Ashes	3.1	6.4
	100.0	100.0

SUMMARY OF THE EXPERIMENT.

	Grains troy.		Carbon.	Hydrogen.	Oxygen.	Azote.	Salts, Earths.
Peas set to germinate	34.534	cont'ng	16.055	2.115	13.843	1.447	1.064
Peas which had germinated	16.595	"	7.292	1.003	6.128	1.111	1.064
Difference	—17.939	"	—8.763	—1.112	—7.715	—0.336	0.000

Peas, during their germination, pushed to this extreme term, therefore, suffered a loss of about 52 per cent., the loss being referable to each of their constituent elements, which are summed up in carbon, water, and ammonia.

7.719 of oxygen taking	0.972	of hydrogen to form water;
0.339 of azote requiring	0.077	of hydrogen to form ammonia ;
	1.049	which represents as nearly as possible the quantity of hydrogen eliminated.

In this experiment, therefore, we see that a seed weighing 3.453 grs. troy, suffered a daily loss of about 0.077 of a grain troy of carbon.

CONTINUED GERMINATION OF WHEAT.

On the 5th of May, 46 corns or grains of wheat, supposed to be quite dry, and weighing 1.665 grm. or 25.704 grs. troy, were set to germinate in the dark.

On the 25th of June, the germinated wheat, when dried, weighed 0.713 grm. or 11.007 grs. troy.

Composition :

	BEFORE GERMINATION.	AFTER GERMINATION.
Carbon	45.5	41.1
Hydrogen	5.7	6.0
Azote	3.4	8.0 supposed.
Oxygen	43.1	39.5
Ashes	2.3	5.4 calculated.
	100.0	100.0

SUMMARY OF THE EXPERIMENT.

	Grains troy.		Carbon.	Hydrogen.	Oxygen.	Azote.	Salts, Earths.
Wheat placed to germinate	25.704	containing	11.704	1.466	11.086	0.879	0.588
Wheat germinated	11.007	"	4.523	0.343	4.373	0.879	0.588
Difference	—14.697	"	—7.181	—0.803	—6.713	0.000	0.000

During the germination, continued for fifty-one days, consequently this wheat lost 57 per cent., and the loss may be wholly referred to the elements of carbonic acid and water, *i. e.* to carbon, hydrogen, and oxygen.*

These results of the elementary analysis of seeds of different kinds, before and after germination, tend, therefore, to show that the chemical phenomena which take place in the earliest periods of germination, continue to go on even after the organic matter of the seed has been changed into a proper vegetable, imperfect, undoubtedly, but still possessing the essential organs of plants,—roots, a stem, and leaves. Deprived of light, the blanched vegetable may be said to vegetate in a negative manner, expending, exhaling the elementary principles contained in the seed whence it sprung.

The general practice of sowing seeds at some depth in the ground, led to the belief, for a long time, that light was prejudicial to germination. Sennebier had even inferred so much from his experiments, which appeared to derive confirmation from those of Ingenhousz, and which were instituted for the express purpose of discovering the comparative influences of sun-light and darkness on the germination and growth of vegetables.† But M. de Saussure showed that the prejudicial influence attributed to the light was connected with the drying of the seed, in consequence of its exposure to a higher temperature. M. de Saussure caused seeds to germinate at the same time under two bell-glasses of equal capacity. One of these shades was opaque, the other was transparent, and so placed as to receive the diffused light of day. The temperature was the same in either. The seeds sprung simultaneously under both glasses.‡ Within a few days, the vegetation under the transparent shade was most advanced; which is exactly what we should have expected from all that has already been said of the functions of the organized parts subjected to the action of light.

We are indebted to M. de Humboldt for a number of very curious observations on the property which chlorine possesses of stimulating or favoring germination. This action of chlorine is so decided, that it is apparent even upon old seeds which will not germinate when placed under ordinary circumstances. The experiments of M. de Humboldt were made, in the first instance, on the common cress, (*lepidium sativum.*) The seeds were placed in two test tubes of glass, one of which contained a weak solution of chlorine, the other common water. The tubes were placed in the dark, the temperature being maintained at about 15° cent. (59° Fahr.) In the chlorine solution, germination took place in six or seven hours; from thirty-six to thirty-eight were required before it was manifest in the seeds in the water. In the chlorine, the radicles had attained the length of .0585 Eng. inch, after the lapse of fifteen hours, while they were scarcely visible at the end of twenty hours in the seeds submerged in water.§

* The small quantity operated on prevented any estimates being made of the azote lost. Its proportion was supposed not to have varied. It is extremely probable, however, that there was some slight disengagement of azote, as in the preceding experiment.

† Saussure, Rech. Chimiques, p. 23.

‡ De Saussure, op. cit. p. 23.

§ Humboldt, Flora fribergensis subterranea, p. 156.

In the botanical gardens of Berlin, Potsdam, and Vienna, this property of chlorine has been made available to excellent ends; by its means many old seeds, upon which a great variety of trials had already been made in vain to make them sprout, were brought to germinate. At Schœnbrunn, for instance, they had never succeeded in raising the *clusea rosea* from the seed; but M. de Humboldt succeeded at once, by forming a paste of peroxide of manganese, with water and hydrochloric acid, in which he set the seeds of the *clusea*, and then placed them in a temperature of from 62° to 75° cent. (143° to 167° Fahr.) It seems very likely that this discovery of M. de Humboldt may yet be taken advantage of in our every-day husbandry. It is quite certain that the whole of the seed which we commit to the ground, does not spring up, especially when we are forced to have recourse to seed that is two or three years old; the loss is then frequently very considerable. But a solution of chlorine, or a mixture which would evolve it, could not cost much, its use would add little or nothing to the very trifling expense which is generally incurred in pickling the wheat that is employed as seed.

---

§ III.—EVOLUTION AND GROWTH OF PLANTS.

As germination advances, we see those organs acquiring shape and size which had appeared at first in the rudimentary state. The roots extend in length, and increase in number, and their extremities become covered with capillary fibres. The stem as it rises puts forth branches in all directions, which become covered with leaves. The cotyledons which had nourished the young plant during the first days of its existence, wither and fall. Under the influence of the solar light, the vegetation progresses amain, and the organic matter, which finally constitutes the plant when it has attained maturity, weighs vastly more than the same matter which existed previously in the seed. To quote a single instance from the family of annual plants, a seed of field beet of the weight of .06175 of a grain, may by the end of the autumn give birth to a root which with its leaves shall weigh 162099grs. or upwards of 28lbs.*

This immense and rapid assimilation can have no other source than the soil, the air, and water. Without, at this time, pausing to consider the useful influence which the soil, and the substances it contains, exert upon the entire development of vegetables, we shall here assume it as a general principle that water and the air of the atmosphere alone, are capable of furnishing them with all the elements which enter into their composition, to wit—carbon, hydrogen, oxygen, and azote. In other words, a seed may germinate, vegetate, give birth to a plant which shall attain to complete maturity, by the

* Actual weight of a beet-root grown at Bechelbronn in 1841

mere concurrence of water and the gases, or vapors which are dif fused through the atmosphere. This fact is demonstrated by the following experiment:—

In a sufficient quantity of properly moistened roughly pounded brick-dust, (which had been heated to redness in order to destroy every trace of organic matter,) a few peas were sown on the 9th of May, and the pot was transferred to a green-house in order to protect the plants from the dust and impurities which always fly about in the open air.

On the 16th of July, the peas, which looked extremely well and healthy, were in flower. Each seed had sent forth one stem, and each stem, abundantly covered with leaves, bore a flower.

On the 15th of August the pods were ripe; no more water was given, and by the end of the month the plants were dry.

The length of the stalks varied from about three feet three inches to five feet; but they were extremely slender, and the leaves not more than one third the ordinary size. The pods were 1.3 inch, by from 0.3 to 0.4 of an inch broad. They generally contained two peas each; one contained a single pea only, but this was almost twice the size of any of the others.

In the course of three months, therefore, these peas came to perfect maturity—ripe seeds were gathered. The analysis of the crop which I shall give by and by, in connection with another question which we shall have to discuss, showed that, the harvest obtained under the conditions indicated, contained a considerably larger proportion of each of the elements found than was originally contained in the seed from which it sprung.

Carbon being the predominating principle in plants, it is our first duty to inquire into the origin of so much of this element as is assimilated in the course of vegetation.

Carbon is met with in very small quantity in the atmosphere in the state of carbonic acid, and as this is one of the most soluble of the gases which enter into the constitution of the air, water always contains a considerable quantity of it in solution. Carbonic acid may therefore be in relation with plants by the medium of the air amidst which they live, and of the water which is no less indispensable to their existence. We have now to ascertain in what way this gas evolves and sets free its carbon in favor of living vegetables.

Bonnet, having put some fresh leaves at the bottom of a jar containing spring water, observed that when exposed to the rays of the sun, they gave off bubbles of air. He sought to ascertain whether this disengagement of gas was due to the leaves, or to the liquid in which they were contained. For the spring water, he therefore substituted water deprived of its air by boiling, and he found that the leaves exposed to the sun's light in this water, no longer gave off any bubbles of air. Bonnet, therefore, concluded that the gas which he collected in his first experiment, proceeded from the water.

In 1771, Priestley discovered, that by emitting oxygen, plants had the property of ameliorating atmospherical air, which had been vitiated by the respiration of animals or by combustion.* This unexpected discovery immediately arrested the attention of vegetable physiologists. Nevertheless, Priestley was not yet master, so to speak, of the capital experiment which he had announced to the world of science. He had not seized all the circumstances which assure its success. Occasionally the leaves which were the subjects of experiment did not cause the disengagement of any gas; occasionally, too, the air disengaged, far from being oxygen—far from ameliorating the atmosphere, was found to be carbonic acid gas. It was Ingenhousz who made out the influence of the solar light upon the phenomenon in question. He proved, by a vast number of distinct experiments, that leaves exhale oxygen when they are exposed to the light of the sun. He perceived, moreover, that in the dark they vitiate the air, rendering it improper for respiration and combustion.†

But the origin of the oxygen disengaged from water by leaves exposed to the light of the sun still remained to be discovered. It was Sennebier who took this important step, by showing that it was to the carbonic acid generally contained in water that leaves exposed to the sun's light owed their faculty of evolving oxygen gas. With this interesting fact, it was easy to render an account of all the anomalies that had been successively announced: boiled water, as Bonnet had observed, could not afford any air, and spring water should usually give more than river water, as Ingenhousz had noticed, for the simple reason that boiled water neither contains carbonic acid gas nor any other kind of air; and that well water generally contains a larger quantity of carbonic acid in solution than river water.

In giving the grand features in the history of this brilliant discovery of the eighteenth century, it may be said that Bonnet was the first who observed the phenomenon of the gaseous evolution effected by the leaves of vegetables:‡ that Priestley announced that the gas disengaged was oxygen; that Ingenhousz demonstrated the necessity of the solar light to the production of the phenomenon; finally, that it was Sennebier, to whom was reserved the honor of showing that the oxygen gas obtained under these circumstances is the product of the decomposition of carbonic acid.

It was, however, matter of supreme interest to study this decomposition of carbonic acid in its last details. It was imperative, for instance, to ascertain what relation existed between the volume of the oxygen disengaged and the volume of the carbonic acid decomposed. This was admirably accomplished by M. Theodore de Saussure in a long series of remarkable experiments, of which I shall here endeavor briefly to state the main results.

The conclusion which follows naturally from the discovery of Sennebier, was that carbonic acid exercised a favorable influence on vegetation by supplying plants with the carbon which enters into

* Experiments and Observations, vol. ii.
† Experiments on Vegetables.
‡ Sur l'usage des feuilles dans les plantes, p. 31.

their constitution. Percival ascertained by direct experiment the accuracy of this inference by placing plants in a current of atmospheric air, mixed with a pretty large proportion of this gas. By means of a comparative experiment, he saw that a plant in such circumstances made much greater progress than one subjected to a current of ordinary air.* The researches of Saussure, in confirming in all respects those of his predecessors, added this farther very important fact: that to act beneficially upon vegetables the carbonic acid must be mixed with oxygen.

Under a bell glass of the capacity of 398 cubic inches, placed over mercury, with a delicate film of water swimming on its surface, he introduced three young peas, which displaced about $\frac{4}{100}$ of the included air. The atmosphere was composed of common air and carbonic acid gas in different proportions.

The experiments were conducted successively in the sunshine and in the shade. In the sun, the apparatus received daily the direct action of the light during five or six hours: when the light was too vivid it was somewhat lessened by shading. In the sunlight the plants lived for several days in an atmosphere composed of equal parts of air and carbonic acid; they then faded. But they died much more speedily in atmospheres which contained two-thirds, or three-fourths, or *a fortiori* which consisted entirely of carbonic acid. The young plants throve decidedly when the atmosphere contained about $\frac{1}{11}$th of carbonic acid; their growth was evidently more vigorous here than it was in simple air; and at the conclusion of one experiment which extended over ten days, almost the whole of the carbonic acid was found replaced by, or changed into oxygen: the peas had assimilated the carbon.

The smallest quantity of carbonic acid added to the air, was found injurious to the plants when they were kept in the shade. Young peas lived only six days under such circumstances, when the atmosphere around them consisted of a quarter of its volume of carbonic acid. They lived ten days when the proportion of this gas did not exceed a twelfth; but then they scarcely grew at all in the mixture; they certainly made much less progress than they would have done in common air. Saussure concluded, from these experiments, that carbonic acid was useful to growing vegetables only when present along with oxygen, and that it ceases to be so when the atmosphere contains more than $\frac{1}{12}$th of its volume of the gas.

To determine the proportion of oxygen set at liberty during the decomposition of carbonic acid by plants, Saussure composed an atmosphere of common air and carbonic acid, the latter in the proportion of 0.075; the mixture was confined under a bell-glass of the capacity of 5.746 litres or 10.112 pints, standing over mercury as in the former experiments. Seven plants of the periwinkle were introduced into the apparatus, their roots dipping into 15 cub. centim. or 5.895 cub. in. of water—the water was limited as much as possible,

* Manchester Memoirs, vol. ii

in order that the absorption of carbonic acid, which must, of course, take place, might be thrown out of the reckoning. The experiment was continued for six days, during which the plants received the direct rays of the sun from five to eleven o'clock in the morning. On the seventh day, the plants were withdrawn. They had preserved their freshness. All the corrections made for temperature and pressure, the volume of the atmosphere in which they had lived was not found changed by more than about 20 cubic centimetres, 7.8 c. in., a quantity which is within the possible errors of computation; but the composition of the air had undergone very notable changes: the carbonic acid had disappeared, and the eudiometer proclaimed 0.24 of oxygen instead of the 0.21 which it contained originally.*

RESULTS OF THE EXPERIMENTS.

	c. inches.		Azote.	Oxygen.	Carb. acid.
Before: Volume of atmosphere........	2257	containing	1650	438.5	169.3
After: " "	2257	"	1704	553	0
	0		+54	+14.8	—169.3

The periwinkles, consequently, had caused 169.3 cubic inches of carbonic acid to disappear, and given off upwards of one hundred and fourteen cubic inches of oxygen. Had the whole oxygen of the carbonic acid been set at liberty, this volume would have been precisely equal to that of the acid gas decomposed; but as no more than one hundred and fourteen cubic inches of oxygen were obtained, it must be inferred that the periwinkles had fixed 54.6 cubic inches of this gas.

This is the conclusion, indeed, to which M. de Saussure came, and subsequent experiments have confirmed its accuracy. The following table contains a summary of five experiments that were instituted:

	c. inches.		c. inches.
Exp. 1. Carbonic acid disappearing...	169.3	Oxygen disengaged...	114.7
		Azote disengaged.....	54.6
			169.3
Exp. 2. " " " ...	121.4	Oxygen disengaged...	88
		Azote disengaged....	33
			121
Exp. 3. " " "	58.5	Oxygen disengaged....	47.5
		Azote disengaged.....	8.2
			55.7
Exp. 4. " " " ...	120.2	Oxygen disengaged....	96.6
		Azote disengaged.....	7.8
			104.4
Exp. 5. " " "	72.3	Oxygen disengaged....	49.5
		Azote disengaged.....	22.4
			71.9

There is one remark which it is impossible to avoid making in surveying this table; it is to the effect, that the azote disengaged rep-

* Saussure, Recherches chimiques, p. 40.

resents almost exactly the volume of oxygen which it would be necessary to add, in order that the oxygen collected should represent the whole of that which entered into the constitution of the carbonic acid decomposed. It is probable that the excess of azote which appeared in all these experiments was present in principal part in the air contained and condensed within the interstices of the plants, or held in solution in the water which bathed their roots. It would be difficult to assign it any other origin; such, for instance, as that from changes in the azotized principles of the plants that were the subject of experiment. In his first experiment, in fact, M. de Saussure fixes the weight of the dry matter of the seven periwinkle plants at 41.6 grains. Now, from numerous determinations of azote which I have had occasion to make in regard to plants of very different ages and species, I think I can say that these periwinkles, taken as dry, did not contain more than .385 of azote; this, in reference to the weight assumed by M. de Saussure, would be 1.042 grs. or 20.8 cubic inches of azote; and the volume of azote disengaged in this first experiment was 54.6 cubic inches. It is proper further to observe, that the state of health which the plants presented on the conclusion of the experiment does not allow us to suppose a total decomposition of the azotized matters which entered into their constitution. These various considerations lead us to infer that the excess of azote collected must have been displaced by oxygen. We are, therefore, at liberty to presume, from the experiments now referred to, that the volume of oxygen produced probably represents the volume of carbonic acid decomposed.

The necessity of oxygen gas in the decompounding action which plants exposed to the light exert so energetically upon carbonic acid, leads us to study particularly the phenomena which oxygen exhibits in connection with growing plants. When a number of freshly gathered and healthy leaves are placed during the night under a bell-glass of atmospheric air, they condense a portion of the oxygen; the volume of the air diminishes, and there is a quantity of free carbonic acid formed, generally less than the volume of oxygen which has disappeared. If the leaves which have absorbed this oxygen during their stay in the dark, be now exposed to the sun's light, they restore it nearly in equal quantity, so that, all corrections made, the atmosphere of the bell-glass returns to its original composition and volume.

Leaves in general have the same effect when they are placed alternately in the dark and in the light; there is, however, a very obvious difference in the intensity with which the phenomenon is produced, according to the nature of the leaves. The quantity of carbonic acid formed during the night is by so much the less, as the leaves are more fleshy, thicker, and therefore more watery. The green matter of fleshy leaved plants, of the *cactus opuntia*, to quote a particular instance, does not produce any sensible quantity of carbonic acid in the dark: but these leaves condense oxygen, and exhale it again like those which are less fleshy, when they are brought into the sun, after having been kept for some time in the dark.

Saussure applied the names of inspiration and expiration of plants to these alternate effects, led by the analogy—somewhat remote, it must be confessed—which the phenomenon presents with the respiration of animals.

The inspiration of leaves has certain limits; in prolonging their stay in the dark, the absorption becomes less and less: it ceases entirely when the leaves have condensed about their own volume of oxygen gas. And let it not be supposed that the nocturnal inspiration of leaves is the consequence of a merely mechanical action, comparable, for example, to that exerted by porous substances generally upon gases. The proof that it is not so is supplied by the fact that the same effects do not follow when leaves are immersed in carbonic acid, hydrogen, or azote. In such circumstances there is no appreciable diminution of the atmosphere that surrounds the plant. The primary cause of the inspiration of oxygen by the leaves of living plants is, therefore, obviously of a chemical nature.

With the facts which have just been announced before us, it seems very probable that during the nocturnal inspiration, the carbonic acid which appears is formed at the cost of carbon contained in the leaves, and that this acid is retained either wholly or in part, in proportion as the parenchyma of the leaf is more or less plentifully provided with water. A plant that remains permanently in a dark place, exposed to the open air, loses carbon incessantly; the oxygen of the atmosphere then exerts an action that only terminates with the life of the plant: a result which is apparently in opposition to what takes place in an atmosphere of limited extent. But it is so, because in the free air the green parts of vegetables can never become entirely saturated with carbonic acid, inasmuch as there is a ceaseless interchange going on between this gas, and the mass of the surrounding atmosphere; there is, then, incessant penetration of the gases, as it is called. There is a kind of slow combustion of the carbon of a plant which is abstracted from the reparative influence of the light.

The oxygen of the air also acts, but much less energetically, upon the organs of plants that do not possess a green color.

The roots buried in the ground are still subjected to the action of this gas. It is indeed well known, that to do their office properly, the soil must be soft and permeable, whence the repeated hoeings and turnings of the soil, and the pains that are taken to give access to the air into the ground in so many of the operations of agriculture. The roots that penetrate to a great depth, such as those of many trees, are no less dependent on the same thing; the moisture that reaches them from without brings them the oxygen in solution, which they require for their development. It is long since Dr. Stephen Hales showed that the interstices of vegetable earth still contained air mingled with a very considerable proportion of oxygen. The roots of vegetables, moreover, appear generally to be stronger and more numerous as they are nearer the surface. In tropical countries various plants have creeping roots which often acquire dimensions little short of those of the trunk they feed.

If a root detached from the stem be introduced under a bell-glass full of oxygen gas, the volume of the gas diminishes, carbonic acid is formed, of which a portion only mingles with the gas of the receiver, a certain quantity being retained by the moisture of the root. The volume of the gas thus retained is always less than that of the root itself, however long the experiment may be continued. In these circumstances, whether in the shade or the sun, roots act precisely as leaves do when kept in the dark. Roots still connected with their stems, give somewhat different results.

When the experiment is made with the stem and the leaves in the free air, while the roots are in a limited atmosphere of oxygen, they then absorb several times their own volume of this gas. This is because the carbonic acid formed and absorbed is carried into the general system of the plant, where it is elaborated by the leaves, if exposed to the same light, or simply exhaled if the plant be kept in the dark.

The presence of oxygen in the air which has access to the roots is not merely favorable; it is absolutely indispensable to the exercise of their functions. A plant, the stem and leaves of which are in the air, soon dies if its roots are in contact with pure carbonic acid, with hydrogen gas, or azote. The use of oxygen in the growth of the subterraneous parts of plants, explains wherefore our annual plants, which have largely developed roots, require a friable and loose soil for their advantageous cultivation. This also enables us to understand wherefore trees die, when their roots are submerged in stagnant water, and wherefore the effect of submersion in general is less injurious when the water is running, such water always containing more air in solution than that which is stagnant.

The woody parts, the fruit, and those organs of plants in general which have not a green color, stand in the same relations to oxygen as the roots: they merely change this gas into carbonic acid, which is then transported to the plant at large, to suffer decomposition by the green parts. In this action we observe a displacement, a kind of translation of the carbon of the lower to the upper parts of plants.

The decomposition of carbonic acid by plants admitted, we have still to examine whether, in the phenomena of vegetation, the leaves decompose the carbonic acid of the atmosphere directly, or the acid gas, previously dissolved in the water, which moistens the ground, be conducted by the way of absorption into the tissues of vegetables, there to suffer decomposition. The quantity of carbonic acid contained in the air is so small, and the growth of plants, on the contrary, is often so rapid, that it might reasonably be suspected that the carbon which they require was introduced in great part by this way of absorption. In that series of beautiful experiments in which M. Saussure exposed plants to the influence of atmospheres more or less charged with carbonic acid, the water in which their roots were plunged was in contact with the mixed atmospheres. It was therefore possible that the carbonic acid gas entered the vegetables in the solution by the roots.

Sennebier made an experiment to show that leaves decompose

both the carbonic acid which is in contact with them externally, and that which is dissolved in the water absorbed by their woody tissue. He took two branches of a peach-tree, and introduced them into a couple of bell-glasses filled with water from the same spring.* The lower end of each branch dipped into a flask. One of the flasks was filled with water charged with carbonic acid; the other contained air: the two bell-glasses were exposed to the light. The leaves of the branch whose extremity dipped into the solution of carbonic acid, disengaged 99.4 cubic inches of oxygen gas under the bell which covered it: the leaves of the other branch only produced 52.2 cubic inches in the same time.

This experiment does not perhaps afford all the sufficient evidence of the decomposition of gaseous carbonic acid as it occurs in the atmosphere, and mixed with a great mass of air. It appears, however, that the leaves of plants have the power of decomposing the gaseous carbonic acid which is mixed with the air, and that even with surprising rapidity.

In the summer of 1840, I introduced into a balloon of the capacity of about twelve quarts and a half, and furnished with three tubulures or openings, the branch of a vine in full growth and bearing twenty leaves. The woody part of the branch was fixed by means of a collar of caoutchouc to the lower orifice of the balloon; a fine tube, intended to establish a communication between the interior of the vessel and the outer air, was introduced into the superior tubulure; the lateral opening communicated by means of a tube with an apparatus which measured with great accuracy the quantity of carbonic acid contained in the atmosphere.

In this experiment the air, before reaching the apparatus for measuring the carbonic acid, passed through the great balloon containing the vine branch. The rate with which the air passed through the apparatus was regulated by the flow of an aspirator, and was at the rate of about twelve quarts per hour.

The apparatus was exposed to the sun; the experiment beginning at eleven and finishing at three o'clock.

In one experiment it was found, all corrections made, that the atmospheric air, after having passed through the balloon, contained in volume 0.0002 of carbonic acid gas; the air of the adjoining court contained at the same moment 0.00045 of carbonic acid.

In another experiment, the air, after having passed over the leaves, contained but 0.0001 of carbonic acid; the air of the court containing 0.0004 of the same gas. In traversing the space in which the vine branch, exposed to the light of the sun, was included, therefore the air was deprived of three fourths of the whole quantity of carbonic acid which it contained.

In operating with the same apparatus during the night, opposite results were obtained; the air in traversing the balloon generally acquired a quantity of carbonic acid, the double of that which the atmosphere contained at the same moment.

* This was common water containing carbonic acid

I conceive that it is by such a method as this, that the genera. phenomena of vegetable respiration in plants still connected with the soil ought to be studied.

The experiments which I have now related must satisfy every one, that the leaves of living plants actually assimilate the carbon which occurs in our atmosphere in the state of carbonic acid; they also explain the well-known fact that plants thrive better in air that is in motion, and frequently renewed, than in a perfect calm.

From all we have seen up to this time, then, we feel authorized to conclude that the greater proportion if not the whole of the carbon which enters into the constitution of vegetables is derived from the carbonic acid of the atmosphere. The experiments cited, show how the vital force acts at first on the oxygen of the air during germination, and next upon its carbonic acid during vegetation properly so called. But in none of the experiments which have been quoted, have we seen any thing which could lead us to suspect that the azote of the atmosphere was absorbed in sensible quantity.

It is true, indeed, that at one time Priestley, and after him, Ingenhousz, thought that they had observed an absorption of azote during the growth of plants in confined atmospheres. But the experiments which have been since performed by Saussure have not confirmed their conclusions upon this point. Saussure even thought that he had perceived a slight exhalation of azote.

Nevertheless, the presence of azote in vegetables being incontestable, and the assimilation of this principle during their growth being in some sort demonstrated by the fact that seeds are multiplied, physiologists were led to imagine that the azote was derived from the soil. And in nature, indeed, the growth of a plant does not take place at the sole cost of water and the atmosphere. The roots which attach it to the earth there also find elements of nutrition. In ordinary circumstances the growth of a plant takes place by the simultaneous concurrence of the food which the roots encounter in the ground, and that which the leaves abstract from the gaseou elements of the air. As it is further acknowledged that the food which is supplied by the soil is for the most part azotized, manures have therefore been regarded as the principal and even as the exclusive source of the azote which is met with in vegetables. The observations of Hermbstædt, in showing that the grain which was grown under the influence of the most highly azotized manures contained the largest quantity of gluten, gave a certain force to this view. Nevertheless, there are facts well established in agriculture which induce us to think that in many cases vegetables find in the atmosphere a part of the azote which is necessary to their constitution.

The majority of crops exhaust the soil; but there are still some which render it more fertile. We shall see, by and by, when treating of the rotation of crops, that if, after having cut a field of trefoil once, the second crop be ploughed down, new fertility is communicated to the ground, in spite of the considerable mass of forage which had previously been taken from it. It appears therefore evident, that in ploughing down this second crop we restore such a quantity of organic or organizable matter, that, all things taken into account, the ground actually receives more from the atmosphere than was taken away from it in the first cutting.

The latest experiments of physiologists would seem to show that plants merely take carbon from the air, and appropriate the elements of water. But the ideas which are now generally adopted in regard to the active principle of manures make it difficult to conceive that the soil, by receiving non-azotized matters only, could acquire the degree of fertility which is certainly obtained from the cultivation of those crops that are called *ameliorating*, a fertility which enables us to follow these crops with others, rich in azotized principles.

There is therefore reason for believing that the ploughing in of certain green crops, and fallowing, are not effectual merely by introducing carbon, hydrogen, and oxygen, but azote also into the soil. And it is absolutely necessary that this should be so, in order that the fertility of those lands may be maintained which, from their position, can receive no manures from without. Let us take, for example, a farm laid out for the growth of white crops, and the rearing of cattle. Every year there is an exportation of grain, of flesh, and of the produce of the dairy; that is to say, there is incessant exportation without any perceptible importation of azotized matter. Nevertheless, the soil maintains its fertility; its losses are repaired by the principles which, in a good system of cultivation, pass from the atmosphere into the earth; and among the number of these fertilizing principles it is beyond all question that azote must be present, in order that so much of this element as has been exported may be replaced.

The best established facts in agriculture, therefore, concurred in showing that azote is among the number of the elements that are fixed by plants during their growth. Still, as this truth had not been proved by the experiments of physiologists, the question had to be considered as yet undecided. It was with the hope of clearing up every thing in connection with it that I undertook the series of experiments, the chief features of which I shall now detail.*

I had necessarily to follow a method of inquiry different from any which had yet been taken; I had no chance of arriving at more definite results than those which had been already come to, had I chosen the old line of investigation. I therefore called in the aid of elementary analysis, with a view of comparing the composition of the seed with the composition of the harvest produced from it, at the sole cost of water and the air. By proceeding in this way I believed that the problem was capable of solution: without flattering myself that I have completely resolved it, I conceive that something has been done in the right direction. The subject is one of the most delicate imaginable, and he who enters it requires indulgence.

For soil, I made use of burned clay or silicious sand freed from all organic matter by proper calcination. In this soil, moistened

* Boussingault, Annales de chimie et de physique, t. lxvii. p. 5, 2e série, année 1838

with distilled water, were sown the seeds whose weight was known By a number of preliminary trials, the quantity of moisture which seed of the same kind, of the same growth, and taken at the same moment, lost by drying, commenced in the stove and finished in an oil-bath, at 110° C. (230° Fahr.) was ascertained. The porcelain vessels, in which the experiment was conducted, were placed in a glass house at the end of a large garden. During the whole term, the windows were kept closed; but the sun shone on the house all day. To remove the produce, the vessels were dried by a gentle heat. The roots of the plants then came out readily; to free them completely from any adhering sand, they were moved about in a little distilled water, but never rubbed or bruised, for fear of loss; it seemed even preferable to leave a little sand adhering. The harvest was then dried in the stove, so that it might be powdered; and the complete desiccation was effected in the oil-bath in vacuo.

In ascertaining previously by muneration the weight of the ashes contained in the seed, that of the produce, freed from all saline and earthy matter, became exactly known.

Elementary analysis then proclaimed the composition of the produce; and it was only necessary now to compare it with the composition of the seed, to have ascertained the proportion and the nature of the elements which had been assimilated during the vegetation.

FIRST EXPERIMENT.

CULTURE OF RED CLOVER DURING THREE MONTHS.

In the beginning of August a quantity of seed was sown, which, being dry and free from ashes, would have weighed 1.586 gramme, or 24.48 grs. troy. The crop presented a very good appearance; the clover was from three to three and a half inches in height. The largest leaves could be included in a circle of about two inches in diameter. The length of the roots varied between two and four inches. Dried and bruised, the color of the produce was a deep green.

The plant gathered quite dry, and supposed free from ash, weighed 4.106 grammes, or 63.38 grs. troy; analysis showed it to consist of—

	In the Seed.	In the Produce.
Carbon	50.8	50.7
Hydrogen	6.0	6.6
Azote	7.2	3.8
Oxygen	36.0	38.9
	100.0	100.0

RESULTS.

	Carbon.	Hydrogen.	Oxygen.	Azote.
24.48 grs. troy containing after the analysis	12.44	1.466	8.815	1.759
63.38 " " "	32.141	4.183	24.155	2.408
38.90 = grs. during cultivation	+19.70	+2.717	+15.840	+0.649

Thus, in the course of three months, the elementary matter of the seed had nearly doubled, and the azote of the plants gathered shows an excess of 0.042 gramme, or 0.649 grs. troy, above the azote of the seed sown.

SECOND EXPERIMENT.

GROWTH OF PEAS.

Five peas, very nearly of the same weight, and together weighing 1.211 gramme, or 18.69 grs. troy, were planted on the 9th of May, in a soil of recently burned clay in rough powder. On the 16th of July, the plants began to bloom, each pea having furnished a stem bearing a single flower.

On the 15th of August the pods were quite ripe; the stems were then from 39 to 40 inches in height. The leaves were smaller than those of the same peas grown in manured earth. The length of the pods was about 1.27 inch, by a breadth of about 0.43 inch. Four of these pods each contained two seeds; the fifth had only one, but it was much longer than any of the others.

The nine peas gathered and dried in the sun, weighed 1.674 gram., or 25.84 grs.; after desiccation in vacuo, at 110° C., (230° F.,) they weighed 1.507 gram., or 23.26 grs. troy; on combustion they yielded 0.9 per cent. of residue.

The roots, the stems, the pods, and the leaves, dried at 230° F., weighed 3.314 gram., or 51.16 grs. troy; and by combustion gave 10 3 per cent. of ashes.

As the result of several experiments, it was ascertained that peas, exactly in the condition of those which had been planted, contained 91.4 per cent. of dry matter, and left by incineration 3.14 per cent. of residue. The five peas planted, taken as dry and free from ashes, would therefore have weighed 1.072 gram., or 16.54 grs. troy.

Analysis showed in the

	Peas sown.	Peas collected.	Straw and roots.
Carbon	48.0	54.9	52.8
Hydrogen	6.4	6.8	6.2
Azote	4.3	3.6	1.6
Oxygen	41.3	34.7	39.4
	100.0	100.0	100.0

RESULTS.

	Carbon.	Hydrogen.	Oxygen.	Azote.
Seeds 16.549, containing	7.950	1.065	6.523	0.710
Crop 68.560, " *	36.680	4.384	25.930	1.559
52.02 grs. by cultivation	+28.73	+3.319	+19.11	+0.849

From this experiment it appears that 16.549 grs. of seed found in the air, and obtained from the water with which they had been supplied during their growth, 52.02 grs. of elementary matter in the course of ninety-nine days' growth, during the warmest months of the year; and that the quantity of azote originally contained in the seed was more than doubled in the produce arrived at maturity.

* Peas	45.89
Straw and shells	22.66
Total weight of the crop	68.55

THIRD EXPERIMENT.

GROWTH OF WHEAT.

Forty-six wheat corns were sown in burnt sand at the beginning of the month of August. At the end of September, the stalks were from fourteen to fifteen inches in height. The greater number of the lower leaves were yellow. The roots were of very considerable length, and formed a kind of mat, which made it difficult to wash and free them from sand.

RESULTS OF THE ANALYSIS.

	Seeds.	Crop.
Carbon	46.6	48.2
Hydrogen	5.8	5.8
Azote	3.45	2.0
Oxygen	44.15	44.0
	100.00	100.0

RESULTS.

	Grs.		Carbon.	Hydrogen.	Oxygen.	Azote
The seed dried	25.38	containing	11.84	1.46	11.19	0.87
The seed dried	46.65	"	22.47	2.67	20.57	0.92
Gain by culture	21.27		+10.63	+1.21	+9.38	+0.05

In the course of three months' growth, therefore, the weight of the seed had, so to speak, doubled; but the grain azote was scarcely appreciable. Nevertheless, this experiment upon the wheat had been conducted under precisely the same circumstances as that made upon the clover. The two crops grew in the same apparatus; they were watered with the same water, which they received very nearly in the same quantity; the seed was even sown in vessels having exactly the same extent of surface, in order that either crop might be exposed to the same chances of error arising from the accidental presence of dust in the atmosphere.

The plants produced under the circumstances indicated were far from presenting the vigor which they would have shown had they been grown in the open field. After three months of growth, the clover was much less forward than some which had been sown, for comparison, in a manured and gypsumed soil at the same time. The wheat showed the same weakness; and after the second month, I observed that each new leaf which was developed upward in the stem, caused one of those at the lower part to droop and grow yellow. The peas, although they reached maturity, had much smaller leaves, and both fewer and smaller seeds than similar plants grown at large.

It is well known that it is in great part due to the fertility of the soil in which seeds are grown that the health and vigor of young plants must be ascribed. A celebrated agriculturist, Schwartz, ascertained, for example, that young coleworts or cabbage plants exhausted in a remarkable manner the soil in which they were raised for transplantation. The good effects of the first nourishment obtained in a well-manured soil must extend subsequently to every part of the vegetable; and it is easily understood that a plant which has languished in its earliest periods of existence can never acquire a good constitution afterwards.

It therefore became interesting to carry out experiments of the nature of those already related, in connection with plants vigorously organized, and which had been raised in the first instance in a fertile soil.

FOURTH EXPERIMENT.

GROWTH OF CLOVER.

In a field of clover sown in the spring of the preceding year, several plants as like one another as possible were chosen. The earth adhering to the roots was removed by careful washing under a small stream of water; the plants were then made dry between leaves of blotting paper, and exposed for a few hours in the air. Three of these plants preserved for analysis weighed when green 6.750 grammes, or 104.20 grs. troy.

Three other plants, weighing 6.820 gram. or 105.28 grs. troy, were set in sand recently calcined and moistened with distilled water. The transplanting took place on the 28th of May, and the plants were forthwith protected from dust.

For some days they seemed to languish, but by and by they became remarkably vigorous. In a month the clover had grown to twice its original height, and the leaves were of the most beautiful green: the plants had in all respects as fine an appearance as the clover of the same age which had been left growing in the field. The flowers showed themselves upon the 8th of July, and by the 15th the flowering was complete: an end was put to the experiment on the 1st of August.

RESULTS OF THE ANALYSIS.

	BEFORE CULTURE.	AFTER CULTURE.
Carbon	43.42	53.00
Hydrogen	5.40	6.51
Azote	3.75	2.45
Oxygen	47.43	38.14
	100.00	100.00

RESULTS.

The trefoil transplanted, weighed when dry and freed from ashes	13.64
After sixty-three days' culture on barren soil, it weighed	34.96
Gained during culture	21.32

		Carbon.	Hydrogen.	Oxygen.	Azote.
The plant contained:	before culture	5.92	0.74	6.46	0.50
"	after culture	18.52	2.23	13.32	0.864
	Difference	+ 12.60	+ 1.49	+ 6.86	+ 0.35

Thus in two months' growth at the cost of the air and water, the clover had, so to say, tripled its quantity of organic matter; and the weight of azote contained in it was very nearly doubled.

FIFTH EXPERIMENT.

VEGETATION OF OATS.

I always failed in my attempts to transfer wheat plants from the ordinary soil in which the grain had been sown to barren sand; they never survived the transplantation. It was not different with oat plants; they also always died. It was at first supposed that the delicate radicles of these plants had been injured in the process of taking them up and freeing their roots from adhering vegetable soil; but I soon saw that this could not have been the case, for the same plants, treated precisely in the same manner, took very promptly when transplanted to garden mould, and even when they were put with their roots in pure water. It was with water, therefore, that the following experiment was conducted.

June 20th, several oat plants were taken up from a field, and their roots were washed and cleansed.

Three plants preserved for analysis, weighed 159.011 grs.

Four plants, the subjects of experiment, weighed 221.844 grs. troy. They were protected from dust, their roots dipping into a vessel containing distilled water, which was regularly kept up to the same level. By the middle of July the stalks of these plants had grown to twice their former length; and at this time it would have been difficult to have distinguished them from those growing in the open field. By the end of July the clusters had formed; and on the 10th of August the grain seemed ripe. It was, therefore, taken up and dried in the stove, and reduced to powder to complete the desiccation at 110° cent. (230° Fahr.)

ANALYSIS OF THE CROP.

	Transplanted.	Gathered from the field.
Carbon	53.0	48.0
Hydrogen	6.8	6.2
Oxygen	36.4	44.0
Azote	3.8	1.7
	100.0	100.0

SUMMARY.

	Carbon.	Hydrogen.	Oxygen	Azote.
The oats when transplanted contained	12.967	1.636	8.770	0.910
After 48 days of growth in distilled water they contained	23.157	2.979	21.180	0.818
	+10.190	+1.343	+12.410	−0.092

The analysis, therefore, indicates a trifling loss of azote.

In recapitulating the conclusions obtained from these experiments, we find:

First. That trefoil and peas grown in a soil absolutely without manure, acquired a very appreciable quantity of azote, in addition to a large quantity of carbon, hydrogen, and oxygen.

Second. That wheat and oats grown in the same circumstances, took carbon, hydrogen, and oxygen from the air and water around them; but that analysis showed no increase of azote in these plants after their maturity.

The mode of experimenting followed had it in view simply to determine the assimilation of azote by certain vegetables, without entering into the question of the means by which this was effected; and, indeed, in reference to the point, I can only offer conjectures.

Azote may enter the living frame of plants directly, or, as M. Piobert has maintained, in the state of solution in the water, always aerated, which is taken up by their roots.* The observations of vegetable physiologists are not generaly favorable to this view. It is farther possible that the element in question may be derived from ammoniacal vapors, which, according to some philosophers, exist in infinitely small proportion in our atmosphere. These vapors, dissolved by rains and dews, would readily make their way into plants, and might there undergo elaboration.

It is long since Saussure alluded to the probable influence of ammoniacal vapors upon vegetation. Prof. Liebig has more recently maintained the same opinion, and has taken particular pains to prove that rain-water always contains a very minute quantity of carbonate of ammonia.

To this cause, which must have the effect of infusing an azotized principle into the tissues of plants, must be added another, which is perhaps not the least energetic. It is this, that under certain electrical influences, of which M. Becquerel has made a particular study, hydrogen in the nascent state, in contact with azote, may actually give rise to ammonia. By means of this view, it becomes easy to conceive how non-azotized organic substances, under the mere influence of the putrid fermentation, might give origin to ammoniacal salts, which would then exercise a fertilizing action on the soil.

During the growth of plants, a portion of the water absorbed by the roots is evidently assimilated; and this circumstance enables us to conceive the formation of many of the immediate principles of vegetables, the chemical composition of which is precisely represented by carbon and the elements of water; such as starch, sugar, etc. We can also understand the presence of those principles, which have further a certain proportion of oxygen in excess, inasmuch as we have ascertained that during the decomposition of carbonic acid by the green parts of vegetables, the whole of the oxygen is not eliminated. But there are substances elaborated by plants which, with reference to oxygen, contain a quantity of hydrogen much greater than is requisite to form water; such are the resins and other carburets of hydrogen in the cone-bearing trees, and the fat oils in the oleaginous seeds. This excess of hydrogen led several physiologists to conclude that water was decomposed in the course of vegetation,—that there was fixation of its hydrogen and disengagement of its oxygen gas.

* Piobert, Mém. de l'Académie de Metz, 1837.

Nevertheless, the presence of hydrogen in excess in certain immediate vegetable principles is no decisive proof of the disjunction of the elements of water; and if no definitive conclusion has been come to on the point, up to the present moment, it is because these hydrogenized principles are produced in plants which live under the influence of certain organic substances that are met with in the soil, where they act as manures, their composition being always complex, and often highly hydrogenized.

The experiments of M. de Saussure do not lead us to suspect the decomposition of water; inasmuch as by keeping plants for a whole month, under receivers filled with atmospheric air freed from carbonic acid, no apparent evolution of oxygen was observed. Operating in the same manner with air containing a certain proportion of carbonic acid, the quantity of oxygen disengaged was always less than that which entered into the constitution of the acid decomposed.

This is the place to observe, and in connection with these very experiments of M. de Saussure, how little satisfactory this partial decomposition of carbonic acid, which corresponds to no definite proportion, appears. We already feel the difficulty of conceiving that this acid should be completely reduced by a living plant; that is to say, that the whole of its carbon should become assimilated. The entire separation of a body so greedy of oxygen as carbon from its most highly oxygenated compound, must needs excite the greatest astonishment.

The readiest conception suggested by the facts is this; that by the agency of the solar light, and under the influence of the green matter, carbonic acid is turned into carbonic oxide by losing a portion of its oxygen. This modification appears more in conformity with the ascertained principles of chemical and physiological science. Still it must be allowed, that facts agree as little with this mode of viewing the question as with that which assumes the entire decomposition of the carbonic acid. On the first assumption, the proportion of oxygen set at liberty is too small; in the second, it is too great.

The negative results of M. de Saussure, in relation to the separation of the elements of water during vegetation, were obtained in the absence of carbonic acid, whilst the experiments which established the decomposition of this latter body, were necessarily made under the influence of moisture. It is possible, therefore, that the water and the carbonic acid underwent simultaneous decomposition; and it becomes interesting, taking this view, to inquire whether the hypothesis according to which carbonic acid undergoes transformation into carbonic oxide does not acquire a certain degree of probability by calling in the effect of the decomposition of water in the phenomena observed.

One volume of the gaseous oxide of carbon takes half a volume of oxygen gas to form one volume of carbonic acid. Reciprocally, one volume of carbonic acid gas, in undergoing transformation into the oxide of carbon, will give one volume of the oxide, $+\frac{1}{2}$ a volume of oxygen gas.

Thus, in the hypothesis which we now discuss, for each volume of carbonic acid that is modified by the vegetation, there will be half a volume of oxygen gas disengaged. Any oxygen more than this half volume which appears, must be regarded as proceeding from the decomposition of water, the hydrogen of which will have been assimilated by the plant at the same time as the carbonic oxide derived from the carbonic acid; and this view would perhaps enable us to conceive how the volume of oxygen which is disengaged during the process of vegetation, may exceed the volume which ought to be produced, if the carbonic acid decomposed really passed into the state of carbonic oxide.

We may perchance obtain a more convincing proof of the separation of the elements of water, in analyzing plants grown in a soil absolutely without any organic matter capable of affording them hydrogenous elements.

In fact, if a plant, which is grown under such circumstances, contains hydrogen in any larger proportion than that which were necessary to transform its oxygen into water, we might conclude, with some certainty, that the elements of water had been separated; the objection made on the score of the presence of manure would then be got rid of entirely. The analyses which have already been laid before the reader supply data for this investigation; it has only to be ascertained whether, in the elements gained in the course of vegetation, the hydrogen is in excess with reference to the oxygen or not. The following table presents a summary view of our experiments:

		Oxygen assimilated.	Hydrogen assimilated.	Hydrogen forming water.	Hydrogen in excess.
Experiment 1.	Trefoil	18.926	2.717	2.362	0.355
Experiment 2.	Peas	19.096	3.319	2.392	0.926
Experiment 3.	Wheat	9.386	1.204	1.173	0.030
Experiment 4.	Transplanted Trefoil	6.854	1.495	0.849	0.646
Experiment 5.	Oats	12.410	1.343	1.343	" "

In the four first experiments, the hydrogen gained evidently exceeds very sensibly the quantity required by the oxygen to form water. The experiment with the oats, indeed, presents an exception; but it must be remembered that here a loss of azote was ascertained. These analyses, therefore, appear to indicate an assimilation of hydrogen in the course of vegetation, in consequence of a decomposition of water analogous to that of carbonic acid, and very probably effected by the same means.

§ III.—OF THE INORGANIC MATTERS CONTAINED IN PLANTS—THEIR ORIGIN—OF THE CHEMICAL NATURE OF SAP.

When a plant is burned, there always remains a residue, which is commonly designated as the ash. Every part of a plant gives a residue of the same essential kind; but it varies in its quantity and somewhat also in its composition. Equal weights of dry herbaceous plants leave more ashes than woody plants.* In a tree, the trunk gives more ash than the branches, and these give less than the leaves.† The residue left by the combustion is commonly composed of salts—alkaline chlorides, with bases of potash and soda, earthy and metallic phosphates, caustic or carbonated lime and magnesia, silica, and oxides of iron and of manganese. Several other substances are also met with there, but in quantities so small that they may be neglected.

The principles usually met with in the ashes of vegetables are always found in the soil which exercises the greatest influence upon the nature and quantity of the saline and earthy matters which remain after the combustion of plants. Those which grow in a soil derived from silicious rocks, yield ashes that are richer in silica than those that are produced in a calcareous soil. But, according to M de Saussure, the quality of the manure has a still more decided influence on the nature of the ash than the geological constitution of the soil; according to this observer, plants of the same species, which have grown upon a calcareous sand, and upon a granitic sand, contain the same kind of ashes, if they have been manured with the same dung; and different species, although growing in the same earth, do not contain the saline and earthy constituents of their ashes in the same proportions.‡

* Kirwan, Memoirs of the Royal Irish Academy, vol. v.
† Pertuis, Annales de Chimie, 1re série, t. xix.
‡ Saussure, Recherches chimiques, p. 283.

QUANTITY OF ASHES CONTAINED IN THE DIFFERENT PARTS OF VEGETABLES, ACCORDING TO M. DE SAUSSURE.*

NAME OF THE PLANT.	Times when taken for analysis.	Ashes.
Oak leaves	10 May	0,053
Ditto do.	27 September	0,055
Oak branches barked	10 May	0,004
Bark of these branches	. . . .	0,060
Oak wood distinct from alburnum	. .	0,002
Alburnum from same wood	. .	0,004
Bark of same tree	. . . .	0,060
Liber of the preceding bark	. . . .	0,073
Leaves of poplar	May	0,066
Ditto ditto	September	0,093
Trunk of ditto	. . . .	0,008
Bark of trunk of ditto	. . . .	0,072
Spanish mulberry-tree wood	. . . .	0,007
Alburnum of mulberry	. . . .	0,013
Bark of ditto	. . . .	0,089
Liber of ditto	. . . .	0,088
Chestnut-tree leaves	10 May	0,072
Ditto ditto	23 July	0,084
Flowers of chestnut-tree	10 May	0,071
Ripe chestnuts	5 October	0,034
Peas flowering	. . . .	0,095
Peas in pod	. . . .	0,081
Beans in flower	. . . .	0,122
Ditto in pod	. . . .	0,066
Bean straw	. . . .	0,115
Beans	. . . .	0,033
Jerusalem artichoke in flower	. . . .	0,137
Ditto in seed	. . . .	0,093
Wheat straw	. . . .	0,043
Wheat	. . . .	0,013
Bran	. . . .	0,052
Indian corn straw	. . . .	0,084
Indian corn	. . . .	0,010
Barley straw	. . .	0,042
Barley	. . . .	0,018
Oats	. . . .	0,031
Pine-tree leaves (Jura)	20 June	0,029
Ditto ditto (Brocken)	20 June	0,029
Pine-tree branches without leaves	20 June	0,015

All these estimates of ashes refer to plants dried during several weeks in a stove heated to 25° cent. (77° Fahr.) By such drying, however, vegetable substances are very far from losing the whole

* Saussure, Recherches chimiques, p. 283.

of the water which they contain. The quantities of ashes, therefore, mentioned by M. de Saussure, if they be referred to vegetables absolutely dry, are somewhat too small.

I present a few estimates of ashes from analyses which I have had occasion to make of some of those plants which are the usual subjects of cultivation with us. The drying here was always performed with care in an oil-bath heated to 110° cent. (230° Fahr.)*

Substance dried at 230° Fahr.	Ashes.	Substance dried at 230° Fahr.	Ashes.
Wheat straw	0,070	Turnip	0,076
Wheat	0,024	Jerusalem artichoke	0,060
Rye straw	0,036	Stems of ditto	0,028
Rye	0,023	White peas	0,031
Oat straw	0,051	Pea straw	0,113
Oats	0,040	Clover hay	0,077
Potatoes	0,040	Meadow hay	0,090
Beet-root	0,063	After grass (meadow)	0,100

We owe to M. Berthier† the following results of the incineration of different kinds of wood burned in the state in which they are generally used.

Kind of wood.	Ashes.	Kind of wood.	Ashes.
Fir	0,0083	Poplar bark	0,0020
Birch	0,0100	Box wood	0,0036
False ebony	0,0125	Oak barked, ash, pine, birch, &c.	0,0040
Hazel	0,0157	Thorn	0,0050
White mulberry	0,0160	Aspin tree	0,0060
Saint Lucia wood	0,0160	Oak bark	0,0120
Elder	0,0164	Black wood	0,0149
Judea-tree	0,0170	Mahogany	0,0160
Oak (branches)	0,0250	Ebony	0,0160
Oak bark	0,0600	Oak (fagot)	0,0220
Lime-tree	0,0500	Fearns	0,0450

We possess several analyses of ashes from different parts of the same plants in the researches of MM. de Saussure and Berthier. As the knowledge of these saline substances may prove highly important in our agricultural applications, and as it further completes, in some sort, the facts that bear upon the chemical phenomena of vegetation, I here add a table of the results obtained by the skilful analysts just quoted:

* Ann. de Chimie, t. i. page 234. 3e. série.
† Traité des Essais, t. i, page 259.

COMPOSITION OF THE SUBSTANCES FOUND BY M. DE SAUSSURE.

No. of analysis.	Plant, or particular part which yielded the ash.	Earthy phosphates.	Phosphate of potash.	Sulphate of potash.	Chloride of potash.	Carbonate of potash.	Earthy carbonates.	Silica.	Metallic oxides	Loss.
1	Chestnuts	0,120	0,280	given along with the chloride.	0,030	0,510	"	0,005	0,003	0,002
2	Menyanthes trifoliata in flower (buckbean)	0,150	"	Idem.	0,120	0,572	0,050	0,020	0,005	0,083
3	The same cleared of seed	0,060	"	0,020	0,140	0,310	0,375	0,028	0,007	0,060
4	Beans	0,259	0,439	0,020	0,009	0,225	"	"	0,005	0,022
5	Wheat straw	0,062	0,050	0,020	0,030	0,125	0,010	0,615	0,010	0,078
6	Selected wheat	0,445	0,320	traces.	0,002	0,150	"	0,005	0,002	0,076
7	Wheat bran	0,465	0,300	"	0,002	0,140	"	0,005	0,002	0,086
8	Indian corn straw	0,050	0,097	0,013	0,025	0,590	0,010	0,180	0,005	0,030
9	Indian corn	0,036	0,475	0,002	0,003	0,140	"	0,010	0,001	0,083
10	Barley straw	0,078	"	0,035	0,005	0,160	0,125	0,570	0,005	0,022
11	Barley in the husk	0,325	0,092	0,015	0,003	0,180	"	0,355	0,003	0,027

In his researches upon the same subject, M. Berthier determined the relation of the insoluble to the soluble matters in each species of ash examined; and the two kinds of salts were then analyzed separately.

ALKALINE SALTS AND INSOLUBLE SUBSTANCES CONTAINED IN THE ASHES OF:

	Alkaline salts.	Insoluble salts.	Nature of soil.
Horn-beam	0,189	0,811	Clayey, sandy and ferruginous.
Oak	0,120	0,880	Very dry, argillaceous, calcareous soil.
Oak bark	0,050	0,750	
Lime-tree	0,108	0,892	Sandy, somewhat calcareous.
Wood of St. Lucia	0,160	0,840	The same.
Elder	0,315	0,685	
Judea-tree	0,190	0,810	Ditto.
Hazel	0,154	0,846	Ditto.
Chinese mulberry	0,189	0,811	Ditto.
White mulberry	0,150	0,850	Ditto.
Ditto ditto	0,250	0,750	Calcareous and argillaceous soil.
Orange or lance-wood	0,096	0,904	Grown in the open field.
White oak	0,075	0,925	
Birch	0,160	0,840	Sandy clay.
False ebony	0,315	0,685	
Fir	0,500	0,500	From Norway, part of a plank.
Wheat straw	0,090	0,810	Grown in a strong and calcareous soil.
Potato stems	0,042	0,958	Silicious and calcareous sand.

ALKALINE SALTS AND INSOLUBLE SUBSTANCES OF ASHES

ACCORDING TO M. BERTHIER.

		Horn-beam.	Oak.	Oak-bark.	Lime-tree.	St. Lucia-tree (Prunus.)	Elder.	Judea-tree.	Hazel.	Chinese mulberry.	Ditto. ditto.	White mulberry.	Orange or lance-wood.	White oak.	Birch.	False ebony.	Firwood.	Wheat straw.
Soluble compounds.	Carbonic acid		0,240	0,232	0,274	0,200	0,240	0,249	0,202	0,226		0,230	0,370		0,170	,,	0,302	
	Sulphuric acid		0,081	0,060	0,075	0,060	0,064	0,031	0,051	0,080		0,083	,,		0,023	0,080	0,031	0,020
	Hydrochloric acid		0,001	0,007	0,018	0,100	0,004	0,005	0,005	0,004		0,040	0,040		0,002	0,020	0,003	0,130
	Silica		0,002	0,008	0,016	0,010	0,002	0,010	0,005	0,010		,,	,,		0,010	0,017	0,010	0,350
	Potash											0,520				,,		
	Soda		0,676	0,693	0,607	0,630	0,670	0,705	0,732	0,680		0,115	0,590		0,793	,,	0,654	0,500
	Water															,,		
			1,000	1,000	1,000	1,000	1,000	1,000	1,000	1,000		0,988	1,000		1,000	,,	1,000	1,000
Insoluble compounds.	Carbonic acid	0,332	0,396	0,385	0,398	0,340	0,314	0,340	0,370	0,187	0,271	0,420	0,335	0,414	0,310	0,190	0,230	,,
	Phosphoric acid	0,100	0,008	,,	0,028	0,063	0,083	0,075	0,048	0,054	0,116	0,018	0,019	0,030	0,043	0,184	0,042	0,012
	Silica	0,050	0,038	0,011	0,020	0,018	0,032	0,024	0,044	0,131	0,077	0,029	0,060	0,033	0,055	0,080	0,080	0,750
	Lime	0,386	0,548	0,501	0,518	0,488	0,492	0,460	0,424	0,556	0,467	0,461	0,450	0,503	0,522	0,456	0,398	0,058
	Magnesia	0,078	0,006	0,008	0,012	0,070	0,025	0,072	0,044	0,072	0,052	0,046	0,070	0,010	0,030	0,090	0,044	,,
	Oxide of iron	0,016	,,	,,	0,001	0,005	0,011	0,013	0,040	,,	0,003	0,005	0,010	0,010	0,005	,,	0,141	0,025
	Oxide of manganese	0,034	,,	0,074	0,006	0,008	0,018	0,007	,,	,,	0,005	0,013		,,	0,035	,,	0,060	,,
	Carbon and loss	,,	,,	0,021	,,	,,	,,	,,	,,	,,	,,	,,	0,056	,,	,,	,,	,,	0,155
		0,996	0,996	1,000	0,993	0,992	0,995	0,991	0,868	1,000	0,991	1,000	1,000	1,000	1,000	1,000	0,995	1,000

COMPOSITION OF THE ASHES OF SEVERAL PLANTS ANALYZED BY M. BERTHIER.

	Fern.	Wheat straw.	Horse-tail grass.	Heath.	Tansy.	Observations.
Sulphate of potash...	0,007	0,004	0,120	0,050	0,033	
Chloride of potassium		0,032	0,114	0,012	0,090	
Carbonate of potash..	,,		,,	0,068	0,167	The wheat straw was from a strong calcareous soil.
Silicate of potash....	,,	0,130	,,	,,	,,	
Silica...............	0,730	0,715	0,505	0,375	0,165	
Carbonate of lime....	0,248	0,096	0,062	0,280	0,434	
Sulphate of lime.....	,,	,,	0,144	,,	,,	
Phosphate of lime ...	0,010	0,023	0,022	0,130	0,100	The tansy was from a sandy garden soil.
Magnesia............	0,005	,,	0,030	0,010	0,002	
Oxide of iron........	,,	,,	,,	0,014	0,007	
Oxide of manganese..	,,	,,	,,	0,061	0,002	

A remark made by Berthier, and arising out of the preceding analyses, is the absence of alumina in the constituent principles of the ashes examined. The results previously obtained by M. de Saussure fully confirm this remark; and if in some cases traces of alumina were detected, the circumstance was attributed to the clay which might accidentally have adhered to the plants. According to M. Berthier the absence of alumina is probably owing to its insolubility in water, and its weak affinity for the organic acids. The soluble salts of alumina with mineral acids are, it is well known, unfavorable to vegetation, and in an arable soil they could not exist along with calcareous or alkaline carbonates: they would be immediately decomposed.

However, alumina appears actually to have been observed in the state of salt in the juices of certain plants: *lycopodium complanatum*, an infusion of which is employed as a mordant in dyeing, contains tartrate of alumina;* the same salt has been detected in verjuice; and as we shall see presently, Vauquelin found acetate of alumina in the sap of the birch-tree. I may add, that in a considerable number of analyses of ashes, produced from plants and seeds of my own growing, I always obtained traces of alumina: but I would not venture to affirm that the earth here was not accidental.

Silica is met with in only very small quantity in the ashes of wood. It is found, on the contrary, in considerable proportion in the ashes of several annual and biennial plants, and more especially in those of the cereals. Sir Humphrey Davy found silica in the epidermis of the Indian rush.

If we compare the ashes of the same species of wood grown in soils of different kinds, we see, says M. Berthier, that they may differ very perceptibly; which seems to establish the fact that the soil exercises a certain degree of influence on their constitution. Thus oak-wood from Roque des Arcs, grown in a decidedly calcareous soil, yielded ashes almost entirely consisting of carbonate of lime,

* Berzelius, *Traité de Chimie*, t. v. p. 130, French translation.

while those left by an oak from the department of la Somme, contained much magnesia and phosphate of lime.* The ashes from a white mulberry of Nemours contained more than 0.10 of phosphoric acid, while scarcely any traces of it were found in those of a similar mulberry from the calcareous soil of Provence. The most remarkable inference deducible from the analyses of M. Berthier, is that which is connected with the composition of the ashes yielded by trees growing in the same soil. It is observed that, for analogous species, the ashes also bear the closest analogy; and on the contrary, it is found that trees of very distinct genera yield ashes of quite a different quality; results which lead to this important conclusion, that plants possess the faculty of selecting in the soil the substances which are best suited to their special organizations. This is a point which we shall have an opportunity of discussing, when we come to treat of rotations of crops.

The substances composing the ashes of vegetables, are not all in the state in which they existed in the vegetable tissue. In plants there constantly exist organic acids, which, in general, are combined with mineral bases. During the incineration of plants these organic acids are destroyed, and the result of their destruction is an alkaline carbonate, if the pre-existing acid was united with soda or potash; a calcareous vegetable salt, again, yields carbonate of lime; and a magnesian salt gives magnesia, from the well-known inability of this earth to retain carbonic acid at a high temperature. Thus, the greater part of the carbonates which enter into the composition of vegetable ashes, are formed by the mere fact of incineration. The salts which resist the action of a strong heat, as the phosphates, sulphates, and chlorides, are the only ones which in the ashes retain the state in which they existed in the living plant.

Water being the vehicle which must convey the mineral salts from the soil into the vegetable, we do not always perceive how they can penetrate. To explain the presence in plants of a salt so insoluble as the neutral phosphate of lime, M. de Saussure admits, from satisfactory experiments, that vegetable juices contain the double phosphate of potash and lime, and of potash and magnesia.† Besides, several bodies considered in chemistry as insoluble, are not so absolutely. Silica seems to possess a certain degree of solubility,—at least, M. Payen has met with it in considerable quantity in the water of the Artesian well of Grenelle, and in the water of the Seine. We know, moreover, that several insoluble earthy salts are dissolved in virtue of the carbonic acid always contained in the waters with which the soil is soaked. Lastly, it is not improbable that certain insoluble salts have their origin in the plant itself, engendered there by the successive arrival and reciprocal action of soluble salts.

It now remains for us to examine by what means saline substances are introduced into the tissues of vegetables, and within what limits

* These ashes were from the carbon of the oak; the insoluble part gave 0.14 of phosphate of lime, and 0.08 of magnesia.

† Saussure, Recherches chimiques, &c. p. 321.

the water, which is essential to living plants, may be charged with them; for it is within what may be called common experience, that saline solutions of certain degrees of concentration, oftentimes act injuriously on vegetation.

The spongioles which terminate roots have too close a tissue to allow any thing but fluids to pass through them. All attempts to make them absorb solid bodies in a state of minute division, and held in suspension in water, have been ineffectual. In these attempts the spongioles have acted precisely like perfect filters, with which those that we employ in our laboratories cannot be compared. Further, the weakest solutions are not entirely absorbed by certain roots; a kind of separation takes place; a portion of the dissolved salt appears to abandon the water at the moment of its penetrating the spongiole. This follows from the researches of M. de Saussure, instituted with a view to ascertain, 1st. If plants absorb substances dissolved in water in the same proportion as they absorb water;* 2dly. If plants make a selection among different substances held in solution in the same liquid.†

In solutions severally containing eight ten-thousandths (0.0008) of each of the following substances—chloride of potassium, chloride of sodium, nitrate of lime, sulphate of soda, hydrochlorate of ammonia, acetate of lime, sulphate of copper, sugar-candy, gum arabic, and extract of humus,‡—several entire plants with their roots, of the *polygonum persicaria*, (lakeweed or redshanks,) which had lived for some time in distilled water, until their roots had commenced growing, were immersed.

The plants lived in the shade during five weeks, throwing out roots in one of the solutions mentioned. They languished, without showing any appearance of growth in the solution of hydrochlorate of ammonia, and could only be kept alive in the sugared water, which soon became changed, by renewing it frequently. They died at the end of from eight to ten days, in gum water, and in the solution of acetate of lime. They held out but for two or three days in the water which contained sulphate of copper.

Observations precisely similar made on the *bidens cannabina* presented the same results, with the sole difference, that this plant lived for much shorter times than the redshanks.

To estimate in what proportion the substances dissolved were absorbed, relatively to the water, M. de Saussure made use of the solutions previously employed; but he brought the experiment to a close when the plants had taken up precisely half the liquid which was feeding them. Each solution fed a sufficient number of plants to allow of this condition being fulfilled in two days. Half of the liquid remaining after each experiment was analyzed, and the quantity of salt found therein, showed, by the difference between this and the quantity originally contained, the amount which had penetrated the vegetable. Representing by one hundred parts the whole

* Saussure, Recherches chimiques, &c. p. 247. † Ibid. page 253.

‡ This entered into the solutions only in the proportion of two ten-thousandths 0.0002.)

of the substance originally dissolved, it is evident that fifty of those parts must enter the plant, if the absorption of the saline substances be in proportion to that of the solvent. But the experiment proved, that in taking up half the volume of the liquid, the polygonum had absorbed but—

15 parts of chloride of potassium,
13 " chloride of sodium,
4 " nitrate of lime,
14 " sulphate of soda,
12 " hydrochlorate of ammonia,
8 " acetate of lime,
29 " sugar,
9 " gum,
5 " extract of humus,
47 " sulphate of copper.

Under the same circumstances the bident took—

16 parts of chloride of potassium,
15 " chloride of sodium,
8 " nitrate of lime,
10 " sulphate of soda,
17 " hydrochlorate of ammonia,
8 " acetate of lime,
32 " sugar,
8 " gum,
6 " extract of humus,
48 " sulphate of copper.

It follows from these experiments that the plants absorbed some part of the different substances presented to them; but without exception, they took up the water in greater proportion than the matters dissolved.

On multiplying and varying these experiments, M. de Saussure always arrived at the same general results. The plants uniformly took up more of the alkaline than of the calcareous salts, and more sugar than gum, though the quantities of the different substances absorbed varied considerably.

The sulphate of copper presented, in the course of these researches, an anomaly which is readily explained. We see that this salt, evidently injurious to vegetation, was taken up in a large dose. This arises from its corrosive action on the roots: sulphate of copper disorganizes the spongioles; and these organs once destroyed, absorption takes place with more rapidity and in greater abundance. A root deprived of spongioles is in the condition of a stalk, or branch, the fresh section of which is immersed in a liquid. Observation proves, in fact, that substances in a state of solution, which by reason of their viscidity are incapable of making their way into a healthy root, are, on the contrary, readily taken up by a cut stalk or branch, in quantity sufficient to dye it deeply, if it was a coloring matter that was presented for absorption.

In the preceding experiments, the solution contained only a single substance. In those which follow, M. de Saussure dissolved in the water two or three salts, a mixture of sugar and gum, &c., in order to ascertain whether the plants would make any selection from mixed solutions.

In 25 fluid ounces of water two or three species of salt were dissolved, the weight of each species being nearly 10 grains troy. Each ounce of water would therefore contain either $\frac{4}{5}$ths or $\frac{6}{5}$ths of a grain of saline or soluble matter. As in the preceding experiments, the plants were made to absorb precisely one half of the solutions. Analysis pointed out the quantity and the nature of the substances which remained in the liquid not absorbed, and consequently the salts which had penetrated the vegetable.

In reducing this table, which exhibits the results obtained, the weight of each particular salt in the solution is represented by 100 parts.

Substances in the solution with which the experiment was made.	Weight of the several substances taken up by the Polygonum in imbibing one half of the solution.	Weight of the several substances taken up by the Bident in imbibing one half the solution.
100 parts by weight.		
Sulphate of soda effloresced .	12	7
Chloride of sodium . .	22	20
Sulphate of soda effloresced .	12	10
Chloride of potassium . .	17	17
Acetate of lime . .	8	5
Chloride of potassium . .	33	16
Nitrate of lime . . .	4	2
Hydrochlorate of ammonia .	16	15
Acetate of lime . . .	31	35
Sulphate of copper . .	34	39
Nitrate of lime . . .	17	9
Sulphate of copper . .	34	56
Sulphate of soda . . .	6	13
Chloride of sodium . .	10	16
Acetate of lime . . .	traces	traces
Gum	26	21
Sugar	34	46

M. de Saussure confirmed these results in experimenting on the common peppermint, (*mentha piperita*,) Scotch pine, and common juniper. The substances absorbed in greatest proportion by the polygonum and bident were also those that were taken up in largest quantities by these plants.

The section of the roots, even their destruction, favors, as we have already said, the introduction of the matters dissolved. Plants whose roots had been removed, no longer selected the substances dissolved in so striking a manner as they did previously; after mutilation they absorbed them almost indifferently, in larger doses, and perceptibly in the same proportion as the water which held them in solution.

Roots with their spongioles entire, therefore, suffer one substance in solution to penetrate the plant in preference to another. The chlorides of potassium and of sodium, for instance, find entrance more readily than the acetate and nitrate of lime; sugar more readily than gum; and precisely as when isolated, are these substances, when combined, absorbed in much less proportion than the menstruum or water of solution.

M. de Saussure is not disposed to admit that the preference which plants evince for certain salts, for certain dissolved substances, results from any particular faculty, from any special affinity. He rather inclines to believe that it should be attributed to the degree of fluidity, or of viscidity communicated to the water by the different substances dissolved; thus the acetate and nitrate of lime, with the same proportion of liquid, form more viscid solutions, which pass with more difficulty through a filter, than the alkaline sulphates and chlorides; and these latter salts in solution were absorbed by vegetables in greater abundance than the calcareous salts. Gum, which imparts more viscidity to water than sugar, is also less capable of being absorbed. Finally, pure water, more fluid than any of the solutions tried, was also that which vegetables preferred to any other.

In the results of the incinerations which we have mentioned, it is obvious that in many plants salts are met with in very small proportion. This circumstance has induced several physiologists to consider the mineral substances found in vegetables as purely accidental, and consequently unnecessary to their existence. M. de Saussure, however, observes that this scantiness is no true indication of their inutility, and he mentions that the phosphate of lime, which forms an element in the organization of an animal, does not probably amount to one five-hundredth part of the entire mass. We shall add, that as the phosphate of lime was met with by M. de Saussure in the ashes of all vegetables which he examined, and as all the analyses performed since the original labors of this celebrated chemist have tended to confirm the accuracy of his general conclusions, we have no ground for supposing that plants could exist without the intervention of saline matter. There are annual plants which, when burned, leave more than 10 per cent. of residue; and vegetables cultivated in soils free from saline or alkaline matter, and watered with distilled water, though they will live and ripen their seeds in some instances, still they never acquire the vigor which they possess when grown in a fertile soil.

Duhamel ascertained that marine plants languish in soils destitute of chloride of sodium; and this is so much the more readily conceived, as those plants which furnish ashes abounding in carbonate of soda, always contain organic acids combined with the alkaline base. Borage, the nettle, &c., thrive only in places where they meet with nitrates ; and it is easy to discover that plants when dried contain a notable quantity of either nitrate of potash or of lime. The vine more especially requires alkaline dressings, in order that the large quantities of potash taken from the soil in the tartrate of potash of the grape may be replaced.

The organic acids, so different in their composition and in their properties, which are met with in the different vegetable families, are always found combined in the state of neutral or acid salts. The proportion of base combined in a plant with a vegetable acid may be readily ascertained from the ashes ; for by the effect of incineration the base passes into the state of an alkaline or earthy carbonate. The vegetable acids undoubtedly perform important functions in the organism of vegetables, and their formation probably depends on the influence of the bases with which they form salts. The nature of the oxide or base itself appears to be of little importance ; it is enough that it be present in the plant. It is known that certain bases may mutually replace each other, equivalent for equivalent.

These principles assumed, Prof. Liebig draws a remarkable inference from the composition of the ashes of different kinds of wood ; namely, that for each vegetable family the sum of the oxygen of the bases combined with the organic acids will be a constant number ; or, in other words, the species of one and the same family will contain the same number of basic equivalents combined with vegetable acids.

Thus, 100 parts of the ashes of a Breven pine-tree, analyzed by Saussure, contain :

Carbonate of potash	3.60	Oxygen of the potash	0.41	9.01 ox.
" lime	46.34	" lime	7.33	
" magnesia	6.77	" magnesia	1.27	
Carbonates	56.71			

The ashes of a pine from Mount La Salle yielded :

Carbonate of potash	7.36	Oxygen of the potash	0.85	8.95 ox.
" lime	51.69	" lime	8.10	
" magnesia	0.00			
	58.55			

M. Berthier found in the ashes of a fir-tree from Allevard the following bases :

Potash and Soda	16.8	Oxygen	3.42	12.82
Lime	29.5	"	8.20	
Magnesia	3.2	"	1.20	
	49.5			

One part of the alkalies containing 1.20 of oxygen was combined with mineral acids, forming sulphates, phosphates, and a chloride The oxygen of the bases combined with the carbonic acid is consequently reduced to 11.62.

The ashes of a Norway fir, according to the same analyst, containing :

Potash	14.10	Oxygen	2.40	12.84 oxygen.
Soda	20.70	"	5.30	
Lime	12.30	"	3.45	
Magnesia	14.35	"	1.69	
	51.45			

In this ash the bases belonging to the inorganic salts contain 1.37 of oxygen. The oxygen of the bases of the carbonates, or in other words of the bases which formed organic salts in the tree, therefore, becomes 11.47. The numbers 9.01 and 8.95 on the one hand, and 11.62 and 11.47 on the other, which represent the quantity of oxygen of the whole of the bases in the ashes obtained from plants of the same species, differ so little, that they may be considered as identical.

Accurate analyses of ashes of plants of the same species grown in soils of different kinds, will determine, says Prof. Liebig, whether the fact, deduced from the composition of the ashes of the pine and fir-tree, constitutes a definitive law.*

Be this as it may, the utility of alkalies in vegetation cannot be a matter of doubt; many usages in agriculture prove it in the clearest manner; and, according to M. Liebig, the fact of the formation of the organic alkaloides in plants affords an additional proof of it. M. Liebig thinks that the organic alkalies have a particular tendency to form in the absence of mineral bases; thus potatoes which germinate in cellars, under conditions where the soil cannot supply them with potash, soda, or lime, develop an organic alkali, solanine, which is not found in the tubers of this vegetable as usually cultivated.† In the cinchonas, quinine and cinchonine are combined with quinic acid; but there is frequently found quinate of lime also. According to the same chemist, the latter base holds the place of a vegetable alkali in the organism; the more prevalent it is in the soil, the less rich will the cinchona plant be in quinine and cinchonine.‡ These ingenious views certainly deserve the careful attention of physiologists; they are calculated to add new interest to the study of the chemical constitution of the ashes of vegetables.

The inorganic substances contained in vegetables evidently come from the soil. By growing seeds, as M. Lassaigne did, in flowers of sulphur, moistened with distilled water, the plant produced contained neither more nor less saline and earthy matter than was originally present in the seed.

The water absorbed by the roots, then, becomes charged during its stay in the ground with the various soluble substances which may be met with there, and which generally contribute to its fertility; such especially are the salts derived from decomposed organic substances. Water charged with small quantities of the soluble substances diffused through the soil, constitutes the ascending sap. When it has penetrated the plant, immediately after its passage by the spongioles of the roots, perhaps even while traversing these

* Liebig, Chimie Organique, Introduction, p. cxi.
† Liebig, idem. p. cxiv.
‡ Liebig, idem.

parts, the organic matters dissolved in the fluid appear to undergo important modifications; for in the sap substances are detected which could not have existed in the water which moistened the soil. During its ascent the sap increases in density, as was ascertained by Mr. Knight, according to whom, the sap of an *acer platanoides*, taken at the level of the ground, has a density of 1.004; at 6½ feet above it this density becomes 1.008, and at 13 feet, 1.012. From this Mr. Knight concluded that the sap took up nutritive matter deposited in the vegetable tissues which it traversed in its ascent.* We have already seen that the sap, after being elaborated in the green parts of trees, takes a route the reverse of that which it followed at first, and we therefore spoke of this modified sap as the descending sap. It is very possible that in Knight's observations the liquid examined was a mixture of the two saps.

We should not be over hasty in concluding that the action of the two species of sap was exerted separately in promoting the development of the plant; it is very probable, as Dutrochet thinks, that the modified sap, by insinuating itself into the permeable tissue of the vegetable, is continually mixed with the ascending sap, in order to concur in promoting the growth of the buds.† The difficulty of obtaining each particular sap separately, if such a separation is really possible, prevents the analytical conclusions we have from possessing all the accuracy that seems desirable.

Vauquelin has studied the sap of the birch-tree, of the hornbeam, of the beech, of the chestnut, and of the elm.

The *sap of the hornbeam* (*Carpinus sylvestris*) was obtained in the months of April and May. At this period it is colorless and clear as water; its taste is slightly saccharine; its odor resembles that of whey; it reddens turnsole paper. The sap of this tree contains water in very large quantity, sugar, extractive matter,‡ and free acetic acid, acetate of lime and acetate of potash, in very small quantity.

This sap left to itself presents in succession all the phenomena of the vinous and then of the acetous fermentation.§

The *sap of the birch-tree* reddens turnsole intensely; it is colorless, and has a sweet taste. The water which forms the greater part of it, holds in solution sugar, extractive matter, acetate of lime, acetate of alumina, and acetate of potash.

When properly concentrated by evaporation, it ferments on the addition of yeast, and then yields alcohol on distillation. The presence of the acetate of alumina may appear extraordinary in this sap, for this reason, that alumina has not yet been discovered in the ashes of the birch-tree.

Sap of the beech, (*Fagus sylvestris.*) The analysis was made in March and April. The color of the sap was a tawny red; it had the taste of an infusion of tanner's bark: it reddened turnsole slight-

* Decandolle, Physiologie, t. i. p. 204.
† Dutrochet, sur la Structure, &c. p. 36
‡ Probably azotized.
§ Vauquelin, Annales de Chimie, t. xxxi. p. 20, 1ère série.

y. It contained, in very small quantity, acetate of lime, acetate of potash, acetate of alumina, extractive matter, tannin acetic acid, and gallic acid.

The *sap of the chestnut-tree*, according to Vauquelin, who, for want of a sufficient quantity of the fluid, was able to study it but very superficially, contains mucilage, nitrate of potash, and the acetates of potash and lime.

The *sap of the elm* was examined at three periods; first, at the commencement of April, then some days after, and lastly, a month later. At the beginning of April its color was yellow, its taste sweet and mucilaginous; it was scarcely acid. Analysis indicated: water 1027.90, acetate of potash 9.23, organic matter 1.06, carbonate of lime 0.80.

At the second period it contained a little more extractive organic matter, and a little less carbonate of lime and acetate of potash. The last examination showed that these two salts had still further diminished in quantity. When exposed to the air, the sap of the elm undergoes decomposition, and becomes alkaline: the acetate of potash passes into the state of carbonate.

M. Regimbeau found in the *sap of the vine*,* bitartrate of potash, tartrate of lime, mucilage, and free carbonic acid.

The *sap of the maple-tree* contains a very considerable quantity of sugar. In Canada, this sap, properly treated, yields sugar which is identical with that of the cane. The nature of the sap is subject to variations; and Duhamel states, that at a certain season it loses its saccharine taste, and acquires an herbaceous flavor.†

Liebig and Will detected the presence of ammoniacal salts in the sap of the maple and birch-tree, and in the tears of the vine. M. Biot examined the sap of a considerable number of trees, and ascertained that the sugar in them often exists in two different states; in that of cane-sugar, properly so called, and in that of grape-sugar, which, as chemists admit, differs from the former only in the possession of an additional equivalent of water. The saps which M. Biot examined, contained besides some animal matter (*albumine*) and a gummy matter; he found no free carbonic acid. The object which he had in view, namely, to study the changes which occur in the nature of sugar, did not lead M. Biot to notice the minute quantities of salts with organic acids which Vauquelin met with in saps.

The trunk of a walnut-tree, tapped on the 11th of February, yielded a sap containing some cane-sugar. The saps of the sycamore, of the acer negundo, and of the lilac-tree, contained the same species of sugar; but that of the birch-tree held in solution some grape-sugar. In the sycamore and birch-tree, M. Biot observed an extremely interesting fact. He ascertained, on felling these trees, that the greater portion of the descending sap was accumulated towards the middle of the trunk. That of the birch-tree was acid and saccharine; the sap of that portion of the trunk which was

* Journal de Pharmacie, t. xviii. p. 36.
† Annales de l'Agriculture, Française, t. v. 2ème série, p. 339.

buried in the ground, contained no sugar, but a substance possessing the principal characters of gum.* It was probably an effect of the season; for Knight states, that he never could discover the least trace of saccharine matter during winter in the alburnum either of the stem or of the roots of the sycamore.†

SAP OF THE BAMBUSA GUADUAS.

The guaduas grows in the hot and marshy countries of the tropical regions; this grass frequently attains the enormous height of from 65 to 100 feet. Its stem, which is hollow, is divided through its entire length into joints spaced rather regularly at distances of from 11 to 12 inches. Each joint indicates the presence of a woody partition, which seems to divide the stem of the guaduas into so many super-imposed tubes. On perforating it immediately above a knot, a clear limpid fluid flows out, which cannot be distinguished from the purest water. This indeed is a store of water of which travellers have frequently availed themselves. This sap, as I have been assured by the inhabitants of the countries where I observed the guaduas, never completely fills the hollow space included between two joints. Analysis satisfied me that the sap of the guaduas is almost pure water. Re-agents detected merely traces of sulphates and chlorides. On evaporating a considerable quantity of it, I was able to discover, independently of these traces of soluble salts, a very small proportion of organic matter and of silica; the latter substance is probably the element which predominates in the sap of the guaduas.

SAP OF THE BANANA PLANT, (MUSA PARADISICA.)

The sap of the banana possesses a well-marked astringent taste; it reddens tincture of litmus. Immediately after escaping from the plant, it is limpid and colorless, like water; nevertheless, it possesses the property of imparting a yellow color to stuffs immersed in it. Exposed to the air it becomes turbid, and throws down flocculi of a dirty rose color. It is to the action of oxygen that this deposite is owing; for it takes place only in contact with the air. After the formation of this deposite, the sap no longer colors stuffs immersed in it. From a chemical examination which I instituted of the sap of the banana, during my sojourn on the banks of the Magdalena, I think I may state that it contains gallic acid, acetic acid, chloride of sodium, salts of lime and potash, and silica.

The sap of vegetables, elaborated during its passage through the leaves, acquires additional consistence. It generally contains peculiar principles, which are the result of this elaboration, and these now constitute the liquid which is usually designated by the name of the particular juice of the plant from which it is procured. This

* Annales du Muséum d'Histoire Naturelle, t. ii.
† Knight, quoted in Annales de l'Agriculture Française, t. v. 2e série, p. 338

proper juice or sap is generally obtained by making an incision which penetrates a little below the bark.

The characters and properties of the elaborated or descending sap, are extremely various. It may, however, be divided into milky sap, saccharine sap, gummy sap, and resinous sap, according to the nature of the juices dissolved or suspended in the liquid. As several of the peculiar juices of vegetables contain principles employed in the arts or in medicine, they have been more carefully studied, and their history is more complete than that of the ascending saps. I do not propose to give a monograph of these juices; in this place I shall only mention those which have been examined with some care.

MILKY SAPS.

The milky saps, as their name indicates, have the appearance of milk; they owe this milky appearance to globules of insoluble matter, minutely divided, and suspended in a liquid.

SAP OF THE PAPAW-TREE, (CARICA PAPAYA.)

The carica papaya grows in tropical regions. The sap, which is extracted from the fruit by incision, is white, and excessively viscous. In a specimen of this sap, which came from the Isle of France, Vauquelin found water in large quantity, and also a matter having the chemical properties of animal albumen,* and lastly fatty matter.

I took occasion to verify the correctness of the results obtained by Vauquelin, on the milk of the fruit of the carica papaya, during my sojourn at Caraccas, where I examined the sap which flowed from the trunk of the tree itself. This sap is less milky, and much more fluid than that which flows from the fruit; it had the appearance of milk-and-water. Its odor is rather nauseating, even when coming from the plant; its taste slightly sour. When exposed to the air it soon coagulates. It contains a considerable portion of matter, which may be compared to animal fibrine, and sugar, wax, and resin, in small quantities.

Evaporated and burnt, it leaves a saline residue. This juice is employed by the inhabitants for medical purposes.

SAP OF THE COW-TREE.

Among the number of astonishing vegetable productions observed in the equinoctial regions, is a tree which yields a milky juice in abundance, similar in its properties to the milk of animals. At the time I left Europe, M. de Humboldt expressly recommended me to direct my attention to the milk of the cow-tree. A short time after my arrival in the Cordilleras, on the shore of Caraccas, M. Rivero and myself were able to comply with the wishes of the distinguished traveller.†

The milk we examined came from the *Palo de Leche*, the milk-tree, which is extremely common in the environs of Maracaibo.

* Vauquelin, Annales de Chimie, t. xlix. p. 219, 1re série.

† Rivero and Boussingault, Annales de Chim. et de Phys. t. xxxiii. p. 229, 2e série.

Vegetable milk possesses the same physical characters as that of the cow, with this sole difference, that it is, in a slight degree, viscous; its flavor is agreeable, slightly balsamic. With respect to chemical properties, these differ perceptibly from those which are peculiar to animal milk. Acids do not curdle it; alcohol scarcely coagulates it.

Under the action of gentle heat, light pellicles are seen to form on the surface of vegetable milk. On evaporating it over a water bath, an extract is obtained resembling fritters; and if the action of the fire be continued for a certain time, oily drops are observed, which increase in proportion as the water is dissipated, and ultimately form a liquid of an oily appearance, in which a fibrinous substance floats, which dries and becomes tough in proportion as the temperature increases. An odor is then diffused, exactly like that of meat frying in fat.

By the mere action of heat, then, the milk of the *Palo de Leche* is separated into two distinct portions: the one fusible, of a fatty nature, the other fibrinous, and presenting all the characters of animal substances.

If the evaporation of vegetable milk is not carried too far, the fatty matter may be obtained unchanged; it then possesses the following properties;—it is white, translucent, sufficiently solid to resist the impression of the finger; it fuses at 140° (Fahr.;) boiling alcohol dissolves it completely; it is equally soluble in potash.

The fibrinous matter presents all the characters of fibrine, obtained from the blood of animals; for this reason we have called it fibrine. In fact, when put on a hot iron, it swells up, fuses, and becomes carbonized, exhaling the odor of grilled meat. Treated with weak nitric acid, it gives out nitrogen gas; by distillation, it disengages ammoniacal vapors in abundance.

The presence and nature of this animalized matter in the milk of the cow-tree, explains how this milk acquires the odor of old cheese on becoming changed. We considered the fatty matter of the milk as analogous to beeswax; I may even say that we made wax-candles of it. However, the property of being completely dissolved in hot alcohol, combined with its ready solubility in potash, establish a well-marked difference between it and the wax of insects. This is a question which can only be completely cleared up by elementary analysis, and we were altogether without the means of making any minute examination of the wax of vegetable milk.

In the water which holds the wax and animal matter in suspension, we met with some saline substances and a free acid, the nature of which we were unable to determine. We did not succeed in detecting the presence of caoutchouc in vegetable milk. According to our researches this milk should contain:

1. A fatty substance similar to beeswax;
2. An animal substance, similar to animal fibrine;
3. Water, salts, a free acid, and a little sugar.

By incineration, we obtained ashes from the milk in which were found phosphate of lime, lime, magnesia, and silica.

During their excursions in the Cordilleras, the inhabitants frequently drink the milk of the cow-tree. M. de Rivero and myself also used it during our sojourn at Maracaibo.

The tree which produces the milk which we examined, is, according to M. de Humboldt, the *galactodendron dulce*, of the family of the verticas, or fig-trees. But several trees are known in the mountains along the coast, which yield a milky juice, and which are often confounded with that just described. For instance, in the environs of Maracaibo, according to M. Desvaux,* the *clusia galactodendron* yields an abundance of very pleasant vegetable milk; this milk, however, does not seem to contain much animalized matter; at least it does not putrefy perceptibly, and instead of the waxy matter, a substance much less fusible and of a resinous character is procured from it.

MILKY SAP OF THE HURA CREPITANS, (AJUAPAR.)

The sap of the *hura crepitans* is dreaded, and not without good reason; it is enough to be exposed to the emanations of this milky juice, when recently extracted, to be seriously affected by it. The use which is made of it, to poison the water of rivers, in order to obtain the fish, is a sufficient indication of its pernicious qualities.†

This vegetable sap would perfectly resemble that of the cow-tree, if it were not slightly yellowish. It has no smell; its taste, which is very little marked at first, soon causes very violent irritation. It reddens the color of turmeric; mineral acids produce in it a white and viscous curd; the surrounding fluid is clear and of a yellow color. Left to itself, the milky sap of the *hura crepitans* yields all the products of the putrefaction of caseum. It contains: 1. An azotized substance similar to gluten, or caseum. 2. A vesicating oil. 3. A crystallized substance, having an alkaline reaction. 4. Malate of potash. 5. Nitrate of potash. 6. A salt of lime, (the malate ?) 7. An odorous azotized principle.

MILKY SAP OF THE POPPY, (OPIUM.)

The milky sap which, by concreting, furnishes the opium of commerce, is obtained by making longitudinal incisions in the capsules of the poppy. The operation takes place before the fruit is ripe, and after the fall of the flower.

* Renseignements communiqués par M. Adolphe Brongniart.

† Rivero et Boussingault, Annales de Chim. et de Phys. t. xviii. p. 430, 2e série. What I shall now state may give an idea of the energy with which this milky juice acts on the animal economy: when M. Rivero and myself examined the milk of the *hura crepitans*, we became affected with erysipelas; the affection continued for several days. The milk had been sent to us in guaduas by Dr. Roulin; the messenger who brought it was seriously affected by it; and along the road the inhabitants of the houses where he lodged felt the same effects.

The concrete sap is brown, firm, of an acrid and bitter taste, and of a peculiar sickening odor. Opium contains a number of principles, the study of which has exercised for a considerable time the ingenuity of the most skilful chemists. It was in this substance that Sertuerner found the first vegetable alkali which was discovered, morphine. After numerous trials made on opium, it was found to contain :—1. Morphine, (vegetable alkali.) 2. Codeine, (the same.) 3. Narceine. 4. Meconine. 5. Para-morphine. 6. Pseudo-morphine. 7. Meconic acid. 8. Resin. 9. Fatty matters. 10. Caoutchouc. 11. Gum. 12. Bassorine. 13. Ulmine. 14. Woody substance. 15. Mineral salts with bases of lime, magnesia, and potash.

MILK OF THE PLUMERIA AMERICANA.

The plumeria, when one of its branches is broken, yields a considerable quantity of milky juice. At the time when I examined this juice, the tree was entirely destitute of leaves. The milk of the plumeria is perfectly white; it is very fluid when it flows from the plant, but soon after deposites a crystalline sediment. The taste is slightly bitter, and it has an acid reaction. The milk of the plumeria appears to contain no animalized matter. I was only able to detect a very large proportion of resinous matter held in solution or suspended in water; and indications of potash, lime, and magnesia, combined with an organic acid.

SAP OF THE CAOUTCHOUC TREE.

Caoutchouc is found in the sap of many trees, and in that of a great number of herbaceous plants; but it is the *hœvea caoutchouc*, the *jatropha elastica*, peculiar to South America; the *ficus Indica*, and the *artocarpus integrifolia*, which grow in the East Indies, that yield the caoutchouc so well known in commerce, and which has been converted to so many useful purposes in the arts.

The caoutchouc tree is particularly common in Choco and forests near the equator. To obtain the elastic gum, the Indians incise the tree below the bark, when there issues a copious discharge of milky sap, which will remain fluid for a considerable time, if it be kept from contact with the air. I have seen it carried to great distances, in wooden vessels hermetically closed. When spread out in thinnish layers, it soon coagulates, and acquires the singular elasticity which characterizes caoutchouc. The action of the oxygen of the air may possibly have some influence in producing this coagulation, unless what I am about to state be the effect of a prompt evaporation of the water of the sap. I have often made a small incision in the trunk of an hœvea from which milk immediately flowed, and by reason of its viscidity, trickled down the tree in a stream of a certain thickness; this milk was at first extremely fluid, but after from one to two minutes' exposure to the air, it suddenly coagulated, so that on raising the drop from the lower end, I obtained a long string or riband of perfectly elastic caoutchouc.

In Guiana the Indians fashion the caoutchouc into the bottles which are so common in trade: they make a clay mould, and this they cover by immersing it in the milk freshly drawn from the tree; they allow it to coagulate, which it does very speedily, especially if it be exposed to the smoke of a wood fire. This first layer being coagulated, they continue the same process until the desired thickness is attained. The mould is then broken and taken out piece-meal from the interior of the caoutchouc bottle which has been formed.

The workmen of Quito, who are very dexterous in manufacturing caoutchouc, make shoes and buskins of it, by applying it in the milky state over moulds of the proper fashion. They also render tissues impervious by spreading it in the same state between two pieces of stuff or cloth; the interposed milk becomes coagulated, and forms a thin elastic lamina, very preferable to the caoutchouc applied by the aid of solvents.

The Indians of Choco sometimes procure this substance by felling the tree, and receiving the milk, which then flows in a stream, into large wooden moulds, generally formed from a hollow stem of the guaduas. By keeping the mould open, the milky mass coagulates after some time. Several of these masses of caoutchouc, which were brought to me by the Indians of the *Chami* nation, were but slightly elastic; their color also was extremely deep. It is probable, that by proceeding in this way, the milky juice is mixed with large quantities of the internal sap which is much less milky.

Several trees in the valley of the Magdalena which bear the name of caoutchouc, which, however, are neither the hœvea, nor the jatropha, also yield a coagulable juice, which may be confounded with the elastic gum; it is, I believe, caoutchouc combined with a large quantity of wax, and probably also of resin; this caoutchouc possesses but little elasticity.

M. Faraday found in the milk of the hœvea, in 100 parts:

Water	56
Caoutchouc	32
Bitter azotized matter soluble in water and alcohol	7
A substance soluble in water and alcohol (sugar)	3
	100

As this milk will remain fluid for a considerable time, provided it be protected from the air, advantage has been taken of this property to convey it to Europe. It is sent in well-filled, hermetically-sealed bottles.

GUMMY AND RESINOUS SAPS.

I place under this head the saps of those trees which yield gum from incisions in their trunk, as the *acacia vera* and *acacia Arabica*, which grow in Arabia, and from which gum-arabic is obtained; acacia Senegal, which also furnishes a species of gum. In general, in very warm countries, the mimosas produce gummy matters in abundance.

The elaborated sap of the coniferæ and terebinthaceæ consists chiefly of resinous matter, dissolved in an essential oil composed of carbon and hydrogen, similar to the essence of turpentine. The balsams of Peru and Tolu are obtained by incising the bark of the trees which produce them. In Choco, where I have seen numerous incisions made in the lower part of the trunk of the Tolu trees, the balsam flows slowly, on account of its thickness; it does not, apparently, contain any water.

SACCHARINE SAPS.

The sap of the *fraxinus ornus*, and that of the *fraxinus rotundifolia*, yield manna on drying or becoming thick. The sap of several palms contains a considerable quantity of saccharine matter. At Java, for instance, crystalline sugar is extracted from the *arenga saccharifera*. In several places, the sap of palm-trees is subjected to fermentation in order to prepare vinous liquors.

The *Cocos butyracea* (*palma de vino*) is very common in the valley of the Rio-Grande de la Magdalena. From a superficial examination which I made of it, its sap contains sugar, an azotized matter, and some soluble salts.

By fermentation, it produces a vinous liquor sufficiently alcoholic to produce intoxication. In order to procure it, the natives of Benadillo first fell the tree, taking care, when it is down, to give the trunk a slight inclination from the summit towards the lower extremity or foot. They then make a hole towards the base of the trunk sufficiently large to hold from fifteen to eighteen pints, the orifice of which they plug up with leaves. The woody tissue, to all outward appearance, contains but little moisture; but in ten or twelve hours after the operation, the cavity is found full of a liquid, of a well-marked vinous odor, and of a sourish taste, owing probably to the carbonic acid which is disengaged in large quantity. The wine thus obtained is rather agreeable. A palm-tree of from 50 to 60 feet in height, and of which the trunk towards the base is from 20 to 24 inches in diameter, will yield from twenty to thirty pints of wine in twenty-four hours during ten or twelve days. The wine must not be allowed to remain too long after it has collected, otherwise it becomes sour.

Sugar is far from being the only useful substance afforded by palms. There are several of these trees which are truly astonishing by reason of the many important uses to which they may be applied; and it is not without reason that the missionaries have styled the palm, the tree of Providence, the bread of life. Such more especially is the *Cocos mauritia*, which grows in the plains of the Apure and Oronoko; its young shoots serve as aliment; from its fruit, while still green, a farinaceous food may be obtained; and when perfectly ripe, it yields oil in abundance. Hammocks and various kinds of cloth are made of the fibrous portion of the bark of this tree; the young leaves serve to make hats, mats, and sails for ships; the tissue which surrounds the fruit furnishes the Indians with clothing; the sap ferments and yields wine; the trunk before fructification contains an amylaceous marrow, of which bread is made; this marrow, on becoming putrid, produces a vast multitude of large white worms which the Indians value as a most delicate dish; finally, the woody part of the mauritia affords excellent timber for building.

It is not necessary to enumerate farther the principles produced by vegetables; we must now study them in reference to their elementary composition.

CHAPTER II.

OF THE CHEMICAL CONSTITUTION OF VEGETABLE SUBSTANCES.

FROM the very first period of vegetable life, during germination, the immediate principles which constitute the seed are destroyed or changed. The young plant, in developing its organs, creates new substances, which are added to the tissues already existing, so as to complete or extend them. In order to account for thé productions or changes which take place in the organism of vegetables, it is expedient first to study the intimate nature and general characters of the materials which compose them. Unfortunately, in the present state of science, this study is as yet but little advanced; and, notwithstanding the efforts which chemical physiology has made in recent times, there still remain numerous and important questions to be solved.

Carbon, hydrogen, oxygen, azote, combined in some cases with minute quantities of sulphur or phosphorus, are the only elements required by nature to give rise to that almost endless variety of vegetable substances, so different in their properties, as well as in their uses. In the food which sustains the life of animals, as in the virulent poison which destroys it, these same elementary bodies are always found combined in various and dissimilar proportions.

The immediate principles of the vegetable kingdom may be divided into three groups, if we look to the number of the elements which constitute these principles as they exist in the several bodies:

1°. *Quarternary*, containing carbon, hydrogen, oxygen, azote.

2°. *Ternary*, containing carbon, hydrogen, oxygen.

3°. *Binary*, containing carbon and hydrogen, or carbon and oxygen, or carbon and azote.

It is by the examination of the immediate principles which exist in the seed, that we should approach the study of the composition of vegetables; and this the more, as we shall find these principles diffused throughout the organs of plants. Once we shall have fully considered their properties and their elementary composition, it will be sufficient merely to indicate where they are to be met with in the organism.

§ 1. QUARTERNARY AZOTIZED PRINCIPLES OF VEGETABLES.

It has now for some considerable time been ascertained that several seeds contain azote, inasmuch as azotized matters, nearly similar to those obtained from the tissues of animals, can be extracted from them. M. Gay-Lussac expressed this fact in the most general manner, by laying it down as a law that every seed contains a principle abounding in azote.*

Azotized animal matters, when heated in close vessels, yield an ammoniacal product; and to satisfy ourselves of the generality of the law laid down by Gay-Lussac, all that is necessary is to subject any seed whatever to dry distillation.

We do not always, indeed, obtain an ammoniacal liquor immediately in this way; rice, for instance, when heated in a retort, yields a product having an acid reaction; but it is easy to demonstrate in the acid liquor, the presence of ammonia by the addition of lime, which at once sets it free. Peas, kidney-beans, in a word all the legumens hitherto experimented on, yield a liquor directly, having a highly alkaline reaction. These differences, in the products of the dry distillation of seeds, are explained in a very natural way. Throwing the husk out of the question, we may consider a seed as formed of two parts; one, non-azotized, possessing a ternary composition, and yielding by the action of heat a liquid with an acid reaction; the other having a quarternary composition, consequently azotized, and yielding an ammoniacal liquor, so that the acid or alkaline reaction of the product, really depends on the predominance of one or other of these two distinct ingredients.

M. Gay-Lussac subjected every kind of seed he could procure to distillation, and all yielded ammonia either directly or indirectly. I shall add, that the numerous analyses which I have had occasion to make for several years back support the generality of the principles laid down by the above celebrated chemist. M. Payen has come to the same conclusion, and has further shown that at the period of germination the azotized matter of seeds is determined towards the parts that are most recently organized. Thus the spongioles situated at the extremities of the radicles constantly produce ammoniacal vapors during their destructive distillation by heat, even though proceeding from seeds which, when distilled, yield an acid liquor wherein ammonia only becomes sensible on the addition of lime.†

The animalized or azotized substance is extracted readily enough from certain seeds, and has consequently been known to exist in them for a very long time. It is found in wheat, for example, in different states, and is obtained with great ease by simply kneading a mass of dough under a small stream of water by which the starch is carried off, and by and by there remains in the hand a grayish highly

* Gay-Lussac, Annales de Chimie et de Physique, t. liii. p. 110, 2e série.
† Payen, Mémoire sur la composition chimique des végétaux, p. 7.

elastic substance of a peculiar heavy odor; this is the gluten of chemists. By this simple process of analysis, however, we are enabled in many cases to estimate the quality of a sample of flour with reference to its richness in gluten, a substance which is rightly considered as the most essential among the nutritive elements of the cereals.

The washings collected and allowed to stand, soon become clear: the starch which was suspended in the liquid subsides, accompanied by flakes of an animalized matter. If the clear liquor be decanted and boiled, a white froth appears upon its surface, which, skimmed off, is found to have the appearance of coagulated white of egg, and which, in fact, has all the characters of animal albumen. The water from which the albumen is solidified, necessarily contains all the soluble substances of the flour. On evaporation, it leaves substances similar to gum and sugar, and traces of saline matters.

With the exception of the starch, which contains very little foreign matter, the different substances obtained by this process of washing are far from being in a state of purity. I have said that all seeds contain fatty substances, but in the products of the operation just described, no oily matter was detected. As it cannot be discovered in perceptible quantity in starch, nor in the substances soluble in water, it must remain attached to the gluten; and this is actually the case. The gluten, the coagulated albumen then, are not pure proximate principles; fat or oil may be obtained from them; and further, by examining common gluten carefully, we learn that it contains several azotized substances, which differ from one another. By boiling crude gluten with alcohol we ultimately obtain a fibrous grayish residue, called by M. Dumas vegetable fibrine. On cooling, the alcoholic liquor lets fall a substance which in its properties resembles the caseum or curd of milk. Lastly, if the cold alcoholic solution be concentrated, a pultaceous substance separates from it, called by Messrs. Dumas and Cahours glutine.

Analysis, accordingly, indicates the presence of four azotized substances in wheat; and when these are all combined in the mass of gluten obtained by washing a lump of dough, they retain fatty matters, from which they may be freed by means of alcohol and ether. The following, according to MM. Dumas and Cahours, is the composition of the azotized principles of wheat, dried at 140 centig. (284° F.)*

	Carbon.	Hydrogen.	Azote.	Oxygen, Sulph. and Phosphorus.
Fibrine	53.2	7.0	16.4	23.4
Albumen	53.7	7.1	15.7	20.5
Caseine (caseum)	53.5	7.1	16.0	23.4
Glutine	53.3	7.2	15.9	23.6

Legumine. Some vegetables, particularly some seeds, contain a substance different from any of those just described. This M. Braconnot was the first to notice in the seeds of the family of the Leguminosæ, and it has been since detected by Dumas and Cahours in

* Dumas et Cahours, Annales de Chimie et de Physique, p. 390, 3e série.

many different seeds. *Legumine*, which plays an important part in the nutrition of animals, is obtained by digesting a quantity of pea or bean meal, or crushed peas or beans in tepid water for two or three hours; the pulp is then pounded in a mortar, and afterwards mixed with its own weight of cold water; after one hour's maceration it is pressed through a cloth. On standing, the liquid throws down some fecula. Filtration is employed to have the liquor perfectly clear; upon which a quantity of acetic acid diluted with from eight to ten times its weight of water is gradually added, when a snowy flocculent precipitate of legumine falls. This is collected in a filter and washed with water; the legumine is then treated with alcohol, dried, and pulverized, preparatory to digestion in ether, in order to free it from fatty matters.

Legumine thus prepared has a pearly or lustrous appearance. It is insoluble in alcohol and ether. Cold water dissolves it in large quantity. On boiling the watery solution, legumine is coagulated and falls in flocculi analogous to those formed by albumen under the same circumstances. Weak hydrochloric acid throws down legumine from its watery solution like the acetic acid; the concentrated acid, again, dissolves it, acquiring a violet tint, a character which also belongs to albumen; but that legumine is actually distinct from albumen is proved by the circumstance of its being precipitated by phosphoric acid with three atoms of water, which albumen is not. The alkalies dissolve legumine at common temperatures.

COMPOSITION OF LEGUMINE, OBTAINED FROM DIFFERENT SEEDS.*

	Sweet Almonds.	Plum Kernels.	Apricot Kernels.	White Mustard.	Hazel Nuts.	Peas.	Lentils.	Kidney Beans.
Carbon	50.9	50.9	50.7	50.8	50.7	50.5	50.5	50.7
Hydrogen..	6.7	6.7	6.7	6.7	6.7	6.9	6.7	6.8
Azote	18.8	18.6	18.8	18.6	18.8	18.2	18.2	17.6
Oxygen....	23.6	23.8	23.8	23.9	23.8	24.4	24.6	24.9
	100.0	100.0	100.0	100.0	100.0	100.0	100.0	100.0

These same azotized compounds, or substances differing but little from them, are very probably those that are now recognised as distributed through the whole body of every vegetable. M. Payen, after having ascertained the presence of these substances in the radicles and spongioles, proved it in nearly all the organs. The examination was extended to a great number of species of different families. The ascending sap of a fig-tree, (*ficus carica*,) that of the lime-tree, of the black poplar, of the vine, have all yielded ammoniacal vapors under the influence of fire; so also do the buds, the young leaves, the stigmas, the anthers, &c.† Thus, according to M. Payen, the nutritious juices which ascend from the extremities of the radicles to the terminal points of the leaves, carry an azotized principle

* Dumas et Cahours, Annales de Chimie et de Physique, t. vi., p. 423, 3e série.
† Payen, Mémoire sur les développemens des végétaux, p. 36.

which accumulates in all the growing organs, at the same time that it is deposited within the entire extent of the canals which the sap traverses. It might, therefore, be supposed that in the latter situation the azotized substance was associated with matters of ternary constitution, so as to form membranes and tissues. But from the various organs of the many species studied, M. Payen succeeded in dissolving out, by means of alkalies, and entirely eliminating the animalized substances, without causing the slightest rent or erosion in the tissues perceptible with the microscope; whence it may fairly be concluded, that if these substances everywhere and always accompany the young tissues of plants, they still form no integral part of them.* The animalized matter seems consequently to preserve a kind of independence with reference to the organs which secrete, which convey, and which contain it; it preserves a sort of mobility which allows of its displacement. And it was in fact necessary that this should be so; for as the period of maturity approaches we see the azotized substance carried more particularly towards the generative organs, and finally become fixed, as it were, and accumulated in the seeds. I have had frequent occasion to satisfy myself that the trefoil, the red-beet, the turnip, &c., contain much less azote after ripening their seeds than they did previously; and all husbandmen know that the straw or refuse of plants that have run to seed, forms very indifferent fodder for cattle.

The *cambium*, that peculiar globulo-cellular matter which is constantly found accumulated where the vegetable is forming woody tissue, contains, according to MM. Mirbel and Payen, the same azotized principle of an animal nature, in conjunction with ternary substances, whose composition, as we shall presently see, is represented nearly by carbon and water.† As the cellular tissue is evolved at the expense of the cambium, the animalized matters show a tendency to quit the consolidated organ. The departure of these matters at the epoch of the growth of the cells, explains satisfactorily wherefore the interiors of old trees contain but a few thousandths of azote, while all the organs of recent formation always contain several hundredths. With the assistance of chemical analysis it is possible to follow the appearance and the removal of the azotized matter; thus in the alburnum and wood it is observed to diminish in quantity from the circumference to the centre; this diminution is also observed in the branches, proceeding from their extremities to their point of junction with the trunk.

* Payen, Mémoire sur les développemens des végétaux, p. 42.
† De Mirbel et Payen, Comptes rendus de l'Académie des Sciences, t. xvi. p. 98.

§ II.—PROXIMATE PRINCIPLES WITH A TERNARY COMPOSITION.

OF STARCH.

Starch is contained in the cells of vegetables under the form of small white granules which have no crystalline structure.

In the year 1716, Leuwenhoeck ascertained that these granules were globular bodies more or less regular in their contours. He believed that he could perceive each globule enclosed in an envelope, a kind of sac different in its nature from the matter which it contained. M. Raspail, a few years ago, confirmed by his own researches the observations of Leuwenhoeck; he further attempted to measure the diameter of the globules in different kinds of starch, and came to the conclusion that their capsule is insoluble, and that it is the internal part alone which is soluble in hot water.* Since then MM. Payen and Persoz have ascertained that if the globules of starch be really surrounded by a capsule, it must be present in a quantity scarcely appreciable—a quantity not exceeding $\frac{1}{4000}$th of the weight of the starch. These first researches were followed by the subsequent observations of M. Payen, who has devoted himself to the study of the amylaceous principle with a zeal and perseverance which must secure him the gratitude of chemists and physiologists.

M. Payen has examined a vast number of fœculæ microscopically; the largest granules he observed were obtained from one of the varieties of potato, from the *menispermum palmatum*, and the *canna gigantea.*

The globules of starch frequently exhibit a polyhedral appearance, a figure which evidently results from their mutual pressure as they have lain in the cells of the vegetable. Notwithstanding a great general analogy of form, the granules of the starch of different species of vegetables still present peculiar physiognomies, so that they can be distinguished in many instances by the practised eye. A character common to the majority of fœculæ, however, is roundness of contour, when their particles have not been compressed by their contact in contiguous cells.

Microscopical and chemical researches alike show that starch is homogeneous in properties, as in composition; that its globules are composed of concentric layers, the external of which have exactly the same characters as the internal layers.† In the natural state, starch is insoluble in water and in alcohol; it is very ductile, and under the influence of certain agents it exhibits a great degree of contractility.

Feculas retain water with considerable force; the quantity re-

* In 1812, Villars, in a paper on the structure of the potato, had already estimated the volume of the globules of different kinds of starch.

† Fritzche, Annales de Poggendorf, t. xxxii. p. 129.

tained varies with the temperature at which the drying was accomplished. Thus the fecula of the potato, which is moist and porous, even when subjected to strong pressure, still retains 45 per cent. of water. This is the green or raw starch of manufacturers. Dry starch is very hygrometric. If after being dried it is placed in an atmosphere saturated with moisture, at 20° centig. (68° Fahr.) it will absorb nearly 36 per cent. of water, and its bulk increases in the ratio of one to one and a half; in this state starch is brilliantly white, and its grains adhere so closely that they form a mass of sufficient firmness to take the impress of a seal; starch in this state, however, pressed upon paper yields no perceptible trace of moisture; it is too hard and adherent to pass through a sieve; and when thrown on a metal plate heated to 125° (257° Fahr.) its particles immediately unite and form a cake. The starch of commerce, in the state in which it is usually found in shops, contains 18 per cent. of water; it is either pulverulent or readily reducible to powder, though by slight pressure in the hand, it may be formed into a mass or ball. After drying in vacuo at the ordinary temperature, starch retains no more than 10 per cent. of moisture; a temperature not less than 140° (284° Fahr.) is required to dry it completely; the water which it retains at this temperature belongs to its constitution, and cannot be taken from it except by combining it with bases.*

MM. Collin and Gaultier de Claubry discovered the important character of starch, that of yielding a fine blue or violet color on combining with iodine.† According to M. Payen, the color is more intense, nearer to blue and more lasting, in proportion as the starch is more strongly compressed; the effect of separation is to turn the blue to shades of violet which approach redness as the substance is looser. The same fecula, according to the degree of its aggregation in plants, is seen to assume shades, which are first reddish, then violet, and eventually of a more decided blue color, under the action of iodine.‡

M. Lassaigne has noticed a very curious property of the combination of iodine and starch: if an amylaceous fluid, having the decided blue color, be heated to 89° or 90° C. (193° or 194° Fahr.) the solution becomes completely blanched; but it resumes its former tint as the liquid cools.§

This property which starch possesses of striking a blue color with iodine, renders one of these bodies an excellent test for the other. However, as the iodine must exist in the free state to produce its effect, it is necessary, when the blue color does not show itself at once, in a solution in which iodine is suspected, and to which starch has been added, to add a few drops of sulphuric acid, so as to decompose the hydriodic acid in cases where it may exist.

It is familiarly known that if raw starch be mixed with boiling vater, the result will be a thick, paste-made starch. According to

* Payen, Mémoire cité, p. 88.
† Collin et Gauthier de Claubry, Annales de Chimie, t. xc. p. 92
‡ Payen, Mémoire cité, p. 105.
§ Lassaigne, Journal de Chimie Médicale, t. ix. p. 510.

M. Payen, the change that takes place in the state of the fecula is owing to a swelling, a rupture, or disgregation of its granules. By heating a drachm of starch, mixed with about a couple of ounces of water, to about 60° cent. (140° Fahr.) the microscope shows us that the smallest or youngest grains—those possessed of the least cohesion—have absorbed a considerable quantity of water, and that the expansion of the contents has caused a certain number of the globules to burst; at this temperature, however, some grains of fecula are observed, which do not appear to have yet attained their maximum of enlargement, and whose contents consequently are not yet diffused through the liquid; it is only between 72° and 100° cent. (161.6° and 212° Fahr.) that the maximum of expansion becomes general, and that the solution acquires its greatest consistency.*

The remarkable property possessed by starch of making a glutinous solution or thick paste with water under the influence of heat, led M. Payen to conjecture that a contrary effect would be produced by lowering the temperature—that the starch might be recovered in its original state of distinct globules by suitable management; and this he in fact accomplished by an ingenious procedure. Starch appears to suffer no actual change when diffused in water by exposure to a temperature of 212° Fahr.; the granules have only swollen to about thirty times their original dimensions by the imbibition of a large quantity of water.

We have already seen how starch may be extracted from wheaten flour; this method, however, is not the one that is usually followed to procure this useful substance, so large a quantity of which is consumed in the arts. Formerly, starch was universally obtained from grain,—wheat; at present the potato furnishes a still larger quantity than grain. In the equatorial regions of South America, starch is abundantly prepared from the *Yuca*, (*Jatropha manihot*,) and from several species of palm.

To obtain starch from wheat, the grain is either coarsely ground and mixed with water in large tubs; or it is put to steep in sacks until it is so soft that a process of kneading suffices to set the starch at liberty.

Starch from potatoes. The potatoes are grated after having been well washed, and the pulp being thrown on a sieve, the starch is carried off by the water and deposited in suitable vessels. The washings in the manufacture of potato starch soon become putrid by reason of the azotized matter which they contain, and until lately occasioned much annoyance, until M. Dailly conceived the happy idea of turning them to account as liquid manure.

Starch of the Yuca, or Jatropha manihot. The manihot yields very large roots, rich in starch. These are taken up a little after the flowering, when the fecula is most abundant. To extract the starch, precisely the same process is employed as in the case of the potato. In South America the manioc is distinguished into *yuca dulce* (mild) and *yuca brava*, (malignant;) the latter epithet applying

* Payen, Mémoire cit. p. 96.

to the jatropha containing poisonous juice. The two yucas are, however, but one and the same species; at least a skilful botanist, M. Goudot, who resided for several years in America, could not perceive any specific differences between them. The poisonous principle of the yuca brava must be very volatile, or readily destroyed by heat, for the root may be eaten with impunity after it has been roasted, while the animals who eat it in the raw state soon experience the most distressing effects.

The Indians seldom prepare starch from the jatropha; but the root frequently constitutes the staple of their food. It is from the *yuca brava* that they obtain the *cassava*, which supplies the place of bread with them. Among the Indians in the country near the river Malta, one of the principal tributaries of the Oronoko, I have seen the cassava prepared in the following manner: the roots of the manioc were scraped on a sort of rasp formed of small fragments of flint stuck into a plank; the pulp was then put to drain in a long strainer made of the entire bark of a species of fig; the juice having drained away, water was added to finish the washing; the liquid came out nearly clear and without bringing away any perceptible quantity of starch. To form the pulp into cakes of cassava, it was spread out on an earthen dish placed over the fire; the process was complete when the cassava was dry, and slightly toasted on the outside. Cassava bread is not very palatable, but it possesses the property of keeping for a long time in spite of heat and moisture, and is frequently an indispensable article of provision with the South American traveller. The Indians say that they cannot obtain cassava from the *yuca dulce*.

Starch from palms. In the Moluccas and Philippine Islands, and in the plains of Apure, there are certain palms which yield a species of fecula. This fecula is found in a soft substance, generally situated in the centre of these trees. The marrow of these palms is dried, and when sifted presents itself in the form of grains, which in commerce bear the name of sago.

None of the amylaceous principles or feculas obtained by the processes which I have mentioned are absolutely pure; even supposing all the soluble substances to have been removed by washing, they still retain fatty matters, azotized principles, and coloring substances. Starch is purified by following up the water washings by the action of alcohol, of acetic acid, and of ammonia. Starch in its state of greatest purity, and dried at 100° cent. (212° Fahr.) contains, according to the analysis of M. Jacquelain:

Carbon	44.9
Hydrogen	6.3
Oxygen	48.8
	100.0*

By slight roasting, amylaceous feculas undergo considerable changes; they become soluble in water, and then present the properties of gum.† Starch thus roasted, supplies the place of gum in

* Jacquelain, Annales de Chimie et de Physique, t. lxxiii. p. 181, 2e série.
† Vauquelin and Bouillon Lagrange, Bulletin de Pharmacie, t. iii. p. 54.

various manufacturing processes; still it should not be confounded with gum in a chemical point of view. The acids act with more or less energy on starch, and give rise to different products. Nitric acid, when it is diluted with water, merely dissolves fecula; but at a certain degree of concentration it exerts a destructive action. In this reaction several acids are formed, among others oxalic acid By employing very dilute sulphuric acid, Kirchhoff succeeded in changing starch into a saccharine substance similar to the sugar of the grape. The operation may be performed in a leaden or silver pan, or, what is preferable, especially when the process is carried on upon the great scale, in wooden vessels, in which the liquid mass is heated by steam. According to M. Couverchel, several organic acids are capable of changing fecula into sugar in a similar manner; such are oxalic, tartaric, and malic acids.

The artificial conversion of starch into grape-sugar has not yet been satisfactorily accounted for. The acid employed does not seem to undergo any change; it is found in its original state and quantity after the operation. M. de Saussure thinks that the effect of the reaction is the fixation of water; thus 100 parts of fecula yielded him 110.40 parts of sugar.*

M. Couverchel and M. Guérin, on the contrary, state that the quantity of sugar obtained was less than that of the starch they employed.

Gluten exerts a reaction on starch similar to that produced by acids; Kirchhoff discovered, that under the influence of the azotized matters which are met with in flour, the fecula is converted into sugar.† Two parts of starch being mixed with four parts of cold water, on adding twenty parts of boiling water, a thick paste is produced; if into this one part of dry powdered gluten be introduced, and the mixture be kept at the temperature of 60° cent. (140° Fahr.) the paste becomes more and more liquid, so that the mixture may be filtered at the end of from six to eight hours. By concentration a sirup is obtained, in which small crystals of sugar are perceived. It is well known that during the act of germination, fermentable saccharine matter is produced. Kirchhoff concluded, from his experiments, that this production of sugar in germination is attributable to the reaction of the gluten on the starch. Germinating grain, barley-malt, for instance, reacts rapidly and powerfully on any fecula with which it is brought into contact; a fact well known to, and constantly taken advantage of, by manufacturers of spirits from potatoes and raw grain, large mashes of which are rapidly converted into sweet fermentable liquids under the action of a little malt.

These facts, it is evident, cannot be explained by Kirchhoff's experiment; in the fermentation of the potato, the mass of fecula to be converted into sugar is too great compared with the quantity of gluten which exists in the malted barley. Further, the gluten in grain which has not germinated, scarcely exerts any appreciable action.

* Saussure, Bibliothèque britannique, t. lvi. p. 333.
† Kirchhoff, Journal de Pharmacie, t. ii. p. 250.

The principle which, in the preceding operations, converts the starch into sugar, must therefore become developed during germination. This important point in the art of the distiller has been investigated with great ingenuity by M. Dubrunfaut;* and MM. Persoz and Payen succeeded in separating the peculiar matter in barley-malt which possesses the property of converting starch into sugar. This matter has been called *diastase*.

Diastase exists in the seeds of all the cereals which have germinated; it is met with more especially near the germs, it seems even that the radicles contain none of it. Nor is diastase observed in the shoots or roots of the potato; it is to be met with only in the tubers, around the eyes or points where the young sprouts are developed, precisely as M. Payen has remarked, in the place where we should conceive its presence to be necessary for effecting the solution of the fecula. It is also found to exist in the bark and beneath the buds of trees, always in contact with starch.† Diastase is generally obtained from malt, and when carefully prepared, its peculiar power is such, that one part by weight is sufficient completely to liquefy two thousand parts of starch. Diastase is solid, white, amorphous, insoluble in pure alcohol, soluble in water and weak alcohol The solution very readily undergoes change; it becomes acid, and then no longer exerts any action on fecula. When dried, it keeps much better; still, at the end of two years, it seems to have lost its distinguishing properties. Diastase has no action on vegetable tinctures, on albumen, gluten, cane-sugar, gum-arabic, or the woody fibre. That which more especially characterizes it, is its powerful action on fecula; it may be advantageously used to separate and purify the preceding substances, when they are mixed with starch. The presence of diastase in malt explains the phenomenon of the liquefaction of starch effected by the action of a small quantity of that substance. This solution is not effected by gluten, nor by hordeine, as M. Dubrunfaut had imagined.

By the action of diastase, or of malted barley, the starch on being liquefied is not entirely converted into sugar; there are other distinct products to be considered in this change. The sirup obtained by concentrating the liquefied starch, contains sugar capable of undergoing the vinous fermentation, and a gummy matter, dextrine. These two substances may be separated by means of dilute alcohol, which dissolves the sugar and leaves the gum untouched. The relative quantities of dextrine and sugar produced by the action of diastase are variable, and depend both on the temperature at which the process is conducted, and on the continuance of the reaction. In the first period of the process, the dextrine predominates; but it becomes less and less by degrees, and finally gives place to sugar.

M. Guérin ascertained a curious fact, which shows how the diastase developed in plants may act on their starch: reaction takes place even at ordinary temperatures. In one of M. Guérin's experi-

* Dubrunfaut, Mémoires de la Société Royale d'Agriculture, année 1823, p. 146.

† Payen and Persoz, Annales de Chimie et de Physique, t. liii. p. 73: t. lvi. p. 337, 2e série.

ments, at a temperature no higher than 20° cent. (68° Fahr.) a quantity of starch, at the end of twenty-four hours, was converted into sirup, which yielded 77 per cent. of saccharine matter.*

Pure dextrine. M. Payen freed dextrine from the sugar which usually accompanies it by precipitating a sirup of fecula previously dissolved in dilute alcohol, by means of alcohol nearly free from water. Dextrine well dried, and reduced to powder, has a specific gravity of 1.51. The specific gravity of pure starch is 1.51, that of the sugar of starch 1.61.†

M. Payen found dextrine dried at a temperature of 212° Fahr. to consist of—

Carbon	44.3
Hydrogen	6.0
Oxygen	49.7
	100.0‡

a composition identical with that of starch.

We have seen that water, acidulated with sulphuric acid, transforms starch into sugar; and that in this respect, the acid acts precisely in the same way as malted barley, like which, the acid first causes the fecula to pass into the state of dextrine: by checking the reaction at the proper moment, this substance may thus be obtained, as was shown by Messrs. Biot and Persoz.§ When starch, for instance, is triturated with concentrated sulphuric acid, if the mixture be diluted with half its volume of water, and be left at rest for an hour, alcohol will throw down almost the whole of the starch employed in the state of dextrine.

M. Payen has remarked that starch is never met with in the vegetable tissues while in the rudimentary state; the spongioles, the radicles, the foliaceous buds, the interior of the ovules, contain none of it. Nor is starch found in the epidermis, nor in the primary cells of the subjacent tissues. This proximate principle seems to be excluded from those parts of vegetables that are more directly exposed to atmospheric influences: it is only met at a certain depth; and the globules which constitute starch increase in number and in size in the cells most remote from the surface. The subterraneous organs of plants,—certain bulbs, most tubers, abound in amylaceous matter. It might be maintained that light modified this substance, at the very moment that it was subjected to the vital influence, and that it was only preserved in the dark.

On the globules of some species of fœcula there is found a point or hilum, which, according to some observers, serves to fix them to the parietes of the cells which enclose them. It often happens, however, that no hilum can be distinguished, even by the help of the most powerful microscopes; to render it apparent, recourse must be had to desiccation, which, by causing the globular mass to shrink, allows the part carrying the hilum to project, by reason of its

* Guérin, Annales de Chimie, t. lx. p. 42, 2e série.

† Payen, Mémoires cités, p. 169.

‡ Idem, p. 157.

§ Biot and Persoz, Annales de Chimie et de Physique, t. lii. p. 73, 2e série.

stronger cohesion. M. Payen does not regard the hilum as a point of permanent attachment, connecting the grain of starch to the interior wall of the cell. He considers it as the orifice of the duct by which growth is effected by intersusception. In support of this view, M. Payen observes, that in a great number of vegetable cells, especially in those of the potato, and of the rhizomas, the globules of starch are developed in such quantity, that it is actually impossible that each of these should be united directly to the inner wall of the cell.*

INULINE.

This substance, discovered by Rose in the Inula helenium, presents certain analogies with starch. It forms the greater part of the solid matter of the tubers of the Jerusalem Artichoke and Dahlia, which do not contain starch. Inuline is dissolved in boiling water; on cooling it is deposited in globules, which, under the microscope, appear diaphanous, adhering to one another like strings of beads; exposed to a temperature of 367° Fahr. it melts completely and acquires new properties, becoming soluble in cold water and in alcohol. Inuline is transformed into dextrine and sugar by the mineral acids; but it possesses certain properties which show it distinct from true starch. In the first place, it is not colored by iodine; and then acetic acid, which is without action on starch, produces with inuline precisely the same effects as the sulphuric, phosphoric, and hydrochloric acids; finally, diastase, whose reaction upon starch is so peculiar, so prompt, and so powerful, does not cause any change in inuline. It is therefore easy to separate these two substances when they are mingled, by treating the mixture either with acetic acid, which dissolves the inuline, or with diastase, which liquefies the starch. Inuline has been analyzed by M. Payen, after having been dried at 253° Fahr. and having been melted at 367° Fahr. In both cases it has the same composition.

Carbon	46.6
Hydrogen	6.1
Oxygen	49.3
	100.0

The composition here is obviously the same as that of starch and dextrine.

OF WOODY MATTER AND CELLULAR TISSUE.

The most solid part of plants, that which forms in some sort their skeleton, is the wood in trees, the woody fibre in herbaceous plants. Woody fibre, as it used to be prepared and considered, viz. by the reaction of certain agents which have the property of dissolving the gummy, resinous, and saline substances which are commonly associated with it, consists, in fact, of two substances, one the cellular substance, constituting the tissue of wood and of all the organs of

* Payen, Mémoires cités, p. 183.

plants, the other the woody substance, properly so called, filling, and in some sort consolidating the cells. This distinction between these two elements of wood was first made by M. Mohl; but M. Payen was the first who fixed the opinion of chemists and of vegetable physiologists upon the true nature of these immediate principles.* By treating the vegetable tissue in its nascent and still gelatinous state—the unimpregnated kernel of the almond, of the apricot tree, &c., the membranous matter of the cambium of the cucumber, the spongioles of radicles, leaves, wood, &c.—with different menstrua, M. Payen obtained the cellular tissue in the state of purity, and having an elementary composition almost identical, from whatever source derived; a fact which may be seen from the following table, which gives the composition of cellular tissue from different sources after having been dried at 352° Fahr.

	Carbon.	Hydrogen.	Oxygen.
Ovules of the almond-tree	43.6	6.1	50.3
" of the apple and pear . .	44.7	6.1	49.2
" of the helianthus annuus . .	44.1	6.2	49.7
Pith of the elder	43.4	6.0	50.6
Cotton	44.4	6.1	49.5
Endive	43.4	6.1	50.5
Banana	43.2	6.5	50.3
Leaves of the agave	44.7	6.4	48.9
Cotton of the Virginian poplar . .	44.1	6.5	49.4
Heart of oak	44.5	6.0	49.5
Pine-tree	44.4	7.0	48.6
Perisperm of the phytelaphas . .	44.1	6.3	49.6
Mushroom	44.5	6.7	48.8

The primary tissue, consequently, which constitutes the skeleton of wood, is still isomeric or identical in elementary composition with starch. With mineral acids the cellular tissue further undergoes changes which assimilate it with starch; for on treating it with sulphuric acid it is changed into dextrine and sugar.

The composition of the cellular tissue differs considerably from that of the woody fibre as it has hitherto been obtained after the action of solvents, and been examined by preceding chemists. Pure wood or woody tissue consists of the following proportions of elements:

* Dumas, Comptes rendus, vol. viii. p. 53.

	Carbon.	Hydrogen.	Oxygen.	Authorities.
Woody tissue of the oak	41.8	5.7	52.5	Gay-Lussac and Thénard.
" of the beech	42.7	5.8	51.5	" "
" of the box	44.4	5.6	50.0	Prout.
" of the willow	44.6	5.6	49.8	"
" of the oak	49.7	6.0	44.3	"
" of the beech	44.3	6.0	49.7	Payen.
" of the aspen	44.5	6.1	49.4	"
Wood in the natural state:	45.6	6.4	48.0	"
" of the oak	39.4	6.2	54.4	"
" of the beech	39.3	6.3	54.4	"
" of the herminiera	46.9	5.3	47.2	"

From these analyses it appears that wood in the natural state contains more carbon than woody tissue obtained in the way of purification, and that this latter substance is also richer in carbon than the cellular tissue which necessarily forms part of it. In the purified woody tissue, therefore, the cellular tissue is associated with the principle which fills its cells, or which incrusts it, and it is to this matter that M. Payen has applied the name of incrusting matter; it is wood properly so called; it is that which gives to wood its hardness, its tenacity; it predominates in hard wood and in knots; it corresponds with the duramen of physiologists; it constitutes almost the whole of the hard particles which are met with in woody pears and in cork, and which are hard enough to blunt well-tempered steel instruments. As this incrusting matter is friable, in many instances it may be pulverized and separated from the tissue which surrounds it, this last tearing or yielding in shreds under the pestle. By means of the sieve the incrusting matter may in this simple way be obtained nearly in a state of purity. The analysis of M. Payen shows it to consist of:—

Carbon	53.8
Hydrogen	6.0
Oxygen	40.2
	100.0

Deducting resinous matters susceptible of solution in alcohol or ether, and of gummy and other substances which are soluble in water, the tissues of vegetables must consequently possess an elementary composition which varies between that of the cellular tissue and that of the incrusting matter; these are the extreme terms, and the entire composition of the mixed tissues will be by so much the richer in carbon as they contain less cellular tissue. The incrusting matter being soluble in alkaline leys, it was by treating wood with solutions of soda and potash that M. Payen succeeded in obtaining the cellular tissue, which is much less susceptible of the action of these agents. But treatment of different kinds, which it is not necessary to enter upon in this place, is required to procure the substance in a state of perfect purity.*

The facts which have just been exposed, in regard to the chemical composition of wood, corroborate the observations of physiologists. We now understand much better than we did formerly the changes which the cells of vegetables experience as they grow and become aged : it is by the appearance of the incrusting woody matter that their walls, thin, transparent, and colorless at first, get thick, become opaque, and acquire consistence. By means of the dissections effected by M. Payen with the aid of purely chemical means, we may obtain assurance that the tissues of all vegetables, whether phœnogamous or cryptogamous, may be reduced to a single substance, cellular tissue, having an invariable composition, and forming the vesicles or bladders of the cellular mass of plants.

This matter exists nearly in an isolated state in the thick walls of the cells of the perisperms of various seeds, those of the date for example. From the microscopic researches of M. Payen and A. Brongniart, it appears that the matter which is added to the young cells is not deposited upon the inner surface of their walls, but that it penetrates and insinuates itself into their tissue. The relation of the cellular to the woody matter in the development of the walls of cells varies very much, some perisperms containing nothing but pure cellular tissue, while the stony concretions of the pear and of cork consist almost entirely of incrusting woody matter.

Wood, in the general acceptation of the word, is the solid part of the trunk and branches ; the properties and aptitudes of the substance vary greatly, according to the plant which has produced it. Wood is of higher density than water, and if it floats in this fluid it is only because of the air with which its pores are filled. Saw-dust, chips, and larger pieces of wood sink when the air which they contain is expelled and replaced by water. The specific gravity of the white woods, such as those of the willow and pine, is about 1.46, that of the heaviest woods, such as those of the oak and the beech, 1.53.

DENSITY OF DIFFERENT KINDS OF WOOD ACCORDING TO BRISSON.

Wood	Density	Wood	Density
Pomegranate	1.35	Orange	0.70
Guaiac, Ebony	1.33	Quince	0.70
Box	1.32	Elm, the trunk	0.67
Oak of 60 years old, the heart	1.17	Walnut	0.67
Medlar	0.94	Pear	0.66
Olive	0.92	Spanish Cypress	0.64
Spanish Mulberry	0.89	Lime	0.60
Beech	0.85	Hazel	0.60
Ash	0.84	Willow	0.58
Hornbeam	0.80	Thuya	0.56
Yew	0.80	Pine	0.55
Apple	0.79	Spanish white poplar	0.52
Plum	0.78	Pine	0.49
Maple	0.75	Poplar	0.38
Cherry	0.75	Cork	0.24

* For an account of these, see Payen in proceedings of the Academy of Sciences vol. vii. p. 1055.

It must not be forgotten, however, that age, climate, and soil exert a marked influence upon the specific gravity of the same species of wood.

Wood, according to the use for which it is intended, is distinguished into fire-wood, building timber, and dye-wood. When first cut down, all timber contains a considerable quantity of water; 100 parts of walnut-tree dried at 212° Fahr. lost 37.5 parts by weight; of white oak, 41 parts; of maple, 48. On an average, the quantity of water contained in green wood may be estimated at about 40 per cent.; and drying or seasoning for eight or ten months will not cause the loss of more than about 25 per cent. of water. The wood which is used for burning almost always contains about a quarter of its weight of moisture, which not only does not assist in producing heat, but actually dissipates a great deal during its conversion into vapor. It is, therefore, highly advantageous in all operations where wood is the fuel, only to employ that which is thoroughly dry. So well is this fact ascertained, that in some manufactories the wood is previously dried in stoves before being consumed in the furnace.

The composition of woody matter may be represented by carbon and water: of carbon the mean may be stated at 52, of hydrogen and oxygen, in the proportions which form water, at 48. The definitive products of its combustion ought consequently to be carbonic acid and water. The heat disengaged during this combustion, necessarily proceeds from the union of the combustible elements of the wood with the oxygen of the atmosphere. But in this particular case, the hydrogen being already present with the proportion of oxygen required for its combustion, it may be regarded as already burned, the state of condensation in which the oxygen exists being considered. The heat produced by the wood, therefore, depends solely on the quantity of carbon which it contains.

Natural philosophers in France agree in designating as unity, in reference to caloric, the quantity of heat necessary to raise a kilogramme, or 2.2 lbs. avoirdupois of water, one degree of the centigrade thermometer, (1°.8 F.) The following table, by Rumford, is intended to show the different calorific or heating powers of different kinds of wood, and its interpretation is this: since 1 kilogramme or 2.2 lbs. avoird. of lime-tree gave out 3460 units of heat, it follows that this quantity of the combustible would suffice to raise by 1 degree centigrade, (1°.8 F.,) for example, from 10° to 11° cent. 3460 kilogrammes, or 7612 lbs. avoirdupois of water

Kinds of wood.	Units of heat evolved.
Lime-tree, dry	3460
The same, thoroughly stove dried .	3960
Beech, dry, four years seasoned . .	3375
The same, well dried in a stove . .	3630
Elm, from four to five years seasoned .	3037
Oak, fire-wood	3550
Ash, dry	3075
Wild cherry	3375
Fir, dry	3037
The same, well dried in a stove . .	3750
Poplar, seasoned	3450
The same, well dried in a stove . .	3712
Hornbeam	3187
Oak, dry	3300

From the experiments of Clement, it appears that the heating power of charcoal is equal to 7050 units. Dry wood containing, as we have seen, 52 per cent. of charcoal, its heating power has been deduced theoretically, as equal to 3666. Mr. Marcus Bull in America, made a series of experiments to determine the relative quantities of heat given out by different kinds of wood, from which M. Peclet has been led to conclude that the same weight of dry wood of every kind has the same heating power, and that this for a kilogramme, or 2.2 lbs. avoird. of wood dried by artificial means, is equal to 3500 units, while the same quantity of the same wood having been cut and seasoned during from ten to twelve months, and containing from 20 to 25 per cent. of water, is no higher than about 260 units.

By way of comparison, I shall here add the heating power of the several combustibles in general use, in contrast with that of wood:

1 kilogrm. or 2.2 lbs. avoird. of		wood-charcoal produces	7226	units of heat.
"	"	coal	6000	"
"	"	peat	3005	"
"	"	peat charcoal	6400	"

Although the same quantities of wood, brought to the same degree of dryness, appear to have the same absolute calorific power, all are not alike adapted to the same purposes. Hard woods burn slowly, and give out less heat in a certain time than the less compact kinds of wood. This is the reason why fir is preferred to oak in furnaces where the object is to obtain the most intense heats. It were foreign to our object to enter upon any consideration of the various qualities, or of the adaptation to particular uses, of different species of timber. I may, however, add a table of the ordinary dimensions of well-grown trees of different kinds, such as are commonly found in these countries.

Trees.	Usual height of Trunk.	Usual Diameter.
	Feet.	Inches.
The spruce fir	26 to 100	47.1
Larch		39.3
Poplar	19 " 65	31.8
Pine	16 " 65	34.1
Plane	16 " 48	36.1
Oak, Elm		31.4
Birch		29.4
Beech		28.2
Lime		25.9
Ash		23.5
Willow		11.7
Chestnut	13 " 48	36.1
Chestnut (another variety)		28.2
Maple	10 " 48	28.2
Service	13 " 39	17.6
Acacia	13 " 26	19.2
Hornbeam	10 " 23	21.2
Mulberry		16.5
Wild Pear		14.1
Crab	6 " 20	12.9
Walnut	6 " 16	36.1

These may be taken as the measurement of trees at their full growth, and fit for felling. The soil being of the same quality, the dimensions of trees depend especially upon their age; individual trees of the same species, however, occasionally acquire extraordinary dimensions.

Every one must have noticed the rapidity with which young trees grow; but is the growth the same for every period of the existence of trees, or do they attain a certain determinate size like animals, and then cease from further increase? We have found that in those climates where vegetation is suspended for a portion of the year, the increase in the diameter of trees takes place periodically by the addition of a concentric layer of woody tissue; so that it is possible to determine the age of a dicotyledonous tree by the number of its concentric rings, counted at the bottom of the trunk. With a view to ascertain the amount of increase in the woody layers at different periods of vegetable life, De Candolle measured their thickness, and found that if the annual increase presented a certain regularity, it was still very far from being absolute even in the case of a single species. The oak especially offered striking anomalies; thus a trunk which had grown slowly in diameter was found to have increased more rapidly as it got older. He found young trees of the same species, the growth of which, very slow at first, by and by became accelerated, and then fell off in a third period of their existence. From the whole of his observations, De Candolle concludes that the growth of our common European trees having gone on with a certain rapidity to the age of from about fifty to seventy years, then became slower, but continued regular to extreme age. The inequalities of growth, conspicuous in the different thicknesses of different rings, he thinks are mainly due to the kind of soil which the mass of the roots encountered in their progress, or to the removal of other trees which grew in the vicinity. The diminished thickness of the rings, after trees have passed a certain age, he ascribes to the depth to which the roots have now penetrated, and their consequent remoteness from the air; and further, to the resistance opposed to the expansion of the trunk by the bark, which has now become thick, hard, and unyielding. Mr. Knight found that old pear-trees, relieved of their outer bark, formed more wood in a couple of summers afterwards, than they had made in the twenty years that preceded the operation.*

The forests of intertropical countries produce a vast number of gigantic trees, many of which might doubtless be turned to excellent use; but the information we have on the trees of these latitudes is very imperfect. In New Granada, the wood which is known under the name of wood of St. Martha, (*astroneum graveolens?*) is frequently employed for building purposes as well as for making furniture. It is very hard, and more beautiful than mahogany, its color being deeper. M. Goudot measured a tree of this kind, which was 1.6 metre or nearly 4½ feet in diameter, including the alburnum, and had 32 centimetres or upwards of 18 inches of heart-wood. Belfries having supports of this wood are met with, which have stood for more than a century exposed to all the inclemencies of the weather. This tree grows in the dry soils of the hottest regions of South America, and seldom at an elevation of more than about fifteen hundred feet above the level of the sea.

Cedar (*cedrela odorata*) is never attacked by insects, doubtless because of its aromatic odor; this valuable property makes it invaluable as building timber. The tree attains to large dimensions. M. Goudot measured one in the forest of Quindiu in South America, which was upwards of 150 feet in height by more than 6½ feet in diameter. It grows freely through a zone of considerable breadth, from a height of about 3280 to 6560 feet above the level of the sea, a circumstance which, according to my own observations, would indicate the extreme temperature of the district which it inhabits to be between 66° and 76° Fahr.

There are several other beautiful and useful timber trees of the Cordilleras—the *Nogal* (*juglans . . . ?*) which grows between 6500 and 9800 feet above the sea line; the *escobo*, the *pino* (*taxus montana Willd.*) whose region lies between the 2.800 and 11.400 feet of elevation; the *arayan* and the *guayacan*,—all are serviceable in one direction or another. The *caracoli* (*anacardium caracoli*) and the fig (*Ignerones*) are trees which attain to extraordinary sizes, and afford light woods that prove useful in various circumstances. Ur.

* De Candolle, Vegetable Physiology, p. 975.

der the tropics, indeed, the trees generally exhibit a luxuriance of vegetation which strikes European travellers with amazement; M. Goudot, for example, measured a *bombax* (*B. pentandrum*) no more than sixty years old, the trunk of which was 8 metres or 26¼ feet in circumference, and whose boughs covered a circular area of 39 metres or 120 feet in diameter.

There is a beautiful tree which grows in the valleys of Arragua in Venezuela, the *Zamang*, a species of mimosa, according to Humboldt, one of which, in particular, is greatly celebrated, and under the shade of which I rested on the 24th of January, 1823. This magnificent tree is to be distinguished at the distance of a league; its branches form a hemispherical crown of 187 metres or 613 feet in circumference, extending like a vast umbrella, the points approaching to within from 10 to 16 or 18 feet of the ground. The trunk of this extraordinary tree is nearly 65 feet in height and upwards of 9¾ feet in diameter. This tree is an object of veneration with the Indians. It does not seem to have altered in its appearance since it was first particularly noticed; the earliest conquerors of Venezuela seem to have met with it in the same state as it is at the present time. When Humboldt measured the Zamang de Turmero, its branches on one side were entirely stripped of their leaves. Twenty years afterwards I found it green in every part; but the leaves and branches with the southern aspect were not so numerous nor so vigorous as the others.

The dragon-tree of Orotava in the Island of Teneriffe is one of the oldest vegetable monuments of the present world. Humboldt gives it a diameter of 17 feet, and its height, as stated by M. Ledru, is upwards of 65 feet. When Teneriffe was discovered in 1402, this tree appears to have had the same dimensions which it presents at the present time.

The mahogany (*cedrela mahogani*) is a very long-lived tree. In Jamaica it sometimes acquires a diameter of upwards of 6 feet, and Sir W. J. Hooker has calculated that two centuries at least are required to supply timber of the large scantling which we constantly see in the yards of our timber merchants and cabinet-makers.

The *Hymenæa courbaril*, one of the largest trees of the Antilles, yields, like mahogany, a timber that is hard and in great request among cabinet-makers and inlayers. It sometimes grows to 19 feet in diameter.

The Baobab (*Adansonia digitata*) lives for centuries, and acquires extraordinary dimensions. Adanson saw one in the Cape de Verdes, in the trunk of which an inscription was found, which was covered by three hundred layers of wood; it had been cut by two English travellers three centuries before. From positive observations collected by Adanson, a table has been constructed to show the progress and probable age of the baobab:

Age of the Baobab.	Diameter of the Trunk.	Height.
1 year,	0.10 feet	5.25 feet
20	1.04	16.40
30	2.41	23.30
100	9.84	30.84
1000	14.76	61.68
2400	19.03	68.24
5150	31.99	76.76

De Candolle has remarked that this longevity of the baobab is made the more surprising by the softness and liability of its wood to decay. But again, it must be considered that the great diameter of the trunk, in relation to the height, gives the tree a stability which is possessed by no other—by enabling it to resist violent gales of wind.

It strikes me that there may very well be some mistake in Adanson's estimates of the age of the baobab. When we see such irregularity in the growth of trees of the same species planted in the same soil, little reliance can be placed on any deductions drawn from the size of the trunk when the concentric rings cannot be counted. In proof of this I here give the measurements of two baobabs planted in 1821 in the Botanical Garden of French Guiana. In 1842 these trees were found:—

	feet.		feet.
No. 1. Length of stem from ground to first branches	7.70	Diameter of the base	5.41
		Do. at origin of branches	4.23
No. 2. Do.	8.36	Diameter of base	2.63
		Do. at origin of branches	1.48

In the tree No. 2 the branches were puny and nowise in relation with the size of the trunk.

The bald cypress (*taxodium distichum*) is a tree that is very abundant in Mexico, and in the southern parts of the United States. At Chapultepec there is one of these trees called the cypress of Montezuma, which tradition says flourished in the reign of that prince. In 1831 the tree was still vigorous, and its trunk was 41 feet in circumference. There is another cypress near Oaxaca, under the shade of which Fernando Cortez is still reported to have rested ; the trunk of this tree is upwards of 39 feet in circumference, and it is 105 feet in height. Michaux measured several taxodiums in the Floridas which approached these two in their dimensions.

We have only uncertain data in regard to the age which palms may attain to ; their sizes, however, are well known. In Egypt, according to M. Delille, the date-trees are generally about 65 feet in height. In the Andes of Quindiu several ceroxylons were measured, the trunks of which were from 195 to 230 feet in height ! Martius assigns the following as the extreme dimensions of the palms of the Brazils : from 75 to 127 or 128 feet in height, by a diameter of from 6 to about 12½ inches.

Among several palms (*arica oleacera*) planted in the Botanical Garden of Cayenne in 1821, the tallest twenty years afterwards was

48 feet from the ground to the bottom of the crown, and 3 feet $6\frac{1}{2}$ inches in circumference at the base; at $6\frac{1}{2}$ feet from the surface of the ground the circumference was only 2 feet 1 inch, and a small fraction. As the palms and baobabs will be carefully protected in the Botanical Garden of Cayenne, an opportunity will be afforded future observers of following these plants in their growth, with a perfect assurance of being correct as to their age.

Particular trees of different kinds have occasionally acquired remarkable dimensions and lived to great ages in Europe. An elm is mentioned which grew on the promenade of Morges, the age of which, reckoned from the number of concentric layers, must have been three hundred and thirty-five years; its trunk was above 18 feet in diameter. The lime is another tree which in temperate countries sometimes grows to a great size. The one planted at Freiburg to commemorate the victory of Morat in 1476, in 1831 was $14\frac{1}{2}$ feet in diameter. Near the same place there is another tree of the same kind which must be older than the last, inasmuch as it was already celebrated for its size a century ago; in 1831 this tree was upwards of 36 feet in circumference, and about 72 feet in height. The lime-tree of Neustadt is scarcely less curious for its size and the immense spread of its branches than for the historical circumstances connected with it. Looking back to old documents, this tree must already have been of great size in 1229; in a poem written in 1408 we are told that this tree was then supported by sixty-seven props; in 1654 it had eighty-two stone pillars to support its branches, and in 1831 the number had increased to one hundred and six. The circumference of the trunk at $6\frac{1}{2}$ feet from the ground measured very nearly 39 feet. An old measurement made one hundred and fifty years before corresponds very nearly with this, a fact which shows that in the course of a century and a half the trunk of the lime-tree of Neustadt had not grown perceptibly. It is said to be from seven to eight hundred years old. The old lime-tree of Chaillé in 1801 was upwards of 49 feet in circumference.

The beech grows rapidly while young; but in more advanced age with extreme slowness. In 1818 Deluc saw several beeches near Geneva, the trunk of which was from 14 to 16 feet in diameter.

De Candolle measured a larch two hundred and fifty-five years old, the trunk of which was upwards of $5\frac{4}{5}$ feet (5.84 ft.) in diameter; and a larch of no more than fifty-four years growth has been measured which was more than $3\frac{1}{2}$ feet in diameter.

The celebrated chestnut-tree of Mount Etna has been stated to be upwards of $206\frac{1}{2}$ feet in girth, (about 68 feet in diameter,) and must therefore be the largest tree described up to the present time; but the tree has been supposed to be formed by several trunks springing from a common root which have grown together. Other remarkable chestnut-trees are mentioned.

The plane is one of the largest growing trees of temperate countries. A traveller, who visited the valley of Bujukdere, near Constantinople, met with a plane upwards of 95 feet in height, and the trunk of which, hollow internally down to the level of the ground, was more than 154 feet in circumference. A plane-tree, which grew in Norfolk, and was of the age of thirty-one years, was $7\frac{3}{4}$ feet in circumference, according to Hunter. Cypress-trees often attain to a very great age. In the garden of the palace of Grenada there is one which has stood for more than three centuries. At La Somma, near Milan, a cypress is shown which in 1794 was 17 feet in circumference.*

Tradition has it that an orange-tree of the convent of St. Sabina at Rome, was planted by St. Dominic in the year 1200; this tree still exists. The orange-tree of Versailles, known under the name of the *Francis I.*, is rather more than three hundred years old. In 1804, orange-trees were shown in the green-houses of Bonn three centuries old, and of which the trunks were more than 30 inches in circumference.† In South America I had myself occasion to observe citron-trees of great age and of very considerable dimensions; the trunks of several of these trees were nearly $27\frac{1}{2}$ inches in diameter.

A sycamore-tree of the village of Trons, in the Grisons, more than five hundred years old, is at this time between 8 and 9 feet in diameter.

Many oaks have been described which had survived from eight hundred to one thousand years. Hunter saw one of these trees still extremely vigorous which was $11\frac{1}{2}$ feet in diameter. Evelyn, who, in his delightful work entitled Sylva, has given a list of the largest oaks known in his day in England, speaks of one growing in Welbeck Lane which must have been eight hundred and sixty years old at least, and the diameter of whose trunk at the base was upwards of $12\frac{3}{4}$ feet.

The olive is one of the trees that reaches a great age; Picconi describes one of about seven centuries, and a circumference of about 25 feet.

The cedar of Lebanon grows vigorously and long, especially in soils that are sufficiently loose and permeable. According to M. Paul Vibray, of Sologne, the growth of this tree is more rapid than that of the coniferi in general. The cedars which grew on Mount Lebanon, and were measured by Nauwolff in 1574, and again by Labillardière in 1787, are generally allowed to be about the age of one thousand years. De Candolle, however, thinks that this age is exaggerated, and in contradiction with observations made on trees, the age of which is positively known. The following are a few of the measurements which have been reported by different observers:

	Age.	Feet circumfer.	Observers.
Cedar of Chelsea	83	12	
" of Paris	40	7	Thouin.
" of ditto	83	9.4	Loiseleur.
" Environs of London	200	16	Hunter
" Ditto	113	14	Ditto.
" of Mount Lebanon	600	36.4	Maundrel.
" of Sologne	30	5	of Vibray.

The yew, as is well known, produces a very hard, close, and en-

* De Candolle, Physiologie, p. 994. † Ibid. p. 999,

during wood, qualities which contribute greatly to the longevity of trees. Some of the oldest trees known have been yews. Here are a few that have been particularly described :

Where they grow.	Probable age.	Circumference.	Observers.
County of York....................	1220	28.25	Pennant.
Ditto..........................	1220	13.85	Ditto.
County of Surrey.................	1287	30.12	Evelyn.
Fotheringal (Scotland)............	2580	62.34	Pennant.
County of Kent....................	2800	62.60	Evelyn.

According to Duhamel it is extremely difficult to fix upon any age as the best in a general way for felling trees, with a view to obtaining the largest quantity of sound available timber. When the tree is too young, the timber has not all the excellence which it would have gained with greater age; when too old, the pores are obstructed, and it has begun to decay in the parts of oldest formation, so that it is not uncommon to find wood in the centre of the trunk which is lighter than that of the circumference. In trees which have already fallen into a certain state of decay, the worst timber in them is decidedly that which is taken from the centre at the base of the trunk; and, indeed, the wood of the centre generally is then of inferior quality to that of more recent formation. Very aged timber always perishes first in those parts which have formed the most internal layers of the tree. It is, therefore, an obvious and grave error to suffer any tree to stand that has given the slightest indications of decay, inasmuch as that which is ordinarily the most valuable timber is likely to be altogether lost. Neither the age nor the dimensions are always the indications of the proper period for felling trees; exposure, soil, situation, have immense influence upon their growth, vigor, and general qualities. Trees ought to be cut just when they are on the turn; the proper moment is that which precedes immediately the alteration of the heart; and although the destructive effects of age are principally felt in the interior, this intestine disorder is nevertheless proclaimed externally; the whole tree suffers when it has taken place.*

Duhamel has given the following characters, as indicating incipient decay, or decline of vigor in trees :†

1. A tree, the top of which forms one uniform rounded mass, is not strong; a vigorous tree always throws out certain branches which surpass the others in luxuriance of growth.

2. When a tree comes into leaf prematurely in the spring, and particularly when the leaves turn, and fall prematurely in the autumn, it is a certain sign of weakness.

3. When several of the top or leading branches of a tree die, even at their mere extremities, the wood in the centre is beginning to undergo alteration.

4. When the bark quits the trunk, or becomes cracked here and there, we may be satisfied that the tree is far gone internally.

5. Mosses, lichens, and funguses growing upon the bark, and red

* Duhamel, Exploit. des bois, t. i. p. 126.

† Idem, t. i. p. 133.

or black spots appearing upon it, always lead to a suspicion of change in the wood.

6. When the sap is observed to flow from crevices in the bark, the death of the tree is at hand.

In France, the cutting down of the smaller wood, such as is used for firing, takes place at from twenty to thirty years; in the forests, the trees are commonly felled at from one hundred to one hundred and thirty years old, and a few trees are generally left as reserves, and for special purposes, till they have attained the age of from two hundred to two hundred and fifty years.

The prevalent opinion among foresters, with regard to the proper season for felling, is, that it should be done when the sap is in the state of greatest repose, or when it is present in least quantity in the trees. The season fixed by the old law of France (1669) was from October to March inclusive. But the experiments of Duhamel tend to show that this is not actually the season when trees contain the smallest proportion of sap, and that fellings made at other times of the year have had very satisfactory results. All things well weighed, says the illustrious cultivator, our only safe guide in such matters is observation; and from numerous experiments he concluded that there was actually as much sap in trees in winter as in summer, and that the spring and summer were the seasons most favorable for the speedy drying of the timber. Trees felled in summer were even found by Duhamel to yield timber which stood better and lasted longer than those that were cut down in winter; while he found the wood of equal strength in either case. He concluded, therefore, that the season of the year at which timber was felled, had no influence upon its quality or durability.*

There is, in fact, no general rule observed in different countries as to the period at which timber is felled. The French still go on cutting from October to March; the English fell in the winter. Convenience of different descriptions appears often to decide the question as to season. In order to procure bark for the tanneries, an act was passed by the English Parliament in 1603, prohibiting all felling of oak timber during the dead season, the penalty for infringement of the act being confiscation of the timber felled, or fine to the amount of twice its value. An exception, however, was still made in regard to timber destined for the public service in ship-building, &c. The price of bark afterwards rose to such a height, that it was found most profitable to cut in the spring; and the practice then became so general, that it by and by became necessary to offer premiums to induce proprietors of oak forests to fell timber in the winter season, for the sake of the British navy. The inhabitants of the county of Stafford appear at a somewhat early period to have sought to combine the advantages of the bark trade, with a fulfilment of the conditions that entitled them to the premium on winter-felled timber: they stripped the trees of their bark in the spring, and felled them the following winter. And Buffon and Duhamel

* Duhamel, op. cit., t. i. p. 400.

showed subsequently, that by barking trees two or even three years before cutting them down, the white external wood could be rendered nearly as hard and durable as the heart-wood of the tree. The recommendation of this procedure by these two distinguished men has not been followed in France; but ever since 1770 the Dutch have adopted it, and it is now practised in many parts of England, particularly in the royal forests.

It is quite certain that the nature of the soil exerts a considerable influence on the rapidity of growth and quality of the timber. The oak, the elm, &c., which have been grown in a damp soil, will not be so hard and compact as the same trees reared on a dry plot. Duhamel found, that although the trees which came in swampy bottoms were very sappy and wet, they were still lighter than others of the same kind which had grown on a dry bank. Their white wood is thick in comparison with their hard wood; they are brittle, and do not readily take or keep the shapes into which they are bent for ship-building or for staves; and then their pores being large and open, and the whole wood being without that kind of varnish which impregnates good timber, they are readily permeable and unfit for the manufacture of vats, &c.—to say nothing of their being much more perishable. Such soft and porous timber is altogether improper for out-of-door constructions and for ship-building; but it answers extremely well for indoor and cabinet work; for the latter it has even certain advantages, it is easily wrought; and once fairly seasoned, it is neither so apt to warp nor to crack as harder wood. It was very probably to guard against any excess of sap in trees, so prejudicial in a general way to the timber they yield, that the Romans, according to Vitruvius, surrounded those that were destined to be cut down with a trench six months beforehand.*

Trees which have grown in a good soil sufficiently drained have a fine bark, and their white wood is moderate or small in quantity. Their woody layers, indeed, are apt to be thinner generally, than those of trees that have grown in a wet soil; but they are much harder and tougher, their grain is more even and close, and their pores are filled with an incrusting matter. They are consequently very heavy, even when thoroughly dry, and with time and due seasoning they become extremely hard, and in the same degree acquire durability. Duhamel was led by his experiments to conclude that the difference in point of density of timber grown in a marshy soil, and in one that was well drained and dry, was occasionally in the ratio of five to seven.

The denser, dry-grown timber supports a relatively much greater weight without breaking than the marsh-grown timber; and when it does yield, it gives way by a large and splintering surface, while the softer, less dense wood snaps off short. In brief, there is no question as to which kind of timber is the most valuable; and measures ought to be taken by landed proprietors and timber-growers at all times, not merely to grow trees, but to grow them under such cir-

* Duhamel, Expl. des bois, t. i. p. 46.

cumstances as shall ensure their yielding good available timber when they have come to maturity.

If wet soils then be unfavorable to the growth of timber of the highest value, in ship-building especially, what has been said must be taken as of application to those trees only which will grow in a great variety of soils. Damp and even marshy lands are well known to be favorable and even indispensable to certain trees, which, by their nature, delight in the neighborhood of water; but these are generally kinds which are rather sought after for their height and lightness, than for their strength and durability.*

Excessively dry soils, on the other hand, have also their disadvantages for forest cultivation. In such ground, trees seldom acquire a sufficient growth to admit of their being applied to any important purpose. It is certain, however, that absolute uniformity is never encountered in any piece of timber. The woody layers that have been formed in a wet or a dry year, in a warm or a cold year, feel and manifest the effects of the varying meteorological influences. They are of different thicknesses and densities, and, when carefully examined, are found to present the characters of the timber grown in soils of the most opposite description in point of wetness and dryness.†

The treatment of trees after they are felled, the drying and seasoning of the timber, are points of the highest importance. Standing trees contain a large quantity of water in their composition. After being cut down the moisture is dissipated, rapidly at first, much more slowly afterwards. This drying process is, of course, favored or retarded by the varying states of heat and moistness of the atmosphere. At length there comes a time when the wood no longer suffers any sensible change by longer exposure to the air; or if it does, the change is now on the one side, now on the other, and merely in harmony with the hygrometric state of the atmosphere. Timber has then lost the whole of the moisture which it can get rid of by this mode of drying; it is now fit for use; it is *seasoned*, to use the technical expression.

Timber is sometimes seasoned by previous total immersion in water. It has been held that this process favored the thorough drying, by dissolving out certain deliquescent salts which are found in the sap, and prevented after-shrinking. However this may be, it is quite certain that in warm countries especially, it is advantageous to sink fresh-cut timber in water, with a view to prevent it from splitting, apparently in consequence of drying too quickly. The old Venetians sank for a season in the sea, the oak timber which was destined for the construction of their galleys. Elm and beech, in particular, are said to improve greatly by the process of submersion in salt water, and to dry afterwards perfectly by simple exposure to the air.‡

* We believe, however, that the live-oak, of which the American navy is constructed, and which supplies one of the most imperishable kinds of timber known, grows exclusively in swamps.—Eng. Ed.

† Duhamel, t. i., p. 57.

‡ Knowles, Maritime and Colonial Annals, 1825.

Mr. John Knowles, who made a particular study of the means most generally employed in seasoning timber, has given an account of a series of experiments undertaken in the arsenals of Deptford and Woolwich, to determine the rate of drying and ultimate degree of dryness attained by timber variously treated—unprepared and prepared by previous submersion in water. The pieces of timber were placed vertically, now in the position they had occupied in growing, now in that opposed to this; and it was found that, circumstances the same, they dried more quickly in the former than in the latter. The general results of these experiments were as follows: 1st. That the pieces of timber were best seasoned by being kept about thirty months in the air, but in the shade and protected from wet. 2d. That they lost more of their original weight after six months' alternate immersions and dryings, than by being kept under water for six months and then dried. Ship-builders are generally agreed that it is not expedient to make use of timber until three years after it is cut.*

Duhamel advises strongly, that in ship-building all timber from trees already on the decline should be rigorously rejected; and this the rather, that the most careful examination often fails at first to perceive any alteration in the heart-wood of such trees, although it never fails to show itself by and by at a sufficient interval after the felling. This is undoubtedly a precept which it would be well to bear constantly in mind; but timber does not always carry within itself the germs of its speedy decay; and that which has been seasoned with the most scrupulous care, and was originally of the best quality, does not escape the rot when it is placed under unfavorable circumstances, any more than that which was of inferior worth and less carefully treated.

Wood appears to perish or decay through three principal a d appreciable causes, which all require similar conditions to con e into play, viz., stagnant air, sufficient warmth, and moisture. Like the generality of organic substances, wood, when moistened in contact with the oxygen of the air, and under the influence of a sufficiently high temperature, undergoes decomposition of a kind which has been compared to a slow combustion, upon which we shall find occasion to say more by and by. It is with a view to escape this kind of decay as much as possible that timber is never, or ought never, to be employed in the construction of ships and buildings until it has been thoroughly seasoned.

Besides this first cause of decay, which may be prevented in a great measure by using certain precautions, wood has still two redoubtable enemies, insects and certain plants of the family of the cryptogamiæ. In one case, the wood perishes because it is fed upon by certain animals which live and grow at its expense; in the other it decays because it serves as the soil to one crop of fungus after another which luxuriate on its surface, while their roots penetrate deeply into its interior. There is nothing in either accident which

* Dupin, Ann. de Chimie, t. xvii. p. 277.

excites astonishment, now that we know the intimate constitution of wood. We know, in fact, that among the number of soluble principles which impregnate the woody tissue, there is an azotized matter analogous in its composition to those that exist so abundantly in all the ordinary esculent vegetables. There is, therefore, in wood ample nourishment for the insects which we find living on it; and if I state now (reserving to myself the opportunity of demonstrating the fact) that all organic azotized matter becomes an active manure by decaying, we shall understand how it happens that plants, which have the power of living in dark, warm, and damp places, wax and multiply in the joistings of houses, and in the ribs and planks of ships, causing a *dry rot*, which separates the integral layers of the wood, and reduces the strongest beams to dust.

The rapidity with which wood is, in some circumstances, devoured by insects is almost incredible. Some years ago the thermites, or white ants, spread in such strength through the docks and arsenals of Rochelle and Rochefort, that in a very short space of time serious damage was done. A learned entomologist, M. Audouin, commissioned by the ministry to take information on the subject, reported that the ravages committed by these insects had been very considerable. But it is principally in warmer climates, where the temperature is steady throughout the year, and where there is no winter, that the thermites occasion the most alarming injury. At Popayan, for example, it is difficult to meet in a building, even of recent construction, with a piece of wood which is not gnawed and ant-eaten. The hardest and most compact woods do not always resist the attacks of these insects, which, further, do not spare every kind of odorous wood, cedar for instance. In such countries it is altogether impossible to preserve books and papers. I remember, in connection with this matter, that having received instructions to examine the archives of Anserma, one of the oldest towns in Popayan, in 1830, I found nothing but books illegible and in pieces; nevertheless, the date of the documents, which it was my business to consult, could not have been older than the year 1600.

The dry rot, which results from the development and growth of cryptogamic plants upon wood, is the curse of navies. Mr. Knowles is of opinion that this disease of timber has been known from the most remote antiquity; he believes that he can even recognise dry-rot in the sore called house-leprosy, mentioned in the 14th chapter of Leviticus. A ship attacked by dry-rot, becomes in a very short space of time unfit for sea. The Foudroyant of 80 guns is often quoted as an instance of its destructive powers: launched in 1798, she had to be taken into dock and almost rebuilt so soon as 1802.*

The fungi which induce dry-rot have been studied by Sowerby. Mr. Knowles signalizes two species in particular; one of which he describes under the name of *Xylostroma giganteum*, the other under that of *Boletus lacrymans*. The Xylostroma does not extend beyond the part where it is developed; but the Boletus, on the contrary, is

* Dupin, Ann. de Chimie, t. xvii. p. 290.

propagated with frightful rapidity, and disorganizes deeply and to a great distance around the texture of the wood where it once appears These fungi are generally found on board ship, between the planking and the ribs, in damp situations, and where the air is scarcely, if ever, changed.*

The temperature most favorable to the development of dry-rot has been found to lie between 7° and 32° cent. or 45° and 90° F. These are the extreme limits: below the minimum vegetation languishes; above the maximum, the fungi droop. With this piece of information it was hoped that vessels might be freed from dry-rot by raising the temperature sufficiently. The trials were made in winter in the "Queen Charlotte," the air in the lower part of the ship being raised as high as 55° cent. or 130° F. But the general result did not answer expectations; for although the fungi were destroyed in the lower part of the vessel, it was found that their growth was rather favored in places at a certain elevation above the kelson. The warm air, in fact, as it rose through the timbers became robbed in its course, and deposited the greater portion of the moisture which it had taken up at a lower level. Above the orlop deck, consequently, there was just about the temperature and the quantity of moisture most favorable to the development of the fungi. The evil was therefore only transplanted, not destroyed. It was now proposed to heat the "'tween decks" at the same time as the hold, making use of due ventilation; but this method of proceeding has not been put into practice.

The extreme slowness of the growth of trees stands in strong contrast with the rapidity of their decay when they are reduced to the shape of timber and employed in constructions of almost every kind. In countries well advanced in civilization, every description of industry tends to consume timber, at the same time that an increasing population is every day contracting the extent of forest land, and diminishing the number of trees grown. In some countries, indeed, it is certain that the production of wood for all purposes, firing, &c., &c., is no longer in relation with its consumption. The price of the article, necessarily high, is therefore tending continually to rise; and it is not surprising that various measures have been suggested and essayed of giving this perishable material greater durability.

The well-known great durability of certain trees, the teak, ebony, lignum-vitæ, &c., naturally led to the conclusion that the fatty or resinous matters which they contain have the property of preserving the wood against the greater number of the ordinary causes of decay; and unctuous and resinous matters appear in fact to have been the means most anciently employed to preserve wood from the air, from moisture, and from the attacks of insects. But it is scarcely necessary, at the present time, to say that these varnishes only accomplish the object proposed in their application in a very imperfect way; paint and varnishes crack, rub, or scale off with the slightest

* Dupin, Ann. de Chimie et de Physique, t. xvii. p. 291, 2e série.

friction; nor do they always remove the causes of internal decay on the contrary, by preventing more complete dryness, they sometimes even provoke or favor them, when applied to tender, that is, imperfectly seasoned wood. Merely laid on the surface, indeed, it has always been seen that varnishes of any kind were but indifferent protectors; that a really good preserver ought to penetrate the substance of the wood, and unite with the tissue itself. But herein lay the whole difficulty; how was the needful penetration to be effected? for the number of chemical substances, from which good effects might reasonably be anticipated, is pretty considerable,—unless indeed we find ourselves prevented from using them by the consideration of the price; for it is imperative that any preservative proposed be extremely cheap.

For a long time the only process for effecting the penetration of timber by substances proposed for its preservation was to macerate them for a longer or shorter time in a solution of the substance. But this means was found as tardy of accomplishment as it was ordinarily imperfectly effected; to have got to the heart of logs of large scantling, years would have been required. Any delay, however, in such circumstances, is of itself a cause of enhanced price of the article. By and by a variety of processes, the element in one being pressure, in another exhaustion, were put in practice, and very satisfactory results obtained. M. Breant showed, that by means of strong pressure he could fill the largest logs from one end to the other with any unctuous or resinous substance proposed, in the course of a few minutes. M. Moll, a learned German, proposed creosote introduced in the state of vapor by forcing, as an effectual means of preserving timber, which it probably would be found; but the high price of the antiseptic, were there no other objections, would necessarily be an obstacle to its general employment. The same objection applies to the bichloride of mercury, (Kyan's patent;) and arsenic is inadvisable from its deleterious effects upon the animal economy. Some workmen are said to have lost their lives in consequence of working timber which had been impregnated with a solution of white oxide of arsenic.

It had been observed that vessels engaged in the lime-trade lasted long; and then it was naturally thought that by impregnating the wood to be used for ship-building with lime it would be rendered more durable. But the result did not answer expectation; the timber treated with lime did not even seem to last the usual time.*

Such was the state of the question when Dr. Boucherie made a highly important communication to the Royal Academy of Sciences on the preservation of timber.† Some estimate of its nature may be formed from the list of subjects discussed in this remarkable paper.

1. To protect timber against dry-rot and the ordinary wet-rot.
2. To increase its hardness and strength.
3. To preserve its flexibility and elasticity.

* Dupin, Ann. de Chimie, t. xvii. p. 286

† Idem, t. lxxiv. p. 113.

4. To counteract its alternate contraction and expansion in consequence of the varying state of moistness and temperature of the atmosphere.

5. To diminish its inflammability and combustibility.

6. To give it a variety of permanent colors and odors.

In the whole of his experiments M. Boucherie set out from this proposition, the truth of which appears indisputable and to require no comment, viz : *That all the changes which wood undergoes proceed or depend upon the soluble matters which it contains.* In conformity with this idea, the first step towards giving durability to timber was, either to render these matters insoluble and inert, or to remove them entirely. M. Boucherie, therefore, in his first trials sought to render the matters insoluble by charging the wood with a substance capable of combining chemically and forming a precipitate with the soluble matter left by the sap. To resolve this problem, M. Boucherie investigated the reactions between the soluble matter of wood, which it was his object to precipitate, and a variety of low-priced chemical agents. He found that the pyrolignite of iron combined the greatest number of desirable properties: it is very cheap, the oxide of iron forms stable compounds with the greatest number of the organic substances which are found in the sap of vegetables, and, to conclude, the crude pyrolignite contains a notable quantity of creosote.

The facts upon which M. Boucherie relies as proving the preservative powers of the pyrolignite of iron flow from numerous experiments performed either on vegetable substances which in themselves readily and rapidly undergo changes; or upon billets of wood of different kinds. A quantity of flour, the pulp of carrots, beet-roots, &c., impregnated with the pyrolignite resist decomposition in a very remarkable manner in contrast with the same substances when they have not been prepared in any way.

The wood which was selected for trial, was generally of the most perishable kind. In December, 1838, several empty hogsheads and barrels made of the best timber unimpregnated and impregnated with the pyrolignite were placed together in the dampest parts of the great cellars of Bordeaux. In August, 1839, it was easy to see that the unimpregnated tubs were already deeply stricken, and after from two to three years they fell to pieces with the slightest force; the casks made of the prepared wood, however, were as sound as on the first day of the experiment.*

M. Boucherie concluded from his experiments instituted with a view to the settlement of the question, that about $\frac{1}{40}$th of the weight of the wood in its green state of the pyrolignite was adequate to precipitate and render insoluble all the principles obnoxious to change, which were contained in the woody tissue.

M. Boucherie, while he regards the pyrolignite of iron as at once the most powerful, and one of the cheapest preservatives of timber, nevertheless indicates several soluble salts, which are readily available in consequence of their low price, and also very effectual when

* Comptes Rendus, t. ii. p. 896.

the wood, which they are to preserve, is not kept constantly wet. Solutions of common salt, of chloride of lime, the mother-water of salt-marshes, &c., were all tried and found useful: casks, the wood of which had been prepared with the chlorides, after having been long kept in very damp cellars, came out as fresh as those which had been impregnated with the pyrolignite of iron; the flexibility of the wood preserved with these alkaline and earthy salts was further as great as at the beginning of the experiment.

Having now come to a conclusion in regard to the substances most effectual in preserving wood, the next business was to make them penetrate its tissue most intimately. Maceration, M. Boucherie soon found, like his predecessors in the same path, to be insufficient, the substances in solution only penetrating a very little way. He then tried various processes of injection; but all inferior to that imagined by M. Breant, and therefore less effectual. He then bethought him of effecting the needful penetration of the wood in the green state, and before it had been sensibly altered by drying and seasoning; he asked himself if the force which determines the ascent of the sap might not be taken advantage of after the tree was cut down, as a means of determining the entrance of a solution of pyrolignite of iron? And all his trials in this direction answered his expectations fully. M. Boucherie had, in fact, discovered a means of securing the penetration of the minutest pores of the largest log by a substance capable of rendering it incorruptible. No one before M. Boucherie thought of taking advantage of an admitted physiological fact for such a purpose. He announces the principle upon which he proceeds in these terms: "If a tall tree be cut down at the proper season, and the bottom of the trunk be then immersed in a saline solution, weak or strong, the liquid is powerfully drawn up into the tree, penetrates its most intimate tissues, rises to its smallest branches, and even to its terminal leaves."*

In the month of September, a poplar, upwards of 90 feet high and nearly 16 inches in diameter, was cut, and the bottom of its bole plunged in a vessel containing a solution of pyrolignite of iron marking 8° of the areometer of Beaumé; in the course of six days it had absorbed upwards of 66 gallons of the fluid.

In his first experiments, M. Boucherie procured the needful absorption by placing the bottoms of his trees in vessels containing the solution; but this mode of proceeding was obviously full of difficulties and open to many objections: the weight of a green tree of large size, with the whole of its top and branches, is often enormous, and to raise a mass of the kind once down again into the perpendicular was no easy task; it implied recurrence to certain mechanical means which are not always at hand, and necessarily expensive. M. Boucherie, therefore, tried other modes of making the trees absorb; he adapted a sac of impermeable material to the bottom of the trunk laid on the ground, and into this sac he poured his solution, and this method answered very well. He next took advantage of

* Ann. de Chimie, t. lxxiv. p. 132.

one or more of the roots to effect the imbibition. He next bored a hole into the bottom of the trunk, still erect; and having brought the cavity thus made to communicate with a reservoir, he still succeeded. This last plan was still further simplified in proceeding as follows: the trunk of the tree is pierced by an auger through nearly the whole diameter. Into the auger-hole thus made, a narrow saw is passed, by working which on either side, the trunk is divided internally to a very considerable extent, and the majority of its sap-vessels are thus cut across and made accessible. An impervious cloth is then tied round the trunk, below the opening, and this is made to communicate with the reservoir of liquid.*

M Boucherie was almost necessarily led, in the course of his experiments, to inquire whether the absorbing power of trees differed at different seasons or not. He ascertained by trials made in the months of December and February, that though in the oak, the hornbeam, and the plane, the solution of pyrolignite of iron always rises several feet, and even several yards, yet that in the colder season of the year, it never rises so high as it does in summer, in spring, and especially in autumn, the season in which the power of ascent is most remarkable. This conclusion is obviously of interest physiologically. It proves that if winter be a season of repose for the sap, it is not so absolutely. There is one remarkable exception to the general fact now announced, and this occurs among the resinous trees that keep their leaves till the spring. It has been ascertained, by direct experiment, that the ascent of the sap continues through the whole course of the winter in the cone-bearing trees, and this to such an extent, that it is always possible to impregnate every part of their trunk by the way of simple absorption at any period of the year. As M. Boucherie remarks, this fact might even have been foreseen from the fresh and green state of the leaves of these trees.

It now became important, in connection with the practical application of M. Boucherie's views, to ascertain whether or not the penetration was energetic in the ratio of the vigor of the tree itself, in proportion as it was more numerously provided with branches, more thickly covered with leaves. Experiment showed that the penetration still takes place after the removal of the greater number of branches, provided only the leading bough or terminal crown be left. A stem furnished with a number of leafy branches continues, as has been said, to imbibe, though separated from the roots; but for how long a time will it continue to do so? This was a capital point to determine. At the end of September, the bottom of a pine-tree, about 14 inches in diameter, was first put into the solution 48 hours after it had been felled; nevertheless the imbibition was complete. In June the same success attended the experiment made on a plane that had been cut for thirty-six hours. Still it is certain that the penetration takes place with so much the more energy as it is arranged close upon the time of the felling. The power by which it is determined declines rapidly after the first day is passed, and by

* Boucherie, Ann. de Chimie, t. lxxiv. p. 134, 2e série

the tenth day it is almost entirely gone. In favorable circumstances these ten days suffice to effect the complete impregnation of the largest stem. In one of his experiments upon a poplar, M. Boucherie saw the absorbed liquid reach the height of about 95 feet in seven days.

In the white woods it is found that there is an axis of variable diameter in different cases which escapes or rather which resists impregnation. In hard woods the parts which are not penetrated are the inner or undermost circles of the heart. M. Boucherie, after having ascertained these facts, explains them thus: in the white woods, according to the testimony of the workmen, the central part which resists the penetration is at once the weakest and the most perishable portion of the log; there is no longer any circulation, any life there; it is dead wood interred in the midst of the living woody layers. This absence of penetration of the woody tissue appears, on some occasions, elsewhere than in the centre of the trunk and branches; it presents itself under the most various forms and in different parts of the trunk: it appears to depend, as has been said, on the presence of wood abstracted from the influence of vital phenomena, and which, impenetrable itself, presents a barrier or an obstacle to the passage of the solution to other parts; it is thus that a knot, or a piece of rotten wood, is generally found as the starting point of the zones that have escaped imbibition. As to the non-penetration of the most central parts of the heart of oak, elm, &c., M. Boucherie views it as affording unquestionable proof of the fact that there the living juices of the tree had long ceased to circulate.

The distinction generally drawn between the white or soft, and the perfect or hard wood, rests on the differences of color presented by a transverse section of the trunk. In the oak, for example, the external and nearly white concentric layers are held as the soft and valueless portion of the log, and are commonly hewn away in squaring it; the darker, more central portions constitute the heart-wood, the valuable timber. But, according to M. Boucherie, the distinction is different when the fact of penetrability is taken as the guide, and all that portion of the trunk which imbibes is considered as alburnum, or soft wood, and all that does not imbibe is regarded as hard wood. The alburnum in this way is so much extended that it may be found constituting three-fourths of the whole mass of the trunk. Once introduced, the pyrolignite of iron, according to M. Boucherie, is not only useful in preserving the wood, it also increases the density of the timber. Impregnated with this salt of iron, wood becomes so hard as powerfully to resist the tools of the carpenter and joiner, who even complain of the increased difficulty with which it is worked.

Flexibility and elasticity in timber are qualities in request for certain purposes, particularly for ship-building. The fir timber of the north of Europe is much more prized than that of the south, especially for masting, on account of its greater flexibility and elasticity, qualities which appear to depend in a great measure on the quantity of moisture retained; to increase these qualities M. Boucherie has even introduced by imbibition a deliquescent salt, such as the muriate of lime, which retains moisture powerfully, as is well known, and seems to have the power of giving a remarkable degree of suppleness to wood. The experiments, contrived to show the effects of deliquescent salts, were made upon deal, which is allowed to be one of the most brittle woods. After having impregnated it with concentrated solutions it was sawed into very thin veneers, some of which I have seen in the possession of M. Boucherie, which after being strongly twisted and bent in various senses, immediately regained their original flatness and evenness when they were left free.

Warping, or shrinking, is occasioned by alternate shrinking and swelling in consequence of varying hygrometric states of the atmosphere. When timber is worked before it is thoroughly seasoned,—and this is apt to happen in regard to pieces of large scantling especially—the shrinking is of course extremely conspicuous when the time necessary to complete desiccation has elapsed. It is this inconvenience which makes it imperative on builders of all kinds, ship-builders more especially, to keep stocks which necessarily absorb a considerable amount of capital. It has long been a question with engineers to find a remedy for this state of things. Seasoning, indeed, is now effected somewhat more quickly by squaring the logs at the time the trees are cut down; but the loss of time is still very considerable. The mode of seasoning by the stove or vapor has been abandoned as too costly.

After having found that the shrinking and separation of pieces of carpentry did not begin to take place until the timber was upon the point of losing the last third of the moisture which it contained at the time of being cut, M. Boucherie thought that to prevent all warping and shrinking it would be enough to retain this quantity of water in combination with the woody tissue; in other words, to prevent complete desiccation. Facts have proved the correctness of this view. Pieces of wood kept at a certain unchanging degree of moistness by means of a deliquescent salt infused into their pores, do not change their bulk or form, in spite of extreme variations in the hygrometric state of the air. Such pieces of wood, however, exhibit great differences in point of weight under the influence of different circumstances.

Several planks of great breadth and extremely thin were prepared with chloride of lime and joined together; some of them were left unpainted, others were painted on one side, or on both sides; after the lapse of a year these planks were found not to have shrunk or warped, while similar planks of the same thickness and kind of wood, but unprepared, were found to have cast in an extraordinary way.*

M. Boucherie has done more than this; he has not only had it in view to preserve wood and to prevent it from warping, qualities so desirable,—he has made use of the same faculty of imbibition to im-

* Boucherie, op. cit. p. 151.

pregnate the wood with a variety of beautiful colors, and thus to give even the most common kinds tints that will admit of their being used in the construction of costly furniture. The pyrolignite of iron alone gives an agreeable brown tint that harmonizes excellently with the natural color of the harder parts of so many trees which usually resist penetration. By following up the pyrolignite with an infusion of nutgalls or oak-bark, the mass of the wood is penetrated with ink, which presents a black, blue, or gray color, according to circumstances; a solution of another salt of iron succeeded by one of prussiate of potash will cause a precipitate of prussian blue in the wood, &c.; in short, by the numerous reactions of this kind with which chemistry is familiar, a great variety of colors may be obtained.

Among the number of useful properties communicated to wood by impregnation with saline solutions, that of being rendered little apt for combustion ought not to be omitted. M. Gay-Lussac was the first who thought of rendering vegetable tissues incombustible by means of saline impregnations.* By incombustible, we are not to understand unalterable by a red heat; for every one must see that the protecting power of no salt can extend so far as this; but tissues which take fire very readily, and burn with great rapidity, cease from giving any flame, and merely smoulder, after they have been impregnated with certain salts; they take fire with difficulty, go out of themselves, become charred, and are incapable of propagating fire. And this is exactly what happens with wood which has been properly charged: it burns, and is reduced to ashes with extreme slowness, so that two huts exactly alike, built one of charged wood, and the other of ordinary wood, having been set fire to at the same moment, the latter was already burned to the ground, when the interior of the former was scarcely charred.†

The ingenious process of impregnating wood by the way of vital inspiration is not without certain objections. In the first place, it can only be performed at those periods of the year when the sap is in motion, and the trees are covered with their leaves. This time, however, is limited to a few months of the year, and the usual practice being to fell timber in the winter, wont and usage are opposed to cutting down trees in the spring and autumn. To meet these objections, M. Boucherie engaged in new experiments, which led him to a means of impregnating timber at all seasons, in winter as well as spring and autumn, and in a very short space of time; this second method is applicable to wood that has already been squared as well as to the round trunk, provided it has been recently felled.

To impregnate timber by this process, the logs are placed vertically, and the upper extremities are fitted with an impermeable sack for the reception of the saline solution destined to charge them; the fluid enters from above, and almost at the same moment the sap is seen to begin running out below. There are some woods which

* Ann. de Chimie, t. xviii. p. 211, 2e série.
† Idem, t. lxxiv. p. 152, 2e série.

include a large quantity of air in their tissues; in this case the flow does not go on until this air has been expelled; once begun, it goes on without interruption. The operation is terminated when the fluid, which drips from the lower part, is of the same nature as that which is entering above. In my opinion this method must be preferable to that by aspiration. In the second mode of proceeding, in fact, we accomplish our object by a true displacement; almost the whole of the sap is expelled, and the saline solution introduced has only to subdue or neutralize the very small quantity of soluble organic matter which may remain adhering to the woody tissue. By accomplishing such a displacement by means of simple water we should undoubtedly obtain results favorable to the preservation of timber, inasmuch as we should have freed it from almost the whole of those matters which are regarded as the most alterable themselves, and the first cause of rotting in timber. The rapidity with which the fluid introduced is substituted for the sap which it displaces, and the quantity of this expelled sap which may be readily collected, exceeds any thing that could have been imagined before making the experiment; thus the trunk of a beech-tree about 52½ feet in length by 33¾ inches in diameter, and consequently forming a cube of somewhat more than 29 feet and a half, gave in the course of twenty-five hours upwards of 330 gallons of sap, which were replaced by about 350 gallons of pyroligneous acid. The liquid which penetrates in this way acts so effectually in displacing the sap, that M. Boucherie says we can readily procure or extract by its means the saccharine, mucilaginous, resinous, and colored juices contained in trees. It would, perhaps, be possible, and I beg to suggest this idea to colonial planters, to apply the method of displacement to the extraction of the coloring matters of dye-woods. The trade in dye-woods does not extend beyond localities favorably situated for exportation, so that at a certain distance from the shores of the ocean, or the banks of rivers, it is found absolutely impossible to carry on a trade, the material of which is so heavy and bulky as timber. The greater number of the coloring matters found in wood being soluble, it is possible to export them in the state of extract. Various attempts of this kind have already been made, and if they have not been successful, the obvious cause of this lies in the method which has been followed, and which has hitherto consisted in treating the wood reduced to chips by means of boiling water, and then reducing the colored solution obtained; but it is obvious that in the remote forests of America, or of Africa, where all mechanical means are wanting, nothing but failure could attend upon such a procedure. By the method of M. Boucherie, the main difficulties appear to be got over; there is nothing more to be done, in fact, than to get the trees into the state of logs, and these are generally readily transportable, after which one or more evaporating pans seem all that are further necessary.

Dye-woods.—The greater number of these woods belong to the family of leguminosæ; the principal kinds met with in trade are:

1. Mahogany wood, (*hæmatoxylon campechianum*,) of a reddish

yellow, which becomes brown with age ; this wood, besides a variety of alkaline and earthy salts, of volatile oil and unazotized matter, contains a particular coloring principle, called hematine, discovered by M. Chevreul.*

The mahogany grows in the hot intertropical regions of America ; Mexico and some of the West India islands export considerable quantities.

Pernambuco or Brazil-wood is the name given in trade to the trunks of several trees of the genus *Cæsalpinia.* The *Cæsalpinia crista* of Jamaica, the *C. sappan* of Japan, the *C. echinata* of Santa Martha, afford kinds that are very much prized. In point of chemical composition Brazil-wood agrees with Campechy wood ; the coloring matter which characterizes it has been named Braziline by M. Chevreul ; it is obtained in small crystals of an orange color.

This wood comes to Europe in fagots of about 39 inches in length. Red Saunders-wood is furnished by the *Ptœrocarpus santalinus ;* it contains a peculiar dye-stuff, santaline, observed by M. Peltier.†

To conclude, the yellow dye-woods of commerce are Fustic, *Rhus cotinus*, of the family of turpentine trees, a native of the south of Europe, and the Cuba and Tampico woods, which are probably varieties of the *Morus tinctoria.*

OF SUGAR.

Sugar is met with in almost every part of vegetables ; it has been found in flowers, in leaves, in stems, and in roots. It is less abundant in seeds ; and it may even be said that the quantity of saccharine matter contained in vegetables in general is invariably diminished at the period of formation of the seed. Sugar, consequently, as well as starch, appears to contribute to the production of the seed.

The very characteristic taste of sugar generally suffices to proclaim its presence ; nevertheless, it would be a great mistake did we rely upon this character alone for discovering the presence of sugar ; several substances possess a very decided sweet taste, without being on that account sugar, in the sense which chemists attach to the name. True sugars, according to chemists, have one property which distinguishes them from all substances with which they may have, in other respects, the greatest analogy ; this characteristic property is that of becoming changed, under the influence of water, a suitable temperature, and contact with yeast, into alcohol and carbonic acid. It is certain, nevertheless, that certain bodies which do not belong to the chemical genus, sugar, may, under the influence of fermentation, yield alcohol. I have already quoted starch as coming under this head ; but it has been distinctly ascertained, as I have also said, that such substances, under the influence of the ferment itself, are first changed into sugar, which subsequently undergoes the vinous fermentation.

* Chimie appliquée à la teinture, 30e leçon, p. 88.
† Chevreul, Chemistry applied to dying, 30th lecture, p. 94.

It is admitted at the present time that fermentable sugars must be divided into two principal species, in harmony with characters which are most easily appreciated. One of these presents itself in the shape of hard, transparent crystals, and is met with in sufficient quantity to be profitably extracted from the juice of the cane and the beet, the sap of the maple and of certain palms; the other is obtained with some difficulty in the solid state, being most frequently and readily procured in the form of sirup; the taste of this is less sweet, less decided; it exists in the grape and the greater number of fruits. The chemical characters of these two kinds of sugar, which are designated cane-sugar and grape-sugar, are somewhat different; and the elegant researches of M. Biot have shown, that from some of their physical properties, particularly the action of their solutions upon polarized light, they cannot be regarded as constituting one and the same species. In the vegetable kingdom, these two kinds of sugar are frequently met with mixed; and there are certain chemical means which enable us readily to transform cane-sugar into grape-sugar. The inverse transformation has not yet been accomplished; but there is nothing which leads us to conclude that it is impossible; and the time, perhaps, is not very remote when the sugar which is manufactured from potato-starch may be changed into crystallized sugar, similar to that which is obtained from the cane.

Crystallized sugar. Cane-sugar is readily obtained in large transparent crystals, which are known under the name of sugar-candy. Sugar is fusible: under the action of a regulated temperature, it acquires a dark-red color, and passes into the state of caramel; a higher temperature effects its decomposition. It is much less soluble in alcohol than in water; highly concentrated alcohol, indeed, only dissolves an extremely small quantity of sugar.

M. Peligot's analysis of cane-sugar shows it to be composed of—

Carbon	42.1
Hydrogen	6.4
Oxygen	51.5
	100.0*

Such is the composition of sugar dried at the temperature of boiling water; but the substance, like the majority of organic matters, still contains a certain proportion of constitutional water, which it abandons when it combines with certain bases. Thus sugar combines with oxide of lead, and forms a true saccharate, in which the sugar, deprived of its water of constitution, plays the part of an acid; this combination, which presents itself to us under the form of white mammillated crystals, analyzed by M. Peligot, would indicate the following as the composition of anhydrous sugar—

Carbon	47.1
Hydrogen	5.9
Oxygen	47.0
	100.0

* Annales de Chimie, vol. lxviii. p. 124, 2e série.

Ordinary sugar, deprived of its water of composition in any other way, has the same elementary composition; thus caramel obtained by heating sugar to 180° cent., (356° Fahr.,) until it no longer loses watery vapor, has, according to M. Peligot, the composition of anhydrous sugar, such as it is found in combination with oxide of lead.

Setting aside all theoretical considerations, it is obvious that to have anhydrous sugar reconstituted ordinary hydrated sugar, it were necessary to add to 100 parts 11.76 of water, containing 1.3 hydrogen and 10.46 oxygen; the 111.76 parts then contain, in elements:

Carbon	47.1	per cent.	42.1	common sugar.
Hydrogen	7.2	"	6.4	"
Oxygen	57.46	"	51.5	"
	111.76		100.0	

Common sugar may therefore be viewed as composed of 100 anhydrous sugar, and 11.8 water.

The whole of the sugar which comes from South America and the West Indies, and a large proportion of that which comes from the East Indies, is extracted from the juice of the sugar-cane.

In America three principal varieties of sugar-cane are cultivated, the Creole, the Batavian, and the Otaheitan. The Creole cane has the leaf of a deep green, the stem slender, the knots very close together. This species, a native of India, reached the new world after having passed through Sicily, the Canaries, and the West India islands. The Batavian cane is indigenous in the island of Java; its foliage is very broad, and has a purple tint; the sap of this variety is much employed in making rum. The Otaheite cane is that which is most extensively grown at the present time; it was introduced into the West India islands and neighboring continent by Bougainville, Cook, and Bligh, in their several voyages, and is certainly one of the most important acquisitions which the agriculture of tropical countries owes to the voyages of naturalists. This variety of cane grows with extraordinary vigor: its stem is taller, thicker, and richer in juice than that of the other species. I observed it along the whole coast of Venezuela, of New Grenada, and of Peru; far from having degenerated by its transplantation to the American continent, it appears to have preserved all its original qualities without alteration.

The sugar-cane is propagated by cuttings. Pieces of the stem about 18 or 20 inches long, and having several buds or eyes, are placed two or three together in holes a few inches in depth, and are covered with loose moist earth. From a fortnight to three weeks are required for the shoots to show themselves above ground. The space to be left between each clump of plants depends much on the fertility of the soil; in the most fertile soils the distance may be about a yard, or a little more; and along the rows the spaces may be about 18 inches. Where land is of no great value it is found more advantageous to give greater space, and so to favor the access of the air and the light. It is not uncommon to see plantations where the canes are spaced at distances of between 4 and 5 feet. The time at which the setting of the slips takes place cannot be definitively indicated; it depends entirely upon the epoch at which the periodical rains are anticipated. But in places where irrigation is possible, the setting goes on through all the months of the year. The holes for the reception of the slips are usually dug with a hoe, and a negro will make from sixty to eighty holes in the course of a day. When the ground has been previously ploughed, as it is in some of the West India islands, he will make twice as many. Loose rich soils, when they have a certain moisture, are the best adapted to the sugar-cane; it does not thrive in an argillaceous soil, which drains with difficulty. In these moist soils the slips are not laid horizontally and covered, but with one end projecting a little way out of the ground. When the young shoots are covered with narrow and opposed leaves, watering is particularly advantageous, and the plants are repeatedly hoed until they have acquired sufficient vigor to choke noxious weeds. About the 9th month after the plantation of the slips, the shaft of the sugar-cane begins to lose its leaves, the most inferior falling first, the others in succession, so that when arrived at maturity, it only presents a tuft of terminal leaves. The flowering generally takes place with the conclusion of the year; and the cane is held sufficiently ripe in from two to three months after this epoch, when the stem has acquired a yellow or straw color. The planters, however, are by no means agreed as to the proper period of the sugar-cane harvest,—some even insist upon cutting before the flowering, believing that the quantity of sugar diminishes on the appearance of the flower. It is unquestionable, however, that the period that elapses between the planting and the harvest, must vary with the nature of the soil, and especially with that of the climate; while in some places the cane may be cut when it is a year old, doubtless there are others where it requires to stand from fifteen to sixteen months. In Venezuela, where the Otaheite cane is grown at the level of the sea, and where the mean temperature of the year is between 81° and 82° Fahr., the cane ripens, according to Colonel Codazzi, in eleven months. In districts at greater elevations under the same parallels of latitude, where the climate is of course not so hot, the cane requires a longer time to come to maturity; where the mean temperature is about 78° Fahr., twelve months are required; where it is about 74° Fahr., fourteen months become necessary; and where it is no more than about 67° Fahr., sixteen months are requisite.* The Otaheite cane grows to very different heights; in very favorable circumstances it will reach a height of 16 feet and upwards, but its general height may be stated at from 9½ to 10½ feet. Great cane plantations are divided into squares of from 100 to 120 yards on the side, each of which coming to maturity in succession, the labor is easily performed, both in regard to field-work and the manufacture of the sugar.

The cane is cut close to the root, and before being carried to the mill, the terminal tuft of leaves is struck off. These heads in the

* Codazzi, Geography of Venezuela, p. 141.

green state afford excellent food for horses and cattle; when dry they are used for thatching houses. After the first cutting, fresh sprouts arise, which require no other attention than hoeing. In good soils one planting will yield five or six harvests by successive shoots; but I have heard planters affirm, that the produce in sugar diminishes from year to year. In Venezuela, cane-pieces are re-planted every five or six years.

The cane with its top struck off is carried to the mill, where the juice is expressed, and the stems, which are spoken of under the name of trash, are dried and used as fuel.

The expressed juice contains crystallizable sugar, an azotized substance analogous to albumen, and some saline matters dissolved in a large quantity of water, which is dissipated by boiling, and the sugar finally won by crystallization. The manufacturing process is conducted with very different degrees of perfection in different places. In some the produce is obtained almost without admixture of molasses, in others the quantity of this article which drains away from the sugar is very large. It is now generally agreed that molasses proceeds in great part from imperfections in the manufacturing processes employed, especially to changes which the sugar undergoes in the course of its concentration by boiling at a high temperature. By the employment of what are called *vacuum pans* of various construction—pans from which the pressure of the atmosphere is removed either by the air-pump, or the condensation of the vapor as fast as it is formed, rapid evaporation is effected at a temperature much below that of boiling water, by which it is found that the relative quantity of sugar to that of molasses is greatly increased. It was long believed, indeed, and that on the authority of the first chemists, that there were two kinds of sugar contained in the sugar-cane, one crystallizable, the other uncrystallizable, and constituting the molasses or treacle. The researches of M. Peligot* have shown definitively that this conclusion is erroneous, that the cane contains no sugar that is not crystallizable, and that the pre-existence of uncrystallizable sugar or molasses is entirely chimerical. M. Plagne had indeed come to the same conclusion some considerable time ago—as far back as 1826; but his labors were not made known by publication till 1840. M. Casaseca, professor of chemistry at Havana, has very lately confirmed these conclusions, so important for the sugar husbandry of the world.† The composition of the juice of the sugar-cane is therefore less complex than it was once believed to be; making abstraction of very minute quantities of an albuminous azotized substance, of several salts and a little silica, substances which altogether do not amount to more than two or three hundredths, cane juice may be said to consist of water and of crystallizable sugar in the proportion of from 17 to 20 per cent.‡ The Otaheite cane analyzed by M. Peligot actually yielded:

* Ann. Maritimes et Coloniales, Aug. 1842.

† Vide Comptes Rendus, 1844. ‡ Peligot, op. cit.

Water	72.1
Woody matter	9.9
Soluble matter (sugar)	18.0
	100.0

This conclusion was verified by M. Dupuy at Guadaloupe in 1841, who, operating on the spot, found the composition to be as follows:

Water	72.0
Woody matter	9.8
Soluble matter (sugar)	17.8
Salts	0.4
	100.0

The analyses of the Creole cane made by M. Casaseca at Havana appear to indicate a larger quantity of woody fibre:

Water	65.9
Wood	16.14
Sugar	17.7
	100.0

The quantity of sugar yielded by the cane, differs considerably. M. Codazzi assigns 6 and 15 per cent. as the extremes, and 7½ per cent. as the mean. M. Dupuy gives 7.1 per cent. as the average. The quantity is, of course, first and most intimately connected with the quantity of juice obtained. But the produce of juice is extremely variable. In Guadaloupe, the juice varies between 56 and 62 per cent. of the cane subjected to pressure. The generality of mills do not, in fact, enable us to obtain more than about 56 per cent. At New Orleans the usual quantity obtained is said to be 50, and in Cayenne only 36 per cent. At Havana, according to M. Casaseca, the riband cane yields 45, the crystalline 35, and the Otaheitan 56 per cent. of juice.

The Otaheite cane was examined by M. Peligot, under a variety of circumstances of age, growth, part of plant, &c. &c. The following table contains the condensed results of his experiments:

	Water.	Soluble matters (sugar.)	Woody fibre.
First shoots	73.4	17.2	8.9
Second do. from original sprouts	71.7	17.8	10.5
Third do. from second do.	71.6	16.4	12.0
Fourth do. from third do.	73.0	16.8	10.2
Inferior part of cane	73.7	15.5	10.8
Middle part of do.	72.6	16.5	10.9
Superior part of do.	72.8	15.5	11.7
Knots	70.8	12.0	17.2
Cane of eight months	73.9	18.2	7.9
Cane of ten months	72.3	18.5	9.2

It would therefore appear, making exception always of the knots which occur in the course of a cane, that the composition of the plant in its various states and conditions, is almost identical. M. Peligot's important paper, while it informs us of the average composition of the Otaheite cane, satisfies us that the gummy and mucilaginous substances and the uncrystallizable sugar, the existence of which was held as demonstrated, are, in fact, nowise constituents of the sugar-cane. Whence we may conclude, with M. Peligot, that every drop of molasses which drains from the sugar is the produce of the manufacture; an opinion to which I assent the more readily from having myself seen oftener than once the juice of the cane yield nothing but crystallizable sugar These analyses further demonstrate, more powerfully than could any discussion, the imperfection of the processes usually followed in manufacturing sugar. They prove, in fact, that in the mill rather more than a third of the whole juice contained in the cane is left in the trash. This loss might be considerably diminished were more perfect pressure employed in extracting the juice. But it appears that the planters are indisposed to crush the trash too much, as by this it is rendered less fit for fuel, a considerable quantity of which, by the present mode of manufacture, is indispensable. M. Dupree, however, says that by insisting on obtaining from 65 to 66 per cent. of juice in all cases, the trash is still left with all its value as a combustible. The trash on coming from the mill appears quite dry. I have seen some which, after having been pressed twice consecutively, looked as if it were impossible by any further amount of pressure to express more liquid. Nevertheless, it was enough to taste this pressed cane, to be satisfied that it still contained a considerable quantity of sugar. To procure this without using more powerful machinery, M. Peligot proposed to steep the trash in water, and to press it a second time. By this means a weak juice is obtained, which, added to the first pressings, raises the produce of sugar from 7 to 10 per cent. upon the whole amount of cane employed. By following this process, suggested by theory, upon the great scale, M. Dupree has succeeded in obtaining $\frac{1}{5}$th more than the usual quantity of sugar without making any change in his apparatus, and without finding the trash too much shaken to be burned under his coppers.* In some circumstances the increase in the quantity of juice which this procedure implies, might be found an objection on account of the larger quantity of fuel required for its evaporation; but wherever a supply of wood is to be had, M. Peligot's method ought undoubtedly to be applied.

The very dissimilar quantities of crystallizable sugar obtained from canes, which as we have seen all contain very nearly the same quantity of this substance, prove that the processes of concentration and purification of the sap also contribute to the loss which has been indicated. M. Peligot has pointed out several causes which concur to deteriorate sugar; among the number: 1. A viscous fermentation which renders the sap thick and stringy, like mucilage, by which the boiling becomes difficult and the crystallization of the sugar which has escaped change, is rendered imperfect. 2. An acidity which takes place when the juice is not run at once into the coppers and boiled, an acidity which requires the addition of lime to destroy

* Peligot, Maritime and Colonial Annals, August, 1842.

or to prevent it. The alkaline earth, as I have had occasion to say, is by no means indispensable; its utility under ordinary circumstances is probably confined to assisting the defecation by forming an insoluble precipitate with some of the organic substances which are always met with in small quantities in cane juice; perhaps also to making an earthy soap with the fatty matters which adhere to the cane and are expressed in the crushing. When lime is added, to correct acidity, it forms an acetate or a lactate, salts which are peculiarly soluble, uncrystallizable, and which necessarily retain a quantity of sugar in the sirupy state. 3. The presence of certain mineral salts in the cane. Common salt, for instance, in combining with sugar forms a deliquescent compound, in which one part of salt is united with six parts of sugar; such a compound as this of course renders a large quantity of sirup indisposed to crystallize. It is therefore impossible to be too cautious, according to M. Peligot, in the choice of manure for a cane-field; that which contains any common salt must needs be injurious in one way, however advantageous it may be in another. The entire absence of this salt in the soil of plantations which are very remote from the sea shore is perhaps one of the causes which increases the quantity of sugar obtained from the crop, and makes it more easily manufactured in such districts.

M. Codazzi reckons the quantity of white sugar produced by a hectare of land, (2.473 acres,) planted with the Otaheite cane in the province of Caraccas, at 1875 kilogrammes, or 36 cwt. 3 qrs. 9 lbs. avoir.; which is at the rate of 15 cwt. 1 qr. 10 lbs. per acre. Taking $7\frac{1}{4}$ per cent. as the average quantity of sugar obtained, the weight of cane brought to the mill must obviously have amounted to 19134 kilog. or 18 tons, 15 cwt. 3 qrs. 10 lbs.; or 7 tons, 11 cwt. 3 qrs. 25 lbs. per acre. Assuming the average composition of the plant to be—

Wood (dry)	11.0
Sugar (minimum)	15.5
Water	73.5
	100.0

One acre of land will consequently yield a crop of—

	Tons.	Cwt.	Qrs.	Lbs.
Wood (dry)	0	16	2	24
Sugar	1	3	2	6
Water	5	11	2	12
	7	11	3	25

The trash of the sugar-cane undergoes rapid fermentation: it soon exhales a distinct smell of vinegar, and almost the whole of the sugar which is left in it is destroyed.

BEET-ROOT SUGAR.

The presence of sugar in the beet was observed by Margraff; and Achard of Berlin by and by attempted the extraction of this sugar on the large scale; but it was only during the period of the continental system that the manufacture of sugar from the beet acquired such perfection in France as made it profitable. The beet so generally cultivated at the present time is derived, according to Thaer, from the *Beta vulgaris*. The two principal varieties of this root are the red beet, which has been grown for a very long time in kitchen gardens, and the white beet. Between these two extremes there are numerous varieties having a flesh color of various intensity, a yellow tint, &c. The seeds of the same plant in fact frequently produce varieties of decidedly different shades of color; the red and the white beet, however, appear to be the most constant; and Thaer has said that the intermediate varieties are crosses between them.

The field beet has a large root which grows in great part above the ground; it is a very hardy plant, which has been cultivated for a very long time in various parts of the continent as food for cattle, and is now also very common in England. The root, which has hitherto been preferred for the manufacture of sugar, is conical, of a rose color without, and its concentric external layers are also colored; but it appears that the white beet of Silesia is the more productive. The beet thrives in almost all kinds of soil, provided only they be sufficiently manured. In Alsace it succeeds in light, and in strong argillaceous soils indifferently. Another precious quality which this root possesses is that of succeeding in the most dissimilar climates; it is grown to purpose both in the north and in the south of France.

The beet is sown at once in the field, or in a bed and transplanted; the latter method appears now to obtain a decided preference, inasmuch as it leaves plenty of time for the preparation of the soil, and especially for accumulating and carrying out manure.

In a piece of ground well broken up by delving or ploughing, and highly manured, which need not be of greater extent than $\frac{1}{10}$th of the entire surface to be planted, the seed is sown in lines or drills as soon as the spring frosts are no longer to be apprehended. The transplanting in the east of France takes place about the middle of May, and even in the beginning of June. The plants are generally set about 15 inches apart. In the north the beet harvest does not begin before the end of September, and generally ends in the course of the month of October. The gathering is delayed as long as possible, inasmuch as the roots increase visibly to the very end of the season. But gathering the beet at a very late period in those countries where the winter seed has to follow this crop, is attended with more than one disadvantage. Without speaking of the difficulties that are incidental to wet seasons, a late seed-time is generally unfavorable for wheat. To meet this difficulty, I have been accustomed for some time to take up my crop of beet at the period when it became necessary to prepare the land for winter seed, that is to say, more than a month before the general harvest of the root. In doing so I relied upon the interesting fact ascertained by M. Peligot in the course of his chemical inquiries, viz: that the composition of the beet is identical at every age. In this premature or anticipated beet harvest, a less weight of root is of course gathered than would have been obtained at a later period; but the nutritious powers of these beet roots are the same as they would ever have been. The grand questions to be determined were, whether the roots would keep or not, and whether the cattle would eat them from the pile as freely as from the field. All this was ascertained in the course of the winter: the beet kept perfectly, the cattle ate it as freely as ever. The procedure to be adopted therefore to secure a crop of beet of average weight, storing nevertheless some considerable time before the usual period, is simply to transplant somewhat more closely, and to put less space between the drills. If experience decides in favor of this method, the sole inconvenience which attends the cultivation of the beet in a freshly manured soil, and as the first crop in the rotation, that, namely, of causing a late and unfavorable seed-time for winter corn, will be completely got over.

The beet which grows above the ground is best gathered with the hand; kinds that grow under ground require to be loosened by running a plough along the drill, &c. In Alsace it is the custom to take away the leaves, and to trim the roots upon the ground; the refuse thus obtained constitutes a considerable mass of manure, which it is well to plough in immediately.

To extract the sugar of the beet the plant is washed and rasped, and the pulp is then subjected to the action of a powerful press. Like the juice of the cane, the juice of the beet speedily undergoes a change; it is therefore immediately heated to 70° cent. or 158°Fahr., and a little lime is mixed with it to neutralize acid and favor the clarification, by combining with albumen. The liquor is skimmed, and in the course of an hour becomes quite limpid, and of a pale yellow color. The liquor is then run upon a filter containing animal charcoal, and from that is transferred to a boiler where it is properly reduced, the process being in all respects the same as in the manufacture of cane sugar.

In France, the produce of each 110 lbs. weight of beet is estimated at 4.56, or somewhat more than 4½ lbs. of white sugar. The amount of loss in the manufacture may be conceived from the actual composition of the beet, which, by the process followed by M. Peligot,* and which consists essentially in drying a certain weight of the root, cut into thin slices, and then exhausting the matter with boiling alcohol of moderate density, appears to contain from 4 or 5, up to 9, 10, 11, and even nearly 12 per cent. of sugar. This analysis of M. Peligot has been confirmed by the experiments of M. Braconnot,† who found the white beet of Silesia to have a very complex composition, comprising as many as twenty-one different ingredients, among the number crystallizable sugar, albumen, woody matter, phosphate of magnesia, phosphate of lime, oxalate of potash, and oxalate of lime, oxide of iron, an ammoniacal salt in small quantity, &c. On an average, the analysis of M. Peligot would lead us to conclude that the beet contained in one hundred parts—

* Recherches sur l'analyse de la betterave à sucre.
† Annales de Chimie, vol. lxxii. p. 442, 2d. series.

Water	87
Matter soluble in water (sugar)	8
Insoluble substances, (woody tissue)	5
	100

from which it appears that no more than about $\frac{3}{5}$ths of the sugar contained in the beet-root is extracted. As in crushing the cane, so in squeezing the rasped pulp of the beet, a part of the loss is owing to a certain quantity of sugar being left in the expressed pulp. In fact, with the presses generally in use, while from 60 to 70 per cent. of juice is obtained, the root actually contains 95 per cent. The loss here, however, is of less consequence than it is in the cane, the trash of which is used for fuel, while the pulp of the beet serves as food for cattle. The pulp, indeed, is found to possess very nearly the same amount of nutritive power as the root which produced it.

One of the considerations which is perhaps of highest importance in connection with the production of sugar from the beet, is inherent in the difficulty of preserving the root after it is full-grown. Gathered at the end of autumn the root suffers no less from severe frost, than it does from mild open weather: frost destroys its organization, and in mild winters vegetation continues at the expense of the sugary principle, which had been formed during the growth. If the beet actually contains at every period of its existence the same quantity of sugar with reference to its weight, there would probably be a great advantage in not waiting for the period of complete maturity, by sowing somewhat thicker than wont; any difference of weight would probably be made up, and then there would be no risk of loss from keeping.

The quantity of beet gathered from a given extent of land necessarily varies with the soil, the pains bestowed upon the crop, and the quantity of manure that has been used; the following are a few particulars from official documents:

	PRODUCE PER ACRE.			
	Ton.	Cwt.	Qrs.	Lbs.
Pas de Calais	12	9	1	19
Department of the North	12	10	2	25
Department of Cher	15	1	39	

but in other departments the produce is considerably smaller, so that the average for the whole country has been estimated at not more that 10 tons 9 cwt. 1 qr. 13 lbs. per acre; an average which approaches very closely to that which I have obtained from my own farm at Bechelbronn, calculated during a period of seven years.

Assuming $4\frac{6}{10}$ths lbs. of sugar to be obtained from every 110 lbs. of beet, the produce in sugar from an English acre in the course of seven months will amount in the present state of things to 9 cwt. 2 qrs. and 7 lbs. By way of comparison I shall here remind the reader that an English acre of land laid out in Otaheite sugar-cane yields in the course of about fourteen months, 15 cwt. 1 qr. 10 lbs. I find from my accounts for 1841, that to manage an English acre of land under beet-root in Alsace, 45.6 days of a man, and 14.1

days of a horse was the amount of labor expended. In a document upon the sugar plantations of Guadaloupe which I have seen, it is stated that a domain of 150 hectares, or 370 acres, is worked by 150 negroes, which, reckoning the time that the crop is on the ground at fourteen months, would bring the number of days labor by a man, to 171.8 per English acre. Such an expenditure of labor must in the nature of things absorb the greater part of the profits; and, indeed, in a commission of inquiry into the laws connected with the sugar trade, it was shown in reference to the plantation in question, that the cost of cultivation and manufacture was equal to the value of the produce. Still the cane presents one considerable advantage over the beet, that, namely, of furnishing the fuel necessary to the boiling, an advantage which will be better understood, when it is known that in the manufacture of every 110 lbs. weight of beet sugar, the consumption of coal amounts to 22 lbs.

In countries where sugar is cheap, it becomes an ordinary article of diet; in the public market-places of the great towns of South America, one of the rations commonly exposed for sale consists of brown sugar and cheese. M. Codazzi estimates the quantity of sugar consumed by each inhabitant of Venezuela at 110 lbs. In England it amounts to about 22 lbs.; in Ireland, to no more than 4 lbs. and $\frac{4}{10}$ths; in Holland it is $15\frac{4}{10}$ lbs.; in France it is $8\frac{8}{10}$ lbs.; in Italy, $2\frac{2}{10}$ lbs.; and in Russia, but $1\frac{4}{10}$ lb. per head.

Maple sugar. (*Acer saccharinum.*) The maple is very common in the east of the United States of America. The tree is occasionally met with in clumps of several acres in extent, but it is more commonly found dispersed in the forest, growing among pines, poplars, ashes, &c. The tree grows particularly in rich soils, and attains the height of the oak; the trunk being often more than three feet in diameter. The maple becomes covered with flowers in the spring before the appearance of the leaves. It is supposed to be in its prime at the age of about twenty years. The sap of the maple is obtained by piercing the trunk to the depth of from six to ten inches. A piece of wood to serve as a gutter is placed in the hole, and the sap is received in a vessel placed underneath. It is usual to pierce the tree first on the side that is towards the south; when the flow of sap begins to lessen, it is tapped upon the north side. The best season for making maple sugar is the beginning of spring, February, March, and April; the sap continues to flow during five or six weeks. The quantity of sap obtained is found to be largest when the days are hot and the nights cold; the quantity collected in the course of twenty-four hours will vary from about half a pint to thirty pints and more; the temperature of the air has the most marked influence upon the flow of the sap; it ceases completely, for instance, in those nights when it freezes after a very hot day.

The maple does not appear to suffer from reiterated perforation; trees are mentioned which were still flourishing after having yielded sugar for forty-two consecutive years. In certain cases, which, however, must be held as exceptions to the rule, as many as 183 pints of sap have been tapped from a maple in the course of twenty-four hours, which yielded $4\frac{4}{10}$ lbs. of crystallized sugar. A maple of ordinary dimensions, in a good year, will yield, on an average, about 198 pints of sap, producing $5\frac{1}{2}$ lbs. of sugar. The sap of the maple must therefore contain about 2.2 per cent. of its weight of marketable sugar. It has been found, that with care and attention the maple becomes more productive ; maples around which other forest trees have been felled, or which have been transplanted into gardens, yield a sap which is not only more abundant, but also richer in sugar, which, in fact, contains about three per cent. of sugar.

The manufacture of maple sugar presents no peculiarity ; precisely the same process is followed as in the case of the cane and beet. Unless very speedily boiled down, the sap ferments, and undergoes change ; in some parts of the United States, indeed, a vinous liquor is made of the sap, by allowing it to run into spontaneous fermentation.

PALM SUGAR.

The palm which in the southern parts of India furnishes crystallized sugar in large quantity, is the cleophora of Gaertner, and reaches a height of about 100 feet. Its fruit hangs in clusters upwards of a yard in length. The natives procure the sap by cutting short one of the shoots that is about to flower and carry fruit, and hanging under the cut part of a calabash or other vessel, into which the fluid distils ; in a large plantation such an apparatus is seen connected with each palm-tree ; the sap is removed every morning, and it is enough to reduce it by evaporation to obtain the sugar, which differs in no respect from the finest sugar of the cane ; in the unrefined state it is known over the whole of the East under the name of jaggery,* and is then a kind of moist and sticky muscovado sugar. The sap of the palm-tree obtained in the way above indicated, is often turned into a vinous liquor, which is much prized in many places. The pith of the tree yields sago. The palm-trees cultivated in India consequently yield three most useful products—sugar, oil, and the farinaceous article of diet called sago. In rearing the cocoa-nut palm, those nuts are selected for seed which fall naturally, and they are dried in their husk. The ground which is to be sown is dug to a depth of eighteen or twenty inches, and it is left to settle for three or four days. Some portion of the surface is then taken away, and the fresh soil is covered, to the depth of about six inches, with sand. The nuts are then placed upon the ground so prepared, and covered over with a little sand and a light stratum of vegetable mould ; they are then watered for three days consecutively. In the course of three months the young palms are fit to be transplanted, and they are set at the distance of about twenty feet every way from one another. For their reception in the permanent

* This is the generic name for sugar, and is obviously either the Latin word saccharum, or from the same root as the Latin word. The cocoa-nut tree treated in the same way as the cleophora yields abundance of sugar, which is also known under the name of jaggery.—Eng. Ed.

plantation, holes are dug of about two feet in depth, in which a layer of sand, about six inches in depth, is put, upon which the young plants, still adhering to the fruit, are placed; the hole is then filled with sand, and the surface is covered with a little earth. The young trees require watering every day during about three years. The palm begins to be productive at the age of seven or eight years, and it continues to yield fruit, or sap for the manufacture of sugar, during a very considerable period, without causing any further cost for cultivation.* The sap of the greater number of the palms appears to be rich in saccharine matter; it is obvious, indeed, that every sap that is capable of supplying a vinous liquor by fermentation, may also furnish sugar; and if the palms have not generally been grown with a view to this product, it is because the fruit must then be given up, and, both in India and South America, the produce in the shape of oil from the nuts of the palm, is almost always more valuable than that which can be had in the shape of sugar.†

GRAPE SUGAR.

We have already said that starch acted upon by acids, and by malted barley, is changed into a saccharine fermentable substance, which, both in regard to flavor and physical properties, differs in many respects from the sugar which we have hitherto been engaged in studying. As this substance exists naturally in the grape, it has been called *grape sugar*, a name for which the generic term glucose has been lately substituted in France, this term being used to include all the sugars that are analogous to grape sugar. Grape sugar occurs in the form of small white and very soft crystals, grouped in tubercular masses; it softens at 60°, (140° Fahr.,) and becomes quite sirupy at 90°, (194° Fahr.) Alcohol free from water dissolves none of it; but diluted alcohol takes up a considerable quantity.

In the grape this sugar is associated with cream of tartar, tartrate of lime, and several other saline matters. It is easily extracted from the fruit; but the grape sugar of commerce is now universally prepared from starch; large quantities, indeed, are manufactured on the continent for the preparation of spirit, and for the amelioration of wine, beer, cider, &c., in short, to supply sugar wherever it is defective in the natural or artificial musts that are subjected to fermentation. In England considerable quantities are also manufactured; but here the law does not allow it to be used in the same advantageous direction as in France and Germany; all that is made is employed for mixing with adulterating cane sugar, which is an article of higher price.

The sugar that is made from starch, and that is obtained from the grape are identical in composition, as is that also which is found in the urine of persons laboring under diabetes.

* Buchanan. A Journey from Madras, &c., vol. i. p. 155.

† In British India the cocoa-nut palm is beginning to be extensively cultivated as a means of producing sugar. A considerable portion of the East India sugar now brought to market, is manufactured from the palm-tree. It is not improbable, indeed, that the palm of one species or another will one day supersede the sugar-cane and the beet as the source of all the sugar consumed in Europe.—Eng. Ed.

	Grape Sugar (Saussure.)	Sugar of Starch. (Guerin.)	Diabetic Sugar (Peligot.)
Carbon	36.7	36.1	36.4
Hydrogen	6.8	7.0	7.0
Oxygen	56.5	56.9	56.6
	100.0	100.0	100.0

Like cane sugar, grape sugar in combining with certain bases bandons a portion of its constitutional water. In the state in which t is combined with the oxide of lead it contains—

Carbon	43.3
Hydrogen	6.3
Oxygen	50.4
	100.0

From these analyses it appears that crystallized grape sugar consists of—

Anhydrous glucose	100
Water	19

On comparing the two kinds of sugar in the crystallized state, it becomes evident that glucose or grape sugar does not differ from cane sugar, except in containing a larger quantity of water. In fact the composition of grape sugar may be represented in this way :

Carbon		42.2	100 of cane sugar.
Hydrogen		6.2	
Oxygen		51.6	
Water	Hydrogen	1.8	15.8
	Oxygen	14.0	
			115.8 of grape sugar.

The cane, the beet, the palm, the maple, the vine, and starch, turned into glucose, are the sources from whence all the sugar of commerce is obtained at the present day, although attempts more or less successful have also been made to extract sugar from the pineapple, from the chestnut, from the sweet orange, and from the stem of the maize or Indian corn. It appears that before the conquest the Mexicans prepared a sirup from the stem of the Indian corn, which was sold in the market-places. Pallas could not obtain more than about 3 per cent. of crystallized sugar from maize, but in an experiment which I made in South America along with M. Roulin, the quantity of raw sugar obtained from this plant was 6 per cent.

SACCHARINE PRINCIPLES NOT FERMENTABLE.

Manna; mannite. This saccharine principle is met with in different plants ; it has been found in the expressed juice of onions, and in that of asparagus, in the alburnum of several species of pine-trees, and in different mushrooms. Manna, which is an exudation from the *fraxinus ornus* and larch, contains nearly $\frac{4}{5}$ths of its weight of mannite, and it is therefore from this substance that mannite is usually obtained, although it can also be had from the juice of the beet and the onion ; but then it is necessary to destroy the cane or grape sugar which they contain by previous vinous fermentation.

and M. Pelouze has even maintained that the mannite thus prepared is a product of fermentation.*

Mannite crystallizes in very white semi-transparent needles; it has a slightly sweet taste, and is soluble in water. According to Liebig and Opperman it contains:

Carbon	39.6
Hydrogen	7.7
Oxygen	52.7
	100.0

Liquorice. This substance, which is obtained from the root of the *Glycirrhiza glabra*, is too well known to require particular consideration; it is soluble both in water and in alcohol.

GUM.

Gum is a substance very extensively diffused in the vegetable kingdom; there is, perhaps, no plant which does not contain some. Gum is divided into two kinds; gum, properly so called, the type of which we have in gum-arabic, and vegetable mucilage, such as we meet in gum-tragacanth.

Gum in dissolving in water produces a thick and adhesive fluid. It is insoluble in alcohol. Some plants contain such a quantity that upon infusion they seem to give, as it were, nothing else: such are the *althea*, the *malva officinalis*, &c.

Gum does not crystallize, it is met with in concrete masses which result from the solidification of the drops which flow spontaneously from the trees that yield it: by long boiling with dilute sulphuric acid it is changed into glucose. Nitric acid alters it, and several new products are the result, among the number of which is mucic acid. Gum-arabic, according to the analysis of M. Gay-Lussac and Thénard, consists of:

Carbon	42.3
Hydrogen	6.9
Oxygen	50.8
	100.0

To obtain vegetable mucilage, a quantity of linseed is treated with water and expressed. It is also obtained by steeping gum tragacanth in about 1000 parts of water and pouring off the solution which covers the mucilaginous mass. The mucilage then forms a jelly more or less consistent, which diluted with a large quantity of water forms a ropy viscid fluid. Dried again, this mucilage becomes hard and translucid; in water it regains its former state.†

VEGETABLE JELLY—PECTINE AND PECTIC ACID.

It is well known that the juice of all fruits contains a gelatinous substance to which many of them owe the property of forming jellies.

* Annales de Chimie, vol. xlvii. p. 419, 2d series.—The refuse wash of the distiller, appreciated by the taste, appears to contain a considerable quantity of saccharine matter, which is probably mannite.—Eng. Ed.

† Berzelius, Chemistry, vol. v.

This matter may be obtained by means of alcohol. If into a quantity of currant juice lately expressed, a portion of alcohol be poured, a gelatinous precipitate is formed after a certain time; this jelly, subjected to graduated pressure and washed with diluted alcohol, gives the gelatinous principle in a state of tolerable purity: this is pectine, discovered by M. Braconnot.

Pectine dried is in membranous semi-transparent pieces resembling isinglass. Thrown into about one hundred times its weight of water it swells considerably and at length dissolves completely, giving rise to a stiff jelly. By increasing the quantity of water, a mucilaginous solution, having a slightly milky aspect, is obtained.

Pure pectine is quite insipid; it does not affect the color of litmus, the weaker acids have no effect upon it; a slight excess of potash or of soda does not change it obviously, and nevertheless pectine is singularly modified under the influence of these alkalies, being changed into a particular body, having acid reaction; for on saturating the alkali employed, it immediately coagulates into a transparent gelatinous mass—pectic acid. As pectine acted upon by the fixed alkalies undergoes so remarkable a change, we may be allowed to conclude, with M. Braconnot, that the pectic acid which is found ready formed in plants, has a similar origin; a view moreover which tends to confirm that formerly announced by Vauquelin, when he ascribed the development of the acids of vegetables to the presence of alkalies.*

Gelatinous pectic acid immediately becomes defluent upon the addition of a few drops of solution of ammonia. By evaporating this solution in a porcelain dish we obtain an acid pectate of ammonia, which swells in distilled water, dissolves in it, and thickens a large quantity of the fluid. As ammonia has no reaction upon pectine, M. Braconnot has taken advantage of this negative property to determine if pectic acid exists or not, ready formed, in certain plants. Thus in treating carrots with cold water, rendered slightly ammoniacal, a liquid is obtained, from which an acid immediately throws down a precipitate of pectic acid.† Pectine and pectic acid, therefore, may exist together in vegetables, and M. Jacquelain has proved that the acid there is often in a state of combination as an alkaline or earthy pectate. It is to these pectates that M. Payen ascribes the origin of the carbonates of the same bases, which are met with in the ashes of plants, the organic acid having of course been destroyed by the combustion.‡

M. Braconnot has described an easy process for obtaining pectic acid in large quantity from carrots.§

M. Fremy has published analyses of pectine and pectic acid, which present this remarkable peculiarity, that the one has exactly the same elementary composition as the other.

* Braconnot, Annals of Chemistry, vol. xlvii. p. 274, 2d series
† Braconnot, op. cit. vol. xxx. p. 99.
‡ Payen, Proceedings of the Academy of Sciences, vol. xv. p. 907
§ Op. cit. vol. xxx. p. 97.

	Pectine.	Pectic acid.
Carbon	42.9	42.8
Hydrogen	5.1	5.2
Oxygen	52.0	52.0
	100.0	100.0

I have thought it right to speak at some length of these two principles, as they appear to play an important part in the phenomena of vegetable life. A careful study of pectine and pectic acid will very probably aid in throwing light upon the metamorphoses which organic substances undergo in the act of vegetation. Pectic acid has been found in every plant in which it has been sought for; M. Braconnot discovered it in the turnip, carrot, beet, peony, in all bulbs, in the stalks and leaves of herbaceous plants, in the wood and bark of all the trees examined, in all kinds of fruit, apples, pears, plums, cucumbers, &c. M. Braconnot is even very much inclined to think that pectic acid may constitute the essential principle in the cambium or organizable matter of Grew and Duhamel.*

OF VEGETABLE ACIDS.

In the series of bodies which we have now considered, one only, sugar, possesses the property of crystallizing. All the others are amorphous, and their globular disposition and gelatinous qualities have led to the presumption that they form in some sort the line of demarcation between things without and things endowed with life. It was also imagined that these amorphous matters, that these products of the vegetable organization, almost organized themselves, would alone suffice for the nourishment of animals. This idea, however, is not well founded; for if it be true that albumen, caseine, legumine, starch, and gum, are powerful elements of nutrition, it is equally so that sugar may perform an important part in this process, by acting in the same manner as starch, the oils, and other principles of ternary composition, in becoming like them a useful, often an indispensable auxiliary of azotized alimentary matters.

This disposition to consider the amorphous state of the more important immediate principles of vegetables as a special and distinctive character, cannot be maintained beside the recent observations of Mitscherlich. This illustrious chemist has found, that if the mineral precipitates which are deposited in liquids, are in many cases formed of crystals more or less regular, they are also sometimes composed of small spheres or aggregated masses, the particles of which do not unite in a regular way as crystals, but remain separated by a thin layer of fluid. Examined under the microscope these masses present themselves under the form of flocks and of shreds, having a granular or gelatinous appearance, and which remain soft and flexible like fresh vegetable or animal substances, so long as they are kept under water; it is only in drying that they become pulverulent or acquire the vitreous aspect.†

* Braconnot, op. cit. vol. xxviii. p. 171.
† Berzelius, Ann. Report, 1841, p. 20.

The substances, the chemical constitution of which we have still to examine, may in general be obtained in the crystallized state; their individuality seems more decided; they are more stable, better characterized, and their specific properties often assimilate them to inorganic bodies. Such, for example, are the acids formed in the course of vegetable existence.

Vegetable acids present all the general characters of mineral acids, while they participate in the properties inherent in organic substances. Thus they form salts by uniting with bases; with potash, soda, ammonia, they form salts soluble in water; the other bases produce compounds that are soluble or insoluble, according to the nature of the acid. These acids, free or uncombined, are very frequently met with in fruit, sometimes in the leaves, more rarely in the seeds and roots; but in combination with bases they are met with in almost all parts of plants. Already very numerous, they are increasing rapidly with the progress of discovery; with the exception of a very few employed in the arts, their study forms a subject of no great interest. I shall therefore confine myself to a few of the most extensively distributed of these acids.

Oxalic acid. This acid exists free in the hairs of the cicer or chick-pea, and united with potash constituting an acid salt, the binoxolate of potash in the wood sorrel and the common or garden sorrel. It is from the former of these plants that the salt called salt of lemons, but which is, in fact, the binoxolate of potash, is still extracted in some countries. The juice of the wood sorrel is expressed and yields about 0.003 of its weight of the salt, from which, by ordinary chemical manipulation, the oxalic acid is readily obtained. At the present time this acid is prepared artificially by the action of nitric acid upon starch; it is a powerful acid, and its affinity for lime is such that it takes this base even from its union with sulphuric acid.

Tartaric acid is met with above all in the grape in the state of bitartrate of potash, a salt which is deposited upon the sides of the casks in which the wine is kept. After having been properly purified, it is known in commerce under the name of cream of tartar, from which the tartaric acid can readily be obtained. Another particular acid, the racemic acid, the composition of which is identical with that of the tartaric acid, has been discovered in the tartar of the wines grown on the Upper Rhine.

Citric acid. This acid is found in the juice of many plants, and abundantly in the juice of lemons, oranges, currants, &c. It is from the lemon and the lime that the citric acid employed in the arts is generally obtained.

Tannic acid. A certain substance which is met with in the bark of particular trees, and which has the valuable property of rendering the hides of animals with which it is combined insusceptible of putrefaction, is familiarly known under the name of tannin. The art of the tanner is founded upon this property of tannin. A solution of gelatine being poured into an infusion of tannic acid, an insoluble precipitate, formed by the union of the acid with the animal matter, is immediately produced. By macerating a piece of raw hide in a solution of tannin, the same combination takes place even into the very interior of the tissue ; the whole of the tannin quits the solution by degrees to combine with the gelatine of the skin.

It is not in the bark only that tannin is encountered, it has been found in different organs of plants. Sir Humphrey Davy has stated these quantities of tannin as constituents of 100 parts of the following substances :

Nutgalls	27.4
Oak bark	6.3
Chestnut bark	4.3
Elm bark	2.7
Willow bark	2.2
Inner white bark of an aged oak	15.0
The same of young oaks	16.0
The same of the Indian chestnut	15.2
The inner colored bark of the oak	4.0
Sicilian sumac	16.2
Malaga sumac	10.4
Souchong tea	10.0
Green tea	8.5
Bombay catechu	54.3
Bengal catechu	48.1

Gallic acid. This acid is found united with tannin in the greater number of barks, or along with the astringent principles of plants. Gallic acid appears to be the product of a kind of fermentation undergone by tannin, as the process by which it is prepared seems to indicate, and which consists essentially in exposing for about a month a quantity of nutgalls reduced to powder and kept constantly moistened. The solution of gallic acid does not precipitate gelatine.

I have added in a table the composition of the principal vegetable acids. I shall speak of the composition of fat acids when I come to treat of fatty substances.

The different vegetable acids do not vary essentially in composition, save in a single instance. With one exception they consist of definite proportions of carbon, hydrogen, and oxygen. The exception alluded to is the hydrocyanic acid, which contains no oxygen, but a large quantity, nearly 52 per cent., of azote.

OF THE VEGETABLE ALKALIES.

The alkaline bases which are formed in the course of vegetation, always contain a certain proportion of azote. Their general properties are those of alkalies ; their watery or alcoholic solutions restore the blue color of the reddened tincture of turnsole, and they constitute salts by combining with acids. In their manner of behaving they bear a certain analogy to ammonia. Like ammonia, the organic alkalies combine with the hydrates of the oxacids, and when they are deprived of their water of crystallization, they fix the hydracids without losing weight.

12

The discovery of the vegetable bases is due to Sertuerner, who, in 1804, indicated the existence of morphine in opium. The majority of the vegetable alkalies are insoluble, or little soluble in water; all are soluble in alcohol; some of them are sufficiently volatile to be susceptible of distillation.

In elementary composition they are all very much alike, consisting of various, but, in each instance, definite proportions of carbon, hydrogen, oxygen, and azote, the carbon varying from about 50 to 75, the hydrogen from 6 to 12, the oxygen from 8 or 9 to 27 and even 37, and the azote from 1.6 to 12, 28, and even 35 per cent.

OF FATTY SUBSTANCES.

Under this title I comprise all the oily substances, liquid or solid, and those that are analogous to wax, which are found disseminated in different organs of plants. A character common to almost all fatty substances, is insolubility in water. They dissolve in sensible quantity in alcohol, and especially in ether. Fatty substances may be divided into two classes: one including those which are easily modified by the action of alkalies, and which form soaps; the other not susceptible of this action, not susceptible of saponification, or, at all events, that are only attacked by alkalies in very particular circumstances.

When a mixture of fat oil and a solution of caustic alkali are heated, the oil is soon observed to incorporate with the alkaline liquid. After boiling for some time, if the alkali is in excess, clots or flocks appear, and in removing the excess of liquid a white mass is obtained which is soluble in water—the oil is saponified; and the product of the saponification is combined with a portion of the alkali which has been employed. If into a hot solution of this soap a quantity of hydrochloric acid be poured, the acid seizes upon the potash or the soda, setting at liberty the fatty body which had been combined with the alkali, and which collects on the liquid. It is easy to discover that the fatty matter thus collected is no longer the same as that which had been originally employed; for example, it is completely soluble in boiling alcohol, which, on cooling, deposites brilliant pearly crystals of a fatty substance possessing acid properties. By evaporating the alcohol from which these crystals are formed, an additional quantity is obtained, and, when the alcohol is entirely dissipated, another unctuous body is obtained, having also acid properties. Three acids having distinguishing characters are, in fact, obtained by the action of alkalies upon fatty substances: the stearic, margaric, and oleic acids. The alkalies consequently transform neutral oily bodies into acid substances, as first shown by the admirable researches of M. Chevreul, before whose time it was always assumed that soap was the result of a direct union of fatty matters with alkalies. The fatty acids are not the only products of saponification, there are several others, particularly glycerine, which, however, need not occupy us particularly here.

The experiments of M. Chevreul would lead us to view all fatty matters as combinations of glycerine playing the part of a base with particular acids; these combinations, analogous to salts if their constitution be merely considered, are generally mixed together in oils and fats; thus the union of stearic acid and glycerine forms stearine, which is fusible at the temperature of about 62° cent. (144° Fahr.) Stearic acid melts at 72° cent., (162° Fahr.,) oleine remains fluid at 4° cent., (24° Fahr.,) and oleic acid is liquid. An oil is, therefore, by so much the more consistent as a larger quantity of solid fatty acid enters into its composition, and it is, on the contrary, by so much the softer and more liquid as this acid is itself more fluid. The wax of the *Myrica cerifera*, for example, is sufficiently hard to be reduced to powder, and is almost entirely formed of stearine. In the fluid vegetable oils oleine always predominates. It is easy to separate these different fatty compounds from one another.

Besides the solid and liquid acids which are obtained from fatty substances, there are others known which are volatile.

Fatty bodies absorb oxygen from the air. This absorption is at first extremely slow, scarcely appreciable; but once begun, it goes on with great rapidity; so rapidly, indeed, that if a large surface be exposed to the air, if, for example, a quantity of rags or tow be impregnated with oil, the mass may take fire. The consequence of this oxidation is always a thickening of oil, and there are some which become completely solid in its course; these are designated by the title of drying oils, and are in particular request for the manufacture of varnishes. Nut oil which has remained long exposed to the air acquires the consistence of jelly, and its unctuous properties have so entirely disappeared that it no longer stains paper.

The alteration which fatty substances undergo in contact with air and moisture is still more remarkable. The oils which are inodorous and without taste soon acquire under these circumstances a strong smell and a disagreeable flavor. Fleshy fruits which contain a large quantity of oil, such as the olive and the oleaginous seeds, when moistened suffer true fermentation, the result of which is the separation of the fatty acids from the glycerine.

Oils subjected to the action of a high temperature are also greatly modified. The glycerine which they contain is decomposed, and gives rise to various pyrogenous products: stearic acid is changed into margaric acid, and oleic acid into sebacic acid, a crystallizable volatile acid which is soluble in hot water.

The fatty substances of plants are principally accumulated in the fruit, and particularly in the seed. In the herbaceous parts they are less abundant, less perfectly elaborated. Oils appear to be included in the vegetable tissue under the form of globules, or minute drops. In such an oily seed as the common almond, when it is growing, we perceive that the cellular tissue is in the first instance full of a colorless and transparent fluid; but as the seed advances, each cell is seen to become filled with numbers of little oil globules which increase continually in size and number until the kernel is ripe; there is at the same time a quantity of azotized matter deposited in the midst of the liquid, which disturbs its transparency; it is this deposite which thickens the walls of the cells.* The capillary force which retains fatty principles combined with the tissue of certain seeds must be very considerable, for having boiled some rape-seed, which contained 50 per cent. of oil, in water, there was not a trace of oily matter perceptible upon the surface of the liquid. Butter appears to be kept diffused in milk by something of a similar force, for milk when boiled yields but a very small quantity of this substance M. Dumas and I maintain that the oil of seeds is intended for the production of heat by undergoing combustion at the period of germination; a series of experiments performed in my laboratory by M. Letellier supports this opinion.

Having ascertained by a preliminary trial the quantity of oily substance contained in a certain weight of seed, some of the same kind was put to germinate, and the quantity of oil which it contained was tested at two periods of the germination; it was found that in the course of this process a considerable proportion of the fatty substance had disappeared; one gramme or 15.438 grains of rape-seed before germination contained 0.50 of oil; after the first period of germination, namely, when the cotyledons had begun to turn green, the quantity of oil was found reduced to 0.43, and at the end of the second period, when the cotyledons had become quite green and the radicles were from 3.9 to 4.6 inches long, the oil was reduced to 0.28.

It would be extremely interesting to ascertain the extreme loss which the oily principles of seeds sustained in the course of the commencement of vegetation, and to follow the return of the same principles in proportion as the plant advanced towards maturity. M. Letellier is going on with these experiments.

The numberless uses to which oil is put, make its manufacture an object of the highest importance. Vegetable oils are generally obtained from olives, from oleaginous seeds, and from the nut of certain palms. Oil is separated by pressure; it may often be extracted from the seed in the natural state, in which case the produce is of fine quality, but seldom abundant. The castor-oil bean, for example, yields its oil under the simple action of the press. In America, however, to obtain this oil, the seeds are first roasted slightly, and being bruised they are then boiled in water; the oil readily separates from the roasted seed. A similar process is sometimes followed in procuring cacao butter.

In the extraction of oil from the common oleaginous seeds, they are first ground or bruised in a proper apparatus; the paste or powder which they now form is generally heated, and being put into woollen sacks, and these enclosed in hair bags, they are subjected to the operation of the press; after one pressure, the magma which remains in the bags is crushed anew, heated, and pressed again. The oil obtained by the second pressing is never so pure as that procured by the first.

The oil-cake is taken out of the bags, completely dry in appearance, but it still contains a large proportion of oil—from 8 to 15 per

* Dumas, Chemistry, vol. v.

cent. of its weight. It is used in fattening cattle and as manure Oil, when newly expressed, is always turbid and very mucilaginous; it becomes clear by standing; but it always retains certain substances which lessen its quality, particularly when it is intended for burning in lamps.

Greater obstacles are encountered in extracting the oil from some of the pulpy fruits than from seeds. In extracting olive-oil, the olives are crushed under millstones; and the paste which results being put into flat baskets of wicker-work, is subjected to the press. The first pressing yields virgin oil, which is used for the table. Having removed the baskets from the press, their contents are mixed with a little boiling water, replaced, and pressed again, by which a new quantity of oil is obtained. But the pulp is not yet exhausted; by special treatment it still yields a quantity of oil of inferior quality, which is employed in the manufacture of soap.

The fruit of the palm yields the oil which it contains with great readiness. I have extracted a butter of excellent quality and very agreeable taste by simply boiling the nuts or berries of the *Palma real* in water. The cocoa-nut yields two qualities of oil, according to the mode of extraction. To prepare the best kind, the fleshy part of the fruit is grated, and the pulp being pressed, a milky fluid is obtained, which yields the oil by boiling. An inferior quality of oil is obtained by causing the cocoa-nuts to putrefy; when the putrefaction has advanced to a certain stage, the oily pulp is thrown into copper vessels and exposed to the sun, and the oil which then rises to the surface is skimmed off. This oil is brown, and has a strong smell; it contains fatty acids which have probably been set at liberty by the putrid fermentation.

The value of the produce in oleaginous seeds of a given extent of land, and the quantity of oil which these seeds will yield, depend, as may readily be conceived, on a variety of causes which it is not always easy to appreciate with precision; such as climate, the nature of the soil, the system of husbandry followed, &c. The observations of M. Gaujac of Dagny on the various plants usually cultivated for the sake of their oleaginous seeds, will however suffice to give a notion of their comparative productiveness in oil and cake:

Crop.	Seed produced per acre in Cwts.	qrs.	lbs.	Whole quantity of Oil obtained per Acre in lbs. avoird.	Oil obtained per cent.	Cake per cent.
WINTER CROPS.						
Colewort	19	0	15	875.4	40	54
Rocket	15	1	3	320.8	18	73
Rape	16	2	18	641.6	33	62
Swedish turnip	15	1	25	595.8	33	62
Curled colewort	16	2	18	641.6	33	62
Turnip cabbage	13	3	19	565.4	33	61
SPRING CROPS.						
Gold of Pleasure	17	1	16	545.8	27	72
Sunflower	15	3	14	275.0	15	80
Flax	15	1	25	385.0	22	69
White poppy	10	1	18	560.8	46	52
Hemp	7	3	21	220.0	25	70
Summer rape	11	3	17	412.5	30	65

M. Matthew de Dombasle made some comparative experiments at Roville on the cultivation of oleaginous plants. The results obtained by this skilful agriculturist are much less favorable than those of M. Gaujac. Instead of 19 cwt. and 15 lbs. of colewort seed yielding 875.4 lbs. of oil per acre, M. de Dombasle only obtained 11 cwt. 2 qrs. 21 lbs. yielding 392.3 lbs. of oil; and the other kinds of seed in proportion. But as I have said already, the fertility of the soil, and the labor and pains bestowed upon it, may have contributed to the differences observed, because here the influence of climate may be overlooked. There is one circumstance, however, which may explain the great differences in the quantity of oil obtained, which is the perfection of the press employed to extract it. In a general way oil-presses are so imperfect that they all leave a quantity of oil more or less in the cake.

Here are two examples: from 2765 lbs. avoird. of fine colewort seed, gathered in 1842, and weighing $52\frac{1}{2}$ lbs. per bushel, I obtained:

	lbs.
Of oil	1130.5
Of cake	1384.9
Loss	249.6
	2765.0

In other terms, per cent.:

Oil	40.81
Cake	50.12
Loss	9.07
	100.00

but by a careful analysis of the same seed in the laboratory, 50 per cent. of oil was obtained.

2d. In 1840 and 1841, I made some experiments on the cultivation of the madia sativa, intermixed with carrots in a fertile soil, well manured with farm dung. The crop of the year 1840 was excellent; it required one hundred and twenty-seven days to come to maturity.

	lbs.
Seed, husks deducted	2424
Dried leaves employed as litter	7700
Carrots without their leaves	31966

The seed gave:

Of oil	635.8
Of cake	17067.6

100 of seed gave:

Oil	26.24
Cake	70.72
Loss	33.4
	100.00

These results agree pretty nearly with those which have been published by other agriculturists; but the seed of this madia, which in the press gave 26.24 of oil per cent., actually yielded 41 per cent. by analysis in the laboratory; this difference between practical results and those of the laboratory, shows us how large a quantity of oil is generally left in the cake. When the cake is used for feeding cattle, the loss is perhaps less to be regretted, inasmuch as the oily matter evidently assists in the fattening; but when the cake is used as manure, the oil which it contains is almost entirely lost.

It is often of importance to the agriculturist to ascertain precisely the quantity of fatty principles contained in oleaginous seeds. For this purpose, it is enough to bruise a given quantity of the seed and to digest it in successive portions of sulphuric ether. After a first digestion, the seed is bruised or pulverized anew, and the bruising is now accomplished without difficulty. The process may be concluded by boiling with a mixture of equal parts of ether and alcohol. The ethereal solutions are decanted from the seed into a porcelain dish, the weight of which is known. The ether evaporates spontaneously and the oil remains, the weight of which is then taken.

The following sums may be taken as a pretty accurate estimate of the average quantity of oil yielded by the different oleaginous seeds: colewort, winter rape, and other species of cruciferous plants, from 30 to 36 and 40 per cent.; sunflower about 15 per cent.; linseed from 11 to 22; poppy from 34 to 63; hempseed from 14 to 26; olives from 9 to 11; walnuts 40 to 70; brazil nuts 60; castor-oil beans 62; sweet almonds 40 to 54; bitter almonds 28 to 46; madia sativa 26 to 28 per cent.

The quantity of oil yielded by any seed subjected to the press is always considerably less than that which it contains, and the oil retained in the cake appears to be in larger proportion as the starch, the woody tissue, and the albuminous matters are more abundant. Thus maize, or Indian corn, which contains from 8 to 10 per cent. of fluid oil, gives mere traces of its presence under the press.

The oily and fleshy fruits, such as those of the olive and the palm, yield a considerable quantity of oil. In the southern countries of Europe, particularly those which are so well protected that their olive-trees escaped the severe winter of 1789, as many as about 816½ lbs. of oil per acre are obtained, with proper care. The trees which were killed during this memorable winter sprouted again from the roots, and at the present day yield from about one quarter to one half the above quantity, according to the spaces left between them, which vary considerably. Under similar circumstances in regard to climate, it will readily be understood, that the quantity of produce will be influenced by the quantity of manure put into the ground. In some countries the olive is never manured, save indirectly; that is to say, the ground between the trees is only manured with a view to another crop, which is grown between them; in other countries, again, in the neighborhood of Marseilles, for instance, it is the practice to manure the olive plantations, directly, every three or four years.

The olive enjoys remarkable longevity; I have mentioned one more than seven centuries old, and the term of the tree's existence appears only to be limited by the severe winters which cause it to die, from time to time. The produce must of course depend upon the age of the trees which compose a plantation. Up to eleven years, M. Gasparin shows that an olive-tree still remains all but unproductive; and that the capital, and the interest upon the capital expended in this husbandry, must necessarily exceed the value of the produce up to the thirtieth year. Yet there are soils which are favorable to the olive, and which are useful for nothing else; a hole in a rock suffices it, if the climate be favorable and it receive a proper dose of manure. But the grand cause of the disadvantages attending the cultivation of the olive, in France, is connected with the periodical occurrence of severe winters, which kill it; in an interval of one hundred and twelve years, from 1709 to 1821, the olive plantations have suffered three great mortalities, which give a mean duration of about forty years to each planting.

The cocoa-nut-tree is one of those which yields the largest quantity of oil with the least labor. The tree grows vigorously in all hot countries, at no great distance from the sea-shore; wherever the temperature is from 78° to 82° Fahr., there the cocoa-nut thrives. It is also found on the banks of great rivers; and the common practice in planting the cocoa-nut is to put a little salt in the hole. When transplanted far from the banks of rivers, it thrives best in the neighborhood of human habitations, which has led the Indians to say that the cocoa-nut-tree loves to hear men talking under its shade. It is a tree which requires a soil impregnated with saline substances, and these are never wanting near the habitations of man. The tree bears its first flowers at the age of four years; it produces fruit the following year, and continues to fructify until it is eighty years old. The spikes generally bear about twelve cocoa-nuts, and the number of nuts yielded by a tree in the course of a year may be taken at about fifty, which will yield about four litres, or rather more than seven pints of oil. Somewhere about ninety trees are generally found upon the acre of land, and these are capable of yielding about 825 lbs. of oil annually.*

The cocoa-nut-tree must, therefore, be regarded as among the most productive in oil, and also as the plant which requires the least outlay in its cultivation. Many species of palm yield oils of a very agreeable flavor for the table, and the produce of all answers admirably for the manufacture of soap. In the same proportion as agricultural industry extends in the equatorial regions of the globe, will the production of palm-oils increase, and this must necessarily influence the cultivation of the olive in a very serious way. The cultivation of the tree being already threatened in Europe by that of the mulberry, and the prodigious extension in the trade in palm-oil upon the coasts of Africa in the course of the last few years, justify this conclusion. In 1817, palm-oil was considered as among the list of mere medicinal substances. At this period a London perfumer thought of making it into a soap for the toilet-table. From this time it became the staple of a bartering trade, which has been by so much the more profitable to the nations engaged in it, as the purchase is always effected by manufactured articles, such as cotton and woollen goods, hardware and crockery, arms, powder, &c. The future

* Codazzi, Resumen de la Geografia de la Venezuela, p. 133.

extent of this traffic may be imagined when it is known that in 1817 the importation of palm-oil into England did not much exceed 140,000 lbs., and that in 1836 it exceeded 70,000,000 lbs! In taking an acre of surface for unity, I find that on an average the

Spring oleaginous plants yield..................	320lbs. of oil.
Winter oleaginous plants......................	534 "
The olive (south of Europe)....................	534 "
The Palm (America)............................	801 "

OF ESSENTIAL OILS.

Aromatic plants owe the odors which characterize them to certain volatile principles, which by reason of certain properties which they have in common with fat oils, such as insolubility in water, solubility in ether and alcohol, inflammability, &c., are generally designated as essential oils. They are met with in all parts of plants; but in one plant the oil is principally found in the flower, in another in the leaves, in another in the bark, &c. It sometimes happens that different parts of the same plant contain oils of different kinds. From the orange-tree, for instance, three distinct oils are obtained, as the flower, the leaf, or the rind of the fruit is treated. In some cases the volatile principle is so thoroughly imprisoned in the vegetable cells, that drying does not dissipate it; in others, as in the greater number of flowers, the oil is formed on the surface, and is volatilized immediately after its formation.

Essential oils are less volatile than water; nevertheless they rise with the vapor of water, and it is by distillation that they are generally extracted. The plant is put into a still or alembic containing water, and heat is applied: the vapor formed is condensed in the receiver, and the essence, by reason of its less density, is found swimming on the surface of the water which has been distilled. Some volatile oils are obtained by pressure, those of the citron and bergamotte, for example.

The volatile principles of plants present somewhat varied physical properties. They are generally limpid and lighter than water; yet there are some which are more dense, and some, such as camphor, which are solid. With reference to their composition, volatile oils may be divided into three classes; 1st. Oils composed entirely of carbon and hydrogen. 2d. Oils composed of carbon, hydrogen, and oxygen. 3d. Essential oils containing sulphur; in addition to which, the essential oil of mustard seed contains azote.

The essential oils undergo a change by long contact with the air: they absorb oxygen, and many of them become acidified; under the influence of this gas, the oil of bitter almonds is changed into benzoic acid, the oil of cinnamon into cinn-amic acid; in a general way, acetic acid is produced. The votatile oil obtained from any plant almost always contains two distinct principles, which may be separated by careful distillation; one of these principles is a carburet of hydrogen, the other an oxygenated oil. *Camphor* is combined with essential oils in many plants of the labiate family. It exudes from certain laurels; it is from the *Laurus camphora* that all the camphor of commerce is extracted in the East, the extraction being effected precisely by the same process as other essential oils. The chips of the *Laurus camphora* are put into iron stills, surmounted by earthenware capitals, in the inside of which a number of ropes made of rice-straw are stretched; the camphor rises and is condensed on the surface of these cords in the state of a gray powder; it is refined by sublimation.

According to M. Dumas, camphor contains:

Carbon	79.2
Hydrogen	10.4
Oxygen	10.4
	100.0

OF RESIN.

Essential oils almost always hold certain substances in solution which make them viscid or sticky. The balsams which exude from the bark of certain trees are nothing more than solutions of resin in essential oils. When the volatile oil has been dissipated by evaporation, the resin remains in the solid state. There is further a natural relation in point of constitution between essential oils and resins. The greater number of essences absorb, as we have said, oxygen from the atmosphere, and by this absorption they become thick, and are changed into resins; so that in one case the resin may be a product of the oxidation of an essential oil, in another it may merely be set at liberty by the dissipation of the essence which held it in solution.

The resins constitute friable, or soft solids. They are fusible, extremely inflammable, and fixed. The resins are inodorous when pure: any odor which particular resins possess is generally attributed to the essential oil which they still retain. The resins are insoluble, or very sparingly soluble in water; some of them dissolve readily in alcohol and in ether, and there are some also, such as copal, which are only soluble in very small quantity. Some resins show acid reaction; they combine with bases, neutralizing them. The greater number of resinous matters obtained from plants are regarded by chemists as mixtures of several particular resins, the study of which is not yet much advanced. Some resins are much employed in the arts, such as colophony and copal, &c. Several balsams are also in familiar use, particularly as medicines, such as the balsam of tolu, balsam of copaiba, &c.

Colophony, or rosin, is extracted from different kinds of the genus *Pinus*. In the Landes, or sandy plains of Bordeaux, it is the maritime pine which yields it. When the tree is from thirty to forty years of age, incisions are made in the trunk, beginning at the lower part, two or three times a week, and these are continued to the height of from 6 to 10 feet from the ground; the last notch generally reaches this height about four years after the tree has been notched for the first time. After this a new series of notches is begun on the opposite side, setting out from the ground as before, and in this way the whole circumference of the tree finally presents a series of notches, so that a tree will continue to yield turpentine during a period of sixty years. The turpentine which exudes from the notches is collected in a hole dug in the ground.

Crude turpentine always contains a quantity of intermixed foreign matters, earth, stones, leaves, &c. It is purified by being melted, and filtered hot through a bed of straw. By distillation it is separated into essential oil, which is condensed in the receiver, and colophony, or rosin, which remains in the still. From 250 lbs. of turpentine 30 lbs. of essence and 220 lbs. of rosin are generally obtained.

Copal is the produce of a tree which is somewhat common in Madagascar, and which M. Perrotet has determined to be the *Hymenæa verrucosa.* The balsam or sap which exudes from the bark solidifies by contact with the air, and the resin is gathered in the state in which it is met with in commerce.

CAOUTCHOUC.

The caoutchouc which we have mentioned as forming a constituent in the sap of certain trees possesses some properties which assimilate it with the resins. Thus pure ether, free from alcohol, dissolves it. The greater number of the essential oils also dissolve it, particularly when hot. It is a solution of Indian rubber in rectified coal-tar oil or naphtha, which is now used so extensively for making stuffs water-proof. According to Faraday pure caoutchouc is composed of:

Carbon	87.2
Hydrogen	12.8
	100

VEGETABLE WAX.

Some plants produce a considerable quantity of a substance which bears a great resemblance to beeswax, and which in some of its properties approaches fatty bodies. Proust discovered that vegetable wax formed part of the green fecula of a great number of vegetables. In the common cabbage it occurs in large quantity. It is often met with forming a varnish on the surface of leaves, fruit, and barks; the substance, however, is far from being identical; it almost always results from the combination of several distinct principles which have not yet been sufficiently studied, but among which there are obviously some true fatty substances, that is to say, bodies capable of saponification, and matters analogous to the resins. I shall here mention a few of the vegetable waxes which are best known.

Wax of the palm. This is the product of the *Ceroxylon andicola,* which is very abundant on the central Cordillera of New Grenada. I believe that I met with the lower limit of the ceroxylon upon the borders of the torrent of Tochecito, at the height of 7500 feet above the level of the sea, and I followed it to an absolute elevation of about 8500 feet. The extreme mean temperatures comprised between these two limits may be valued at from 11° to 18° cent.; 51.8° to 64.4° Fahr. Towards the superior limit, the ceroxylon is exposed to a cold during the night, which approaches the freezing point of water; it is therefore frequently met with in company with the great oak of America, whose climate it stands very well.

The Indians obtain the wax by scraping the bark of the palm: the scrapings are then boiled in water; the wax swims—without, however, melting; it is merely softened, and the impurities which it contains are deposited. The matter thus purified is formed into balls and set to dry in the sun. It is with this substance, to which, however, a small quantity of fat is often added to render it less brittle, that the loaves of wax and the candles of the country are formed. After it has been melted, the cera de palma is of a deep yellow color, slightly translucid, as brittle as resin, and presenting a waxy fracture well characterized. Its melting point is a little above that of boiling water. Boiling alcohol dissolves it readily; in cooling, the solution sets into a gelatinous mass. Ether dissolves it, as do the alkalies also.

The wax of the palm consists of two principles; one, fusible above the temperature of the boiling point of water, has all the physical properties of beeswax; the other has the properties of resin. The composition of these substances upon analysis appears to be:

	Wax.	Resin.
Carbon	81.6	83.7
Hydrogen	13.3	11.5
Oxygen	5.1	4.8
	100	100

Wax of the Myrica cerifera. This wax is procured by boiling the fruit of several species of myrica in water. The tree is extremely common in Louisiana and the temperate regions of the Andes. The fruit yields as much as 25 per cent. of wax, and a single shrub will yield from 24 to 30 lbs. of berries per annum. The crude wax is green, brittle, and, to be made into candles, requires the addition of a certain quantity of grease. According to M. Chevreul the wax of the myrica is saponifiable.

Wax of the sugar-cane. The sugar-cane, particularly the violet variety, is covered with a powder or bloom of a waxy nature, which melts at the temperature of 82° cent. (180° Fahr.) This wax is so hard that it can be pulverized; it may be made into candles, which, for the brilliancy of their light, are not inferior to those of spermaceti. M. Avequin, who directed attention to this subject, found by his experiments that a hectare (nearly $2\frac{1}{2}$ acres English) of the violet cane would furnish nearly 200 lbs. of wax. This wax is entirely soluble in boiling alcohol; ether does not dissolve it in the cold. It appears to constitute a perfectly defined immediate vegetable principle, the composition of which, according to M. Dumas, is the following:

Carbon	81.4
Hydrogen	14.1
Oxygen	4.5
	100

CHLOROPHYLLE.

The green matter which colors the leaves of vegetables is so designated. The attempts which have been made to isolate this matter, render it probable that it is somewhat of the nature of the vegetable waxes. Pelletier and Caventou endeavored to procure it by treating with cold alcohol, the pulp remaining after expressing all the juices from the leaves of various herbaceous plants. By evaporation of the alcoholic liquor, a substance of a deep-green color was obtained, which is chlorophylle, a matter soluble in ether, in alcohol, the oils, and the alkalies. Heated, it softens and is decomposed before it melts. Acetic acid dissolves it in very appreciable quantities, so do the sulphuric and hydrochloric acids; water precipitates it from these acid solutions. Berzelius says that chlorophylle exists only in very small quantity in plants, the leaves of a large tree will not perhaps contain more than about 100 grains.

OF COLORING MATTERS.

The matters which color the different parts of plants are extremely numerous; they present great varieties of shade, but are in general derived from red, yellow, and green. It is seldom that the coloring matter of a plant exists isolatedly; it is almost always allied with one or several immediate principles, which are themselves frequently colored. Thus red coloring matters are generally combined with yellow principles, which having nearly the same properties, one is with great difficulty separated from another.

Coloring matters are solid, inodorous, and have little taste. Some are soluble in water, others only dissolve in alcohol or in ether. All combine with the alkalies, and several of them unite intimately with acids; the greater number are powerfully affected, undergo a true destruction, on exposure to the sun's rays, especially when in contact with moist air. It is familiarly known that vegetable tissues of all kinds, beeswax, &c., are bleached by exposure to the sun and air; a high temperature acts like light: some vegetable colors are altered, bleached, when they remain exposed for a time to a temperature of from 334° to 424° Fahr. The oxygen of the air, which so quickly destroys certain colors, develops others under particular circumstances.

Alkalies and acids, by uniting with vegetable colors, almost always modify their tints and often change them entirely. Many blues, for instance, become reds, under the agency of acids, greens or yellows under that of alkalies. By neutralizing the acid or the alkali, the color generally resumes its original tint.

Several substances, which are colorless in the state in which they are formed in vegetables, become colored by the united action of oxygen and an alkali, such as orceine, which is oxidated and becomes blue under the simultaneous contact of air and ammonia. The greater number of vegetable coloring matters are destroyed and bleached by chlorine. Many of the same matters unite intimately with alumina and oxide of tin to form lakes, insoluble compounds in which the colors remain fixed; thus a colored liquid often becomes colorless when it is shaken with a hydrate of alumina. Charcoal, in a state of extreme subdivision, acts like alumina, and is a powerful discharger of colors in every-day use in the arts. Coloring matters are generally ternary compounds, though some of them also contain azote; and several of them exhibit the remarkable phenomenon, that in undergoing oxidation in contact with ammonia they assimilate the azote of this alkali. I shall now indicate the origin and the mode of preparing a few of the more important of these coloring matters.

Indigo. This substance, so essential in the art of dyeing, has been one of the great staples of trade with Asia from the most remote times. For a long while indigo was regarded in Europe as a mineral substance found in India; it used to be designated Indian or India stone, whence the name of indigo. It was not until after the discovery of America that the true nature of this dye-stuff was known, although before this period indigo had been made in Arabia, Egypt, and even in the Island of Malta.

Indigo is volatile, so that to obtain it pure, it is enough to put a small quantity into a platinum capsule, to cover it with a lid and to expose it to heat. Indigo is volatilized in the state of violet-colored vapor, and collects in crystals upon the middle part of the sides of the capsule. Indigo gives nothing to water or to ether. Alcohol takes up a very small quantity of it; concentrated sulphuric acid dissolves and modifies it.

All bodies greedy of oxygen appear to reduce or deoxidize this coloring principle; it changes to a yellow, and becomes soluble in water in contact with alkalies; by exposing the alkaline liquor charged with the uncolored indigo to the air, it absorbs oxygen rapidly, and the indigo becomes insoluble and is precipitated with its original blue color. It is most easy, as said, to disoxidize indigo; it is sufficient to bring it into contact with hydrogen gas in the nascent state, a condition which is readily secured by throwing iron or zinc filings into water containing the coloring matter previously dissolved in sulphuric acid. The disengagement of the hydrogen has scarcely commenced before the deep-blue color of the solution declines in intensity, and by and by it becomes of a very pale gray. When the discharge of color is completed, and no more hydrogen is disengaged, the colorless indigo begins to react upon the air, it absorbs oxygen, becomes again oxidized, and by and by the liquid has resumed its deep blue. This property of indigo of becoming soluble in alkaline solutions under the influence of disoxidizing bodies, is taken advantage of in our laboratories to obtain indigo in a state of purity, and in the arts to prepare a dyeing liquid. If a mixture be made of 15 parts of the indigo of commerce reduced to fine powder, 10 parts of the sulphate of the protoxide of iron, 15 parts of lime, and 60 parts of water, and it be left for several days in a closed vessel, a colorless liquid is obtained The liquid decanted and exposed to the air, deposites the whole of its indigo after a time. It is with similar ingredients that the dyer prepares his bath for blue colors. It is into the alkaline liquor so prepared that the stuff to be dyed is dipped; it is then hung up in the air, where it soon becomes blue; it is re-dipped and re-exposed again and again, until it has acquired the depth of tint required, after which it is washed. The indigo, regenerated by the action of the air, remains fixed in the stuff, and proves, as all the world knows, one of the most solid of colors.

Chemists are not agreed as to the true nature of colorless indigo which may be obtained in the solid state. Some regard it as indigo disoxidized, others as indigo hydrogenized. On the latter supposition, the hydrogen fixed would be derived from the water decomposed, the oxygen of which would be transferred to the bodies greedy of this element, which are brought into play. The latter of these views appears to have the ascendant at the present time. However this may be, the following is the composition of indigo in each of its states, as determined by M. Dumas:

	Blue Indigo.	White Indigo.
Carbon	73.1	73.0
Hydrogen	4.0	4.5
Azote	10.8	10.6
Oxygen	12.1	11.9
	100.0	100.0

The plants which have hitherto been cultivated for the production of indigo with any profit are not numerous; they belong to the genera *Indigofera*, *Isatis et Nerium;* it is the genus *Indigofera* which is most generally cultivated, and the species designated *argentea* is found to be the most profitable. M. Chevreul has ascertained that in the living plant the indigo is not colored, and that it is consequently during its extraction that it becomes blue. The experiments of M. Pelletier upon the *Polygonum tinctorium* have confirmed the old researches of M. Chevreul. After having dried a leaf, Pelletier digested it with ether in a closed flask. The whole of the *chlorophylle* was dissolved, and the leaf became completely blanched; by exposing it afterwards to the air, it turned blue if it contained indigo.

In the republic of Venezuela, where I had an opportunity of studying the cultivation of the indigo-bearing plants, I saw that those soils were preferred which were light and susceptible of irrigation, a condition indeed which seems to me all but indispensable to the profitable exercise of agriculture within the tropics. Indigo requires a warm climate; at an elevation of about 3250 feet English above the level of the sea, where the mean temperature is not more than from 72° to 75° Fahr., the indigo husbandry cannot be carried on with advantage. Nevertheless, the *Indigofera sylvestris* is met with at an elevation of about 4900 feet above the level of the sea; but the attempts that have been made to obtain coloring matter from the plant have proved fruitless. In the valley d'Aragua, where the best plantations are met with, the plant is sowed in lines, the holes destined to receive the seed being about $1\frac{3}{4}$ inch in depth, and somewhat more than 25 inches apart. A pinch of seed is dropped into each hole, and is covered with a little earth. The sowing takes place in soils that are moist but well drained, or in situations generally which have no system of irrigation at the period of the first rains. The seeds shoot in the course of the first week; hoeing is performed in the course of the month. The first cutting takes place when the plant is coming into flower; from fifty to sixty days generally intervene between the sowing and this cutting; but the time necessary for the development of the leaves depends of course upon the climate. In the neighborhood of Maracaibo, where the mean temperature is about 78° Fahr., the gathering does not take place before the third month. The second cutting is performed from 45 to 50 days after the first; and in this way several successive crops are obtained, until it is seen that the plant begins to degenerate. In good soils the indigo will last for two years; in soils of inferior quality the crop is generally annual.

The indigo harvest is immediately transported to tanks or large rectangular reservoirs built of masonry, and disposed on different levels, the superior reservoir or steeping tank being much larger than the two others. In the valley d'Aragua, there are some which are upwards of 20 feet long by 15 feet wide, and 20 inches in depth.

The second, or mashing tank, is narrower and deeper than the former. The third reservoir, or depositing tank, receives the liquor from the mashing tank, and in it the indigo subsides. In some manufactories the third tank is not used, the deposition taking place in the mashing tank itself.

The leaves, as the name implies, are thrown into the steeper, covered with water, and kept down by planks loaded with stones; fermentation soon begins, and is allowed to continue during about eighteen hours; and in the management of this first operation lies much of the art of the indigo-maker. By continuing it too long some portion of the coloring matter is destroyed; by stopping it prematurely, a quantity of indigo is left in the leaves. The fermentation judged to be sufficiently advanced, the liquor is run off into the battery, and vigorously stirred until the *grain* is deposited. The fluid is then either let into the subsider, or left in the battery, and the deposition is complete at the end of about twenty hours; the supernatent fluid is drawn off, and the indigo paste is scooped out and placed upon cloths to drain. When sufficiently firm, it is divided into lumps, and these are set in the shade to dry. In the valley d'Aragua it is estimated that with a good soil and careful management, each hectare of surface will yield 280 lbs. of marketable indigo,* which is at the rate of about 112½ lbs. per English acre.

In Carolina the cultivation of indigo appears to be much less productive than in the equinoctial regions, and the produce is of inferior quality. There they sow in drills in the commencement of the rainy season which follows the vernal equinox, and the first crop is gathered about the beginning of July; the second is secured two months afterwards, and when the autumn is mild, a third but insig-

* Cadozzi, Resumen de la Geografia de Venezuela, p. 144.

nificant gathering takes place at the end of September. One negro is allowed to be able to work nearly two acres and a half of ground, from which about 160 lbs. of indigo are obtained.

In the East Indies, upon the Coromandel coast, the growth of indigo takes place upon sandy soils which are not irrigated, and in which vegetation is only possible during the rainy season. The loamy soils that admit of being irrigated are almost always reserved for the growth of rice. Immediately after the rains have set in, in December, the land receives two superficial ploughings ; the indigo is sown broad-cast, and the seed is harrowed in by dragging a fagot of bamboos over the surface, or by treading in by means of a flock of sheep. The first and principal gathering takes place in March ; any other crop that may be won is purely casual, and entirely dependent on the rain that falls. The crop rarely fails to feel the effects of the droughts which so frequently take place upon the Coromandel coast. It is never abundant, and the plants have little vigor. The harvest takes place after the flowering season. The crop is dried in the sun ; the plant is then beaten with switches, by which the leaves are detached from the stems, after which hey are exposed anew to the sun to secure their being perfectly dry. They are then reduced to coarse powder, and handed over to the indigo-maker, for in India the planter is never himself the manufacturer of the dye-stuff.

On the coast of Coromandel, indigo is always extracted from the dried leaves, which, bruised and broken, are infused in three or four times their bulk of cold water during two or three hours ; the infusion is then filtered through a loose stuff made of goat's hair ; the filtered liquor is beaten for two hours, and after this about five gallons of lime-water are added for every 100 lbs. of dried leaves ; the mixture is stirred, and then left to settle. When the deposite has formed, the supernatent liquor is drawn off, the sediment is washed with a little boiling water, and being thrown upon a cloth, the indigo is drained and dried. It is then pressed, and the paste is cut into cubical lumps which are thoroughly dried in the air, and of which each weighs nearly 3 ounces.

In the Indian method of manufacturing indigo, all is accomplished, as appears, without fermentation. This indigo is little esteemed in commerce ; it is heavy, of a pale blue, without much of the coppery aspect, rough on the broken surface, and presents here and there white points and vegetable *débris*. An acre of land on the Coromandel coast will produce from 48 to 49 lbs. of indigo annually.

In spite of the high price of indigo, so small a quantity would scarcely cover the cost of production, were not the wages of the Indian laborer exceedingly low. The whole expense of producing a kilogramme, or $2\frac{2}{10}$ lbs. avoird. of indigo, according to M. Plagne, amounts to 3 francs, 20 cents, or about 2*s*. 8*d*.

The cultivation of the indigo plant has been attempted several times in the south of Europe, particularly in Spain and Italy. There is no doubt but that indigo may be grown in Europe in those situations where for three or four months of the year the temperature is truly tropical; but it seems probable that indigo can never be advantageously introduced into the agriculture of temperate countries.

Before indigo was so extensive an article of commerce as it is now, the south of France used to furnish almost all the markets of Europe with a blue dye, which was the best then known; this was woad or pastel, the produce of the *Isatis tinctoria*.

The isatis is sufficiently hardy to stand the cold of winter. In the south it is sown in March, and the seed springs in from eight to ten days. When the plant has five or six leaves, it is hoed with care. The crop is gathered when the leaves have acquired their greatest size, when they even begin to fade a little. The preparation of woad bears a certain resemblance to that of indigo, and need not detain us here.

The *Polygonum tinctorium* has of late attracted the attention of European cultivators. The plant is a native of China, where it has been cultivated from time immemorial; it was brought into France and propagated under the care of M. de Lille. In the course of three months the plant has thrown out all its leaves, and in the south of France it never fails to ripen its seeds. From some experiments that have been made, the leaves of the polygonum appear to contain about the five-thousandth of their weight of indigo, and as the acre of land will yield between 11,000 and 11,900 lbs. weight of leaves the produce of coloring matter will come to upwards of 56½ lbs.

The indigo obtained from the polygonum by the methods generally practised is not always of fine quality. It contains matters which, having been dissolved in the water used for maceration, had subsequently been precipitated M. Vilmorin proposed to adopt on the great scale and in the manufactory, the methods which are used for purifying indigo in the laboratory, and which consists, as we have seen, in reducing colored and insoluble indigo to the colorless and insoluble state by means of a salt of the protoxide of iron in contact with an alkali, and subsequently to restore its color, and effect its precipitation by contact with the oxygen of the air. It is obvious that this method is perfectly applicable to the treatment of the whole of the indigoferous plants, and I believe that its adoption would be a great improvement.

Orchil. This coloring matter, of a deep purple, is prepared from certain lichens or lung-worts; that which is most prized is the *rocella tinctoria*, a native of the Canaries and Cape de Verd islands. The *variolaria dealbata*, the *var. aspergillia*, and the *lichen corallinus*, which grow upon the rocks of Auvergne, and on the Alps and Pyrenees, yield a produce of inferior quality.

To obtain the orchil, the lichens are steeped for several days in their own weight of stale urine. Into the mixture about 5 per cent. of slaked lime in powder, a small quantity of arsenious acid, and a little alum, are introduced. Fermentation is by and by set up in the mass, which soon acquires the characteristic color of the orchil, but the tint is nèver complete until the expiration of about a month.

Orchil readily communicates its peculiar color to water and to alcohol; the watery solution, which is of a fine crimson, becomes colorless in a few days when it is kept in a flask hermetically sealed; it regains its color by exposure to the air. The color which is acquired during the manufacture of orchil indicates that the lichens which yield it contain principles which are colorless in themselves, but which possess the singular property of becoming tinted under the influence of oxygen and ammonia; for in the preparation of orchil, the addition of urine and of lime has no other purpose beyond the introduction and development of this alkaline base. This view is shown to be correct, moreover, by the facts brought to light by MM. Heeren and Robiquet in their inquiries into the chemical constitution of the lichens. These chemists, in fact, succeeded in extracting from several of the lichens which yield orchil a variety of colorless crystalline principles, in particular orcine, a substance which may be procured in very regular quadrangular prisms, and which is soluble in water and in alcohol. The watery solution of orcine mixed with ammonia and exposed to the air, becomes gradually of a deeper and deeper red until it has the color of blood. The result of this oxidation of orcine under the action of ammonia is a coloring principle, orceine, into the constitution of which azote enters, an element which formed no part of orcine; the analyses of these two substances made by M. Dumas, show this fact very distinctly:

	Dry orcine.	Orceine.
Carbon	67.8	55.9
Hydrogen	6.5	5.2
Oxygen	25.7	31.0
Azote		7.9
	100.0	100.0

The lichens furnish several other principles analogous to orcine in their property of acquiring color under similar circumstances.

Turnsole. This coloring matter is met with in commerce in two states, in mass and in thin cakes, and is procured from various lichens which have not yet been accurately specified; in any case the substance is obtained by a process which differs little from that used in the manufacture of orchil. According to Mr. Kane, the coloring principles of turnsole are naturally red: they only become blue by combining with a base. The coloring matters which predominate in turnsole are erythrolitmine and azolitmine, which are united with lime, potash, and ammonia, and further mixed with a considerable quantity of chalk and sand. The analyses of Kane show these two substances to have the following composition:—

	Erythrolitmine.		Azolitmine.
Carbon	55.6		49.8
Hydrogen	8.4		5.4
Oxygen	36.0	oxygen and azote	44.8
	100.0		100.0

Turnsole occurs in trade in the shape of thin cakes, made up of chips, which are colored by the juice of *chroyophoria tinctoria*, a plant of the euphorbiaceous family, and of which the dyeing properties appear to have been known to the earliest naturalists.

Madder. The root of the madder plant, so commonly employed in dyeing, contains more than one coloring matter; but the most important of these, and that which constitutes the most useful element in the root, is alizarine, which was discovered by M. Robiquet.

This substance is scarcely soluble in boiling water, but it is soluble in alcohol and still more so in ether, to which it imparts a golden yellow color. Alkaline liquors dissolve it and acquire a violet shade extremely agreeable to the eye. Alizarine is sublimed by the action of heat in the form of brilliant red needles.

Madder, (*rubia tinctorum,*) is a native of the south; but as it stands the winter, it is now cultivated almost all over Europe; the plant is propagated by seeds, but there are certain advantages in using the sprouts which it throws out in the spring, and which readily take root. The plant requires a friable soil, sufficiently moist and highly manured to receive it; the soil must be previously trenched or have had a very deep ploughing. In the east of France the planting takes place in April or May. The sprouts which are to be transplanted must be about 6 inches long. When the plant has struck root, the ground is cleaned, and fifteen or twenty days afterwards it is hoed; in the course of the summer, several other hoeings are required. In Alsace, madder is planted in rows and in patches, a certain interval between each patch being left, the earth of which, in the month of March of the following year, is thrown over the ground that is planted.

In the neighborhood of Haguenau, madder occupies the ground during two years; the crop is gathered about the middle of November. In some districts the plant remains in possession of the soil for five or six years. It is generally allowed that the amount of produce increases with time; but in those countries, such as Alsace, where the plant is liable to be attacked by frost, it is generally thought prudent to gather it at the end of two years; the harvest is then profitable, and in the course of the third or fourth or any succeeding winter, it might run the risk of such severe frost as would destroy it entirely. In southern countries the growers say that a crop of the fourth year exceeds very considerably one of the third year; but it might be made a question whether the increase actually compensates for the longer occupation of the soil. And then when the cultivation is too much prolonged, a species of fungus is developed around the root and kills it. The crop of madder is gathered with the hoe, a laborious and costly process; the roots are then dried in stoves and sent to the mill, so that it is in the state of powder that madder is met with in commerce.

In Alsace, where madder remains two years in the ground, the mean produce per acre is estimated at about 3300 lbs. of dried roots, which is equal to an annual crop of about 1650 lbs. In the south of France, the mean annual produce amounts to something less than this, or to about 1560 lbs.;* but the quantity has been estimated a a considerably lower amount still.

* De Gasparin, Agricultural Memoirs, vol. ii. p. 243.

Besides its roots, madder yields an abundance of leaves which are excellent forage.

Reseda luteola, or dyers' weed, is a plant in common use, and owes its properties as a dye-stuff to the presence of a yellow crystalline principle, luteoline, discovered by M. Chevreul. This substance is soluble in ether, alcohol, and alkaline solutions.

Dyers' weed is sown in autumn, stands through the winter, and ripens in the month of August following. The plant is gathered when it begins to turn yellow, and it is in a marketable state after it is dried. An acre of land will produce about 1833 lbs. weight of marketable dye-weed.

Saffron. This plant is cultivated in the south of France and in Austria, but appears to be a native of Asia. Saffron requires a light and yet fertile soil in order to produce abundantly, although it may also be cultivated in soils of middling quality. The ground, trenched one spit deep, is set out with bulbs from an old plantation. In the south the transplanting takes place in the month of June. The first flowers appear towards the middle of October; they are few during the first year. They are gathered, and the pistils removed; the gathering continues for about a fortnight. In the course of the year which follows the planting, the ground receives a surface dressing; it is freed from weeds, and the withered leaves are removed. The next gathering takes place at the same period as the former, but the flowers are now much more abundant, and the same process is continued until the roots are taken up, which they are in France at the end of the second year; but in Austria the culture is continued for a much longer period in the same piece of ground. The extraction of the pistils is an occupation in which the whole family of the saffron-grower take part, and employ their evenings; in the course of an evening of five hours, eight persons will generally have drawn 250 grammes or about 8 ounces of saffron. In some places the pistils are dried in the sun, in others by being exposed in a sieve over a fire of twigs; the latter process appears to be the better one.

M. de Gasparin estimates at about 110 lbs. the saffron which is gathered in the course of two years from about $2\frac{4}{10}$ths acres of land; this would give a mean annual produce of about 43.7 lbs. per English acre, and the price of saffron being from 27 to 28 shillings per pound, the value of the produce may easily be reckoned. In Austria, where the crop is allowed to occupy the ground for three years, the produce has been estimated at about $19\frac{1}{2}$ lbs. per acre per annum.

Roucou is a dye-stuff extracted from the fruit of the *Bixa orellana*, a tree which is extremely common in the hot regions of Southern America.

Chica. This and the former dye-stuff are in use among the native Americans for staining the skin. It is obtained from the leaves of the *Bignonia chica*, which are of a beautiful green when fresh, but become red by drying.

Chica has the color of cinnabar: it is without taste and without smell; a mass of this pigment may be compared to a mass of indigo it only differs from this substance in its color. Like indigo, it acquires the metallic polish when rubbed with a hard body. It dissolves in alcohol, and in alkaline solutions; it mixes readily with grease, and it is with such a mixture that the Indians paint their bodies. Chica has been employed in cotton-dyeing, and the color is found to stand the sun perfectly.

---

§ II.—COMPOSITION OF THE DIFFERENT PARTS OF PLANTS

The immediate principles, the history of which has now been sketched, are met with in greater or lesser quantities in different parts of plants; some of them are accumulated in the roots, others in the seeds, the barks, the leaves, &c. To complete the study of the chemical constitution of vegetables we have still to examine with reference to their composition certain parts or organs which present sufficient interest either from their extensive employment or their importance in an agricultural point of view.

ROOTS AND TUBERS.

The Potato, (*Solanum tuberosum.*) This plant is a native of South America. Two English travellers, Messrs. Caldcleugh and Baldwin, were so fortunate as to meet with it lately in the wild state in Chili, and not far from Monte Video. It is probable that the cultivation of the potato spread from the mountains of Chili to the chain of the Andes, proceeding northward and obtaining a footing successively in Peru, at Quito, and upon the plateau of New Granada. This, as Humboldt observes, is precisely the course which the Incas took in their conquests. The potato does not appear to have been introduced into Mexico until after the European invasion of that country; and it is well ascertained that it was not known there under the reign of Montezuma, although there are not wanting some who maintain that the potato was found in Virginia by the first colonists sent thither by Sir Walter Raleigh. It is said that it was then brought into England by Drake; but it seems well established that long before Drake's time, namely, in 1545, a slave merchant, John Hawkins by name, had introduced tubers of the potato from the coasts of New Granada into Ireland. From Ireland the new plant passed into Belgium in 1590. Its cultivation was at this time neglected in Great Britain, until it was introduced by Raleigh at the beginning of the seventeenth century. When the potato came from Virginia to England for the second time it was already disseminated over Spain and Italy. It has been ascertained that the potato has been cultivated on the great scale in Lancashire since 1684; in Saxony since 1717; in Scotland since 1728; in Prussia since 1738.*

* Humboldt, Essai Politique, t. ii. p. 461

It was about the year 1710 that the potato began to spread in Germany, and that it there became a plant in common use; it had, indeed, before this time been cultivated in gardens; and had even made its appearance at the tables of the rich some time previously. The severe dearth of the years 1771 and 1772* seemed necessary to lead the Germans to cultivate this useful plant upon the great scale. From this time it was shown that it was a substitute for bread; and once fairly introduced, men were not long of perceiving the many recommendations which it possesses as an article of food. In fact, of all the useful plants which the migrations of communities and distant voyages have brought to light, says M. Humboldt, there is none since the discovery of the cereals, that is to say, from time immemorial, which has had so decided an influence upon the well-being of mankind. In less than two centuries it may be said literally to have overspread the earth, or to have been welcomed in every country suited to its cultivation, so that at the present day it is found growing from the Cape of Good Hope to Iceland and Lapland. "It is an interesting spectacle," adds the illustrious traveller quoted, "to see a plant, a native of mountains situated under the equator, advance towards the pole, and growing even more hardily than the grasses which yield us grain, brave the inclemencies of the North."† The potato, like all other tubers, is a collection, an exuberance which is evolved upon the subterraneous stems. Its varieties, which are very numerous, present rather remarkable differences in regard to size, form, color of the surface and of the interior, taste, and the time which they require to come to maturity.

Next to water, fecula or starch is the principle which predominates in the potato, but it also contains a certain quantity of azotized matter. Vauquelin has published a detailed account of the soluble matters which are met with in the potato, and which, strange to say, have been neglected in the greater number of analyses of this useful vegetable which have been published. In 100 parts of potatoes he found: asparagine 0.1, albumen 0.7, azotized matter not defined 0.4, citrate of lime 1.2, and undetermined quantities of citrate of potash, free acetic acid, phosphate of potash, and phosphate of lime.

In examining forty-eight varieties of potato he found that they contained in 100 parts: first, from 1 to $1\frac{1}{2}$ of pulp; second, from 2 to 3 of soluble or extractive substances; third, from 20 to 28 of starch; fourth, from 67 to 78 of water.‡

In a variety grown in the neighborhood of Paris, Henry found the following ingredients, viz: pulp 6.8, starch 13.3, albumen 0.9, uncrystallizable sugar 3.3, acids and salts 1.4, fatty matter 0.1, and water 74.2=100.0.

The proportion of starch varies considerably in the different varieties; M. Payen has ascertained the extent of this diversity in a certain number of varieties grown in the same soil and under the

* Thaer, Principes raisonnés d'Agriculture, t. iv. p. 119.
† Humboldt, op. cit. t. ii. p. 463.
‡ Thénard's Chemistry, vol. v. p. 82.

same circumstances. The results are contained in the following table:

Varieties.	One of seed potatoes produced.	Produce per Acre.				In 100 parts.		Starch.	Starch per Acre.			
						Dry matter.	Water.					
		tons.	cwts.	qrs.	lbs.				tons.	cwt.	qrs.	lbs.
The Rohan	58	14	14	2	16	24.8	75.2	16.6	2	9	0	12
The great yellow .	37	9	8	0	27	31.3	68.7	23.3	2	3	3	4
The Scotch shaws	32	8	3	2	2	30.2	69.8	22.0	1	16	0	1
The late Iceland ..	56	14	6	1	23	20.6	79.4	12.3	1	15	1	2
The Segonzac.....	32	8	3	2	2	28.8	71.2	20.5	1	14	0	5
The Siberian......	40	10	4	2	12	22.2	77.8	14.0	1	8	2	16
The Duvillers.....	40	10	4	2	12	21.7	78.3	13.6	1	8	0	12

In the particular circumstances under which this experiment was made, therefore, it is obvious that the Rohan variety contained the largest quantity of nutritive matter and starch.

Potatoes which have been exposed to a temperature a few degrees below the freezing point of water, undergo so great a change in their texture, that it becomes difficult afterwards to extract the starch which they contain. They besides acquire so disagreeable a flavor, as all the world knows, that cattle sometimes refuse to eat them. After having ascertained that a potato has the same chemical composition before and after congelation, M. Payen examined the starchy substance under the microscope, and found that the starch obtained from a frozen tuber presented itself in compound granular masses, four or five times the size of the largest natural grains of starch. The pulp which remained upon the sieve in the preparation of this starch, was formed by a collection of cells, for the most part full of starch. It would therefore appear, that in consequence of the changes of volume of the fluid successively congealed and liquefied the adhesion between the cells was destroyed: they become separable with the slightest force, and merely part one from another by the action of the grater without being torn; the larger number remain unbroken and still filled with starch. This fact enables us to understand how potatoes, which have been frozen, will yield nearly the whole of their starch if they be treated before they are thawed. The cells then sealed up by the congealed water resist sufficiently to be broken by the teeth of the grater. Potatoes which have been frozen, are generally less farinaceous, at the same time that they have a decidedly sweet taste, which, according to M. Payen, is owing to vegetation having already made some progress in the tubers before congelation; and we know that during germination there is always a formation of sugar at the expense of the fecula. Frosted potatoes have always a disagreeable taste, and a most unpleasant smell, so that in many places they are thrown upon the dunghill. The effect of the frost, in fact, is to set the juices which are enclosed in the tissue of the potato at liberty, and the higher temperature which accompanies and follows a thaw, exposes these juices to be acted upon immediately by the air of the atmosphere, the consequence of which is, that they behave like all other vegetable juices left to themselves, they become putrid. The putrid odor and the acrimony which are developed in the frosted potato, are by so much the more remarkable as a certain layer which exists immediately under the skin, and presents various shades of color, tawny red or violet, is more highly developed. The tissue of this layer, examined under the microscope, was found by M. Payen to be totally without starch, but it contains the greater part of the (strong-smelling) coloring principles.* These principles, which give such unpleasant qualities to frosted potatoes, appear to be soluble, or at least destructible by long exposure to the open air. Thus, if frosted potatoes be spread upon the ground and exposed to the weather, they dry spontaneously, become hard, whitened, and they may then be preserved for a very long time. This method of making use of frosted potatoes has been several times employed in practice, and it might perhaps be recommended for general adoption, were it only ascertained that by such treatment the tubers did not lose a great proportion of their most nutritive principle, viz. albumen. However this may be, it is by a similar process that the natives of the Andes of Peru preserve and render more transportable the tubers which form a principal element in their food. In the steepest parts of the Peruvian Cordillera, nearly at the superior limit of vegetation, where a miserable field of barley and of *Quinoa* is only seen here and there, various tubers are collected in the hollows of the surface, such as the *Maca*, the *Oca*, the *Ulluco*. To preserve these they are exposed for several days to the alternate action of the frost and of the sun. At these great elevations, which are upwards of 13,120 feet above the level of the sea, it always freezes in clear nights when the air is moderately calm. During the day the rays of the sun, which strike with great force, dry the tubers rapidly, the watery juices of which have been shed into the amylaceous tissue by the effect of the preceding night's frost. Thoroughly dry, they may be kept for more than a year by being stored and protected from moisture. Various other modes of preparation are practised in regard to the other kinds of tubers which have been mentioned. By previously boiling the common potato, pealing it, and exposing it alternately to the frost of the night and the heat of the sun, until it is completely dry, the Indians prepare one of their most agreeable and wholesome articles of subsistence.

The potato thrives in soils of very various kinds, provided it be sufficiently fertile, and the climate is favorable. This crop, like the beet, is generally planted in freshly manured ground, and is succeeded in the autumn by a winter crop of corn—wheat or rye. The potato is set when apprehensions of frost are no longer entertained; in the east of France the setting is generally ended about the middle of May. In Alsace the cuttings of the potato are dropped at the distance of about a foot from each other in furrows made by the plough, the furrows being from 18 inches to nearly 2 feet apart.

* Payen, Journal of Practical Agriculture, vol. i. p. 498.

When the plants are from 10 to 12 inches high, and the weather is dry, the furrows are lightly earthed up. In dry soils the earthing plough must not be carried very deeply; and I may say that in the elevated table lands of America, where the natural drought of the climate is often to be apprehended, I have seen very fine crops of potatoes which had never been earthed up at all. The potato, like all plants that are hoed, requires considerable care; but this care, as it is immediately profitable, is still more so remotely upon the white crops which are to follow. M. Crud reckons at 58.3 the number of days work that are required upon an acre of land which has received between 19 and 20 tons of manure. This is very nearly what we have found to be the truth at Bechelbronn, where for the same extent of surface, manured in the same way, we reckon fifty days labor of a man, and rather better than eleven days of a horse.

In Europe the potato harvest takes place at the end of autumn. In the intertropical Cordillera, where the cultivation depends principally upon the heat of a very steady climate, the potato remains in the ground from four to seven months, as it is cultivated at a greater or less height above the level of the sea; it succeeds best where the mean temperature ranges between 13° and 18° centigrade, (56° and 65° Fahr.) In Venezuela, indeed, it is still cultivated in places where the temperature is not far from 24° centigrade, (76.5° Fahr.;) but I am doubtful that the culture is then advantageous. In warm and moist regions the potato yields a large quantity of top, and few tubers. I have gathered some very bad ones at Riosucio de Engurama, a village situate at the distance of about 5900 feet above the level of the sea, where the mean and constant temperature is about 22° centigrade, (72° Fahr.)

The produce per acre, noted by different observers, is as follows:

Countries.	Tons.	Cwts.	Qrs.	lbs.
Prussia	5	18	2	1
Palatinate, mean of ten years	5	7	1	14
Austria, mean of thirteen years	9	11	3	10
Brabant	11	17	0	2
West Flanders . . .	9	13	0	17
Pays de Waes . . .	10	8	3	13
Pays de Tongres . .	6	14	0	25
England	10	2	2	7
England	9	9	0	25
Ireland	9	9	3	14
Alsace	8	14	3	8
Alsace (Bechelbronn) . .	5	8	1	12
Neighborhood of Paris . .	11	2	2	13
Venezuela	9	16	1	20*

There is an obvious relation between the quantity of seed-potato planted and the amount of the crop. In Alsace from 25 to 30 bushels per acre are usually planted. In some places too much seed

* This is the produce of two harvests, which they gather in the same year.

is used, in others not enough. It were very desirable that certain experiments were undertaken which should fix the proper quantity of seed-potato to be used for each variety of soil and situation.

The Jerusalem Artichoke, (*Helianthus tuberosus.*) This plant is generally believed to be a native of South America, but M. de Humboldt never met with it there, and according to M. Correa, it does not exist in Brazil. The property which the tubers of this plant have of resisting the cold of our winters, and several botanico-geographical considerations, lead M. A. Brongniart to presume that the plant belongs to the more northern parts of Mexico.

The Jerusalem artichoke rises to a height of from 9 to 10 feet; it flowers late, and I have not yet seen it ripen its seeds. It is propagated by the tubers which it produces, and which are regarded, for good reason, as most excellent food for cattle; in times when the potato was not very extensively known, it also entered pretty largely into the food of man; when boiled, its taste brings to mind that of the artichoke, whence the name.

The tuber of the Jerusalem artichoke, from an analysis of M. Braconnot, appears to contain in 100 parts:

Uncrystallizable sugar	14.80
Inuline	3.00
Gum	1.22
Albumen	0.99
Fatty matter	0.09
Citrates of potash and lime	1.15
Phosphates of potash and lime	0.20
Sulphate of potash	0.12
Chloride of potassium	0.08
Malates and tartrates of potash and lime	0.05
Woody fibre	1.22
Silica	0.03
Water	77.05
	100.00

M. Payen found a larger proportion of sugar in this tuber than that stated above, and he ascertained that the fatty matter consists chiefly of stearine and elaine. In the Jerusalem artichoke I myself found:

Of dry matter	20.8
Water	79.2
	100.0

One trial for azote would lead me to conclude that M. Braconnot had estimated the albumen too low in his analysis, or, as is more probable, that several azotized principles had escaped him. The dried tuber gave me 0.16 of azote, a number which would indicate 1.0 as the proportion of vegetable albumen. There are few plants more hardy and so little nice about soil as the Jerusalem artichoke; it succeeds everywhere, with the single condition that the ground be not wet. The tubers are planted exactly like those of the potato, and nearly at the same time; but this is a process that is performed but rarely, inasmuch as the cultivation of the helianthus is incessant, being carried on for many years in the same piece; and after the harvest, in spite of every disposition to take up all the tubers, enough constantly escape detection to stock the land for the following year, so that the surface appears literally covered with the young plants on the return of spring, and it is necessary to thin them by hoeing. The impossibility of taking away the whole of the tubers, and their power of resisting the hardest frosts of winter, is an obstacle almost insurmountable to the introduction of this plant, as one element in a regular rotation. Experience more and more confirms the propriety of setting aside a patch of land for the growth of this productive and very valuable vegetable root.

Of all the plants that engage the husbandman, the Jerusalem artichoke is that which produces the most at the least expense of manure and of manual labor. Kade states that a square patch of Jerusalem artichokes in a garden was still in full productive vigor at the end of thirty-three years, throwing out stems from 7 to 10 feet in length, although for a very long time the plant had neither received any care nor any manure.*

I could quote many examples of the great reproductive power of the helianthus; I can affirm, nevertheless, that in order to obtain abundant crops, it is necessary to afford a little manure. I shall show in another chapter, however, that this is manure well bestowed.

Like all vegetables having numerous and large leaves, the helianthus requires air and light; it ought, therefore, to be properly spaced The original planting of course takes place in lines, but in the succeeding crops, and those which are derived from small tubers accidentally left in the ground, the order is of course lost; it is only necessary to destroy a sufficient number of the young sprouts which show themselves in the spring, to leave those plants that are preserved with a sufficient space between them. When the plants are somewhat advanced, the ground should receive one or two diggings with the spade, and a hoeing or two to destroy weeds.

The leaves of the helianthus are used in many places as forage, the stems being cut a few inches from the ground; the gathering takes place at different periods of the year, but probably to the detriment of the tubers; it may be lucrative to destine the leaves for the nutriment of cattle, but I believe we have to choose between the green crop and the crop of tubers. It is unquestionable that the premature removal of the green stems must prove injurious to the roots; in my own farm the leaves are never removed, and my opinion is, that it is vastly more advantageous to depend upon the crop of tubers alone. The tubers are gathered as they are wanted, for, not dreading the frost, they remain in the ground the whole of the winter; they do not require, like the potato, to be collected and pitted at a certain period; they require no particular situation, no particular care for their preservation; the only disadvantage that accompanies their being left in the ground, is that during very hard frosts the labor required to get at them is very great. During win-

* Schwertz, Culture of Forage Plants.

ter the woody stems of the plant die and dry up, they are then useful as combustible matter; but a better use of them perhaps is to make them enter in certain proportions into the litter of the hogstye; the pith there absorbs a large quantity of the liquid manure. Schwertz estimates the mean quantity of dry leaves and stems at 3 tons, 1 cwt. 1 qr. and 15 lbs. per acre. The following quantities of tubers have actually been gathered in Alsace.

	Tons.	Cwts.	Qrs.	lbs.
Sandy soils	4	3	3	6
Soils of the best quality	10	8	3	13
At Bechelbronn (mean)	10	16	0	8
Bechelbronn crops of 1839–40	15	16	1	16

The Carrot, (*Daucus carota.*) This root is frequently cultivated, particularly by intercalation; it is frequently grown along with the poppy, where the seed is raised for the sake of its oil, occasionally also being sown with white crops in the spring, it comes to maturity after them in the autumn; it is a plant that is much liked by animals, but which by no means possesses the very high value as an article of food which is generally ascribed to it by husbandmen. The carrot requires a deep, somewhat loose and homogeneous soil, fresh manure, and much care in the cultivation. Schwertz, taking the mean of three years, estimates the produce per acre at 13 tons, 18 cwt. 1 qr. and 2 lbs. of roots, and about one-third the same quantity of green leaves, which are valuable as fodder and as elements of manure. In a field at Bechelbronn where this vegetable had been intercalated with the madia sativa, we obtained upwards of 5¾ tons of roots in addition to our principal crop of oleaginous seed. The carrot contains a large quantity of water in its constitution—87.6 per cent., according to some of my experiments.

The juice of the carrot contains sugar, albumen, a crystallizable coloring principle, called carrotine, a volatile oil, fatty matters, pectic acid, pectine, starch, malic acid and alkaline, and earthy phosphates.

The parsnip, (*Pastinaca sativa.*) This plant is not very extensively cultivated, yet it has the advantage of standing the winter in the open field. It has been recommended as very useful in fattening cattle. In its composition it must assimilate with the carrot and beet. Drappier obtained as much as 12 per cent. of cane-sugar from the parsnip.*

BARKS.

Cinchona barks. The barks of cinchona, which are employed with so much success in medicine, are the produce of different species of a family of trees which grows in the mountains of South America; the active principle of all the varieties of bark resides in the vegetable alkalies, quinine, cinchonine, and cinchovatine.

The medicinal properties of bark were made known to Europeans in 1638, on the occasion of an obstinate fever from which the Countess of Chincon, vice-queen of Peru, suffered at Lima in that year. A corregidore of Loxa, who had been cured by the Indians while

* Berzelius, Chemistry, vol. ii. p. 199.

affected in the same way, recommended the bark. The medicine was completely successful; and to show her gratitude, the vice-queen had large quantities brought down from the mountains for distribution among persons affected with fever. It was from this circumstance that the bark was at first known under the name of the Countess's powder. By and by, the members of the college of Jesuits having been charged with its distribution, it of course became the Jesuits' bark or powder. Lastly, the Cardinal de Lugo having brought it to Rome, the new medicine was known under the name of the cardinal's powder.

The cinchonas are met with principally in forests at a considerable elevation above the level of the sea, in a temperate climate, and growing in a stony soil. The proper period for gathering is known by the circumstance of the inner surface of the bark, when detached from a branch, acquiring in a few minutes a red, yellow, or orange tint, according to the species.

The trees are cut down one or two days before the process of barking begins, by which the operation is rendered more easy, and the cuticle is no longer liable to be rubbed off. The bark of the trunk and branches is removed by means of a large knife, in strips, or bands, which are kept as broad as possible. The bark is placed upon cloths and put to dry in the sun, each piece being kept isolated, in order to facilitate the drying, and especially to favor the *quilling* or *rolling* up; when the bark is dried in a heap, and when the pieces touch, it often acquires a most disagreeable odor in consequence of incipient putrefaction, and the quilling does not take place.* The bark, when thoroughly dry, is packed in bullocks' hides and sent to Europe. From my own observations, and those that have been supplied me by M. Goulot, the different species of bark appear to be distributed upon the mountains of New Granada in the following order:

	Heights where most abundant.	Temperature.
Gray bark, C. lancifolia..............	6560 feet	19° (66½ F.)
White bark, C. ovalifolia............	4264 "	21° (70 F.)
Red bark, C. oblongifolia	2296 "	24° (75½ F.)
Yellow bark, C. cordifolia	1968 "	25° (77 F.)

Pelletier and Caventou discovered in the gray bark, 1st, cinchonine in combination with quinic acid; 2d, a fatty substance; 3d, red and yellow coloring matters; 4th, tannin; 5th, quinate of lime; 6th, gum; 7th, starch; 8th, woody matter. In the yellow and red bark these learned chemists found the same principles, and, moreover, quinate of quinine.

Barks of the willow and poplar. Decoctions of these barks are often employed with success in the treatment of intermittent fever. In searching after the active principle of these medicines, M. Roux discovered the particular substance, salicine, in the bark of the willow, (salix helix,) the medicinal effect of which is analogous with that of the febrifuge principles of the true barks. M. Braconnot has

* Ruiz, Quinologia.

further succeeded in obtaining another crystalline matter from the leaves of the aspen (*populus tremula*) populine.

Cork. The oak which yields cork is known in Spain under the name of the *alcornoque*. It forms extensive forests upon the abrupt slopes of the Pyrenees, where it is often seen growing upon arid and stony soils, that seem doomed to eternal sterility. The cork-tree has flexible and strong roots, which creep over the naked surface of the granitic masses, turning round blocks, and searching everywhere for fissures and collections of sand and alluvium, into which they penetrate deeply, in search of the nourishment necessary to the tree. At maturity, the alcornoque rises to a height of sixty-five feet, and its trunk may be three feet and a quarter in diameter.

In the Spanish Pyrenees, the superior limit of the cork-tree region, is that of the vine, about 1640 feet above the level of the Mediterranean. In France this tree grows luxuriantly in the communes of Passa, Lauro, &c., the mean elevation of which is 1148 feet. In Spain, as in France, the soils on which the cork forests grow are of primitive origin; and it is said, on good authority, that the cork-tree only grows on soils derived from granite, gneiss, mica-slate, or porphyry, and never on soils of calcareous origin.

The cork-tree is reproduced spontaneously on these silicious soils, among cistuses and heaths; but the reproduction in this way is so slow, that art often interferes advantageously to aid it. There are many varieties of cork oak; and as that which is covered with a smooth and grayish cuticle, yields the article which is most prized in commerce, the seeds of this variety ought to be selected for sowing. The acorns of the cork-oak are tumid, of considerable size, and a sweet taste; the acorns ripen from October to December, and are much employed as food for hogs. The Catalonians sow the acorns in a cultivated soil at the same time that they plant the vine, and for twenty or twenty-five years the produce of the vine compensates the outlay upon the young cork-trees; but the produce of the vine diminishes as the cork-tree overshadows it, and finally there comes a time when the vines die out completely. The cork-tree is of slow growth, and at four years of age it may be from thirty-six to forty inches in height, and requires incessant care until the trunk is from seven to ten feet high, at which time it may be about twenty years of age; and its total height, including its branches, may be about twenty-two feet.

The barking of the cork-tree begins about the middle of July, and may be continued so long as the sap is in motion. When stripped off, good cork is formed of from ten to twelve layers, each of which indicates an annual deposition. The two outer layers constitute the cuticle; the others adhere closely together, and although of variable thickness they present a homogeneous mass. The time for removing the cork is indicated by the interior acquiring a slightly rosy tint, which happens about the tenth year. The barking is performed by means of an axe, with which a cut is made the whole length of the trunk, care being taken not to wound the woody layers; two other cross cuts are then made at the top and bottom of the trunk.

By means of the handle of the axe, which is shaped like a wedge forced into the vertical cut, the cork is then loosened and stripped from the living bark beneath it, the whole covering of the tree being often taken away in a single piece, although it is more commonly removed in two pieces. The process of barking is very easy when the sap is abundant.

The cork-tree must be about forty years of age before its bark has any commercial value; that of a tree of twenty years is always treated as rubbish. An oak a century old may furnish 200 lbs. of marketable cork; as many as 480 lbs. however have been taken from a single tree; the mean produce may be reckoned at about 106 lbs. per tree; to fit it for the market, cork undergoes a variety of preparations which need not detain us here.

LEAVES.

The herbaceous parts of vegetables have all a very similar composition, if they be regarded in the most general point of view. The leaves and green stems, along with the woody fibre which forms in some sort their skeleton, always contain albumen or an analogous azotized principle, saccharine and gummy substances, chlorophylle, wax, fatty and resinous substances, free or combined acids, and frequently also essential oils. Such is the general constitution which chemists agree in assigning to clover, hay, leaves, in a word to green forage of all kinds; nevertheless, to this constitution, which may be regarded as standard, we have frequently other particular matters added, some of which we have already studied, and which by their medicinal properties, or the economic uses they possess, render the plants that contain them of high importance in an agricultural point of view. I shall here only speak of two of these plants, tobacco and tea, the leaves of which, almost in universal use, are a source of great commercial prosperity to the people who cultivate them.

Tobacco, (*Nicotiana tabacum*,) a native of America, appears to have been introduced into Spain and Portugal about the middle of the sixteenth century by Fernandez de Toledo. Its name is generally believed to be derived from that of the Island of Tobago, one of the West India islands, at no very great distance from the coast of Venezuela, whence the first importations were made. Nicot, the French ambassador to Portugal, first made its use known in France, whence the name *nicotiana*. At the present time, the cultivation of tobacco appears to have spread almost over the whole surface of the globe.

Tobacco requires a somewhat friable soil, rich in humus; it consequently succeeds in lands just broken in. In America the mode of cultivation and of preparation are almost everywhere the same.

In Venezuela the seed is sown in a very rich loam, and after from forty to fifty days, the young plants are transplanted in rows, distant a little more than three feet from one another, the plants being about two feet apart; the transplanted plant is generally covered with a banana leaf for a few days to preserve it from the burning rays of the sun. When the plant is about eighteen inches high, a bud is formed at the superior extremity ; this bud and any others that may appear are removed, as well as any sprouts which show themselves on the stem. By this treatment the tobacco becomes bushy and thick ; by and by the leaves acquire a decidedly blue tint, and the time of gathering is indicated by the appearance of a stain of a deep-blue color near the pedicle. The leaves do not all ripen simultaneously, so that the business of the planter, in gathering those that appear ripe, is incessant for a certain time.

After they are gathered, the leaves are carried under sheds, where they are disposed two and two upon hurdles arranged for their reception. The tobacco soon becomes yellow and pliant, and the ribs of the leaves having been removed, they are twisted into a rope which is coiled up into a mass of the weight of from sixty to eighty pounds. These coils are placed upon a bed made with damaged leaves and the ribs which have been removed. The whole is covered, and left to ferment during forty-eight hours, a little water being supplied if the tobacco appears too dry ; during the fermentation the temperature rises, and the process having been carried sufficiently far, the coils are exposed separately to the air; they are then unrolled, and hung up under sheds to dry completely.

The vertical zone in which the cultivation of tobacco within the tropics is carried on, is extensive ; it reaches from the level of the sea to an elevation of about 5,900 feet above it. The time during which the crop remains on the ground varies according to the mean temperature of the place ; according to M. Codazzi, the leaves are gathered one hundred and fifty days after the sowing in the hottest regions of the coast of Venezuela. In more elevated situations, where the thermometer ranges from 65° to 68° Fahr., the first leaves are not fit to be gathered until after about seven months and a half from the sowing.

In Ceylon, tobacco is cultivated almost precisely as in America. There they also prevent the plant rising in height, and they limit the number of leaves upon each stem according to the quality of the tobacco which they desire to grow. By leaving the plant with no more than from ten to twelve leaves, the most esteemed quality is obtained ; if eighteen or twenty leaves be left, the tobacco is far from having the same strength. Lastly, in leaving the plant to itself, by suffering the stem to run up and to flourish, a large crop is obtained, but the produce is not esteemed. The leaves gathered from the plant in this state of maturity are often held fit for consumption after being simply dried without further preparation. The tobacco is then yellow, extremely mild, and perfectly suited to the immoderate use made of it by the Cingalese.

If the mode of cultivation enables the tobacco-grower to obtain a superior quality at the cost of quantity, it is still indubitable that climate exercises the chief influence on the quality of the article. That which is grown in the temperate regions of the Andes, in Virginia, and in Europe, can in no way be compared with the tobacco of the Havana, of Varinas, of Giron, of the valley of Cauca.

The cultivation of tobacco has appeared to me more especially advantageous in localities where the mean temperature does not fall below 75° Fahr.

Tobacco is decidedly a plant of a hot climate; there only does it yield a produce of the best quality. In Venezuela, where its cultivation is followed with great skill, and in situations where the temperature of the climate keeps from 77° to 80° Fahr., about five plants are held necessary to produce 1 lb. of tobacco; and as a mean, 11 cwt. 1 qr. 23 lbs. of the prepared article are produced from an acre of unmanured land.

In Alsace, tobacco is sown about the middle of March; the transplanting takes place in the beginning of June, and the harvest follows in autumn. The husbandry is very similar to that which has been already described, each plant being left with eight or ten leaves. Schwertz reckons the produce at about 15 quintals per hectare of two and a half acres, which is equal to about 12 cwt. 1 qr. per English acre. Thaer estimates the produce in Prussia at 11 cwt. 1 qr. 23 lbs. per acre.

The consumption of tobacco has lately increased considerably throughout the whole of Europe. In France, government sold tobacco in one form or another to the extent of 31,116,340 lbs. in the course of the year 1837. Public documents show that in 1841 there were in France 8,158 hectares, or 20,175 acres under tobacco, which yielded 21,261,064 lbs. of the article; the difference between the quantity produced and the quantity consumed is of course supplied by importation.

The virtues of tobacco very probably reside in the volatile vegetable alkali, nicotine, which it contains. The analyses of M. Posselt and Kiemann show the leaf of tobacco to be composed as follows: Nicotine 0.07, extractive matter 2.87, gum 1.74, a green resin 0.27, albumen 0.26, gluten 1.05, malic acid 0.51, malate of ammonia 0.12, sulphate of potash 0.05, chloride of potassium 0.06, nitrate and malate of potash 0.21, phosphate of lime 0.17, malate of lime 0.72, silica 0.09, woody matter 4.97, and water 86.84,—100.00. During the fermentation of the leaves, there is always a formation of ammoniacal salts.

Tea, the use of which is and has so long been universal in the Chinese empire, began to be known in Europe in the seventeenth century, when it was imported by the Dutch East India Company. In 1669, the importation of tea into England did not exceed 1 cwt.; in 1833, the East India Company set aside for the consumption of Great Britain alone nearly 24,200,000 of pounds!

The tea plant commonly attains a height of from three to about five feet. In China it blossoms in the early part of the spring, and ripens its seeds in December and January. Its branches are covered with short thick leaves of a deep-green color and elliptical form. It is one of the most hardy plants, and thrives from the equator to the forty-fifth parallel of north latitude; but the districts best adapted to its growth appear to be comprised between the twenty-fifth and thirty-third degrees of latitude.* Tea requires a moist climate, and a light and sandy soil. No manure is given, and no attention is paid to the nature of the soil where irrigation is practicable.

The shrub is propagated from seed. Several seeds are dropped into holes at the distance of from three to six feet apart, and the plant begins to produce from the third year: the gathering is done by the hand, the leaves being picked off; but a few are always left upon each branch. The number of gatherings made in the same year varies from one to three, according to the age of the plant. It is very seldom indeed that a fourth gathering is practised. In China the tea harvest begins about the middle of April, a period at which the leaf buds appear surrounded by a slight cottony down. The first gathering is very small, but it constitutes the highest-priced tea, the Shou-chun or tea of the first growth. The second gathering takes place in June, when the branches are covered with leaves of a pretty deep color; these leaves are very abundant, but inferior in quality to the buds of the former gathering; they constitute the tea called Urh-chun, or tea of the second growth. The third gathering is performed a month later, and the produce passes by the name of the San-chun, or tea of the third growth. The leaves are now of a deep-green color, tough, and are manufactured into the most common kinds of tea.

Considerable plantations of tea are now established in Assam, in British India, and in the Brazils, and it seems not improbable that the plant may be cultivated at some future day in Europe.

According to Guillemin, who studied the cultivation and preparation of tea in the Brazils, the leaves are dried as soon as they are gathered. From four to six pounds are thrown into an iron pot, the interior of which is polished, and which may be somewhat more than three feet in diameter, by about a foot in depth. The temperature of the pot is maintained at about the boiling point of water; a negro stirs the leaves in all directions with his hand, until they become quite soft and pliant, so that they can be moulded into pellets by movement between the hands. When the leaves are in this state, they are thrown upon a tray made of bamboo, and strongly kneaded for a quarter of an hour, so as to force out a green sap of a disagreeable taste. The kneaded leaves are then returned to the pot, and dried completely, being all the while stirred about with the hand, being separated when they stick together, and being continually tossed up in order to prevent them adhering or getting scorched by remaining too long in contact with the metal. During this process, which lasts for some half hour, a large quantity of dust is disengaged, which proceeds from the cottony down with which the leaves are covered. By this rapid drying, the leaves crisp and curl up of themselves, and acquire the appearance of the tea that is in every-day use. On being taken from the drying pot, the tea is thrown upon a sieve of a certain mesh, and the leaves which have rolled themselves up into the smallest compass, and which are those

* Robinson, a Descriptive Account of Assam, p. 131.

which proceed from the buds, are separated from the others; these after having been winnowed, receive a new touch of the fire, when they acquire a leaden-gray color, and constitute the tea of the best quality, which in the Brazils passes by the name of Imperial or Uchim tea. The leaves which remain upon the sieve are heated, winnowed, and sifted again, and the produce is fine Hyson tea; and by the same means other varieties are procured, until at length a kind remains upon the sieve consisting of the leaves that have not become rolled up, which being added to the broken particles derived from the winnowing operations, is called family tea, because it is consumed upon the spot.

During, and for some time after the drying, tea exhales an herbaceous and not very pleasant odor, which however becomes modified in the course of time. The aroma of the Chinese teas is said to be communicated to them by a highly odorous plant, which is believed to be the *Olea flagrans*. It is also said that the green tea is colored by means of indigo; but it is possible that the shades of color of the different kinds of tea depend solely upon the degree of roasting which they have undergone. Guillemin has said nothing of the produce of the tea shrub in Brazil. In China, according to a manuscript of M. Carpena, a shrub with care will produce annually during thirty or forty years from 2 lbs. to 2½ lbs. of marketable tea.

From the analysis of M. Mulder, tea appears to contain: 1st. A volatile oil. 2d. Chlorophylle. 3d. Wax and resin. 4th. Gum. 5th. An extractive matter. 6th. A coloring matter. 7th. Azotized substances analogous to albumen. 8th. Woody fibre and inorganic salts. 9th. A particular crystalline principle,—theine or coffeine, which is ranked among the vegetable alkalies, and which is also met with, as implied by the name, in coffee: this new principle crystallizes in colorless needles of a silky aspect and bitter taste. It is little soluble in alcohol and in ether; water dissolves about $\frac{1}{50}$th of its weight, and it sublimes without undergoing decomposition; it is by sublimation, in fact, that Mr. Stenhouse proposes to obtain it from tea.

This is undoubtedly the principle which communicates to tea its bitter taste, and several of its properties; experiment has shown, that when administered even in considerable doses it produces no ill effect on the animal economy; different kinds of tea, as might have been presumed, contain it in different proportions. Mr. Stenhouse obtained from 100 parts of

Hyson	1.09	of Coffeine
Congou	1.02	"
Assam	1.37	"
Twankay, green	0.98	"

SEEDS.

Wheat. This valuable grain is the produce of several kinds of *triticum*—winter wheat, and spring wheat, *T. hybernum*, and *T. æstivum*, spelter, *T. Spelta*, and *T. monocon.*

Wheat is sown either upon a fallow or upon land that has carried some forage crop, or such a crop as beans and peas. It requires a stiff rich soil, containing a certain proportion of calcareous earth, and abounding in organic matter; it does not thrive well in soils where the sandy element predominates over the clayey. For seed the best grain is selected; but this and all other precautions do not suffice to preserve the plant from many diseases, such as smut, rust, mildew. Farmers are wont, before putting their seed wheat into the ground, to prepare it in various ways with a view to destroying the germs of certain parasites which are believed to adhere to it externally. The process is generally called pickling, or liming, because milk of lime, in which the seeds are put to steep for twelve or fifteen hours, is often employed in its course. Means that are said to be more efficacious have also been recommended: some make use of alum, others of sulphate of iron, sulphate of zinc, sulphate of copper, sulphate of soda, and even white oxide of arsenic. All these means appear to conduce to the same result. We employ sulphate of copper, which indeed is the custom in a considerable part of Alsace, and I can assure the reader that our fields of wheat are never infected. 100 grammes, or about 3¼ ounces troy, are allowed to a hectolitre or sack of nearly 3 bushels of wheat; the salt is dissolved in as much water as is held requisite for the submersion of the grain, which is steeped in the solution during about three quarters of an hour, after which it is thrown into baskets to drain, and being then spread out on the floor it is dried before being sown.

The season at which wheat is sown in autumn ought to vary with the climate, and nothing can be more displaced than those precise dates which are set down by the majority of writers. The great point to be held in view is, that the young plant may have got a certain length before the frost sets in, that the roots may have penetrated to a depth which shall protect them from the severe cold of the winter. In each district, experience has already proclaimed the proper time for sowing, and this can rarely or never be departed from without detriment. In the east of France, in Alsace, the sowing of winter wheat generally takes place in the first week in October; in the southern hemisphere, in certain parts of Chili, for example, the wheat is sown in April, and is exposed to the cold weather of June, July, and August. The quantity of seed sown may vary from about 7 pecks to 18 pecks and more per acre. Farmers generally agree, however, that we have seed enough when we employ about 2 bushels to the acre; this is the quantity which is used at Bechelbronn; but in the same district, and even on contiguous fields, we frequently see proportions of seed employed which vary in the ratio of from one half to twice the quantity specified, without, so far as I know, any sufficient reason being given for this parsimony or prodigality. It is, however, a question of the very highest importance to ascertain the proper quantity of seed. The question may be considered in two ways: 1st. with reference to the produce of a given extent of surface, and 2d. with reference to the produce from the grain sown. It is quite certain that in sowing thick, a larger produce per acre will be obtained than by sowing very thin; but on

15

the other hand, thin sowing yields a larger number of times the quantity of seed put into the ground. The reasons which should guide us in determining the dose of seed are numerous and extremely complex; they must evidently be taken in connection with the value of the ground and of cultivation, the price of wheat and of straw, the cost of labor and of manure. Thus in countries where the rent of land is extremely low it may be a good practice to scatter but a moderate quantity of seed over a large extent of surface. I remember a field in the neighborhood of Pampeluna, where the wheat was growing in isolated tufts, all extremely vigorous and very heavy in the ear: the ground had had but very little preparation; nevertheless, they expected to gather from sixty to eighty times the seed. This, without doubt, was a profitable crop; nevertheless, I am satisfied that it could not have yielded more than from 6½ to 7½ bushels per acre.

For the same reasons the first settlers of the United States must have followed a somewhat similar mode of cultivation. "An English farmer," says Washington, in a letter addressed to Arthur Young, "must have a very indifferent opinion of our soil when he hears that with us an acre produces no more than from 8 to 10 bushels of wheat; but he must not forget that in all countries where land is cheap and labor is dear, the people prefer cultivating much to cultivating well."

In Alsace we do not reckon any crop profitable which yields less than from 19½ to about 23 bushels per acre; and in these circumstances we do not receive back more than from 9 to 10 times the seed.

Nevertheless, it must be allowed, even in these extreme cases in which the value of ground is so different, inasmuch as it may vary in the ratio of from 1 to 1000, that there are certain limits with reference to the seed which must not be passed; and there is without doubt an opportunity of making a series of curious and useful experiments, with the view of ascertaining the true ratios which exist between the produce and the seed. I am well aware that the results of experiments of this kind have already been made public; but I know also that these data have not been deduced from a sufficient number of facts perfectly comparable with one another, and noted under a variety of climatic influences; in a word, that they are not such as they ought to be, to put an end to the uncertainty which still exists in the minds of the best-informed farmers and rural economists upon the subject.

In Europe, the wheat that is sown in autumn generally stands upon the ground for from nine to ten months. The time, however, varies considerably with the climate; in the Andes, it is in proportion to the proper temperature of each place.

Wheat, which is now an important article of agricultural produce in America, was introduced from Europe very shortly after the Conquest The first particles of wheat sown in Mexico before 1530, are said to have been found by a negro belonging to Fernando Cortez among the rice destined for provision to the army.* Wheat

* Humboldt's Essay on New Spain, vol. ii.

was brought into Quito by a Fleming, Father Jose Rixi, a monk of the order of St. Francis. I was shown in the Convent of St. Francis, the vessel in which the first seed is said to have come from Europe.

In Mexico, where the ground can be irrigated, and this, all things else being the same, always yields the best crops, wheat is watered at two different periods—when it has sprung and when it is shooting into ear. According to M. de Humboldt, who has collected so much that is interesting upon the agriculture of New Spain, the richness of the harvest is truly surprising; irrigated soils often yield from 40 to 60 times the seed; 16 for 1 is reckoned a middling crop, and, taking the whole of Mexico, the mean produce may be estimated at from 22 to 25 for 1.

The cultivation of wheat is especially lucrative in districts which enjoy a mean temperature of from 18° to 19° C., (65° to 67° F.;) yet it may be pursued amid plantations of coffee-trees and sugar-canes, although I doubt whether it can there be extremely productive. The extreme limits of the growth of wheat in the Cordilleras, according to my own observations, correspond with the mean temperatures of from 12° to 23.5° C. (54 to 75° F.) M. Codazzi estimates at 37 for 1 the mean produce of wheat in Venezuela.

The hectolitre of wheat in France (22.009 gallons, something less than a sack of three bushels English) is held to weigh, on an average, 169.4 lbs., or 61.6, say 61½ lbs. per bushel; but this weight varies according to the quality of the grain between 56 and 64 lbs. the bushel.

The following is a table of the mean produce of the wheat crop in different countries, from documents that may be relied on:

Localities.	Produce per acre (seed deducted.)	Authorities.
	Bushels.	
Germany: Moeglin	26.7	Thaer.
Lavanthal	22.0	Burger.
Saalfelden	18.0	Lurzer.
Lombardy: irrigated lands	25.6	Burger.
" non-irrigated lands	15.9	Dandolo.
" "	11.0	Verra.
Average of Venetian Lombardy	15.9	Burger.
England: the best soils	34.0	Arthur Young.
" average	23.0	Arthur Young.
Brabant and Flanders	25.0	Schwertz.
France: Alsace (after tobacco)	29.0	Schwertz.
" " Bechelbronn	22.3	Lebel and Boussing.
Environs of Paris	25.2	Dailly.
" Oise	21.5	Official statistics.
America: (East of the Alleghanies)		
" rich lands	35.0	Blodget.
" middling ditto	9.0	Blodget.
Mississippi: rich lands	44.2	Blodget.
" middling ditto	27.0	Blodget.
Venezuela (Valley of Aragua)	44.0	Humboldt.
" temperate regions	14.0	Codazzi.

The cereals, besides their principal produce, their farinaceous seeds, yield another, which is of great importance in rural economy; this is straw, which no European agricultural establishment could do without. After having been used as food and as litter for cattle, it is returned to the ground as manure, and contributes powerfully in preventing the exhaustion of the soil, which the cultivation of wheat always produces. The quantity of straw which can be reckoned on in a farm, is, of course, in proportion to the soil under white crops. The relative weight of grain and straw, however, varies considerably according to circumstances; in a wet year, for instance, the wheat crop contains a relatively large proportion of straw, and a small proportion of grain; in dry years the contrary relation obtains. Lands recently and abundantly manured, yield a larger quantity of straw than clover breaks. Thick sowing always yields a large quantity in contrast with the grain; lastly, climate exerts the most marked influence upon the two kinds of produce which we are considering. The differences which are observed between one year and another, in the same districts, in consequence of very different meteorological conditions, are not less remarkable. I shall quote a single instance. The years 1840, 1841, and 1842 gave us crops of grain at Bechelbronn which were far from excellent; in the first the rains were too abundant, and in the second the drought was too long continued. In these opposite circumstances, the weight of the straw to that of the grain was—

In 1840–41 : : 100 : 24
In 1841–42 : : 100 : 90

The latter harvest, in fact, occasioned a complete dearth of litter in our establishment. In ordinary years we procure about 38 of grain for 100 of straw, a relation which agrees with those that have been reported by different observers, who vary in their calculations from 33 and 35 to 41, 44, and 50 of grain to 100 parts of straw.

In the cereals the amylaceous matter, which constitutes the principal part of the seed, is surrounded by a flexible perisperm, of the nature of woody tissue. The object of grinding is to break this case and to reduce the interior of the grain to powder. In France, the grinding of wheat is performed by a succession of operations; in England it is completed at once. The French mode, however, appears to yield the largest quantity of fine flour.

	English.		French.	
Fine flour	58	72	66	74
Second flour	14		8	
Bran	26		23	
Loss	2		3	

The proportion of flour furnished by the cereals does not, however, depend alone upon the mode of grinding, but also upon the nature of the grain. Wheat, for instance, of different kinds, yields 78, 83, and $85\frac{1}{2}$ per cent. of flour.

Spelter. This grain is so firmly enclosed in the husk that it cannot be freed by threshing; so that, in the countries where this grain is grown, the mills are provided with an apparatus for husking it.

Schwertz made many experiments in Wurtemberg to determine the quantity of flour yielded by spelter, and he found that from 100 of grain he obtained 90 of husked grain, and 8.7 of bran; there was a loss of 1.3. The quality of the flour always varies according to the wheat from which it is procured: it contains moisture in variable proportions, gluten in variable proportions, and, finally, various quantities of woody matter. The wheat of the south is harder and tougher than that of the north, and appears richer in azotized principles; as it contains less moisture, it also keeps better; it is, undoubtedly, in consequence of the large quantity of water which our northern wheats contain, that we meet with such indifferent success when we attempt to keep them for any length of time in our granaries. The wheat of Alsace, for example, frequently contains from 16 to 20 per cent. of moisture; and I have ascertained, by various experiments, that it is almost impossible to keep it without change, in vessels hermetically sealed. To secure its keeping, the proportion of water must be reduced to from 8 to 10 per cent., and this is nearly the quantity of moisture contained in the hard and horny wheat of warm countries. I am therefore of opinion that we shall never succeed, in these countries, in keeping wheat for any length of time—in the magazines of fortified towns, for example—whatever care be taken.

The flour of the cereals, particularly that of wheat, absorbs a large quantity of water, and forms a paste, which is by so much the firmer and more elastic, as the flour contains a larger proportion of gluten: the azotized principle of wheat has, in fact, the remarkable property of being extensible like a membrane, when it is moist, and this property it communicates to the whole of the paste or dough. In order to be brought into the state of dough fit for making bread, flour will absorb from 55 to 70 per cent. of water. The quantity of bread obtained necessarily depends upon the heat and length of exposure in the oven; but, in a general way, from 100 of flour, 130 of the best white bread of Paris is procured. In the country, the bread is generally less baked than in Paris or London, and therefore retains more water; so that from 100 of flour, 140, 145, and 146 of bread are made: thus, admitting 16 per cent. of moisture as existing in wheat originally, we have of absolute dry matter 64½, 57, and 56 in different kinds of bread.

Bread is by so much the more nutritious as it is made from flour containing a larger proportion of gluten; to add any starch therefore is to prejudice the interests of the consumer; nevertheless it is the practice to do so almost openly; when potato starch is at a low price, the adulteration frequently begins with the miller and is extended under the baker. The quantity of gluten contained in different kinds of wheat varies greatly. Vauquelin found in the—

	Gluten.	Starch.	Sugar.	Gum or dextrine.	Water.	Bran.
Flour of French wheat	11.0	71.5	4.7	3.3	10.9	
Flour from hard Odessa wheat	14.6	56.5	8.5	4.9	12.0	2.3
Flour from soft Odessa wheat	12.0	62.0	7.6	5.8	10.0	1.2
Flour from the baker's	10.2	72.8	4.2	2.8	10.0	

The method of analysis employed by Vauquelin, whose results

are given above, by means of washing, is however far from being very accurate; it is impossible to prevent the loss of some gluten which passes with the starch, and the vegetable albumen is entirely lost by reason of its solubility in water: and then to dry gluten is a very long and delicate process; and if we would pretend to any degree of accuracy, we must ascertain the quantity of fatty matter contained in the samples. I therefore thought that with reference to the azotized principles particularly, the better way would be by proceeding to ascertain these by immediate ultimate analysis.

The four azotized principles which we have already admitted have very nearly the same elementary composition; the mean proportion of azote in each is 0.16. With this datum, it is evident that if a particular sample of flour is found to contain 0.04 of azote, it may be inferred that this azote represents 0.25 of gluten, albumen, fibrine, and caseine, dried at 140° C. (284° F.,) and as these are the most valuable elements in flour, I took the pains to ascertain their proportion in a considerable number of varieties of wheat, the whole of which were grown in the same year, in the same soil, which was well manured, and under climatic influences that were identical; nor did I restrict myself to the azotized matters of these samples; I also endeavored to ascertain the precise relative quantities of bran and of flour. The following table contains the results of my experiments.

Variety of wheat.	Character of the grain.	In 100 of the grain.		In 100 of the Flour.		Appearance of the flour.
		Bran.	Flour.	Gluten and albumen.	Starch, sugar, gum, water.	
Triticum spelta rufa mutica	small, thin.	21.9	78.1	24.1	75.9	grayish, coarse.
Small spelter, T. monococon	middling.	20.8	79.2	24.8	75.2	soft.
Great spelter	very large.	26.9	73.1	22.1	77.9	very coarse.
Mecca wheat	horny, long.	32.0	68.0	23.8	76.2	yellow, coarse.
Bearded wheat	small, brown.	13.2	86.8	22.7	77.3	very coarse.
Winter wheat, T. hybernum	middling.	38.5	61.5	18.3	81.7	yellowish, soft.
Common wheat (mouret)	reddish.	23.5	76.5	23.5	76.5	coarse.
Reyel wheat	yellow, fine.	14.0	86.0	18.7	81.3	very white, soft.
Red Egyptian wheat	small, hard.	15.0	85.0	21.6	78.4	yellowish, rough.
A large wheat growing in 4 ranks	hard.	15.0	85.0	20.4	79.6	somewhat coarse.
Fine red wheat of Roussillon	reddish.	16.0	84.0	22.6	77.4	white, very soft.
Red Marcel wheat	large.	21.5	78.5	19.0	81.0	yellowish, soft.
Dantzic wheat	soft.	24.0	76.0	22.7	77.3	white, very soft.
Wheat from the North	pretty hard.	20.5	79.5	22.4	77.6	yellowish.
Fine red wheat (pays de Foix)	soft.	18.5	81.5	21.9	78.1	white, very soft.
Smyrna wheat	white, hard.	19.0	81.0	19.9	80.1	white, rather rough.
Bengal wheat	ditto.	21.5	78.5	18.6	81.4	soft, white.
Tangarok wheat	small.	23.5	76.5	24.1	75.9	white, extremely fine.
Hard African wheat	gray, hard.	24.5	75.5	26.5	74.5	yellow, very coarse.
Cape wheat	yellow, large.	19.0	81.0	18.2	81.8	white, very soft.
Russian wheat	wrinkled.	18.0	82.0	22.1	77.9	yellowish, very soft.
Sicilian wheat	small, red.	19.5	80.5	24.3	75.7	yellow, coarse.
Giant St. Helena wheat	hard, very large.	25.0	75.0	20.9	79.1	ditto.
Subernac wheat (Pyrenees)	well-formed.	20.5	79.5	19.0	81.0	white, very soft.

The quantity of gluten and albumen contained in these samples of flour is much larger than that usually indicated; I have given reasons which explain, to a certain extent, this difference. I ought to add, however, that the varieties of wheat, the flour of which was analyzed, were all grown in the rich soil of the garden, a circumstance which, as Hermbstädt has shown, exerts the most powerful influence in increasing the quantity of gluten in wheat.

It was already known, from the experiments of Tessier, that the proportion of gluten in the same species of wheat might vary in the ratio of from 12 to 36 per cent. of the weight of the flour, according to the nature of the soil and the quantity of manure. But it was Hermbstädt who first made truly comparative observations on the action of the excrements of different animals on the culture of the cereals.

The excrements made use of by this able cultivator in his inquiries were always dried in the air at a temperature of $12\frac{1}{2}°$ C. ($54\frac{1}{2}°$ F.,) and equal areas of the same soil were sown with equal weights of winter wheat, and had a similar dose of manure of one kind or another spread over them. One hundred parts of the flour obtained from wheat thus grown yielded:

	Gluten.	Starch.	Bran, soluble matter and moisture.
With human urine	35.1	39.3	25.6
" bullock's blood	34.2	41.3	25.5
" human excrement	33.1	41.4	25.5
" sheep's dung	22.9	42.8	34.3
" goat's ditto	32.9	42.4	24.7
" horse ditto	13.7	61.6	24.7
" pigeon's ditto	12.2	63.2	24.6
" cow's ditto	12.0	62.3	25.7
Soil not manured	9.2	66.7	24.1

It is apparent, therefore, that in general, for the exception onl refers to the pigeon's and the horse dung, the wheat grown in groun manured with the most highly azotized matters yields the larger quantity of gluten.

By way of adding to and confirming these conclusions of Hermbstädt, I shall give the results of an experiment of my own, made i 1836, in which the same variety of wheat was grown in the ope field, and in garden ground very highly manured. The grain wa analyzed after having been dried at 110° C., (230° F.,) and gave:

	From the open field.	From the garden ground.
Carbon	46.10	45.51
Hydrogen	5.80	5.67
Oxygen	43.40	43.00
Azote	2.29	3.51
Ashes	2.41	2.31
	100.00	100.00

In the produce of the garden there were 21.94—very nearly 22 per cent. of gluten and albumen; in that of the open field no more than 14.31 per cent. of the same principles.

Davy was of opinion that the wheat of warm climates was richer in azotized principles than that of temperate lands. Southern countries are known to produce harder, tougher grain, the flour of which contains more gluten than the soft and more friable wheat of the north ; and the inquiries of M. Payen appear to bear out the conclusion of the illustrious English chemist. M. Payen, in fact, found in the hard wheat of Africa 3.00 of azote, equivalent to 18.7 ; and in that of Venezuela 3.50 of azote, equivalent to 21.9 of gluten and albumen. The experiments quoted above, however, prove that we may have wheat grown in Europe fully as rich in azotized elements as any that is grown between the tropics ; the influence of the soil in this direction is probably more than the influence of climate.

In all the analyses of wheaten and other flour published up to the present time, we find no mention made of the fatty matters which they contain ; and late views in regard to the special part which these matters play in nutrition make it very necessary to supply the omission. Along with MM. Dumas and Payen, I therefore determined the quantity of fatty matter contained in a considerable number of the vegetables and vegetable substances used as food, from which it appears that grain of different kinds contains from 2 to 10 per cent. of oil. One hundred parts of winter wheat gathered at Bechelbronn lost 14.5 of water by drying at 110° C., (230° F.,) and therefore contained 85.5 of dry matter. 100 of this dry wheat gave 13.7 of bran and 86.3 of flour.

Various analyses showed the composition of this wheat and its parts to be as follows :

Dry matter.	Gluten and Albumen.	Starch.	Glucose. (Sugar.)	Gum.	Fatty matter.	Woody tissue.	Ashes.
Bran	20.0			28.8	5.5	45.7	
Flour	13.4	73.2	5.6	4.2	2.1		1.5
Wheat	14.3	63.2		12.[illegible]	1.6	7.5	

Rye, (*Secale cereale.*) Rye is an important article ot food, particularly in the north of Europe, where the people live upon it almost entirely. It is a very hardy plant, and will thrive in soils which are altogether unfit to grow wheat. In the husbandry of the north this grain occupies the place of wheat in the south ; it requires much the same treatment, and stands upon the ground for nearly the same length of time. The bushel of rye weighs on an average about 60 lbs. avoird. The usual quantity of seed sown is from 10 to 11 pecks per acre, and the produce per acre, the seed being deducted, has been stated as follows :

	Bushels.
Brabant	23.0
Flanders	32.4
Austria	20.6
England	22.0
France	19.0

The German agriculturists say, that the weight of the straw to the weight of the rye produced is in general as 100 is to 47 ; others say as 100 is to 50, and some have taken it even as high as 100 to 33. The relation seems to differ extremely in different years. At Bechelbronn, for example, in 1840–41 we had 63 of grain to 100 of straw ; in 1841–42 we had but 25 of grain to 100 of straw.

Rye yields flour that is not so white nor so fine as that of wheat, which is in consequence of the woody covering of the grain getting ground, in great part, in the mill. If but from 50 to 65 parts per cent. of flour be taken from rye, it is white and looks well. The dough made with rye flour is not very adhesive; it contains little vegetable fibrine, the azotized principle which gives gluten its elastic properties. It is this want of vegetable fibrine which renders it more difficult to make good light bread of rye than of wheaten flour, although experiment shows that rye flour of the first quality will form as large a proportion of bread as wheaten flour; 100 of rye flour have given 145 of bread.

Rye bread is more hygrometric than that of wheat, and consequently remains for a longer time soft and fresh. Rye generally contains 24 of bran to 76 of flour; by drying at 230 F. it loses about 17 per cent. of water. Analyses of a dried sample grown at Bechelbronn yielded:

Gluten and albumen (azotized principles united)	10.5
Starch	64.0
Fatty matters	3.5
Sugar (glucose ?)	3.0
Gum	11.0
Woody matter and salts (phosphates)	6.0
Loss	2.0
	100.0

Barley, (*Hordeum vulgare.*) The usual produce of barley varies much from 15 or 20 to 50, 60, and even 70 bushels per acre; the average for France is stated at about 43½ bushels; and the weight of the bushel may be taken on an average at about 504 lbs. The ratio of the straw to the grain varies very much, but may be taken generally at that of 100 to 50. Barley contains:

Of flour	68.6
Bran	18.4
Water	13.0
	100.0

Dried, this grain gave 0.0214 of azote, which represents 13.4 per cent. of gluten and other azotized principles.

Oats, (*Avena sativa.*) When oats yield 43 or 44 bushels per acre, the crop is a fair one. At Bechelbronn we have frequently had upwards of 45 bushels per acre.* Schwertz states the relation between the straw and the grain as 100 is to 60.

Some oats gathered in 1841–42 yielded 78 of meal and 22 of husk per cent.

One hundred parts of these oats lost by drying at 230° F., 20.8 of water; thus dried, analysis showed that they contained:

Of starch	46.1
" gluten, albumen, &c.	13.7
" fatty matter	6.7
" sugar (glucose)	6.0
" gum	3.8
" woody matter, ashes, and loss	21.7
	100.0

* This would be reckoned a poor crop in the North of England and Scotland, when 80, 90, and even 100 bushels of oats per acre are frequently grown.—Eng. Ed

Maize, (*Zea mais.*) This is the true wheat of the Americans, and it is now generally allowed that the plant is a native of the New World. It is also well known that maize was introduced into Spain long before potatoes. Oviedo states in his work, printed in 1525, that he had seen it growing in Andalusia and the neighborhood of Madrid. The cultivation of this useful plant was observed everywhere on the discovery of America by Europeans, from the most southern parts of Chili to Pennsylvania in the north; and in the neighborhood of the equator, from the level of the sea to the high table-lands of the Andes. Garcilasso gives a particular description of the procedure followed by the Incas in the cultivation of this plant, the kind of manure, &c. At Cusco the Indians manured with human excrement dried and reduced to powder. On the coasts they employed in one place guano; in others, as the dusty and sterile soils of Attica, Atiquiba, &c., they made use of the offal of fish.

The uses of maize are very numerous. In America it is made into cakes, which are a substitute for bread; by fermentation a vinous liquor is prepared from it called chicha. Before the conquest, the Mexicans manufactured a sirup from the expressed juice of the stems. In describing to Charles V. the various articles of provision that were met with in the march to Tlatclolclo, Cortez says, "They sold us the honey of bees, wax, and honey from the stems of the maize plant." Maize when ground and boiled makes a kind of pudding in universal use, and the ear, when nearly ripe, whether boiled in water or roasted in the ashes, is held a luxury by all classes. In the tropical Cordillera maize is advantageously cultivated from the level of the sea to the height of 9186 feet above it; that is to say, it thrives in temperatures which vary between 14° and 27.5° C. (57.5° and 81.5° F.;) this circumstance explains its very general introduction into Europe.

Maize succeeds on all soils when they are properly manured; I have seen beautiful crops upon the most sandy soils and upon the stiffest clays; it requires much the same management as our ordinary grain crops; the climate alone should decide as to whether its introduction into a particular district is opportune or not; a certain degree of heat is necessary to ripen it, and above all, the cold to which it is exposed must not be too severe. It is for this reason, that in the east of Europe the maize is sown in spring, when there is no longer any apprehension of frost; there would be a real advantage in sowing late, were it not for fear of the frosts of autumn at the season of ripening. The susceptibility of maize to frost and climate generally, appears to me very analogous to that of the vine; and I doubt whether it would be wise to attempt its cultivation on the great scale where the grape does not ripen in ordinary years.

Maize is sown either with the dibble or with the hand, following a furrow opened by the plough; I believe that it ought never to be sown broadcast, for it is a plant that requires room; it is only in the hottest countries that the drill system is less necessary. In Alsace the drills are about 2½ feet apart, and the seeds are sown at the distance of about a foot from each other. This very considerable space left between the maize plants appears to authorize the general custom that prevails of interposing some other crop in the fields under Indian corn; that which is most generally interposed is either the dwarf haricot or the potato. I observed the same custom in the more temperate valleys of the Andes, where it is almost as necessary as in Europe to leave free spaces between the plants to give them air and sun; but the plant is cultivated alone in the hotter regions. Soon after maize has sprung it receives a first hoeing, and after it has got to a certain height, a second; in Alsace, for instance, it is customary to hoe towards the end of June; but I never saw any operation of the kind performed between the tropics: the only care they seemed to take of their fields of Indian corn, was to pull up foul weeds. In Europe it is usual to take away the sprouts which rise beside the principal stem; this precaution is also unnecessary in equatorial countries where the ground is fertile; the more lateral stems that rise, the better, as they all become richly laden with grain. I may also say as much for the system of topping which prevails among us, that system which consists in removing the extremity of the stem which bears the male flowers after the fecundation has been effected. The leaves and heads of stems which are obtained by this operation, compose a forage by no means to be despised.

The time during which the crop of maize remains on the ground, is greatly influenced by the mean temperature of the climate; in hot intertropical countries, the grain ripens in less than three months, and there are even farms upon which four considerable crops are gathered in the course of the year. On the temperate plateau or table-land of Bogota, the plant ripens in six months; in Alsace about the same length of time is required, although at Bechelbronn, in 1836, the maize which was sown on the 1st of June was gathered ripe on the 1st of October. Maize is dried either in exposing the spikes stripped of their covering upon the floor of a well-ventilated granary, or by hanging them up in bunches or sheaves under sheds, or under the eaves of the house. In warm countries the drying is accomplished by one or two days' exposure to the sun, after which the spikes are stored. The maize is freed from the stem with the hand in small farms, with the flail in larger establishments. In America the operation is never done until the moment when the grain is wanted, as it is said that the grain is less subject to be attacked by insects when it is kept in the ear. When animals are fed on maize, they are accustomed to separate it for themselves.

The produce in Indian corn varies greatly, as appears by the following table, in different countries:

Countries.	Produce in bushels per acre.
Lavanthal	81
Carinthia	55
Austria and Moravia	24
Hungary and Croatia	42
Tuscany	96
France (climate of Paris)	29
Alsace	43
Venezuela	147

By far the finest crops of Indian corn in America are obtained upon breaks of virgin soil. I do not hesitate to say that the husbandman gains from six hundred to seven hundred times his seed under such circumstances. The mode of proceeding upon these breaks, which I have frequently witnessed, deserves to fix attention for a moment.

The planter chooses the end of the rainy season for cutting down the trees and the brushwood: every thing remains where it falls until it is sufficiently dry; fire is then set to the heap, and the burning extends and lasts even for weeks; all the smaller branches are completely consumed, nothing but the charred trunks of the larger trees remain. As the rainy season is about to return, a man, with a pointed stick in his hand, goes over the burnt surface, making a hole of no great depth at intervals, into which he throws two or three particles of Indian corn, over which he draws a little earth, or rather ashes, by a slight motion of his foot. This primitive mode of sowing terminated, the planter takes no further heed of the crop; his habitation is often so remote, that he never visits it until harvest time: the rain and the climate do all the work. It is unnecessary to hoe, the burning having destroyed all the plants that were indigenous to the soil; nothing rises but the grain which has been sown. In such fields, stems of Indian corn are frequently seen of the height of from twelve to fourteen feet. It rarely happens that more than three consecutive crops are taken from the burnt soil; and the last, though still very superior to any thing which we can obtain by our regular husbandry, is not to compare with the first. As there is no want of forest, it is held preferable to make a fresh break.

Taking the seed as unity, it is found, from documents now possessed, that 1 of seed will yield—in Mexico (an indifferent harvest) 150; in New California (beyond the tropics) 80; Alsace (the plants very far apart) 190; Venezuela (an ordinary crop) 238. Besides the grain and the straw, the husks and the cores of Indian corn are all extremely valuable upon the farm as forage, and as affording manure.

Maize has been analyzed by M. Payen, and found to contain: starch 71.2; gluten, albumen, &c., 12.3; fat, oil, 9.0; dextrine and glucose 0.4; woody tissue 5.9; and salts 1.2; 100.0. I found 0.02 of azote in a sample of dry maize, which I analyzed, a quantity which indicates 12.5 of gluten and albumen, a result that coincides exactly with M. Payen's analysis.

Rice, (*Oriza sativa.*) Rice is an aquatic plant which can only be grown in low moist lands that are easily inundated. The ground is ploughed or stirred superficially, and divided into squares of from twenty to thirty yards in the sides, separated from each other by dikes of earth, about two feet in height, and sufficiently broad for a man to walk upon. These dikes are for retaining the water when it is required, and to permit of its being drawn off when the inundation is no longer necessary. The ground prepared, the water is let on, and kept at a certain height in the several compartments of the rice-field, and the seedsman goes to work. The rice that is to be used as seed must have been kept in the husk; it is put into a sack, which is immersed in water, until the grain swells and shows signs of germination; the seedsman walking through the inundated field, scatters the seed with his hand as usual, the rice immediately sinks to the bottom, and may even penetrate to a certain depth into the mud. In Piedmont, where the sowing takes place at the beginning of April, they generally use about fifty-five pounds of seed per acre. The rice begins to show itself above the surface of the water at the end of a fortnight; as the plant grows, the depth of the water is increased, so that the stalks may not bend with their own weight. About the middle of June this disposition is no longer to be apprehended; the rice is no longer so flexible as it was, so that the water can be drawn off for a few days to permit hoeing, after which the water is let on and maintained to the height of the plant; in July it is usual to top the stalks, an operation which renders the flowering almost simultaneous. Rice generally flowers in the beginning of the month of August, and a fortnight later the grain begins to form. It is at this period especially that the stalks require to be supported, and this is effectually done by keeping the water at about half their height. The rice-field is emptied when the straw turns yellow. The harvest generally takes place at the end of September. In the Isle of France, rice is cultivated in very damp soils, upon which a great deal of rain falls, but which are not flooded artificially. I have seen the same process followed in other tropical countries which I have visited, but I do not think that the produce is so great, or the crop so certain, as where inundation is employed. In Piedmont, the usual return from a rice-field is reckoned at about 50 for 1 of seed. At Muzo, in New Granada, the paddy fields, which are not inundated, under the influence of a mean temperature of 26° cent. (79° Fahr.) yield 100 for 1.

Three kinds of rice yielded, on analysis, the following quantities of—

	Carolina.	Piedmont.	Rice.
Starch	89.5	90.1	86.9
Gluten, albumen, &c.	3.8	3.9	7.5
Fatty matters	0.2	0.3	0.8
Sugar (glucose ?)	0.3	0.1	0.5
Gum	0.7	0.1	
Woody tissue	5.1	5.1	3.4
Phosphate of lime	0.4	0.4	0.9
Chloride of potassium, phosphate of ditto, &c.			
	100.0	100.0	100.0

M. Payen's analysis indicates a proportion of azote, the double of that found by M. Braconnot. In a trial for azote, which I made myself, I found 1.2 of this element per cent., which would show the amount of albumen and gluten to be 7.5, a quantity that corresponds exactly with M. Payen's valuation.

Coffee, (*Coffea Arabica.*) The habit of using the infusion of coffee appears to have been introduced into Europe about the middle of the sixteenth century. The first public establishments for the sale of the drink were opened in Constantinople, in the year 1554. The use of coffee remained for a long time confined to the East; but by degrees it spread, and at the present day the consumption of the article in Europe exceeds 660,000,000 of pounds annually. The greater portion of coffee consumed in Europe is the produce of America, and yet it is not more than a century since it was first grown in the New World.

The coffee-plant thrives between the tropics in situations where the mean and nearly constant temperature is between 22° and 26° C., (71.5° and 80° F.)

Coffee is rarely sown in a nursery; the seeds are made to germinate still surrounded by their natural pulp, and wrapped up in leaves of the banana. The young plants, after seven or eight days of germination, are put into the ground. In the valley d'Aragua an acre of ground of good quality is generally laid out with about 1040 plants. The coffee-plant flourishes in the course of the second year; when left to grow unimpeded it will attain a height of from 23 to 26 feet, but it is seldom allowed to grow so high, its upward progress being checked by pruning; the planters of Venezuela generally keep it at a height of from five to six feet. The shrub receives the care of the planter during the first two years; the ground must be kept free from weeds, and the growth of parasites must above all be prevented. To thrive, the coffee-plant requires frequent rains up to the time of flowering. The fruit bears a strong resemblance to a small cherry, and is ripe when it becomes of a red color, and the pulp is soft and very sweet. As the berries never ripen simultaneously, the coffee harvest takes place at different times, each requiring at least three visits made at intervals of from five to six days. A negro will gather from ten to twelve gallons of fruit in the course of a day.

Two beans are found in the interior of each berry; in order to free these from the pulp which surrounds them, they are passed through a kind of mill, and the coffee is steeped in water for twenty-four hours in order to free it from the mucilaginous matter which adheres to it; it is then dried by being spread out upon a floor under a shed. In the coffee plantations of Venezuela which I visited, I saw them proceed in another way. The berries were exposed to the sun upon a piece of ground somewhat inclined, and spread out to about three inches in thickness; the pulp soon enters into fermentation, and a very distinct vinous odor is exhaled, and the juice altered either flows away or dries up; at the end of a fortnight or three weeks the berries are all dry and shrivelled, and they then undergo two triturations, one to obtain the seeds or beans, the other to detach a thin pellicle which surrounds them. Three bushels of berries will yield from 85 to 90 lbs. of marketable coffee.

During the destruction of the sugary matter contained in the pulp of the berry, a considerable quantity of spirit is produced and dissipated. M. Humboldt, struck with the readiness with which the berry of the coffee-plant runs into fermentation, expresses his surprise that no one ever thought of obtaining alcohol from it. In an old work, however, I find the following passage: "The inhabitants of Arabia take the skin which surrounds the coffee bean and prepare it as we do raisins; they form a drink with it for refreshment during the summer."* This vinous liquor appears to enjoy all the exciting properties which are esteemed in the infusion of coffee.

The coffee-plant continues to produce to the age of forty to forty-five years; it bears to a considerable extent even in the third year. Some shrubs yield from 17 to 22 lbs. of dry coffee beans; but this is a very large quantity. An acre of land in the valley d'Aragua, planted with about 1040 shrubs, will yield about 940 or 950 lbs., which is at the rate of somewhat less than 1 lb. per shrub.

Coffee contains the same active principle as tea, coffeine, but in less proportion; the researches of different chemists have also shown the presence of a particular acid called coffeic acid, of fatty matters, a volatile oil, a coloring matter, albumen, tannin, and alkaline and earthy salts.

Cocoa, (*Theobroma cacao.*) The ancient Mexicans cultivated the cocoa-tree, and with its seeds prepared tablets similar to the chocolate of modern times. The use of cocoa appears to have been introduced after the conquest into the other parts of the continent; nevertheless, the cocoa-tree is indigenous in the hot and humid forests of South America. M. Goudot discovered several species in New Granada; among others, that which is known at Muso under the name of the *Cacao montaraz*: this cocoa-tree, which attains a height of from 25 to 30 feet, yields a considerable quantity of fruit; the natives prepare a chocolate from its beans, which is extremely bitter, and which they regard as an excellent febrifuge. The wild Indians still appear to be ignorant of the profit that may be made of the seeds of the cocoa-tree; they only eat the pulp of fruit which surrounds them. Cocoa was introduced into Europe by the Spaniards, and in no long space of time this production of the New World became the object of a very considerable traffic.

It is a fact well known to the husbandmen of tropical countries, that a virgin soil is quite indispensable to the success of a cocoa plantation; nothing but failure has followed attempts to replace the sugar-cane, indigo, maize, &c., with cocoa, a plant which to succeed requires a rich, deep, and moist soil, heat and shade; nothing suits it better than a forest brake, the surface of which is susceptible of irrigation.

All the important cocoa plantations which I visited had a common physiognomy: they were all situated in the hottest regions, at a short distance from the sea, near torrents, or on the banks of great rivers. The cocoa husbandry ceases to be profitable in localities which have not a mean temperature of at least 24° C., (75.2° Fahr.,) and I have had occasion to take part in attempts that were as fruitless as expensive to cultivate the cocoa-tree in a brake where the heat of the climate from my own observations did not exceed 22.8° C. (73° Fahr.) Under the influence of this temperature, the trees presented a very good appearance; in the course of a few years they flowered, but the fruit, which was always small, rarely came to maturity. When a piece of land has been selected for a cocoa

* Mem. of the Academy of Inscriptions, vol. xxiii p. 214.

plantation, they begin by establishing a good system of shade. Occasionally a certain number of trees, with large and leafy crowns, are left standing; but in general certain plants, which grow rapidly, are had recourse to as a means of procuring shade. In the neighborhood of Caraccas they shade with the *erythrina umbrosa;* and in some plantations they take advantage of the shade of the banana; finally, the two modes of procuring shade are frequently conioined.

In the province of Guayaquil they plant the beans of the cocoa directly. In Venezuela they prefer raising the plant in a nursery, which is always selected of the most fertile soil, and deeply trenched. The seeds are sown immediately before the setting in of the rains, and germination takes place in from eight to ten days. In a good soil, at two years of age the cocoa-plant will have attained a height of nearly 3 feet; it is then pruned by having two of its upper branches removed, and is transplanted. In the upper valley of the Rio Magdalena, where there are many valuable cocoa-groves, the sowing is performed in ground well prepared and protected by screens made with palm leaves; here the young cocoas are transplanted when they are six months old. During the whole of the time that the plants remain in this nursery they continue to be well shaded; and they are watered once a week by water poured upon the screens.

The tree seldom comes into flower under thirty months old. I have known planters who always destroyed these first flowers, and who never suffered any fruit to ripen before the fourth year, and that too under the most favorable circumstances in regard to climate, in situations where the mean temperature was 27.5° C., between 81° and 82° Fahr. In less favorable situations it is necessary to wait six or seven years before gathering the first fruits of a cocoa plantation. There are few arborescent plants which have so small a flower, and especially a flower so disproportionate to the size of its fruit, as the cocoa-tree. The diameter of a bud, measured at the moment of its expansion, does not exceed 4 millimetres—0.157 of an English inch. The flowers appear principally upon the trunk of the tree itself; they rarely show themselves beyond the middle of the larger branches; occasionally they appear upon the roots which happen to be above the ground.

To receive the young plants grown in the nursery, the ground properly shaded is first freed from weeds. Trenches are then cut, either to season the ground or to irrigate it when requisite. The young plants are set in rows at regular and considerable distances, which vary, however, with the quality of the soil; and the general opinion is, that the better the soil the greater should be the space from tree to tree. Thus in the valley of del Tuy, in the neighborhood of Puerto Cabello, the cocoa-trees are set at the distance of about 16 feet apart in the best soils, and at the distance of about 13 feet only in soils of inferior quality. In the windward islands, where the soil is generally less fertile than on the continent, the trees stand at the distance of from 6 or 7 to 9 or 10 feet apart. A reason for this practice may be readily assigned; in the more fertile soils the trees grow more vigorously, the branches spread further, and consequently require a larger space.

Once the cocoa-tree is in the plantation, it is regularly pruned to prevent its branches becoming too numerous. It sometimes happens that the branches show a tendency to bend down towards the ground, in which case they are fastened up around the trunk, until they acquire strength and a better direction. The soil around the trunk is hoed from time to time to the extent of about a yard in circumference, and the capillary roots, which spring from the base of the trunk, are removed in the course of the operation.

From the fall of the flower to the complete ripeness of the fruit there elapses an interval of four months. The fruit is of an elongated form, slightly bent, and terminated in a point; its length is about 9 inches, and its greatest diameter, which is near the point of attachment, is from 6 to 7 inches. Externally, the cocoa-nut pod is furrowed longitudinally. Its color varies from a greenish white to a reddish violet, the latter being the more common tint. Internally the flesh of the fruit is generally white, although it has sometimes a rose-color; it is sweet and acid, and of a very agreeable flavor. The seeds are generally twenty-five in number in each fruit, and at first are white; they are oleaginous and slightly bitter; in drying they acquire a brown tint. The fruit is known to be ripe by its color, and particularly by the ease with which it is gathered from the tree. There are two grand cocoa harvests in the course of the year, at six months' interval; still, in old and large plantations the harvest is almost incessant, as it is not uncommon to observe, on the same cocoa-tree, ripe fruits and fresh flowers. To obtain the seeds the fruit is opened with a piece of wood, having a rounded extremity. The produce is classed according to its quality, care being taken to throw out all the beans that are not sufficiently ripe or that are damaged; they are then exposed in the sun. Every evening the day's gathering is collected into a heap under a shed, and a brisk fermentation is soon set up, which would become destructive were it suffered to continue. Next day the heap is scattered, and the drying goes on in the sun, several days' exposure being required before the drying is complete. Occasionally the drying is retarded and rendered difficult by the occurrence of rain, and there would certainly be many advantages in effecting it by the stove. It has been found that 100 lbs. of fresh beans give from 45 to 50 lbs. of dry and marketable cocoa. In Venezuela, a cocoa-tree which is over seven or eight years old, will yield annually for more than forty years over 1½ lb. (1.65 lb.) of dry and marketable cocoa. An acre of ground, which in good plantations will be set with about two hundred and thirty-three trees, produces in a middling year about 383 lbs. weight The cocoa-tree appears to yield most abundantly when it is abou twelve years of age, and its produce in the fertile lands of Upper Magdalena, according to M. Goudot, is greatly superior to what it is in Venezuela. At Gigante, for example, each adult tree yields

4.4 lbs. of dry cocoa annually, and the produce of an acre there may be estimated at 733 lbs.

Cocoa beans contain albumen, a particular principle, theobromine, analogous to coffeine, a coloring matter, and a large quantity of oil or fat, which, from experiments made in my laboratory, appears to amount to 43 per cent. The presence of a large quantity of albumen and fatty matter in cocoa explains its highly nutritious qualities. It is indeed one of the most wholesome and restorative articles of sustenance known. Nevertheless, very opposite statements have been made upon the virtues of cocoa or chocolate, of which the bean forms the basis. Benzoni, in his History of the New World, declared chocolate to be a drink that was fitter for hogs than men; and Father Acosta declares the taste for cocoa to be unreasonable. On the other hand, Fernando Cortez and one of his gentlemen followers are perhaps guilty of exaggeration when they say, "that he who has taken a cup of chocolate may march the rest of the day without other aliment!"* Without going the whole of this length with Cortez, I still allow that chocolate is one of the best articles for travelling upon, especially in the uninhabited forests of South America, where it is a matter of the highest moment to have the bulk and the weight of necessary rations as small as possible.

Seeds of leguminous plants. The leguminous plants that are cultivated as food for man are beans, peas, haricots, and lentils; vetches are grown exclusively for the use of cattle.

Leguminous plants scarcely ever open rotations; but they very often wind them up. Speaking generally, however, they may follow any crop. In speaking of the Indian corn, I have said that haricots and beans might be advantageously intercalated.

The meteorological observations I have made in different countries lead me to conclude that to succeed, leguminous plants require a temperature which in the mean does not fall below from 14° to 15° C., (57° to 59° F.) Hot climates agree with them perfectly; I have followed them from the sea-board of the equatorial Andes to a height of from 8200 to 9800 feet above the level of the sea. Schwertz has given the following statement of the produce of the different leguminous plants generally cultivated:

Plants.	Weight per bushel in lbs.	Produce per acre in bushels.	Weight of dry straw or haulm per acre.			
			Tons.	Cwts.	qrs.	lbs.
Haricots . .	47.5	66.7				
Beans . . .	65.5	66.2	2	2	2	17
Peas . . .	57.9	38.5	2	4	2	11
Lentils . . .	62.3	39.8				
Vetches . . .	62.3	41.2	2	4	2	11

The analyses we have of leguminous vegetables show the following proportion of elements:

* Humboldt, Travels, vol. v. p. 285.

	Haricots	Peas.	Lentils.
Legumine . . .	22.0	20.4	22.0
Starch	41.0	47.0	40.0
Fatty matters . .	3.0	2.0	2.5
Sugar (glucose ?) . .	0.3	2.0	1.5
Gum	4.0	5.0	7.0
Woody fibre, pectic acid	8.0	11.0	12.0
Salts, phosphates, &c.	3.2	3.0	2.5
Water and loss . .	17.5	9.6	12.5
	100.0	100.0	100.0

Besides these principles, a quantity of tannin has always been found in the skin of the seed of all leguminous plants.

The Hop, (*Humulus lupulus.*) From its very general use in making beer, the hop has become an object of great importance, both in an agricultural and commercial point of view.

The hop may be cultivated in any soil that is of sufficient depth and fertility; it thrives especially in rich and turfy loams, such as those of Haguenau, where there are many beautiful hop-gardens. The plant is propagated in the spring by setting the sprouts or radicular buds in ground trenched to the depth of 18 inches at least, at intervals of about a couple of yards from one another. Within a few weeks the young hop-plant is growing lustily; and as it is a climber, it is trained upon a pole of from 12 to 20 feet in height. The ground is usually hoed towards the end of June. The first crop from a new plantation is always trifling in amount; the ground is then manured. The following spring all the eyes or buds that have become developed near the root are removed, except six or seven, which are left to shoot. The hop harvest generally occurs about the middle of September: the poles are pulled up, the stems are cut, and the strobiles are picked off into baskets by hand, and immediately carried to the stove or kiln, where they are dried with a very gentle heat, in order not to dissipate their fine aromatic particles.

A hop-garden produces very variously in different countries and districts, and in different years. The produce of an acre in hops has been stated to be:

In Flanders, ...	13 cwt.
Germany (mean of 10 years,) . .	10 "
France (near Paris,) . . .	10 "
" (Roville, mean of 10 years,)	7¼ "
[England, from 9 to 10, and from 12 to 14 cwt.]	

The strobiles of the hop are covered with a yellow pulverulent substance, which has been held to furnish in principal part the extractive matter that is so valuable in brewing. To procure this substance it is enough to sift a quantity of hops after they have been dried by a gentle heat. This yellow powder, which appears to be the useful principle in the hop, and consequently gives it its value is not found in the same proportion in the produce of all hop-gardens. This clearly appears from the inquiries of Messrs. Payen and Chevalier. They found, for example, that while 100 parts of the hops of Belgium contained 18 of yellow substance and 70 of mere leaf, those of England contained no more than 10 of yellow matter and 87 of leaf, and those of Germany the still smaller quantity of 8 of yellow matter to 88 of leaf. This yellow pulverulent matter contains wax, resin, gum, a bitter principle, certain azotized principles, a volatile oil, and salts, among others acetate of ammonia.

FLESHY OR PULPY FRUITS.

The fleshy fruits almost all contain the same principles, but in very different proportions. It is consequently the predominating principle which in some sort characterizes each variety, that gives it its flavor, odor, &c.: sugar, albumen, gum, starch, acids, fixed oils, essential oils, woody fibre, are almost invariably found secreted in their pulps, with a larger or smaller quantity of water. An ingenious classification of fruits has been formed on the basis of the predominance of the different substances which have just been enumerated: thus those fruits in which the starchy principle predominates are feculent or amylaceous fruits; those in which the sugar predominates are saccharine fruits, and so on.

M. Bérard has analyzed a great number of fruits in the course of his researches on their ripening.* It is proper to say, however, that some of the principles brought to light by modern analysis do not figure in M. Bérard's list of elements; among the number, pectic acid, gallic acid, small quantities of volatile oils, and of salts of potash formed by vegetable acids.

	Apricots.	Peaches.	Currants.	Cherries.	Green gage plums.	Pears.
Albumen and gluten	0.2	0.9	0.9	0.6	0.3	0.2
Coloring matter .	0.1				0.1	
Vegetable tissue .	1.9	1.2	8.0	1.1	1.1	2.2
Gum . . .	5.1	4.9	0.8	3.2	2.1	2.1
Sugar . . .	16.5	11.6	6.0	18.1	24.8	11.5
Malic acid . .	1.80	1.1	2.4	2.0	0.6	0.1
Citric acid . .	"		0.3			"
Lime . . .		0.1	0.3	0.1		
Water . . .	74.4	80.2	81.3	74.9	71·0	83.9
	100.00	100.0	100.0	100.0	100.0	100.0

Banana. Of all the pulpy fruits, the banana is that perhaps which is most extensively used as food by man. It is the usual nourish-

* Ann. de Chimie, t. xvi. p. 225, 2d series

ment of the inhabitants of most of the countries between the tropics, where its cultivation is as important as that of the cereals and farinaceous roots in the temperate zone. The ease with which it is cultivated, the small space of ground it occupies, the certainty, the abundance, and the continuance of its produce, the diversity of food it yields according to the degree of maturity, make the banana an object of admiration to the European traveller. In climates where man scarcely feels the necessity of clothing himself, or of raising a shed for his protection, he is seen gathering almost without labor supplies of food as abundant as they are wholesome and varied from the banana-tree. It is the banana which has given rise to that proverb so consoling and so true, which is frequently heard between the tropics, viz. "No one dies of hunger in America;" he who is hungry will be welcomed and fed in the very poorest cabin. Botanists distinguish three principal varieties of the banana: 1st. the *Musa paradisica;* 2d. the *Musa sapientium;* 3d. the *Musa regia.*

The American origin of the banana has been called in question. Oviedo in his natural history of the Indies affirms that it was brought from the Canary Isles to St. Domingo by a monk. Foster adopted this opinion, which is corroborated, says M. de Humboldt, by the complete silence of the first travellers who visited the New World in regard to it. Nevertheless, the testimony of the Inca Garcilasso de la Vega proves obviously that the banana flourished in America before the arrival of the Spaniards; in his royal commentaries he speaks of the banana as constituting the chief food of the Indians in the warmest parts of Peru.

The banana is everywhere cultivated in the neighborhood of the equator, in situations at no great height above the level of the sea. The cultivation is most profitable, the crop is most abundant, and attains maturity in the shortest space of time in low lying districts where the mean temperature is from 24° to 27.5° C., (75.5° to 82° F.) Some estimate may be formed of this from the low price of the banana in such districts; upon the borders of the great river de la Magdalena, I gave one franc or about 10d. for about 220 lbs. weight of the fruit. The day's wages of a man being generally about 1s. 8d., it is beyond all doubt the cheapest food that can be had in the world.

In looking at the cultivation of the banana at different heights in the equatorial Cordilleras, I arrived at the following conclusions:

Temperature 28° C. (between 82° and 83° F.) the cultivation extremely advantageous; at 24° C. (between 75° and 76° F.) the cultivation advantageous; at 22° (71° and 72° F.) the cultivation middling; at 19° C. (or between 66° and 67° F.) the cultivation disadvantageous.

The banana is propagated by means of suckers or offsets. It requires a rich and humid but well-drained soil, the plantation being arranged a little before the setting in of the rains. The earth is freed from weeds, and dug either entirely or more generally only at regular distances here and there, where it is proposed to set the new plant, a space of 6 feet at least being left between each. The plant throws up several shoots, generally 6 or 7, each of which will be allowed to grow and to carry fruit; when a greater number make their appearance, some of them are cut away. The time which passes between planting the slip and gathering the fruit varies according to the situation; in the hottest districts near the level of the sea the banana comes into flower about nine months after it has been planted; and in three months more the fruit has formed and become ripe. In cold situations an interval of four months will elapse between the flowering and the ripening of the fruit. The care required by a banana plantation is not very great, the principal duty being to hoe around the young plants. As the banana is renewed by stems which arise continually from the neck of the root, it is easily understood that the plant will go on yielding fruit for an indefinite length of time; when the fructification is complete in one stem, the leaves, &c., wither and fall, and give place to a new stem. It is thus that the gatherings from the banana go on successively at short intervals, and that the same plant presents at one and the same moment fruit that is ripe, fruit that is half ripe, fruit that is beginning to be formed flowers, and finally young stems, which are rising as preparations for the future. Thus no crop is more assuring to the planter than the banana. Climatic circumstances may sometimes delay, but can never destroy the hopes of the husbandman. The extraordinary droughts which under the burning climates of the equator so frequently interrupt or destroy ordinary herbaceous plants, rarely exert any pernicious influence upon the banana plantation, the thick shade of which presents a constant obstacle to the evaporation of moisture. During the dry season, when for whole months the heavens preserve their purity, and no drop of rain falls to refresh the earth, the soil which surrounds the banana still continues moist. It looks every morning as if it had been watered during the night; this salutary effect is produced by the nocturnal radiation of the leaves into the clear sky. These leaves, whose extent of surface is considerable, always fall several degrees below the temperature of the surrounding air, and thus condense the watery vapor contained in the atmosphere, which drips down to the foot of the plant.

The produce of a banana plantation depends first upon the distance at which the bananas are placed, and next upon the climate. It is generally estimated in the very warm climates, that a crop of bananas will weigh about 44 lbs., and that from an adult plant three crops will be obtained in the course of a year. In temperate countries, and towards the superior limits of the banana plant, they do not reckon on more than two crops. According to M. de Humboldt, the produce per acre, in hot countries where the mean temperature is about 82° Fahr., will amount to 75 tons, 8 cwt. 1 qr. 17 lbs.; at Cauca, where the temperature is about 79° Fahr., the produce amounts to 61 tons, 8 cwt. 0 qr. 2 lbs.; at Ibagué, where the temperature is not higher than about 72°, the produce, according to M. Goudot's estimate, is 26 tons, 17 cwt. 3 qrs. 2 lbs. The pulp of the banana is surrounded by a pod or husk of some thickness, which is easily detached, and of which account must be taken if we would estimate the actual weight of the truly alimentary matter afforded. In a general way, and when the banana is ripe, the shell may be estimated at about 36.8, the edible banana at 73.2 per cent.

The *Musa paradisica* is the variety of banana generally cultivated, and it also yields the heaviest crops. The fruit of the other two varieties mentioned is much smaller; but it is of a much more delicate flavor. The ripe fruit of the banana is of the consistence of a pear; it is very sweet, and slightly acid. In the common variety, I found crystallizable sugar, gum, an acid, (probably the malic,) gallic acid, albumen, pectic acid, woody fibre, and alkaline and earthy salts. Dried in the sun, 1000 parts of ripe banana were reduced to 439 parts; so that they contained 561 parts of water. The green or unripe banana has a white and almost insipid flesh. In this state it scarcely contains any sugar; it is starch that predominates. In this state, therefore, it is made a substitute for bread, for the potato, or Indian corn; it may be considered a farinaceous vegetable. After having removed the rind, the banana is dressed by being roasted under the ashes until the outer part is slightly brown; it is then served up at table, and constitutes a kind of soft bread, very agreeable to the palate, and greatly preferable, in my opinion, to the produce so much vaunted of the bread-fruit tree. In the expeditions which are undertaken into the forest, and when the habitations of man are to be quitted for some considerable time, the green banana is always made a principal part of the provision; but then it is previously dried, first to lessen its weight, and then to destroy its vitality so far as to prevent its ripening. This drying is performed in a baker's oven, into which the green bananas, stripped of their husks, are introduced, and where they are kept for about eight hours. On being taken out, the bananas are hard, brittle, translucent, and present the appearance of horn; 100 lbs. of the green fruit give but 40 of dry substance. The banana thus prepared is called *fifi*, and will keep for a great length of time without change. To prepare it for food, it is put to steep in water, and then boiled; by adding a little salted meat, a very substantial and nutritious meal is prepared. I once made a voyage on the Pacific, in a vessel which was principally victualled with dried bananas, which were served out to the company like biscuit.

When ripe, the banana is no longer farinaceous; as it ripens, its starch is changed into gum and sugar, and an acid is developed. But between the farinaceous and the sugary or perfectly ripe state, there is one intermediate, in which it is generally eaten. Roasted in the ashes, the banana has then a taste which brings to mind that of the chestnut; it is also eaten as a vegetable, boiled in the usual way in water. Completely ripe, the fruit is eaten raw or dressed, it is then extremely sweet; a very common practice is to fry it, cut in slices, in grease.

I have no data upon which to estimate the nutritive value of the banana, still I have reasons for believing that it is more nutritious than the potato. I have seen men do a great deal of hard labor upon an allowance of about 6½ pounds of half-ripe bananas, and two ounces of salted meated per diem.

CHAPTER III.

OF THE SACCHARINE FRUITS, JUICES, AND INFUSIONS USED IN THE PREPARATION OF FERMENTED AND SPIRITUOUS LIQUORS.

THE juice of all the sweet fruits when expressed and left to itself under the influence of a suitable temperature, presents the remarkable phenomenon of fermentation, in the course of which the sugar disappears completely, and is replaced by alcohol, the change from first to last being accompanied by the disengagement of carbonic acid gas.

Sugar alone does not suffice to cause the vegetable juices, which contain it, to ferment: for example, a solution of pure sugar in distilled water will remain for a very great length of time without suffering the least change; exposed to the open air it would evaporate, and the saccharine matter would be found in the same state as it was before solution. If, however, a small quantity of that azotized principle which we have called albumen, gluten, &c., be introduced into the solution, fermentation will speedily be set up, and will run through its usual course; it would, therefore, appear to be upon this principle that the commencement and continuance of fermentation depends. Fermentation is not set up immediately in the juice of fruits; a certain time longer or shorter always elapses before it is manifested; the reason of this is, that the albumen or gluten which always enters into the constitution of these juices, must itself have undergone a certain change in order to act as a ferment. The proof of this is comprised in the fact that all vinous liquors contain a very small but constant quantity of carbonate of ammonia, as was shown by M. Doebereiner. These azotized principles, which in the fresh state remain without action upon sweet juices, act immediately as powerful ferments when they are employed after having been exposed for some days to the contact of air and moisture; after, in a word, they have themselves begun to suffer change. The quantity of ferment used up or consumed in exciting and maintaining the fermentation of saccharine juices is so small, that we are led to believe that it really acts by its presence or contact alone. This view appears the more likely, when we know that, after having added an azotized substance to induce fermentation rapidly in a liquid which, besides sugar, contains albumen, we find from six to eight times the quantity of ferment after the phenomena have ceased, which had been added in the first instance; that is to say, we find the whole, or almost the whole, of the original ferment, and, in addition, that which has been produced by the azotized principles pre-existing in the matter subjected to fermentation; this fact is seen every day in the process of making beer.

The ferment or yeast thus produced is but little soluble in water, and in composition bears a remarkable affinity to the azotized matters from which it is derived; M. Dumas has in fact found it to be composed of:

Carbon	50.6
Hydrogen	7.3
Azote	15.0
Oxygen, Sulphur, Phosphorus	27.1
	100.0

Under the influence of ferment, sugar becomes entirely changed into alcohol and carbonic acid. The composition of grape-sugar—which appears to be the only one that is susceptible of fermentation, for cane-sugar before undergoing this process passes into the state of grape-sugar, as was demonstrated by M. Henry Rose—the composition of grape-sugar is as follows:

Carbon	36.4
Hydrogen	7.0
Oxygen	56.6
	100.0

and the constitution of the substances which are produced in the process of fermentation, viz. alcohol and carbonic acid, being as under:

	Anhydrous alcohol.	Carbonic acid.	Water.
Carbon	52.19	27.27	
Hydrogen	13.02	"	11.1
Oxygen	34.79	72.73	88.9
	100.0	100.0	100.0

It appears that the composition of 100 parts of grape-sugar may be expressed by:

			Carbon.	Hydrogen.	Oxygen.
Alcohol	46.16	containing	24.24	6.05	16.17
Carbonic acid	44.45	"	11.12	"	32.33
Water	9.09	"	"	1.01	8.08
	100.00		36.36	7.06	58.58

by which it would appear that during the transformation of hydrated grape-sugar into alcohol and carbonic acid, the combined water is set at liberty.

The first fermented vegetable juice of which I shall speak is *cane-wine*, or *guarapo* of the South Americans, a drink which is in common use wherever the sugar-cane is cultivated. It is prepared from the juice of the sugar-cane suffered to run into fermentation.

The *chicha* of South America is a fermented liquor prepared from Indian corn, and constitutes the wine of the Cordilleras. The grain is steeped for six or eight hours in water, bruised upon a stone and boiled; the pulp which results is then diffused through 4½ times its volume of water, and the temperature being from 60° to 65° F., a violent fermentation is soon set up in the fluid, which begins to subside after a period of twenty-four hours, when the chicha is potable, and now constitutes a liquor of an agreeable and decidedly vinous flavor, in high repute with those who have acquired a taste for it, although its muddy appearance and the sediment which it alway lets fall in the vessel into which it is received, render it somewhat unpleasant at first to European eyes. The Indians, however, always drink it in the muddy state, and even shake the cask before turning the tap. The truth is, that chicha is at once a drink and a very nutritious food.

Guarazo is another vinous liquor which the Indians prepare with rice much in the same manner as they proceed with Indian corn.

Cider and *Perry*. In countries where the vine is not cultivated, a substitute for wine is found in the fermented juice of a variety of sweet pulpy fruits, more particularly of apples and pears. Of the numerous varieties of apples which are grown in cider countries, the preference is generally given to one which has a rough and somewhat bitter taste. The fruit is gathered by shaking or beating the trees, and the few that remain are taken off by the hand; the fruit is piled up in large backs placed in cellars. It is crushed about two months after it is gathered, and the pulp is left for ten or twelve hours to macerate in the juice, in order to give the rusty or yellow color which is esteemed in cider. The pulp is pressed and the juice is run into large vats or tuns, in which it undergoes fermentation, which having gone on for about a month, the temperature being from 55° to 58° F., the liquor is racked off into smaller vessels, in which the fermentation goes on slowly, and the cider is preserved. The fermentation of cider is, or always ought to be, slow; still, with time, the whole of the sugar is transformed into alcohol, if the process be not interfered with.

Wine. Grape-juice contains—1st. grape-sugar; 2d. albumen and gluten; 3d. pectine; 4th. a gummy matter; 5th. a coloring matter; 6th. tannin; 7th. bitartrate of potash; 8th. a fragrant volatile oil, or cream of tartar; 9th. water. It is obvious, therefore, that grape-juice contains within itself the elements necessary for the production of the vinous fermentation. The relative proportions of these different elements, however, are singularly modified according to the nature of the vine, the quality of the soil, and especially the heat of the climate. There are indeed few crops that are so much at the mercy of the atmosphere as that of the vine; even in the vineyards that are most favorably situated, it is rare that wines of equal quality and flavor are produced in two consecutive years; and in districts upon the verge of the productive limits of the vine, under what may be called extreme climates, where the vine only exists in virtue of hot summers, its produce is still more variable, more inconstant. The limits to the culture of the vine in Europe are generally fixed where the mean temperature is from 10° to 11° C., (50° to 52° F.;) under a colder climate no drinkable wine is produced. To this meteorological datum must be added the further fact that the mean heat of the cycle of vegetation of the vine must be at least 15° C. (59° F.,) and that of the summer from 18° to 19° C., (from 65° to 67° F.) Any country which has not these climatic conditions cannot have other than indifferent vineyards, even when its mean annual temperature is above what I have indicated. It is impossible, for instance, to cultivate the vine upon the temperate table-lands of South America, where they nevertheless enjoy a mean temperature of from 17° to 19° C., (about 62.6° to 66.2° F.,) because that which characterizes the climate of these elevated equinoxial countries is the constancy of the temperature; the vine grows, flourishes, but the grapes never become thoroughly ripe. In these equatorial countries good wine cannot be made where the constant temperature is not at least 20° C., (or 68° F.)

In France the vine begins to sprout towards the end of March, and the vintage generally occurs in the course of October. As the quality of wine depends mainly on the ripeness of the grapes, of course the vintage does not take place until this is complete, or until there is no longer any prospect of improvement.

The must of the grape is procured by treading and pressing the fruit; the juice is run into vats, and the fermentation takes place in cellars; different procedures, however, are followed in different places. The fermentation having subsided in the larger vessels, the wine is drawn off into smaller casks, which are carefully filled up from time to time, and in which it is preserved.

Wine may be defective, especially by wanting strength and being too acid. Sharp wine contains an excess of cream of tartar and free vegetable acids, and is always the produce of grapes which have not been completely ripe. The deficiency of strength is due to the same cause; for it is well known that as the grape ripens its acids disappear and are replaced by sugar. This deficiency of saccharine matter in the must, is now habitually supplied by the addition of a quantity of artificial grape-sugar, prepared from starch. In warm countries, where the grape always ripens, the quantity of tartar is small; the sugar then predominates greatly, sometimes to such an extent that the azotized substance of the must is insufficient as a ferment, and it is then that we have wines of too sweet a flavor, such as those of Lunel and of Frontignac. When these musts, which are so rich in sugar, contain the proper quantity of ferment they produce very strong wines, in which, of course, the sweet flavor no longer predominates; such are the dry wines of southern vineyards, of which that of Madeira may be taken as the type. There are some wines which participate at once in the properties that distinguish the two varieties that I have mentioned, or that show one of them in excess according to circumstances; such are the wines of Xeres, Alicant, Malaga, &c. Some of these wines are what are called boiled wines, that is to say, a portion of the must, as it flows from the press, is concentrated to a fourth or a fifth of its original bulk by boiling, and this being added to the rest, the strength of the resulting wine is increased. Sometimes the concentration of the juice is effected by drying the grapes partially. It is in this way that the celebrated Hungarian wine, called Tokay, is prepared; the clusters are left upon the vines after they are ripe, and alternately exposed to the cold of the night, which probably decomposes to a certain extent the texture of the grapes, and to the heat of the sun. They shrivel and become partially dry. In this state the grapes are subjected to pressure, and a very sweet must, as may be conceived,

flows from them. In less favorable climates, where the rains of autumn prevent the drying of the clusters upon the vine stocks, the same thing is effected by laying the bunches upon straw in open or well-aired granaries or sheds. It is with the must procured from grapes so treated, that the sweet and often strong wines, which are called *vins de paille*, or straw wines, are obtained. Wines when stored in the cask always deposite with time a copious sediment, the lees. This sediment, in which tartar predominates, appears to be the consequence of an increase in the proportion of alcohol in the liquor. The alcohol may increase from two causes: first, by the fermentation which, though nearly insensible, goes on in most wines so long as there is any sugar left unchanged; and next from mere keeping. It is well known, in fact, that wine put into the best casks, and kept in a well-ventilated cellar, loses a very perceptible quantity by evaporation; it is found necessary to fill up the casks from time to time: the loss has taken place through the pores of the wood, in virtue of an attraction exerted between the substance of the wood and the included liquid; and as this attraction is much greater between the organic matter and water, than between organic fibre and alcohol, it is easy to conceive how wine kept in wood should improve. The very same thing, in fact, appears to go on in regard to wine in corked bottles: the cork does not oppose all evaporation, and it seems probable that it is not merely upon some new and little known change of a chemical nature in the constitution of the wine that its improvement and mellowing in bottle depend, but also upon the loss of a certain quantity of its water through the pores of the cork.

Throwing quality, flavor, &c., out of the question, it is well known that a vineyard, culivated in the same way, year after year, receiving the same quantity of the same kind of manure, of which the vintage is managed in the same manner, the wine made by the same method, &c., yields a produce which differs greatly in regard to the quantity of alcohol it contains in different years. The vineyard of Schmalzberg, for example, near Lampertsloch, which has been under my management for several years, yields wines of the most dissimilar characters from one year to another. Some idea of this may be formed from the different quantities of alcohol which the wine of different years contains:

Years.	Mean temperature.						Wine per acre in gallons.	Pure alcohol per cent.	Pure alcohol per acre, in gallons.
	Of the whole term of the growth of the vines.		Of the summer.		Of the beginning of autumn.				
	deg.	deg.	deg.	deg.	deg.	deg.			
1833	14.7C.	58.4F.	17.3C.	63.1F.	11.4C.	51.5F.	311	5.0	11.4
1834	17.3	63.1	20.3	68.½	17.0	63	314	11.2	46.3
1835	15.8	60.2	19.5	67	12.3	54	621	8.1	50.0
1836	15.8	60.2	21.5	71	12.2	54	544	7.1	38.6
1837	15.2	59.5	18.7	66	11.9	54	184	7.7	14.0

If we now inquire how the meteorological circumstances of each of these five years influenced the production of our wine, we see at once that the mean temperature of the days which make up the period of the cultivation of the wine has a perceptible influence. The temperature of the summer was 17.3° C. (63.1° Fahr.) of the year which yielded the strongest wine, and only 14.7° C. (58.4° Fahr.) in 1833, the wine of which was scarcely drinkable.

A hot summer is naturally favorable to the vine: the mean heat of 1833 did not exceed 17½° C. (63½° Fahr. ;) with the exception of this year, which must be regarded as one of the very worst, the three favorable summers, 1834, 35, and 36, show a mean temperature of about 20° C. (68° Fahr.) It is not, however, with the warmest summer that we find the strongest wine to correspond. Besides the sustained heat, which is necessary during the whole year's growth of the vine, it would appear that a mild autumn was a condition necessary to the perfect ripening of the grapes: this is one of the essential conditions. We see, in fact, that in 1834, the months of September and October presented the extraordinary temperature of 17° C., (62.6° Fahr.,) while in 1833, the temperature of the same months did not rise higher than 11.4° C. (51.5° Fahr.) I shall here add, that the year 1811, so remarkable over Europe for the quantity and the excellence of its wines, was distinguished by the high temperature of the early part of its autumn; we find, in fact, from the excellent series of observations with which M. Herrenschneider has presented Alsace, that in this year, after a summer the mean temperature of which was 19.6° C. (67.8° Fahr.,) the heat of the months of September and October was maintained at 15° C. (59° Fahr.,) the usual temperature of the months of September and October not being higher than about 11.5° C. (52.7° Fahr.)

If we deduct from these observations the years 1833 and 1837, which were decidedly bad, it seems that we must conclude that meteorological influences have a greater effect upon the quality of wines, than upon the whole quantity of alcohol formed; thus, although the wine of 1836 was very inferior to that of 1834, it actually yielded a larger proportion of alcohol from the acre.

In Alsace, in order that a year may be favorable to the vine, the temperature of those months during which the plant is alive must be sensibly superior to the mean: a fact which appears from M. Herrenschneider's long series of observations. In a climate where the vine requires such a condition to succeed, it is obvious that its cultivation can never be advantageous; and this, in fact, is the case · the cultivation of the wine would, indeed, be altogether ruinous, were it not for the circumstance that the value of wine increased in a much greater ratio than its quality, so that one good year often indemnifies the grower for many bad years. Another consideration is this, that the vine, like the olive, grows and thrives in situations where it would be difficult to put any thing else.

The produce of a vineyard also depends upon its age; and it would be curious to examine the progressive increase of the quantity of wine yielded. This information I am able to give in connection with a vineyard established in Flanders; I only regret that I have no means of presenting parallel observations from a country more favorable to the vine. The vineyard of Schmalzberg was planted in 1822, with new cuttings from France, and from the borders of the Rhine. The vines are trained as espaliers, and are now rather more than four feet in height. The vineyard began to yield wine in 1825, and the following table shows the results in the successive years up to 1837:

Years.	Wine per Acre in Gallons.	Years.	Wine per Acre in Gallons.
1825	68.75	1832	209.9
1826	192.0	1833	311.6
1827	0.0	1834	413.4
1828	115.0	1835	620.0
1829	55.9	1836	544.5
1830	0.0	1837	184.4
1831	153.0		

The mean quantity of wine furnished by this vineyard from the date of its plantation, is 224½ gallons per acre. M. Villeneuve reckons the mean produce of many vineyards in the southwest of France at from about 146 to 192 gallons per acre, considerably less consequently than our vineyard at Schmalzberg; and official documents, while they give the mean produce of the vine for the whole of France as 170.9 gallons per acre, state the whole of the wine produced over the country at 976,906,414 gallons.

From documents recently published, the whole produce of the vineyards of the German States brought to market appears to be 59,180,000 gallons.

Pulque. This is a vinous liquor, indigenous to Mexico and some parts of Peru, and is prepared from the sap of the *Agave Americana.* When this plant is about to flower, a hole is made into the upper part of its stem, which by and by becomes filled with juice, and is removed two or three times in the course of the twenty-four hours; this sap is very sweet, runs quickly into fermentation, and yields the liquor called pulque. The flow of sap continues for two or three months, and a single plant will yield from six to eight quarts per day. In the neighborhood of a large town, which ensures a ready sale for the produce, a plantation of *Agave* is one of the most profitable possessions; in the neighborhood of Cholula there are single plantations which are worth from £8,000 to £12,000.

CHAPTER .V.

OF SOILS.

The solid mass of our earth does not everywhere present the same physical characters, or the same chemical composition. In traversing a mountainous country of any extent, we seldom fail to observe a notable difference in the nature and relative position of the rocks which compose it; the idea which forces itself upon the mind in such circumstances is, that these mineral masses have not had the same origin, that they have been formed and placed in their several situations at distinct and often distant epochs.

In examining attentively the inequalities which mark the surface of the globe, we soon perceive that those rocks which generally form the most elevated points, the axis or skeleton of mountain chains, result from the agglomeration or intimate mixture of different mineral substances which may be isolated and separately studied.

These crystalline masses are frequently covered to a certain depth, and even completely concealed by rocks of more recent formation, the fragmentary elements of which proclaim their origin from the attrition or breaking down of the strata which support them. The regular stratification of these superimposed rocks, the configuration of their minute particles, the remains of organized beings which are found in them, proclaim them to be deposites which have taken place successively, and from the ocean. The formation of the crystalline rocks probably dates from the period at which the crust of the globe became solid. These elements, intimately mingled by fusion, combined as they cooled, according to the laws of affinity, to constitute the mineral species which we encounter; just as it happens that mineral species, identical with those which we observe in nature, are produced and crystallize during the consolidation of certain scoriæ from our furnaces.

The various circumstances which have accompanied the cooling of the crust of the globe, have doubtless occasioned the differences which we observe in the distribution of the minerals that enter into the composition of rocks. Thus granite and mica schist, which present so dissimilar a structure, are nevertheless, and very certainly, varieties of the same species, and contain quartz, felspar, and mica. In sienite, the mica is replaced by amphibolite, and in protogenite by talc. In trachite, a volcanic rock, both of older and more recent date, quartz is almost entirely wanting; the amphibolite is replaced by pyroxenite, and the felspar which is encountered, is no longer identical in its chemical composition with that which enters into the constitution of granite. The limestone rock, which belongs to the same Plutonic epoch, is granular or saccharoid; occasionally the intervention of magnesia makes it pass into dolomite.

The sedimentary strata do not vary less in their composition. The -

causes which segregated the rocks of igneous origin, appear to have destroyed or removed one or several of their elements before their new consolidation ; one of the most common deposites, sandstone or grit, is almost wholly composed of grains of quartz, amidst which particles of mica are frequently encountered ; but felspar is extremely rare. In the oldest sedimentary strata of the series, as in the greywackes, the igneous elements are met with more complete, and less altered. The structure of the calcareous rocks of this epoch is often compact, clayey ; it becomes porous and friable in deposites of more recent date.

The stratified rocks must have been deposited in parallel superimposed layers, and these strata, horizontal in the beginning, have been forced into the inclined and perpendicular positions which they now occupy by the tumefaction or rising of the masses upon which they rest. The organic remains which they present, frequently in such quantity, proclaim that in the period when the revolutions of the globe took place that gave them birth, there were already animated beings and plants growing upon the surface of the earth. The production of sedimentary strata, is an obvious proof that the igneous rocks of which they are the product, must have been segregated, so as to form beds of gravel, and sand, and clay. The elements of all stratified rocks must necessarily have passed through these different states before the powerful causes which consolidated them, of the nature of which we cannot now form an estimate, came into play. The disintegration of the crystalline igneous rocks proceeds under our eyes, as it were, from the combined actions of water and the atmosphere.

Water, by reason of its fluidity, penetrates the masses of rocks that are at all porous; it filters into their fissures. If the temperature now fall, and the water comes to congeal, it separates by its dilatation the molecules of the mineral from one another, destroys their cohesion, and produces clefts which slowly reduce the hardest rocks to fragments, and then to powder. During the frozen state, the ice may serve as a cement, and connect the disintegrated particles ; but with the thaw, the slightest force, currents of water, the mere effect of weight, suffices to carry the fragments to the bottom of the valley, and the rubbing and motion to which these fragments of rocks are exposed in torrents, tend to break them still smaller, and to reduce them to sand.

The quantity of earthy matter brought down by streams and rivers, is considerable : an idea may be formed of it from the thickness of the slime or mud deposited by a river which has overflowed its banks. In many situations, the arable soil is either formed entirely, or is powerfully ameliorated by such alluvial deposites. The fertilizing powers of the mud of the Nile are well known ; according to Shaw, the waters of this river carry with them about the 132d part of their volume ; those of the Rhine, at the periods of its great increase, bring down more than the 100th part ; and Dr. Barrow, from observations made in China, estimates at the 200th part of the volume of the mass of fluid, the mud and slime which are carried towards the sea by the

Yellow river. These fluviatile deposites accumulate at the mouths of great rivers, and gradually encroach upon the ocean; this is very conspicuous, for example, at the mouths of the Elbe, where, at the turn of the tide, when there is an interval of calm, the earthy matters which are held in suspension are precipitated, and a sediment results, which is thrown up by the next waves upon the beach. By these successive deposites, the beach rises gradually, and an extensive alluvium is formed which remains dry at neap and ordinary tides. These new lands, the fertility of which is truly surprising, constitute the polders of which the Dutch make so much. During spring tides, and storms from particular quarters, these polders would of course be all submerged, had not the active industry of the inhabitants raised dykes, which successfully oppose the waters of the ocean.

Besides the mechanical causes of the destruction of rocks already quoted, there is a chemical action depending upon meteorological influences, which exerts a powerful influence upon the constituent elements of crystalline rocks. Felspar, amphibolite, mica, and the protoxide of iron suffer decomposition in certain circumstances with surprising rapidity, without our being able to foresee, and still less to explain, this singular tendency to destruction. In granite, for example, the felspar and the mica lose their vitreous and crystalline state, they become friable, earthy, and are transformed into an argillaceous substance, which is known in the arts under the name of kaoline, and which is extensively used in the manufacture of porcelain; amphibolite, and pyroxenite, undergo an alteration of the same kind. In these minerals the protoxide of iron passes to the state of the maximum of oxidation. The air and moisture appear to exert a great influence upon this alteration, which frequently extends to a great depth, as we see in the beds of porcelain earth, which are worked in various granite districts, and as I have myself ascertained, in a bed of decomposed syenitic porphyry where there are very extensive subterraneous works. In these works, which are carried on in auriferous strata, the alteration in the felspar and amphibolite can be followed to a depth of nearly 330 feet. In the midst of the rocks so changed, we every here and there meet with masses which have resisted the decomposing action, and still possess all their original hardness and freshness. Historical monuments also show us unalterable granites; such is that, for instance, which now forms the obelisk in the square of San Giovanni di Laterano at Rome, and which was cut at Siena, under the reign of a king of Thebes, thirteen hundred years before the Christian era. Such is further the obelisk of the Place of St. Peter, which was consecrated to the sun by a son of Sesostris more than three thousand years ago.

The schists, by reason of their structure, wear away with much greater facility. Calcareous rocks resist atmospherical agencies somewhat better; but their softness in general suffers them to be readily attacked by mechanical causes, and water even acts upon them as a solvent through the medium of the carbonic acid which it always contains. The resistance of the greywackes, and of the sandstones depends in a great measure on the nature and cohesion of the cement which unites their particles; their power of resisting, however, is generally inconsiderable, and these rocks fall down pretty rapidly into sandy soils.

The modifications experienced by the constituent minerals of rocky masses, do not happen solely from changes in the molecular state of their elements; their chemical nature is further deeply changed, and some of their original principles disappear. The felspars, for example, into the constitution of which potash and soda enter, abandon almost the whole of these alkalies, in passing into the state of kaoline. This is made manifest by a comparison of the analyses of the mineral in its two states. Besides the alkali which is lost, we also perceive that in kaoline, the proportion of alumen relatively to that of silica, is much greater than in the undecomposed felspar, a fact which, according to M. Berthier, demonstrates that the alkali is removed in the state of silicate.

The final result of the disintegration of rocks, and of the decomposition of the minerals which enter into their constitution, is the formation of those alluviums which occupy the slopes of mountains that are not too steep, the bottoms of valleys, and the most extensive plains. These deposites, however formed, whether of stones, pebbles, gravel, sand, or clay, may become the basis of a vegetable soil, if they are only sufficiently loose and moist. Vegetation of any kind succeeds upon them at first with difficulty. Plants which by their nature live in a great measure at the expense of the atmosphere, and whick ask from the earth little or nothing more than a support, fix themselves there when the climate permits. Cactuses and fleshy plants take root in sands; mimosas, the broom, the furze, &c., show themselves upon gravels. These plants grow, and after their death, either in part or wholly, leave a debris which becomes profitable to succeeding generations of vegetables. Organic matter accumulates in the course of ages, even in the most ungrateful soils in this way, and by these repeated additions they become less and less sterile. It is probable that the virgin forests of the new world have thus supplied the wonderful quantity of vegetable mould, in which the present generation of trees is rooted. At Lavega de Supia, in South America, the slipping of a porphyritic mountain covered completely with its debris, to the extent of nearly half a league, the rich plantations of sugar-cane which were there established. Ten years afterwards I saw the blocks of porphyry shadowed by thick groves of mimosas; and the time perchance is not very remote when this new forest will be cleared away, and the stony soil, enriched with its spoils, will be restored to the husbandman.

The chemical composition of the earth, adapted for vegetation, must of course participate in the nature of the rocks and substrata from which it is derived; and the elements which enter into the constitution of mineral species ought to be found in the soils, which, by the effect of time or human industry, may serve for the reproduction of vegetables. It is on this account that it becomes interesting to know the composition of the minerals which are the most abundantly dispersed in the solid mass of the globe.

The solid part of our planet, as is well known, occupies but one-third of its whole surface. The ocean occupies two-thirds, and the majority of the rocks of sedimentary formation must have been primarily deposited at the bottom of the sea. These rocks will therefore be apt to contain the saline substances which are met with in sea-water, and it is a fact that many of the secondary sandstones show unequivocal traces of these substances. Deltas and low downs, left by the ocean, are constantly being brought under tillage, and the fierce winds of the sea frequently carry saline matters to vast distances, even to the centre of great continents; lastly, as we shall see by and by, the ocean supplies agriculture with powerful manures. Analysis shows that sea-water contains, besides chloride of sodium or common salt, hydrochlorate of magnesia, sulphate of soda, sulphate of magnesia, sulphate of lime, carbonate of lime, carbonate of magnesia, and a quantity of carbonic acid, to which must be added the substances discovered in the mother waters of salt marshes, and which occur with reference to the others in quantities so small as to escape direct analyses of any moderate portions of sea-water: these substances are iodides, bromides, and certain ammoniacal salts.

The minerals most generally found in rocks are quartz, felspar, mica, amphibolite, pyroxenite, talc, serpentine, and diallage.

Quartz is frequently composed of silica nearly in a state of purity; but I may save time by presenting in a single table the composition of the principal mineral species such as we find it indicated by the best chemical analysts:

Minerals.	COMPOSITION.										
	Silica.	Alumina.	Lime.	Magnesia.	Potash.	Soda.	Oxide of Iron.	Oxide of Manganese.	Fluoric Acid.	Water.
Felspar of Lomnitz	66.8	17.5	1.3		12.0		0.8			
Ditto Domite	61.0	19.2		1.6	11.5		4.2			2.0
Ditto Albite of Finland	68.0	19.6	0.7			11.1	0.2	0.5		
Ditto Albite of Arendal	68.7	19.9		traces		9.1	0.3	traces		
Siberian Mica	42.0	16.1		26.0	7.6		4.9		0.7	
Mica from the U. States	48.5	33.9			11.3			1.3		3.0
Amphibolite of Pargas .	45.7	12.2	13.8	18.8			7.3	0.2	1.5	
White Pyroxenite......	54.6		24.9	18.0			1.8	2.0		
Green ditto............	54.9	0.2	23.6	16.5			4.4	0.4		
Serpentine	42.3			44.2			0.2			13.3
Ditto, another kind	43.1	0.3	0.5	40.4			1.2			12.5
Spezian Diallage	47.2	3.7	13.1	24.4			7.4			3.2
Talc from St. Bernard..	58.2	traces		33.2			4.6			3.5
Ditto from St. Gothard..	62.0			30.5	2.8		2.5			0.5

If we now compare the analyses of the ashes of vegetables which we have already given with those just indicated, we see that the mineral substances which meet us in plants also exist in the soil independently of any addition from manure. We may therefore lay it down as a principle that the mineral substances encountered in vegetables are obtained in the soil, and that the whole of these substances come from rocks which form the solid crust of our planet. I ought, however, to observe in this place that the phosphates, which are so constantly present in plants that it is to be presumed they are essential to their organization, do not figure among the elements of crystalline rocks; we only meet with phosphoric acid in the strata of a more recent geological epoch,—strata the formation of which has indeed followed the appearance of organized beings; so that it would be quite fair to maintain that this acid had been introduced into these new strata by the animated beings which are buried in them. Still the phosphates are by no means wanting in the rocks of igneous origin. In metalliferous strata, to quote those of more common occurrence only, we find phosphate of lead, of copper, of manganese, and of lime; it is even difficult to discover a ferruginous mineral which does not contain a larger or smaller dose of phosphoric acid. And I must here add, that if phosphoric acid has been rarely indicated as a constituent of mineral substances, this is by no means from its uniform absence there, but because it escaped the researches of the analyst, in the same way as iodine and bromine for a long time escaped notice in all the analyses that were made of sea-water. Chemists, in fact, only discover those bodies readily which exist in some very appreciable quantity in the compounds they examine. The substances whose presence is not foreseen, those which only enter in extremely small quantity into a mineral, are apt to pass the eyes of even the most skilful and conscientious unperceived.

The ashes of every vegetable examined up to the present time show us phosphates, and yet these salts have never been detected in any of the analyses of saps (not very numerous it is true) which we possess; it is, nevertheless, all but certain that the sap must contain phosphoric acid in some state of combination or another.

Thaer compares the soil in husbandry to the raw material upon which the industry of the manufacturer is exercised; the comparison would, perhaps, be more exact were the soil likened to the mechanical agents he uses; and, in fact, even as the prosperity of manufactures and the perfection of their produce depend upon the perfection of the machinery employed, so are the quality and the quantity of crops connected in the most intimate manner with the quality of the soil. The highest skill of the husbandman, even under a favorable climate, and otherwise in the most advantageous circumstances, may all be made nugatory by the incessantly renewed difficulties which meet him in a barren soil.

To be truly fit for agriculture the earth ought to present several essential qualities; a soil, for instance, must be sufficiently open, sufficiently loose, to permit the roots of plants to penetrate it, and to prevent the water from stagnating upon it. The matter of which it is composed must, further, be of such a kind that the air may insinuate itself into it and be renewed, without, however, too rapid a desiccation following.

A great deal has been written since Bergman's time upon the

chemical composition of soils. Chemists of great talent have made many complete analyses of soil ; noted for their fertility ; still practical agriculture has hitherto derived very slender benefits from labors of this kind. The reason of this is very simple ; the qualities which we esteem in a workable soil depend almost exclusively upon the mechanical mixture of its elements ; we are much less interested in its chemical composition than in this ; so that simple washing, which shows the relations between the sand and the clay, tells, of itself, much more that is important to us than an elaborate chemical analysis. The quality of an arable soil depends essentially on the association of these two matters. Sand, whether it be silicious, calcareous, or felspathic, always renders a soil friable, permeable, loose ; it facilitates the access of the air and the drainage of the water, and its influence is more or less favorable as it exists in the state of minute subdivision, or in the state of coarse sand or of gravel.

Clay possesses physical properties entirely opposed to those of sand ; united with water it forms an adhesive plastic paste, which, once moistened, becomes almost impermeable. With such characters, it will easily be conceived how it is impossible to work to advantage a soil that is entirely argillaceous. The proper character, or, if you will, the quality of a soil, depends, then, essentially on the element which predominates in the mixture of sand and clay that composes it ; and between the two extremes, which are alike unfavorable to vegetation, viz., the completely sandy soil and the unmixed clay, all the other varieties, all the intermediate shades can be placed. It is rare, indeed, that arable soils are formed solely of sand and clay : not to mention certain saline substances which are generally encountered, although in small quantity, we always find the remains of organic matters, remains which constitute that part of a soil which has been designated under the somewhat vague name of humus. Although a soil which is entirely without humus may be cultivated by calling in the aid of manure, and as humus, consequently, need not be regarded as indispensable, still this matter generally enters, in certain proportions, into the constitution of soils. The soils of forest lands contain a large quantity of it, and some soils are mentioned which are very rich in this substance, and which yield abundant crops of grain for ages, and with very little attention.

In examining a soil, attention ought to be directed, 1st, to the sand, 2d, to the clay, 3d, to the humus which it contains. It would, further, be useful to inquire particularly in regard to certain other principles which exert an unquestionable influence upon vegetation, such as certain alkaline and earthy salts.

Vegetable earth dried in the air until it becomes quite friable may, nevertheless, still retain a considerable quantity of water, and which can only be dissipated by the assistance of a somewhat high temperature. It is therefore proper, in the first instance, to bring all the soils which it is proposed to examine comparatively, to one constant degree of dryness. The best and quickest way of drying such a substance as a portion of soil, is to make use of the oil-bath ; a quantity of oil contained in a copper vessel is readily kept at an almost uniform temperature by means of a lamp. A thermometer plunged in the bath shows the degree to which it is heated; the substance to be dried is put into a glass tube of no great depth, and sufficiently wide; or into a porcelain or silver capsule, if the quantity to be operated upon be somewhat considerable: these tubes, or vessels, are placed in the oil so as to be immersed in it to about two-thirds of their height. For the desiccation of soils, the temperature may be carried to 150° or 160° C., (334° or 352° F.) The weight of the vessel is first accurately taken, and a given weight of the matter to be dried is then thrown into it, after which it is exposed to the action of the bath. If we operate upon from 600 to 700 grains, the drying must be continued during two or three hours; the weight of the capsule with its contents, after having been wiped thoroughly clean, is then taken. It is placed anew in the bath, and its weight is taken a second time after an interval of fifteen or twenty minutes; if the weight has not diminished, it is a proof that the drying was complete at the time of the first trial. In the contrary case, the operation must be continued, and no drying must be held terminated, until two consecutive weighings, made at an interval of from fifteen to twenty minutes, show any thing more than a very trifling difference. Davy points out another and much more simple method, which, although far from accurate, may, nevertheless, suffice in many general trials. The soil to be dried is put into a porcelain capsule heated by a lamp, and a thermometer, with which the mass may be stirred, is placed in its middle, and shows the temperature at each moment. Lastly, in many circumstances the marine bath may suffice. In drying, the main point is to do so at a known temperature, and one which may be reproduced; for the absolute desiccation of a quantity of soil could not be accomplished except at a heat close upon redness, and this would, of course, alter or destroy the organic matters it contains.

The organic matters contained in ordinary soils consist, in part, of pieces of straw and of roots, which are usually separated by sifting the earth through a hair sieve; the gravel and stones which the soil contains are separated in the same way.

The earth sifted is now washed. To accomplish this, it is introduced into a matrass, with three or four times its bulk of hot distilled water, the whole is shaken well for a time, the matrass is left to stand for a moment, and then the liquid is decanted into a wide porcelain capsule. The washing is continued, fresh quantities of water being added each time, until the whole of the clay has been removed, which is known by the fluid becoming clear very speedily; the sand which remains, is then washed out into another capsule. The argillaceous particles, or the clay and all the matters held in suspension in the water, are thrown upon a filter and dried; the desiccation is completed by the same process, and under the same circumstances as that of the soil had been. The sand is, in like manner, dried with the same care.

If we would ascertain the nature and quantity of the soluble salts, the whole of the water used in the washing must be put together and evaporated, which may be done upon a sand-bath. The evaporation is pushed to dryness, and the salts that remain, having been previously weighed, are thrown into a small platinum capsule, in which they are heated to a dull red by means of a spirit-lamp, in order to burn out the organic salts, and thus distinguish, by means of a subsequent weighing, between them and the inorganic salts.

The sand may be silicious or calcareous. The presence of carbonate of lime is readily ascertained by treating it with an acid which will form a soluble salt with lime, such as hydrochloric, nitric, or acetic acid. Effervescence shows the presence of a carbonate ; the quantity of which may be estimated by weighing the sand dry before and after its treatment with the acid, particular care being of course taken to wash the remaining sand well before setting it to dry. This, however, is an operation of little use, the great object is to ascertain the quantity of sandy matter. Had we a particular interest in ascertaining the presence and estimating the quantity of the earthy carbonates contained in a sample of soil, it would be advisable to make a special inquiry, inasmuch as the finely divided calcareous earth being carried off along with the clay in the course of the washing, the sand obtained never contains the whole of the carbonate of lime.

The argillaceous matter procured by the washing is far from being pure clay; it contains a quantity of extremely fine sand, particles of calcareous earth, and if the soil contain humus, the more delicate particles of this substance will also be included.

To determine the quantity of humus, recourse is generally had to its destruction by heat. A known weight of dried earth is heated to redness in a capsule, and constantly stirred for a time, and when no more of those brilliant points or sparks, which are indications of the combustion of carbon, are observed, it is set to cool and then weighed. This is the method which has been generally followed by Davy and others. It would be difficult to find a method more convenient than this, but it is unfortunntely very inaccurate. Soils dried at a temperature at which organic matter, such as humus, &c., begins to change, still retain a considerable quantity of water in union with the clay. This water is disengaged at the red heat required for the combustion of the organic matters; and as their quantity is estimated by the loss of weight on the subsequent weighing, it is obvious that the loss from the dissipation of water is added to that which proceeds from the destruction of the humus. It is undoubtedly to this cause of error that we must ascribe the large proportions of humus mentioned in the soils examined by Thaer and Einhoff; it is therefore better to restrict the examination to the determination of the presence or absence of humus than to attempt to ascertain its quantity by so imperfect a method.

Priestley and Arthur Young were already aware that a more delicate operation was required to determine the quantity of humus. They recommend calcination of the soil in a close vessel, and that the gaseous products should be collected. This mode of proceeding, however, would have but slight advantages over that which I have just criticised, inasmuch as the volume of gas collected varies with every difference of heat employed.

The only method in my opinion which we have of learning the quantity of humus, of organic debris, which is contained in a soil, is that of an elementary analysis. It is by burning a known quantity of earth thoroughly dried by means of the oxide of copper, aided by a current of oxygen, that the carbon and hydrogen may be determined. But the most important point of all is to ascertain the amount of azote included in the organic remains of the soil; and we have happily precise means in our elementary analysis of ascertaining the quantity of azote, from which the amount of azotized organic matter may be accurately inferred.

It may be very useful to determine the presence or absence of carbonate of lime in a soil; this knowledge would of course guide us in our applications of lime, marl, &c. Two modes may be employed for this purpose; 1st. the soil may be treated by nitric acid slightly diluted with water. Any effervescence will denote the presence, in all probability, of carbonate of lime. I say in all probability, because the disengagement of carbonic acid gas under such circumstances generally indicates the presence of carbonate of lime; it is not, however, a special character, because the disengagement may be due to the presence of any other carbonate. It is well to boil the acid solution upon the sample of soil that is analyzed; the part which is not dissolved is thrown upon a filter and washed with distilled or rain-water boiling hot. Into the clear filtered liquor which results from all the portions of water used in the washing, a little ammonia is added; if any precipitate falls, it is collected upon a filter and washed: to the new liquors obtained by this washing, a solution of oxalate of ammonia is added. If there be any lime present, it is thrown down in the state of oxalate, and the liquor, having been left at rest for five or six hours, becomes completely clear; the addition of a few drops of the solution of oxalate of ammonia to this clear fluid satisfies us whether the whole of the lime has been precipitated or not. The oxalate of lime is received upon a filter, washed, and dried; it is then thrown into a platinum capsule along with the piece of filtering paper upon which it was collected, and is heated to a dull red, until the paper of the filter is completely consumed and no further trace of carbon appears; the capsule is then taken from the fire or from over the spirit lamp, and cooled; when cold, the matter which it contains is moistened with a concentrated solution of carbonate of ammonia.

The matter is then dried, great care being taken that nothing is lost by particles flying out, and the capsule is again heated to a dull red; when cold, it is weighed accurately, and the quantity of matter contained then becomes known. This matter is carbonate of lime, 100 of which represent 56.3 of lime and 43.7 of carbonic acid. I have said that in arable soil other carbonates may be met with besides that of lime; calcareous soils, for example, very commonly contain carbonate of magnesia. If we would ascertain the quantity of this earth, the mode of proceeding which I have just particularly indicated enables us to do so; we have but to evaporate the liquid from which the oxalate of lime was deposited, and then to calcine the product of the evaporation in a platinum capsule. Any nitrate of magnesia which may exist there will be decomposed at a dull red heat, as well as any oxalate of ammonia which may have resulted from ammonia added in excess. By treating the residue of the calcination with water we obtain the magnesia, which being washed, has only to be calcined and its weight ascertained by weighing.

2d. If we would be content with a simple approximation, we may judge of the quantity of calcareous carbonate contained in a vegetable soil by measuring the quantity of carbonic acid which we obtain from it. We counterpoise upon the scale of a balance a phial containing some diluted nitric acid; we weigh a certain quantity of the earth to be analyzed, and this is added by degrees to the acid. If the earth contains carbonates, effervescence ensues. The liquid is shaken with care, and having waited a few minutes in order to let the carbonic acid which is mixed with the air of the phial escape, the phial with its contents is again put into the balance. If there has been no disengagement of carbonic acid, it is clear that to restore the equilibrium it will be sufficient to add to the opposite scale the weight of the earth which was put into the phial; whatever is wanting of this weight represents precisely the weight of carbonic acid which has been disengaged. Presuming this acid to have been combined with lime, the weight of the calcareous carbonate can be calculated exactly.

Sulphate of lime is an occasional constituent of soils; to ascertain its presence and quantity, the following is the method of procedure:

The earth well pulverized is first roasted for a considerable time in a crucible or platinum capsule, until all the organic matter is completely destroyed; it is advisable to operate on about 100 grammes, or about 3.2 ounces troy of soil. After this operation the matter is boiled in 4 or 5 times its weight of distilled water for some time; water being added to replace that which is dissipated by evaporation; we then filter, re-wash, and having added all the liquors, we evaporate in a capsule until the volume of the liquid is reduced to a few drachms. To the liquid thus concentrated we add its own bulk of alcohol. If the solution contains sulphate of lime it will be deposited, and the deposite being received upon a filter and washed with weak alcohol, its weight is taken after having been dried and calcined. This salt is frequently seen deposited in the form of fine colorless needles on the cooling of the sufficiently concentrated solution; but the addition of alcohol is always useful, because the sulphate of lime, which is not very soluble in water, is altogether insoluble in weak spirit, which on the contrary dissolves certain alkaline and earthy salts whose presence would interfere with the accuracy of the result.

It may be matter of great moment to determine the existence and the quantity of phosphates contained in a soil destined for cultivation. Although the search for phosphoric acid may perhaps require a certain familiarity with chemical analysis, I shall nevertheless indicate the method of procedure. It is much to be desired that enlightened agriculturists should not remain strangers to manipulations of this kind.

The soil to be analyzed must be first deprived of all organic matters by calcination. After having reduced it to a very fine powder it is to be boiled for about an hour with three or four times its weight of nitric or hydrochloric acid. The solution is then diluted with distilled water, and filtered; the matter which remains upon the filter is generally silica or alumina which has escaped the action of the acid. After having reduced the washings by evaporation, and added them to the acid liquor, ammonia in solution is poured in. Taking the simplest instance, the precipitate which falls upon the addition of this alkali may contain, 1st, phosphoric acid in union with the peroxide of iron and lime; 2d. oxide of iron and of manganese; 3d. silica. This precipitate, which is usually of a gelatinous appearance, is received upon a filter, well washed and dried, when the precipitate is readily detached from the filter. It is thrown into a platinum capsule which is raised to a white heat, after which the weight of the residue is taken. The precipitate after calcination is thrown into a small glass matrass, and dissolved by hot hydrochloric acid. If there is any silica undissolved, its quantity is merely estimated, if it be very small; if it be a larger quantity, it is to be collected upon a filter and weighed. To the new acid solution, about three times its weight of alcohol is added; the mixture is shaken, and pure sulphuric acid is then instilled drop by drop until there is no longer any precipitate. The precipitate is sulphate of lime, which is thrown upon a filter, where it is washed with diluted alcohol; it is then dried, calcined, and the weight of the sulphate of lime obtained, permits us to calculate that of the lime which formed part of the precipitate thrown down by the ammonia in the first instance. 100 of sulphate of lime are equivalent to 41.5 of pure lime.

The alcoholic liquor is concentrated in order to expel the spirit; as it is acid, it is saturated with ammonia until a slight precipitate begins to be formed, which is not redissolved upon shaking the mixture. A few drops of the hydrosulphate of ammonia are then added, upon which the iron and the manganese fall in the state of sulphurets. As a part of the metals has been precipitated in the state of oxide by the ammonia added in the hydrosulphate, it is well to digest for eight or ten hours, because the hydrosulphate of ammonia always ends by changing the metals present into sulphurets, which being washed, dried, and reduced to the state of oxides by calcination in a platinum capsule, are weighed.

If the first ammoniacal precipitate did not contain phosphoric acid, its weight ought to be reproduced by adding that of the lime to that of the metallic oxides proceeding from the calcination of the sulphurets. Any loss which is noted after this, is due, if the process has been well conducted, to phosphoric acid, which had not been collected, but which has remained in the state of phosphate of ammonia in the liquid treated by the hydrosulphate. To determine with precision the presence of phosphoric acid, the liquid in question must be evaporated to dryness, and the residue heated strongly in a platinum capsule. After the dissipation and decomposition of the ammoniacal salts, there remains watery phosphoric acid, distinguishable by its powerful acid reaction, its sirupy consistence, and its fixity.

By way of example, I shall give the results obtained in an analysis of this kind:

From the acid liquor, ammonia threw down of:	grs. troy.
Phosphates and metallic oxides	8.012
These gave of sulphate of lime	8.768
Equivalent to lime	3.612
Hydrosulphate of ammonia caused a precipitate, which, calcined, gave of metallic oxides . .	1.620
Lime and metallic oxides together . .	5.233
Difference due to phosphoric acid . .	2.789

The analysis for phosphoric acid may be simplified by employing a process conceived by M. Berthier, and which is founded upon the strong affinity of this acid for the peroxide of iron and the insolubility of the phosphate of the peroxide of iron in dilute acetic acid. If to a fluid containing at once phosphoric acid, lime, peroxide of iron, alumina, and magnesia in solution, ammonia be added, the precipitate will contain the whole of the phosphoric acid. The acid will be in great part combined in the state of phosphate of iron, if the peroxide of iron be in quantity more than sufficient to neutralize it, a condition which must be frequently expected in an arable soil; however, to make sure of this point it is well to add a certain quantity of the peroxide of iron to the soil which is to be analyzed. Besides the phosphate of iron, the precipitate may contain phosphate of lime, phosphate of alumina, and certainly ammoniacal magnesian phosphate. Finally, with these phosphates will be found associated alumina and oxide of iron, the latter especially if it has been introduced in excess. The precipitate collected upon a filter and washed, must then be treated with dilute acetic acid, which will dissolve the lime, the magnesia, and the excess of the oxides of iron and alumina, and there will remain phosphate of iron or phosphate of alumina, because the latter salt is as insoluble as the former in acetic acid. Whenever the precipitate in question, therefore, leaves a residue which is insoluble in vinegar, the presence of phosphoric acid may be inferred; this residue may consist of basic phosphates of iron or alumina, or of a mixture of the two salts, and no great error will be committed if one hundred parts of this residue, calcined, be assumed as representing fifty of phosphoric acid.

The presence of silica in the precipitate insoluble in acetic acid may, however, lead to error. To make sure that the precipitate is formed by a phosphate it must be redissolved in hydrochloric acid, and the acid solution evaporated to dryness, so as to render the silica, which may exist in it, insoluble. By treating the residue with hydrochloric acid again, the phosphates alone will be dissolved. The presence of phosphoric acid may otherwise be determined by treating the phosphate of iron in solution in the way which I have already indicated.

From what precedes, it must be obvious that the most carefully conducted chemical analysis of a soil, only leads us to the discovery of certain principles which exist in very small quantity, although their action is unquestionably useful to vegetation. As to the determination of the relative quantities of sand and loam, this rests upon simple washing; and a chemist would spend his time to very little purpose, in seeking by means of elementary analyses to determine the precise composition of these substances. The finest part carried off by the water will always show properties analogous to those of clay; the sand, which is generally silicious, will exhibit the characters of quartz; and the calcareous fragments, which are mixed with it, will exhibit those that belong to carbonate of lime. It will be sufficient then in connection with the mineral constitution of arable soils, to expose very briefly the general properties of clay or loam, of quartz, and of carbonate of lime, substances in fact which form the bases of all arable lands. Pure *clay* composed of silica, alumina, and water, does not contain these substances in the state of simple mixture. The inquiries of M. Berthier have satisfactorily shown that clay is an hydrated silicate of alumina. When we remove a portion of the alumina from clay, for example, by treating it with a strong acid, the silica which is set at liberty will dissolve in an alkaline solution, which would not be the case were the silica present in the state of quartzy sand, however fine.

Pure clays are white, unctuous to the touch, stick to the tongue when dry, and when breathed upon give out an odor which is well known, and is commonly spoken of as the argillaceous odor. This property of dry clay to adhere to the tongue is owing to its avidity for water. It is known, in fact, that dry clay brought into contact with water, first swells, and finally mixes with it completely. Duly moistened it forms a tough and eminently plastic mass. Exposed to the air, moist clay, as it dries, shrinks considerably; and if the drying be rapid, the mass cracks in all directions. It is to an action of this kind that we must ascribe the cracks and deep fissures which traverse our clayey soils in all directions during the continuance of great droughts.

The constitutional water of clays is retained by a very powerful affinity, and does not separate under a red heat; pure clay has a specific gravity of about 2.5; but the weight is frequently modified by the presence of foreign matter, for it contains sand, metallic oxides, carbonate of lime, carbonate of magnesia, and frequently even combustible substances from bitumen to plumbago, all of which admixtures of course modify the properties which are most highly esteemed in clays, such as fineness, whiteness, infusibility, &c.

Quartz is abundantly distributed throughout nature, and is met with in very different states in the form of transparent colorless crystals constituting rock crystals, as sand of different fineness; finally, in masses constituting true rocks. Quartz is the silica of chemists, and a compound, according to them, of oxygen and silicon, in the proportion, Berzelius says, of 100 of the radical to 108 of oxygen.

Silica in a state of purity occurs in the form of a white powder, and having a density of 2.7. It is infusible in the most violent furnace, but it not only melts in the intense heat which results from the combustion of a mixture of hydrogen and oxygen gas, but it is even dissipated in vapor. As generally obtained, silica is held insoluble in water; still, when in a state of extreme subdivision, it is soluble; and then its insolubility is probably not so absolute as is generally supposed, for M. Payen has found notable quantities in the water of the Artesian well of Grenelle, and in that of the Seine. Silica exists especially in very appreciable quantity in certain hot springs where the presence of an alkaline substance favors its solution; the water of the hot springs of Reikum in Iceland contain about $\frac{4}{1000}$th parts of its weight of silica; and the thermal spring of Las Trincheras, near Puerto Cabello, deposites abundant silicious concretions. The water of this latter spring, which is at the temperature of 210° F., besides silica contains a quantity of sulphurated hydrogen gas, and traces of nitrogen gas. Rock crystal when colorless and transparent may be regarded as pure silica; in the varieties of quartz which mineralogists designate as chalcedony, agate, opal, &c., the silica is combined with different mineral substances, particularly oxide of iron and of manganese, alumina, lime, and water.

Carbonate of lime, considered as rock, belongs to every epoch in the geological series, and frequently constitutes extensive masses. When pure it is composed of lime 56.3, carbonic acid 43.7; and its density is then from 2.7 to 2.9. It dissolves with effervescence without leaving any residue in hydrochloric or nitric acid. Exposed to a red heat its acid is disengaged, and quick-lime remains. Carbonate of lime is insoluble in water, but it dissolves in very considerable quantity under the influence of carbonic acid gas. When such a solution is exposed to the air the acid escapes by degrees, and the carbonate is deposited, by which means those numerous deposites of carbonate of lime are produced, which we see constituting tufas and stalactites. The solubility of carbonate of lime in water acidulated with carbonic acid, enables us to understand how plants should meet with this salt in the soil, inasmuch as rain-water always contains a little carbonic acid.

The mineral substances which we have now studied, taken isolatedly, would form an almost barren soil; but by mixing them with discretion a soil would be obtained, presenting all the essential conditions of fertility, which depend as it would seem much less on the chemical constitution of the elements of the soil than on their physical properties, such as their faculty of imbibition, their density, their power of conducting heat, &c. It is unquestionably by studying these various properties that we come to form a precise idea of the causes which secure or exclude the qualities we require in arable soils. This has been done very ably by M. Schübler, and his admirable paper will remain a model of one application of the sciences to agriculture.*

The researches of M. Schübler were directed to the mineral substances which are generally found in soils, viz: 1st. silicious sand; 2d. calcareous sand; 3d. a sandy clay containing about $\frac{4}{10}$ths of sand; 4th. a strong clay containing no more than about $\frac{2}{10}$ths of sand; 5th. a still stronger clay containing no more than about $\frac{1}{10}$th of sand; 6th. nearly pure clay; 7th. chalk, or carbonate of lime in the pulverulent state; 8th. humus; 9th. gypsum; 10th. light garden earth, black, friable, and fertile, and containing, in 100 parts, clay 52.4, quartzy sand 36.5, calcareous sand 1.8, calcareous earth 2.0, humus 7.3; 11th. an arable soil composed of clay 51.2, silicious sand 42.7, calcareous sand 0.4, calcareous earth 2.3, humus 3.4; and 12th. an arable soil taken from a valley near the Jura, containing clay 33.3, silicious sand 63.0, calcareous sand 1.2, calcareous earth and humus 1.2, loss 1.3.

The object of these inquiries was to ascertain, 1st. the specific gravity of soils; 2d. their power of retaining water; 3d. their consistency; 4th. their aptitude to dry; 5th. their disposition to contract while drying; 6th. their hygrometric force; 7th. their power of absorbing oxygen; 8th. their faculty of retaining heat; and 9th. their capacity to acquire temperature when exposed to the sun's rays.

Specific gravity of soils. The weight of soils may be compared in the dry and pulverulent state, or in the humid state, or the specific gravity of the particles which enter into their composition may be determined. This last information is easily obtained by the following method: take a common ground stopper bottle, weigh it stoppered and full of distilled water; let it then be emptied, in order that a known quantity of the soil, in the state of powder and quite dry, may be introduced into it. A quantity of water is now poured in, and the phial is shaken to secure the disengagement of all air bubbles; the phial is then filled with distilled water, and when the upper part has become clear the stopper is replaced; the phial is then wiped dry and weighed again. The difference between the weight of the phial full of water plus that of the matter, and the weight of the phial containing the matter and the water mixed, gives the weight of the water displaced by this matter. Thus:

Weight of the phial full of water	60.0
Weight of the matter	24.0
	84.0
Weight of the phial containing the mingled earth and water	74.4
Difference of water displaced	9.6

which is the weight of the volume of water equal to that of the matter introduced into the phial; we have consequently for the specific gravity of the earth $\frac{2.4}{0.6}=2.5$, the weight of the water having been taken as 1.

* Schübler, Annals of French Agriculture, vol. xl. p. 122, 2d series.

This number represents the mean specific gravity of the isolated particles of the powder which has been examined. But we must not from this density pretend to deduce the weight of a particular volume of soil, a cubic foot or a cubic yard, for instance ; we should come to far too high a number. The weight of a given volume of earth must be determined immediately by ramming it into a mould or measure of a known capacity.

From M. Schübler's experiments it appears, 1st. that silicious and calcareous sandy soils are the heaviest of any ; 2d. that clayey soils are of least density ; 3d. that humus or mould is of much lower density than clay ; 4th. that a compound soil being generally by so much the heavier as it contains a larger proportion of sand, and so much the lighter as it contains a larger quantity of clay, of calcareous earth, and of humus, it is possible from the density of a soil to infer the nature of the principles which prevail in it. In the course of his experiments M. Schübler found that artificial mixtures always gave higher densities than those that ought to have resulted from the several densities of each of the sorts of substance which formed the mixture.

Imbibition of water. The power which soils possess of retaining water or of resisting the too rapid dissipation of their moisture, is highly important in its influence upon their fertility. This faculty is measured comparatively in the following manner: a given quantity of soil is taken, say from 3 to 400 grains ; it is dried until it ceases to lose weight ; it is then made into a thin paste and thrown upon a moistened filter ; when it has ceased to drop it is weighed. The increase of weight is plainly due to the quantity of water retained by the soil, thus :

Weight of the dry soil	300.0
Weight of the moistened filter	75.0
	375.0
Weight of the filter and moistened earth	525
Water absorbed	150

In the experiment quoted, 100 of dry earth absorbed or imbibed 50 of water. The following table contains the result of experiments made on the imbibing power of different soils.

Kind of earth.	Water absorbed by 100 parts of the earth.
Silicious sand	25
Gypsum	27
Calcareous sand	29
Sandy clay	40
Strong clay	50
Sandy clay	70
Fine calcareous earth	85
Humus	190
Garden earth	89
An arable soil	52
Another arable soil	48

It appears, therefore, that the silicious and calcareous soils and the gypsum have the least affinity for water ; the clayey soil retained by so much the more as it contained a smaller quantity of sand ; the fine calcareous earth retained 15 per cent. more than the pure clay; while the calcareous sand retained 41 per cent. less. This fact proves how much the state of subdivision must influence the physical properties of soils; and it is easily to be understood that in noting the presence of calcareous matter in an arable soil we are carefully to indicate the form and degree of subdivision in which it occurs; humus, however, is the substance which shows itself most greedy of moisture, and we perceive from this fact wherefore soils rich in this principle have so strong an affinity for water.

Consistence, tenacity, friability of soils. The consistence or tenacity of soils is an important property which agriculturists indicate when they speak of soils being strong or stiff, and light, the amount of power expended in ploughing being taken as a measure of these qualities. To compare different soils under the point of view of their tenacity in the dry state, M. Schübler moulded various kinds, duly moistened, into equal and similar parallelopipeds. When these solids were completely dry, he placed either extremity upon a fixed support, and by means of the scale of a balance hung exactly from the middle of the prisms, he added weights gradually until they gave way; the weight, supported by each parallelopiped immediately before it broke, expressed its tenacity.

In working a damp soil we have not only to overcome its force of cohesion, but further and principally to get the better of its adhesion to our implements. This consideration led M. Schübler to estimate, always comparatively, the power which it is necessary to expend in working soils of different descriptions. As the material which enters into the construction of agricultural instruments is in general iron and wood, he did no more than ascertain the disposition of the soil to adhere to these two substances. In the experiments, the results of which we shall immediately detail, two discs were employed, one of iron, the other of beech-wood, having equal surfaces. The disc was connected with the extremity of the arm of a very delicate balance; it was then brought into perfect contact with the moist soil, and when it adhered, the opposite scale of the balance was loaded until the adhesion was overcome. In experiments of this kind it is obviously indispensable that the soil in each instance should have the same degree of humidity; they were tried, consequently, at the point of saturation with water.

Pure dry clay possessed the greatest tenacity, and its power was expressed by the number 100; the tenacity possessed by other matters was then compared to that of pure clay. The following table exhibits the results of the two series of experiments, viz. those having reference to the tenacity and those having reference to the force of cohesion.

Kind of soil.	Tenacity of soil, that of pure clay being 100.	Tenacity expressed in weight.	Cohesion in the moist state.	Vertical adhesion to iron and to wood on a surface of 3.937 square inches.
		kil.	kil.	kil.*
Silicious sand	0	0.	0.17	0.19
Calcareous sand......	0.	0.	0.19	0.20
Fine calcareous earth.	5.0	0.55	0.65	0.71
Gypsum	7.3	0.81	0.49	0.53
Humus	8.7	0.97	0.40	0.42
Sandy clay...........	57.3	6.36	0.35	0.40
Stiff clayey soil.......	68.8	7.64	0.48	0.52
Strong clay	83.3	9.25	0.78	0.86
Pure clay	100.0	11.10	1.22	1.32
Garden earth.........	7.6	0.84	0.29	0.34
Earth from Hoffwyll ..	33.0	3.66	0.26	0.28
Earth from the Jura ..	22.0	2.44	0.24	0.27

M. Schübler finds, from his experiments, that a dry soil is very easily worked when its tenacity does not exceed 10, that of pure clay being 100 in the moist state. Soils are further worked with ease when their adherence to a surface 3.937 inches square is represented by a weight of from 0.15 to 0.30 kil.; *i. e.* from 0.380 to 0.760, or nearly ⅓d to ¾ths of a lb. avoird.; the latter term passed, the difficulty of working increases rapidly, and a very considerable force is required when the adherence to the same surface amounts to 0.70 kil. or 1.840 lbs. avoird.

The tenacity of a wet soil is not, however, in the direct ratio of its faculty of imbibition. Loams and loose calcareous soils, which absorb much more water than clay, are nevertheless much less tenacious; and then water actually makes sandy soils stiffer than they are when dry.

Every practical farmer knows how much more friable stiff wet soils become from the effects of frost. The water in expanding as it becomes solid pushes apart the molecules of the soil, and it is to this action that the advantages of autumn ploughing are with justice ascribed. M. Schübler found that the cohesion of a stiff clay which was equal to 68 fell to 45, when before it was tried the clay was exposed to the frost.

Disposition of the soil to become dry. The faculty of throwing off by evaporation any excess of water with which it may be charged, is as essential to constitute a good soil as is that of retaining moisture in due proportions. Those soils which throw off too slow ly the excess of moisture they have acquired during the winter occasion much trouble to the husbandman. They are perfectly unworkable in the spring, and consequently can only be sown very late. M. Schübler tried the retentive powers of the soil by the following method. A metallic disc, furnished with a narrow rim, was suspended to the arm of a balance. Over this disc, the soil to

* The abbreviate kil. in the above table signifies killogramme, a weight equal to 2.5 lbs. avoirdupois. As the weights are principally interesting in their relations to one another, it has not been thought necessary to reduce them to English weights.

be tried and previously brought to the point of saturation with moisture, was spread as evenly as possible. The weight of the disc thus charged was noted, and it was weighed anew, after having been kept for four hours in a temperature of 18.75° cent. (65.75° Fahr.) The weight of the water lost by evaporation was obtained by a second weighing; the complete desiccation of the soil was then completed in the stove. The following is the detail of one operation :

Weight of the moist earth	310
Weight after four hours' exposure to the air	260
Water evaporated	50
Weight of the moist earth	310
Weight after complete desiccation	200
Whole quantity of water contained in the soil tried	110

Thus 100 of water of imbibition lost 45.5 during exposure to the air for 4 hours at a temperature of about 66° Fahr. A more severely accurate method might readily be contrived, but that employed by M. Schübler appears sufficient for ordinary purposes. His results, in regard to the different kinds of soil he tried, are these :

Kinds of soil.	100 parts of the water contained in the soil lose in the course of 4 hours at 66 deg. Fahr.
Silicious sand	88.4
Calcareous sand	75.9
Gypsum	71.7
Sandy clay	52.0
Stiffish clay	45.7
Stiff clay	34.9
Pure clay	31.9
Calcareous soil	28.0
Humus	20.5
Garden earth	24.3
Arable soil of Hoffwyll	32.0
Arable soil of the Jura	40.1

Of all the substances examined, sand and gypsum are obviously those which allow the water to pass off most rapidly by evaporation. The calcareous or chalky soil again has a high retentive power: but it varies much in different instances, apparently in consequence of different degrees of fineness ; it is however surpassed by humus, and the garden soil which was tried. Humus is therefore at the head of the list of substances in reference to retentive properties.

All soils *shrink* more or less in drying, and form cracks, in the way already indicated ; the shrinking has been estimated by means of prisms of soils measured in the moist state, and after being dried in the shade :

Kinds of soil.	100 parts cube shrink to.
Carbonate of lime in fine powder	950
Sandy clay	940
Stiffish clay	911
Stiff clay	886
Pure clay	817
Humus	846
Garden earth	851
Arable soil of Hoffwyll	880
Arable soil of Jura	905

Gypsum, silicious, and calcareous sand do not appear in this table, because they do not shrink in drying. The humus appears to have shrunk the most; and dry humus is liable to swell in the same proportion when it is moistened. This property explains the obvious elevation of certain turfy or mossy soils at the period of the rains.

Hygrometric property of soils. Agriculturists allow, that those soils which have the property of attracting moisture from the atmosphere are generally among the most fertile. This hygrometric property must not be confounded with that in virtue of which moisture is retained. It appears to depend especially on the porousness of a soil, and, probably, also, in some degree, on the deliquescent salts which it contains, even in very small quantity. Davy was disposed to regard the hygrometric property of soils as a certain index of their good quality; and the experiments of M. Schübler upon the point, all tend to confirm the accuracy of this view. In M. Schübler's experiments, the increase of weight of dry soils was ascertained by exposing them for a certain time in an atmosphere kept at the point of saturation with moisture, and at the same temperature, between 60° and 65° Fahr.

Kind of soil.	500 centigrammes, or 77.165 grains troy, of soil, spread upon a surface of 36,000 millimetres, or 141.48 square inches, absorbed in—			
	12 hours.	24 hours.	48 hours.	72 hours.
	Grains.	Grains.	Grains.	Grains.
Silicious sand	0	0	0	0
Calcareous sand	.154	.231	.231	.231
Gypsum	0.077	0.077	0.077	0.077
Light clay	1.617	2.002	2.156	2.156
Stiffish clay	1.925	2.310	2.618	2.695
Strong clay	2.310	2.772	3.080	3.157
Pure clay	2.849	3.234	3.696	3.773
Chalky soil in fine powder	2.002	2.387	2.695	2.695
Humus	6.160	7.469	8.470	9.240
Garden earth	2.695	3.465	3.850	4.004
Arable soil of Hoffwyll	1.232	1.771	1.771	1.771
Arable soil of Jura	1.078	1.463	1.540	1.540

From the results comprised in the preceding table, we may conclude, first, that the faculty of absorbing lessens as soils acquire moisture; second, that humus is the most hygrometric of all the substances examined; third, that the clays which absorb the largest quantity of moisture are those which contain the smallest proportion of sand; and fourth, that silicious sand and gypsum do not absorb moisture in any appreciable quantity.

Absorption of oxygen gas by arable soils. Humboldt had already observed, before the year 1793, that argillaceous soils, the lydian stone, certain schists, and humus, deprived the air of its oxygen. He had also observed that the sides of the large cavities dug in the salt mines of Saltzburg, absorbed this gas, and thus rendered the stagnant atmosphere of the workings irrespirable and incapable of supporting combustion. Finally, this illustrious observer had satisfactorily ascertained, at the same period, that earth taken from the galleries of these mines, only became fertile after having been exposed to the atmosphere for a considerable length of time. I have quoted these curious observations because they are, so far as I know, the first which established the necessity of the presence of oxygen in the interstices of the soil, or, as M. Humboldt then said, and, indeed, as may still be maintained, the utility of a previous oxidation of the soil.

All our agricultural facts, indeed, confirm this view of the necessity of air in the interstices of the soil that is destined for the growth of vegetables. When, by ploughing very deeply, for example, we bring up a portion of the subsoil into the arable layer, in order to increase its thickness, we always lessen the fertility of the ground for a time; in spite of the action of manures, and of any treatment we may adopt, a certain time must elapse before the subsoil can produce an advantageous effect; it is absolutely necessary that it have been exposed to the atmospheric influences, and it is then only that deep ploughing, which gives the arable layer a greater thickness, pays completely for the expense it has occasioned.

I am disposed to ascribe the absorption of oxygen gas by clayey soils, to the oxide of iron, which they almost always contain, and which is in the minimum state of oxidation, when the clay lies at a certain depth. In the performance of some soundings in a tertiary soil of the department of the lower Rhine, which I performed in 1822, I had occasion to observe that the clays brought up white by the borer, very speedily became blue by exposure to the air; and that in gaining color they condensed oxygen. I propose returning upon this fact to show the important part which this simple superoxidation probably plays in the amelioration of soils.*

M. Schübler, again, has studied the action of oxygen gas upon the component parts of arable soils, and, according to him, the absorption of this gas cannot be doubted; it is very trifling in connection with sand and gypsum, very decided as regards clay, loam, and humus. As M. de Humboldt and M. de Saussure had already done, M. Schübler observed humus to change a portion of the oxygen which it fixed, into carbonic acid; but, in general, the other substances, or soils, or elements of soils, upon which he experimented, appeared to absorb the oxygen by the intermedium of the protoxide of iron, from which they are never altogether free. Besides this cause, due to the superoxidation of a metal, M. Schübler thinks that a certain portion of the oxygen disappears by condensation within the pores of some soils; and in support of his opinion he appeals to the admirable observations of M. de Saussure, on the condensation of the gases by porous bodies. Starting from the fact that the roots of plants require the presence of oxygen in order to thrive, he

* Austin proved, that during the oxidation of metallic iron under water, there is a constant production of ammonia. Certain experiments commenced some time ago, and which I still continue, will establish in the most precise manner, as I hope, the fact that this formation of ammonia likewise takes place during the passage of the protoxide of iron to the state of hydrated peroxide. The theoretical conclusions deducible from this fact, and the economic applications which may flow from it, must be obvious.

ascribes a greater power to the gases compressed or condensed within the interstices of the soil. But the action of the air upon the roots of vegetables is readily conceivable in soils of a loose nature, especially if they have been sufficiently worked, without the necessity of having recourse to such an explanation.

Capacity of soils for heat. The quantity of heat which a soil will receive, retain, or throw off in a given time, depends upon the conducting power which it possesses. M. Schübler endeavored to measure this power comparatively by measuring the rates of cooling. In a vessel of the capacity of 595 cubic centimetres, or 234.2 cubic inches, filled with the substance to be tried, a thermometer was placed, with its bulb in the centre. The temperature having been brought up to 62.5° C., (144.5° Fahr.,) the time was noted which each substance required to fall to 21.2° C., or about 70° Fahr., the temperature of the surrounding air being 16.2° C., or about 61° Fahr.

Kind of soil.	Power of retaining heat, that of calcareous sand being 100.	Time which 234.2 cubic inches of soil required to cool from 144° to 70° Fahr., the temperature of the surrounding air being about 61° Fahr.
		h. m.
Calcareous sand	100.0	3.30
Silicious sand	95.6	3.27
Gypsum	73.2	2.34
Sandy clay	76.9	2.41
Stiffish clay	71.1	2.30
Stiff clay	68.4	2.24
Pure clay	66.7	2.19
Calcareous soil	61.8	2.10
Humus	49.0	1.43
Garden earth	64.8	2.16
Arable soil of Hoffwyll	70.1	2.27
Arable soil of the Jura	74.3	0.36

The general observations which these experiments suggest, are that, for equal volumes, calcareous or silicious sand possesses greater powers of retaining heat than any of the other substances tried. This fact explains the high temperature and the dryness which sandy soils maintain even during the night in summer. Humus is obviously the substance which possesses the highest conducting powers.

Degrees in which soils become heated under exposure to the sun. There is no one who has not had occasion to observe the high temperature which bodies acquire when exposed to the rays of the bright sun. There are some, such as dry sand, slates, and certain colored rocks, which become burning hot. It is by the heat of the sun that the soil, before it is shaded by the leaves and stems of plants, rises in temperature, and throws off the excess of moisture which it had imbibed in the winter. Agriculturists all know how different the degree in which this heating takes place, even in soils that are close to one another. A light-colored, moist, clayey soil will heat much less than a dark-colored, calcareous, or sandy soil. The differences in the heat acquired in various soils depend, 1st, on the state of their surface ; 2d, on their composition ; 3d, on the quantity of water which they contain ; and, 4th, on the angle of incidence of the sun's rays. M. Schübler, by a method which is far from being unobjectionable, but which may be excused, considering the difficulties of the subject, measured the temperature acquired by different soils exposed to the sun for the same length of time, and in circumstances as nearly alike as possible ; the numbers obtained are given in the following table :

Soils	Highest temperature acquired by the upper layer, the mean temperature of the atmosphere being 25° C. (77° F.)	
	The soil moist, degrees centigrade.	The soil dry, degrees centigrade.
Silicious sand, yellowish gray ...	37.25 (99° F.)	44.75 (112°.5 F.)
Calcareous sand, whitish gray....	37.38	44.50
Bright gypsum, whitish gray......	36.25	43.62
Poor clay, yellowish	36.75	44.12
Stiff clay....	37.25	44.50
Argillaceous earth, yellowish gray	37.38	44.62
Pure clay, bluish gray	37.50	45.00
Calcareous earth, white..........	35.63 (96°.1 F.)	43.00 (109°.4 F.)
Humus, blackish gray............	39.75 (103°.5 F.)	47.37
Garden earth, blackish gray......	37.50	45.25
Arable earth of Hoffwyll, gray....	36.88	44.25
Arable earth of the Jura, gray.....	36.50	43.75

In comparing the circumstances which concur in assisting the action of the sun's rays in raising the temperature of the soil, it appears that the color and moistness of the soil and the angle of incidence of the sun's rays are the most influential ; they may occasion differences in the temperature acquired of from 14° to 15° C., (25° to 27° F.) The nature of the surface and the composition of the soil are far from producing such marked differences ; although, according to M. Schübler, the effect of inclination is very decided, and may occasion a difference to the amount of 25° C., (45° F.)

CLASSIFICATION OF SOILS.

Agriculturists class soils according to their fertility, and the cropping which they will stand to advantage. In practice two grand divisions have been adopted : *strong* soils, and *light* soils ; every soil belongs wholly or in part to one or other of these divisions.

In strong soils clay is the predominating element ; in light soils it is sand which prevails. The first are stiff, little permeable, and slow in drying ; the second are loose, dry speedily and readily, are permeable, and less difficult to labor. Humus always adds to the qualities of these two kinds of soil, though possessed of properties so opposite ; but its utility is especially remarkable in argillaceous or clayey soils, the extreme stiffness of which it diminishes.

Stiff or strong soils share in the advantages and disadvantages peculiar to clay; they absorb a great deal of moisture, and they do not dry readily, retaining obstinately a considerable quantity of water. The humus which they contain, and the manures which are spread upon them in the course of cultivation, remain with them for a long time, preserved as it were from the too active agency of atmospheric influences; the fertilizing power of these substances is further rarely interfered with by too great a degree of dryness in the soil. Nevertheless, in very wet seasons, and in years of extraordinary drought, the advantages which I have enumerated disappear. In wet seasons clay lands become immoderately humid, sometimes they approach the state of mere puddle; and on the contrary, under severe and long-continued drought, they become so hard that the roots of vegetables can no longer penetrate them, and then they crack in all directions, and the roots perish for want of being properly covered. I might add that severe frost is the cause of effects disadvantageous in the same degree; so that very stiff clays are liable to the same bad effects under the influence of two causes diametrically opposed: the great heat of summer and the severe cold of winter.

In such soils all agricultural operations are often impracticable; changed into a liquid mud, neither horse nor plough can be put upon them, or baked into a mass having the hardness of stone, the share will not penetrate them.

Light soils rarely accumulate an excess of moisture in their interstices, so that they are liable to suffer under want of rain of even short continuance. They are worked with infinitely greater ease, and at much less expense; vegetation upon them is quicker, and harvests earlier; but manure is less profitable than in clayey soils, because the rains dissolve and carry it away.

The defects of these two kinds of soils are precisely of a nature to compensate one another, and it is in fact by a mixture, or that which is equivalent to a mixture of these two extreme kinds of soil, that those lands are formed which are admitted to be the best adapted to cultivation, and the most fertile of all. Messrs. Thaer and Einhoff, in submitting to mechanical analysis an immense number of arable soils, and in studying at the same time the system of culture best adapted to these soils, and to their relative fertilities, have given us results of great importance, and which may be made the basis of a practical classification of arable soils.*

An argillaceous or clayey soil properly so called, generally contains about 40 per cent. of sand. If the quantity of sand be less than this, the crop from such a soil will be more or less precarious, and the tenacity will be such, that considerable difficulty will be experienced and necessary expense incurred in working it; such a clayey soil, (having at least 40 per cent. of sand,) when it contains a sufficient quantity of humus and is properly treated, may be regarded as favorable for wheat. Barley succeeds better than wheat, when

* Thaer's Rational Principles of Agriculture, (in French,) vol. ii. p. **115.**

the quantity of sand is as low as 30 per cent. With less than 30 per cent. oats will thrive. Wheat may still be advantageously cultivated upon lands that contain from 40 to 50 per cent. of sand; beyond this term, when the soil contains from 50 to 60 per cent. of sand, it is more advantageous to grow barley. Such a soil will not be completely pulverized by reiterated ploughing, as will that which contains a larger proportion of silicious matter, and it does not become hard and cracked under drought like lands that are more essentially clayey, because it retains a sufficiency of moisture; it is equally well adapted for trefoil of all kinds, for tubers, for plants with tap roots, and for many other crops of great marketable value, such as cabbage, flax, tobacco, &c. It is almost always accessible, a circumstance which allows of the greatest care being bestowed upon the crops which are raised upon it. In soils which yield on washing from 60 to 80 per cent. of sand, we cannot reckon securely on the success of wheat. At 70 of sand, it ceases to be well adapted to the cultivation of this grain, except with especial precautions; but it is still well adapted to barley, and it is in such a soil especially that rye succeeds best.

Land with such a dose of sand is always easily labored, but it is more apt to be overrun by foul weeds than a soil that is decidedly argillaceous. Manures are speedily consumed in it for the reason already given; it is, therefore, advantageous to manure such land frequently, laying on less dung at a time. A soil having 75 per cent. of sand, is qualified by Thaer as an oat soil, and even up to 85 per cent. of sand it may be regarded as suitable to this grain; this term passed, nothing but rye or buckwheat ought to be sown upon it, and that only after it has had a sufficient dose of manure. The reiterated ploughings which some of these sandy soils require to get rid of the foul weeds which rush up in such quantities upon them, sometimes render them so open that rye will not succeed. The best course is then to lay them down in grass, and allow them to become consolidated by rest.

It is extremely difficult, at least in this climate of ours, to make any thing of soils that contain 90 per cent. of sand; in times of drought they become true moving sands. As we have already shown, calcareous matter may replace silicious sand in the part which it plays in an arable soil; like sand, calcareous matter tends to destroy the strong cohesion of the particles of clay: but it appears that chalk or lime, especially when it is in a state of minute subdivision, besides this effect, really contributes to the amelioration of wheat lands.

The following table comprises the results obtained by Thaer and Einhoff. I must observe, however, and from causes which have been already explained as influencing the determination of the humus, that this substance is evidently estimated at much too high a figure in several of these analyses, which deserve to be made anew under the precautions that are now familiarly known.

Soils according to composition.	Usually designated.	Clay.	Sand.	Lime or chalk.	Humus.
Clay with humus	Rich wheat land.........	74	10	4	11.5
ditto..............	ditto	81	6	4	8.5
ditto..............	ditto	79	10	4	6.5
Marly soil	ditto	40	22	36	4
Light soil, with humus..	Meadow land	14	49	10	27
Sandy soil, humus......	Rich barley land	20	67	3	10
Argillaceous land.......	Good wheat land.........	58	36	2	4
Marly soil	Wheat land..............	56	30	12	2
Argillaceous land.......	ditto................	60	38		2
Stiffer argillaceous land.	ditto................	48	50		2
Clay	ditto................	68	30		2
Stiff argillaceous land...	Barley land of the 1st class	38	60		2
ditto..............	ditto 2d class	33	65		2
Sandy clay	ditto ditto	28	70		2
ditto..............	Oat land................	23.5	75		1.5
Clayey sand	ditto	18.5	80		1.5
ditto..............	Rye land	14	85		1
Sandy soil	ditto	9	90		1
ditto..............	ditto	4	95		0.75
ditto..............	ditto	2	97.5		0.5

Schwertz has given a summary of the opinions of Thaer upon the value of different soils from an eminently practical point of view. Agreeing with this distinguished agriculturist, that it is well to judge of the soil by its produce, he also forms a scale of comparison after the different kinds of grain, taking as extreme terms wheat and barley, the first succeeding in bad argillaceous soils, the second still growing in sandy soils of the poorest description. In these extreme or boundary soils, wheat and barley succeed very indifferently indeed; but between the two extremes are comprised every variety of soil which results from the fusion of the strongest or stiffest with the lightest soils, from the most tenacious clay up to loose sand. In these mixed soils of intermediate qualities, wheat and barley gradually approach one another, taking the place successively of barley, oats, and buckwheat, until they meet in the middle of the scale in a kind of neutral soil, upon which every variety of grain may be grown.

Schwertz arranged his scale in the following manner :*

0. Moving sand........................0. Stiff clay.
1. Rye land..........................1. Wheat land.
2. Rye and buckwheat land.............2. Wheat and oat land.
3. Rye, buckwheat, and oat land.........3. Wheat, oat, and barley land.
4. Rye, oat, and small barley land.......4 Wheat and large barley land.

5. Wheat, rye, barley, and oat land.

The species of soil which suit these different crops are :

1. Light dry sand.......................1. Cold stiff clay.
2. Moist, very slightly argillaceous sand...2. A lighter moist clay.
3. Argillaceous sand....................3. A warm dry clay
4. Sandy clay..........................4. Rich clay.

5. Clay.

* Precepts of Practical Agriculture, (in French,) p. 49.

The preceding considerations are more than sufficient to give a precise idea of what is to be understood in regard to the composition of arable soils. Nevertheless, with a view to making the subject more complete, I shall quote a few of the analyses of arable soils, published by different chemists at a time when a certain importance was attached to researches of this kind. I may remark generally, that from the whole of the analyses of good wheat lands which have hitherto been made, it appears that carbonate of lime enters in considerable quantity into their composition; and theory, in harmony with practice, tends to show that it is advantageous to have this earthy salt as a constituent in the manures which are put upon soils that contain little or no lime.

Analysis of a soil under the variety of rape called colza, by M. Berthier:

Silica	78.2
Alumina	7.1
Peroxide of iron	4.4
Lime	1.9
Magnesia	0.8
Carbonic acid	1.4
Water	5.8
	99.6

This soil was dried in the air, after having been reduced to powder; it lost 34 per cent. by drying. It is remarkable that it contains no trace of organic matter, the rather as it was held favorable for colewort. M. Berthier believes that this soil would gain in fertility by the addition of a certain quantity of calcareous matter, and M. Cordier* explains its inability to grow grain to advantage from the deficiency in lime. The stalk of the grain grown in this soil is weak, especially in wet seasons, and the seed is particularly apt to shake out when it is ripe.

If the presence of lime in a wheat soil is a guaranty against loss by shaking in harvest, M. Berthier's analysis is still far from proving that the presence of lime in a soil is indispensable, inasmuch as beautiful wheat crops are grown in the neighborhood of Lisle without lime. In proof of this fact, I shall here cite the analysis of one of the most fertile soils in the world, the black soil of Tchornoizem, which Mr. Murchison informs us constitutes the superficies of the arable lands comprised between the 54th and 57th degrees of north latitude, along the left bank of the Volga as far as Tcheboksar, from Nijni to Kasan, and stretching over a still more extensive district upon the Asiatic side of the Ural mountains. Mr. Murchison is of opinion that this land is a submarine deposite formed by the accumulation of sands rich in organic matters. The Tchornoizem is composed of black particles mixed with grains of sand; it is the best soil in Russia for wheat and pasturage; a year or two of fallow will suffice to restore it to its former fertility after it has been exhausted by cropping; it is never manured.

M. Payen found in this black and fertile soil:

* On the Agriculture of French Flanders, p. 232, (in French.)

Organic matter	6.95	containing 2.45 per cent. of azote.
Silica	71.56	
Alumina	11.40	
Oxide of iron	5.62	
Lime	0.80	
Magnesia	1.22	
Alkaline chlorides	*1.21*	
Phosphoric acid	*a trace.*	
Loss	*1.24*	
	100.00	

Bergman gives the following as the composition of a fertile soil in Sweden:

Carbonate of lime	30	
Gravel	30	Clay.
Silicious sand	26	Clay.
Alumina	14	Clay.
	100.0	

Several fertile soils of Senegal, examined by M. Laugier, contained:

	LOCALITIES.				
	Rawei.	Doukitt.	Diague.	Roso.	N'Dick.
Silicious sand and silica	87.0	72.0	89.0	78.0	91.0
Alumina	3.6	10.0	3.0	7.0	1.8
Oxide of iron	3.4	8.0	3.6	5.2	3.0
Carbonate of lime	trace	trace	0.5	trace	0.5
Organic matter and water	4.4	10.0	3.6	9.0	3.0
Loss	1.6	"	0.3	0.8	0.7
	100.0	100.0	100.0	100.0	100.0

M. Plagne, who has studied the agriculture of the Coromandel coast, divides the soils he met with there into argillaceous, or clayey, sandy, and mixed, and gives their several compositions as follows:

	Argillaceous.	Sandy.	Mixed.
Silica	22.0	82.0	64.0
Alumina	59.0	6.5	19.5
Carbonate of lime	3.5	3.5	2.5
Oxide of iron	2.5	4.0	4.0
Phosphate of magnesia	"	2.0	trace
Sulphate of lime	"	2.0	trace
Azotized organic matter	5.0	"	7.0
Water and loss	8.0	2.0	3.0
	100.0	100.0	100.0

The soils in which the tea-plant is grown in Assam and China, have been examined by Mr. Piddington;* they contain respectively:

	Chinese soil.	Assam soil.
Silica and sand	76.0	84.8
Alumina	9.0	4.5
Oxide of iron	9.9	7.0
Phosphate and sulphate of lime	1.0	traces
Organic matter	1.0	1.5
Water	3.0	2.3
	99.9	100.1

Sir Humphrey Davy found the various soils most generally cultivated in England, to have the following composition:

* Robinson, Account of Assam, p. 130.

Districts.	Silicious sand and gravel.	Silica.	Alumina.	Carbonate of lime.	Carbonate of magnesia.	Oxide of iron.	Salts and organic matter.	Sulphate of lime.	Water.	Loss.	Remarks.
County of Kent	66.3	5.2	3.3	4.8	0.8	1.2	8.0	0.5	4.9	5.0	Rich soil, under hops.
Norfolk	88.9	1.7	1.2	7.0	..	0.3	0.6	..	0.3	..	Ditto under turnips.
Middlesex	60.0	12.8	11.6	11.2	..	..	4.4	..	..	..	Very good soil, wheat.
Worcestershire	60.0	16.4	14.0	5.6	..	1.2	2.8	..	..	..	Very fertile soil.
Vale of the Teviot	83.3	7.0	6.8	0.7	..	0.8	1.4	..	..	..	Good soil.
Salisbury	9.1	12.7	6.4	57.3	..	1.8	12.7	..	..	..	Excellent grazing soil.

M. Gasparin has published analyses of the soils of the south of France. Instead of destroying the humus and organic matter by calcination, as Davy and the generality of analysts did, M. Gasparin dissolved out the humus by means of a strong alkaline solution. This method of procedure is, however, at least as liable to objection as the other.

Districts.	Humus.	Calcareous matter.	Clay.	Sand.	Remarks.
From Thor (Vaucluse)	7.5	92.5	6.0	1.5	Middling wheat soil.
Alluvium of the Rhone	3.4	2.3	53.5	42.7	Well adapted for madder, wheat & lucern.
From Palus, near Orange	2.5	55.5	43.5	1.0	Bad wheat land.
Old deposite of the Rhone	5.0	32.5	56.0	11.5	Good wheat land, very indifferent for madder.
From the plains near Orange	4.0	50.0	48.0	2.0	
Neighborhood of Auch	1.5	3.5	73.0	23.0	Ditto, bad for madder.

There is an important element which must always be taken into the account in estimating the value of soils, no matter what their special composition; this element is their depth, or thickness. In running a deepish furrow in a cultivated field, we generally distinguish at a glance the depth of the superficial layer, which is commonly designated as the mould or vegetable earth; this is a layer generally impregnated with humus, and looser and more friable than the subsoil upon which it rests. The thickness of this superficial layer is extremely variable; it is frequently no more than about 3 inches; but it is also encountered of every depth from 3 or 4 to 12 or 13 inches. It must be held an exceptional and unusual case when it has a depth of 3 feet or more. Nevertheless we do meet with collections of vegetable soil of great depth, deposited by rivers, washed down into the bottoms of valleys, or accumulated on the surface, as in the virgin forests or vast prairies of America. Depth of mould, or vegetable soil, is always advantageous; it is one of the

20

best conditions to successful agriculture. If we have depth of soil, and the roots of our plants do not penetrate sufficiently to derive benefit from the fertility that lies below, we can always, by working a little deeper, bring up the inferior layers to the surface, and so make them concur in fertilizing the soil. And, independently of this great advantage, a deep soil suffers less either from excess or deficiency of moisture; the rain that falls has more to moisten, and is therefore absorbed in greater quantity than by thin soils, and, once imbibed, it remains in store against drought.

The layer upon which the vegetable earth rests, is the subsoil, which it is of importance to examine, inasmuch as the qualities, and, consequently, the value of an arable soil, have always a certain relation with the nature and properties of this subjacent stratum. Frequently, and especially in hilly countries, the mineral constitution of the subsoil is the same as that of the soil, and any difference that the former may present is owing especially to the presence of humus, and to the looser condition which results from the growth of vegetables, from ploughing, &c., and not from atmospheric influences. By deep ploughing done cautiously, the thickness of the layer of arable land may be increased at the expense of the subsoil, and, when plenty of manure can be commanded, the operation will go on with considerable rapidity. Still it is maintained, and indeed in many cases it is unquestionable, that the soil loses temporarily some portion of its fertility by the introduction of a certain quantity of the subsoil, and that, under ordinary circumstances, several years elapse before any amelioration becomes perceptible.

In plains, in high table-lands, the analogy, in point of constitution, between the soil and subsoil is not so constant. In such situations the arable land is frequently an alluvial deposite proceeding from the destruction or disintegration of rocks situated at a great distance. When the superior strata possess properties that are entirely different from the subsoils, it may be understood how the vegetable earth may be improved by the addition of a certain dose of the subsoil, and this is the case in which the amelioration is the least expensive. The impermeability of the subsoil is one grand cause of the too great humidity of a cultivated soil. A strong soil, very tenacious through the excess of clay which it contains, has its disadvantageous properties considerably lessened, if the subsoil upon which it rests is sandy, 1st, from the evident amelioration which must result from an admixture of the two layers, and, next, because it is always a positive advantage in having a soil which has a strong affinity for water superposed upon a subsoil which is extremely permeable. The inverse situation is scarcely less desirable; a light friable soil will have greater value if it lies upon a bottom of a certain consistency, and capable of retaining moisture; with this condition, however, that the clayey layer shall not be too uneven in its surface, that it shall not present great hollows in which water may collect and stagnate; an impermeable subsoil, to act beneficially in such circumstances, must have a sufficient inclination to admit of its draining itself. The most essential distinction, then, in regard to the nature of subsoils is, into permeable and impermeable. Acquainted with the nature of vegetable earth, it is easy to judge of the advantages or disadvantages which will be presented by subsoil having the faculty of retaining or of permitting the escape of moisture.

In some situations, particularly upon the slopes of hills, the layer of arable land is of very limited thickness, and it is not uncommon to see it lying upon rocks of the most dense description, such as granite, porphyry, basalt, &c.; in such circumstances the substrata are unavailable, and there is nothing for it then in the way of amelioration except to transport directly vegetable earth from other situations. Mica schist is perhaps the least intractable rocky subsoil; the plough often penetrates it, and in the long run it becomes mingled with the arable layer. It is generally agreed that limestone rocks form a less unfavorable substrate. There are in fact some calcareous rocks which absorb water, and crumble away, and the roots of various plants, such as cinquefoin, penetrate them deeply; but there are many limestone rocks so hard that they resist all decomposing action for a very long period of time.

The qualities which we have thus far sought to determine in soils, do not depend solely on their mineral constitution or their physical properties, nor yet on those of the subsoils which support them. These qualities to become obvious require that the soils shall be placed in certain conditions which must not be left out of the reckoning. Such are those of the climate enjoyed and of the position more or less inclined to the horizon in one direction or another. The precepts which we have laid down are especially applicable to the arable lands of Germany, England, and France. But in generalizing it would be proper to say that clayey lands answer better in dry climates, and light sandy soils in countries where rains are frequent. Kirwan made this remark long ago in connection with numerous analyses of wheat lands. The conclusion to which this celebrated chemist came was this, that the soil best adapted for wheat in a rainy country must be viewed in a very different way with reference to a country where the rains are less frequent. The fertility of light sandy soils is notoriously in intimate relationship with the frequent fall of rain. At Turin, for example, where a great deal of rain falls, a soil which contains from 77 to 80 per cent. of sand is still held fertile, while in the neighborhood of Paris, where it rains less frequently than at Turin, no good soil contains more than 50 per cent. of sand. A light sandy soil which in the south of France would only be of very inferior value, presents real advantages in the moist climate of England.* Irrigation supplies the place of rain, and in those countries or situations where recourse can be had to it, the question in regard to the constitution of soils loses nearly the whole of its interest. Land that can be irrigated has only to be loose and permeable in order to have the whole of the fertility developed which climate and manure can confer. Sandy deserts are sterile because it never rains. Upon the sandy downs of the coasts

* Sinclair's Practical Agriculture.

of the Southern Ocean, a brilliant vegetation is seen along the course of the few rivers which traverse them; all beyond is dust and sterility. I have seen rich crops of maize gathered upon the plateau of the Andes of Quito in a sand that was nearly moving, but which was abundantly and dexterously irrigated.

A sandy and little coherent soil is by so much the more favorably situated as it lies in the least elevated parts of a district; it is then less exposed to the effects of drought; any considerable degree of inclination is unfavorable to such a soil, inasmuch as the rain drains off too quickly, and because it is itself apt to be washed away. It is to prevent this action of the rains, that the abrupt slopes of hills are generally left covered with trees; and the deplorable consequences which have followed from cutting down the woods in mountainous countries are familiarly known.

Strong soils, on the contrary, are better placed in opposite circumstances. A certain inclination is peculiarly advantageous to them; and, indeed, in working clayey lands that stand upon a dead level, we are careful to ridge them in such a way as to favor the escape of water.

In countries situated beyond the tropics, where consequently shadows are cast in the same direction throughout the whole year, the exposure of a piece of land is by no means matter of indifference. In our hemisphere, the lands which have a considerable inclination and a northern exposure, receive less heat and light, and remain longer wet than those that slope towards the south; vegetation consequently is less forward upon the former than the latter lands: but, on the contrary, the latter are less exposed to suffer from want of rain; and it is a fact, now well ascertained from data collected in Switzerland and in Scotland, that the slopes which descend towards the north, if they be only not too abrupt, are actually the most productive. This kind of anomaly is explained by the frequence and rapidity of the thaws which take place upon slopes that lie to the south. Frost, when not too intense, is certainly less injurious to vegetables than too rapid a thaw; and it is easy to understand that in situations where, from the mere effect of nocturnal radiation, vegetables are covered almost every morning through the spring with hoar frost, a rapid thaw must take place every day immediately after the rise of the sun. With a northern exposure, the frost occurs in the same measure; but the cause of its cessation does not operate so suddenly, the fusion of the rime being effected by the gradual rise in temperature of the surrounding air. In other respects, it is obvious that the advantages and disadvantages of different exposures are connected with the nature and the constitution of soils. The same may be said with reference to means of shelter from the action of prevailing winds. Stiff wet lands are much benefited by the action of free currents of air; our stiff soils at Bechelbronn remain impracticable for our ploughs during but too long a period of the spring, when they have not been well dried in the months of March and April by strong winds from the east. Light and sandy soils, again, require to be well sheltered. The whole object of studying the soil, is its amelioration; the industry of the agriculturist is, in fact, more effectually bestowed, and exerts a greater amount of influence upon the soil than upon all the other and varied agents which favor vegetation.

To improve a soil is as much as to say that we seek to modify its constitution, its physical properties, in order to bring them into harmony with the climate and the nature of the crops that are grown. In a district where the soil is too clayey our endeavor ought to be, to make it acquire to a certain extent the qualities of light soils. Theory indicates the means to be followed to effect such a change; it suffices to introduce sand into soils that are too stiff, and to mix clay with those that are too sandy. But these recommendations of science which, indeed, the common sense of mankind had already pointed out, are seldom realized in practice, and only appear feasible to those who are entirely unacquainted with rural economy. The digging up and transport of the various kinds of soil according to the necessities of the case, are very costly operations, and I can quote a particular instance in illustration of the fact: my land at Bechelbronn is generally strong; experiments made in the garden on a small scale showed that an addition of sand improved it considerably. In the middle of the farm there is a manufactory which accumulates such a quantity of sand that it becomes troublesome; nevertheless, I am satisfied that the improvement by means of sand would be too costly, and that all things taken into account, it would be better policy to buy new lands with the capital which would be required to improve those I already possess in the manner which has been indicated. I should have no difficulty in citing numerous instances where improvements by mingling different kinds of soil were ruinous in the end to those who undertook them.

A piece of sandy soil, for example, purchased at a very low price, after having been suitably improved by means of clay, cost its proprietor much more than the price of the best land in the country. Great caution is therefore necessary in undertaking any improvement of the soil in this direction,—in changing suddenly the nature of the soil. Improvement ought to take place gradually and by good husbandry, the necessary tendency of which is to improve the soil. Upon stiff clayey lands we put dressings and manures which tend to divide it, to lessen its cohesion, such as ashes, turf, long manure, &c. But the husbandman has not always suitable materials at his command, and in this case, which is perhaps the usual one, he must endeavor, by selecting his crops judiciously, crops which shall agree best with stiff soils, and at the same time meet the demands of his market, to make the most of his land. In a word, the true husbandman ought to know the qualities and defects of the land which he cultivates, and to be guided in his operations by these; and in fact it is only with such knowledge that he can know the rent he can afford to pay, and estimate the amount of capital which he can reasonably employ in carrying on the operations of his farm.

In an argillaceous or clayey soil, which we have seen above is the best adapted for wheat in these countries, it would be absurd to persist in attempting to grow crops that require an open soil. Clayey lands generally answer well for meadows, and autumn ploughing is always highly advantageous to them by reason of the disintegrating effects of the ensuing winter frost.

Chalk occupies a large space in recent formations ; as a general rule, the soil it supports immediately is of no great fertility. Sir John Sinclair proposed to improve such soil by growing green crops and consuming them upon the spot. Properly treated, the chalky soils of England produce trefoil, turnips, and barley, and they are particularly adapted to cinquefoin. It is doubtful whether in France, where the climate is not so moist as in England, chalky lands could be treated to advantage on the English plan. Recent inquiries have shown that chalk contains a small quantity of phosphate of lime, a salt, as we shall see by and by, whose presence is always desirable in arable lands.

Turf or turfy soils yield rich crops when we succeed in converting the turf into humus. The grand difficulty in dealing with turf is to dry it properly, inasmuch as it is generally found at the bottom of valleys or of old lakes and swamps. By a happy coincidence, turfy deposites frequently alternate with layers of sand, of gravel, of clay, and of vegetable earth, which have been accumulated at the same epoch. By a mixture, by a division of these different materials, preceded in every case, however, by proper draining, mere peat bogs may be turned into good arable soil. Pyritic turf, however, shows itself more intractable, it rarely yields any thing of importance. To improve such a soil it is absolutely necessary to have recourse to substances of an alkaline nature, such as chalk or lime, wood-ashes, &c., which have the property of decomposing the sulphate of iron which is formed by the efflorescence of the pyrites. Turfy lands can also be brought into an arable state, with the help of paring and burning. Scotch agriculturists, who are very familiar with reclaiming land of this kind, hold, that the best method of improving turf or bog lands, is to turn them into natural meadows. Where the wet and soft state of the soil does not allow cattle to be driven upon it, the crop of hay should only be cut once, the second crop should be left standing. By proceeding in this way mere bogs have been turned into productive meadows.* Turfy lands thoroughly drained and improved, present many advantages connected with their natural but not excessive moistness. In the neighborhood of Haguenau, magnificent hop-gardens are found upon bottoms of this kind ; madder also thrives in it equally well, and for certain special crops it is in my opinion one of the richest soils.

Sandy soils do perfectly well in countries which are not exposed to long droughts ; their cultivation is attended with little expense, and they grow excellent crops of turnips, potatoes, carrots, and rye ; but it is well to exclude clover, oats, wheat, and hemp, which require a soil of greater consistence. In southern countries, a system of irrigation is absolutely necessary, in connection with the

* Sinclair, Practical Agriculture

cultivation of sandy soils if they are not watered, they remain nearly barren; the only mode of making them productive is to lay them out in plantations of timber.

Those moving sandy plains of great extent, which are found in the interior of many continents, seem at first sight stricken with eternal barrenness. Nevertheless, the mobility of the sand of the desert, which permits it to be swept hither and thither, and to be tossed about like a liquid mass, depends less upon the total absence of argillaceous particles than upon the want of the moisture necessary to agglutinate or to fix its grains. The burning steppes of Africa and America have their oases here and there, the surface of which, moistened by a spring, is green with vegetation; and whenever sandy plains are bathed by a river, it is possible to render them fit for cultivation. In Spain, for instance, in the neighborhood of San Lucar de Baromeda, a powdery soil of extreme dryness has been fertilized by the hand of man. The mammillated downs of San Lucar are covered on the surface by a layer of quartzy sand, so loose that it is blown about by the wind; but by a happy disposition of things, a lower stratum of these downs is kept constantly moist by the waters of the Guadalquiver, and it is only necessary to remove the superficial sand, and to level the surface, in order to have a loose soil which unites in the highest degree two essential conditions of fertility, viz: openness, and a constant supply of moisture, which penetrates the soil in virtue of its permeability; under the influence of a fine climate and manure, the market gardens established in the midst of this desert are remarkable for the rapidity and the vigor of their vegetation. To avoid great expense, the labor of removing the sand is only undertaken in places where the layer is least thick; and what is removed being heaped up as a mound around the soil which is cleared, a kind of boundary wall is formed, which is not without its use in affording shelter, and which becomes productive itself by the plantations of vines and fig-trees that are made upon it with a view mainly to its consolidation. In the same way in Alsace, in the plains of Haguenau, the soil which was a moving desert of sand, has, in the course of less than forty years, become one of the most fertile under the influence of incessant cultivation; in the same way also it is that in Holland, mountains of sand, which had been accumulated by the winds, have been fixed. This sand, which rests upon a wet bottom, draws up the moisture by capillary attraction, and so becomes fit to support certain vegetables. These downs, which may be said to have come out of the sea, have a constant tendency in many places to encroach upon the cultivated lands. To oppose their progress, the Dutch sow them with the arundo arenaria, the long and creeping roots of which bind together the moving mass and imprison the particles of sand within a kind of net-work. These masses of sand become fixed in this way; but they remain nearly or altogether unproductive.

It is therefore a problem of the highest importance in many instances to fix permanently masses of sand blown up from the sea, bv covering them with productive plantations. This problem was studied and successfully resolved by M. Bremontier, a French engineer, who by sagacity in the choice of means and perseverance in their employment gave a complete and practical solution of the question among the downs of the Gulf of Gascony.*

The downs formed by the sand which is thrown up by the ocean between the mouths of the Adour and the Girond, occupy a surface of 75 square leagues and have a mean elevation of from 60 to 70 feet. They form a multitude of hillocks, which appear connected by their bases, the crowns of many of them rising to a height of 160 feet and upwards. Under the influence of the prevailing west winds, these masses of sand move with a mean celerity of about 80 feet per annum, covering forests and villages in their progress. A part of the little town of Mimizan is already invaded, and it has been calculated that in the course of twenty centuries, things proceeding at their present rate, the rich territory of Bordeaux will have completely disappeared. In their progress these moving masses of sand choke up the beds of rivers, and increase the disastrous effects they produce otherwise by causing formidable inundations.

The sands of the Gulf of Gascony, like those of Holland and the Low Countries, are not altogether without moisture; a very short way below the surface they are moist, and even present a certain degree of cohesion. This, in fact, might have been predicated, for otherwise the wind which brings them from the sea would have dispersed them in clouds of dust and to great distances; but no such dispersion takes place. Downs advance slowly, at the rate already indicated, and by rolling over, as it were, upon themselves. The sand driven by the wind creeps up on the flanks of the ridges as upon an inclined plane; after having got over the summit of the hillocks already formed, it falls down the opposite slope, and accumulates at the base. The action of the wind is only exerted upon so much of the sand as is rendered loose and moveable by its dryness; but the moist part is exposed, dried, and swept away in its turn; in this way the whole mass of sand which was at first deposited upon the west aspect of the hillocks is carried to the east, where it is in the shelter. By this process, under the influence of a wind which blew steadily for six days, a hillock has been seen to advance towards the interior of the country through a space of 3¼ feet.†

The moisture contained in the sand proceeds from the rains, from the surface water that filters through it and displaces the salt water which impregnated it originally. The very slight trace of sea-salt that finally remains it it has no unfavorable influence on vegetation.

Once aware of the fact that certain plants throve in the sands of downs, Bremontier saw that they alone were capable of staying their progress and consolidating them. The grand object was to get plants to grow in moving sand, and to protect them from the violent winds which blow off the ocean, until their roots had got firm hold of the soil.

* Annals of French Agriculture, vol. xxvii. p. 145.
† D'Aubuisson, Geognosy, vol. ii. p. 467.

Downs do not bound the ocean like beaches. From the base of the first hillocks to the line which marks the extreme height of spring tides, there is always a level over which the sand sweeps without pausing. It was upon this level space that Bremontier sowed his first belt of pine and furze seeds, sheltering it by means of green branches, fixed by forked pegs to the ground, and in such a way that the wind should have least hold upon them, viz., by turning the lopped extremities towards the wind. Experience has shown that by proceeding thus, fir and furze seeds not only germinate, but that the young plants grow with such rapidity, that by and by they form a thick belt, a yard and more in height. Success is now certain. The plantation so far advanced, arrests the sand as it comes from the bed of the sea, and forms an effectual barrier to the other belts that are made to succeed it towards the interior. When the trees are five or six years of age, a new plantation is made contiguous to the first and more inland, from 200 to 300 feet in breadth, and so the process is carried on until the summits of the hillocks are gradually attained.

It was by proceeding in this way that Bremontier succeeded in covering the barren sands of the Arrachon basin with useful trees. Begun in 1787, the plantations in 1809 covered a surface of between 9,000 and 10,000 square acres. The success of these plantations surpassed all expectation; in sixteen years the pine trees were from thirty-five to forty feet in height. Nor was the growth of the furze, of the oak, of the cork, of the willow, less rapid. Bremontier showed for the first time in the annals of human industry, that moveable sands might not only be stayed in their desolating course, but actually rendered productive. Like all other inventors, this benefactor of humanity was soon the object of jealousy among his contemporaries. Doubts were of course entertained at first of the possibility of consolidating the moving sands of downs; and when this was demonstrated, the honor of originality was denied him. The ingenious engineer defended himself with moderation, and demanded an inquiry; in the course of which it was satisfactorily proved that nothing of the same kind had been attempted by others previously to the year 1788. The labors of Bremontier must be regarded as another of those remarkable struggles which the industry of man has successfully waged with the elements.

CHAPTER V.

OF MANURES.

WHATEVER may be its constitution and physical properties, land yields lucrative crops only in proportion as it contains an adequate quantity of organic matter in a more or less advanced state of decomposition. There are favored soils in which this matter, designated by the name of *humus*, or mould,* exists by nature, while there are others, and they form the majority, which are either totally destitute of it, or contain it but in insignificant proportion. To become productive, these soils require the intervention of manure; for this there is no substitute, neither the labor which breaks them up, nor the climate which so powerfully promotes their fecundity, nor the salts and alkalies which are such useful auxiliaries of vegetation. Not that land entirely destitute of organic remains is incapable of producing and developing a plant. We have already seen that the atmosphere, light, heat, and moisture, suffice for its existence; but in such a condition, vegetation is slow and often imperfect; nor could agricultural industry be advantageously applied to a soil which approached so near to absolute sterility.

Plants, considered in their entire constitution, contain carbon, water, (completely formed, or in its elements,) azote, phosphorus, sulphur, metallic oxides united to the sulphuric and phosphoric acids, chlorides, and alkaline bases in combination with vegetable acids; many of these elements form no part of the atmosphere, and are necessarily derived from the soil. Moreover, the manures most generally made use of are nothing but the detritus of plants, or the remains or excretions of animals, including by the very fact of their origin the whole of the elements which constitute organized beings; and although it is very probable that certain tribes of plants are more adapted than others to appropriate the azote or the ammoniacal vapors of the atmosphere, experience proves that azotized organic remains contribute in the most efficacious manner to the fertility of the soil. Besides, we are far from being able to affirm that the carbon of plants is derived from the carbonic acid of the atmosphere. Doubtless this acid is its principal source; but it is possible that certain elements of carburetted dungs may be directly assimilated.

The writers who have treated of manures, have generally formed them into two grand classes:

1st. Manures of organic origin, in which are again found all the elements of the living matter.

2d. Mineral manures, saline or alkaline, which are particularly designated under the name of stimulants, thus ascribing to them the faculty, purely gratuitous, of facilitating the assimilation of the nutriment which plants find in dung, and of stimulating and exciting their organs. Such a distinction has no real foundation, and nothing shows so much how scanty our knowledge upon this subject has hitherto been as this tendency in the ablest minds to connect vegetable nutrition with the feeding of animals.

All the agents employed by the agriculturist to restore, preserve, and augment the fecundity of the soil, I shall term Manures. In my view gypsum, marl, and ashes are manures, as much as horse-dung, blood, or urine; all contribute to the end proposed in employing them, which is the increase of vegetable production. The best

* *Mould*, or vegetable earth, as the word is generally used, is not exactly *humus*; but as it derives its principal qualities from the presence of the humus of the chemist, I shall generally employ the terms as synonymous.—Eng. Ed.

manure, that which is in most general use, is precisely that which by its complex nature contains all the fertilizing principles required in ordinary tillage.

Particular cultures may demand particular manures; but the standard manure, such as farm dung, for example, when it is derived from good feeding, supplied to animals with suitable and abundant litter, affords all the principles necessary to the development of plants; such manure contains at once all the usual elements which enter into the organization of plants, and all the mineral substances which are distributed throughout their tissues; in fact, carbon, azote, hydrogen, and oxygen are found therein united with the phosphates, sulphates, chlorides, &c.

In order to be directly efficacious, every manure must present this mixed composition. Ashes, gypsum, or lime spread upon barren land, would not improve it in any sensible degree; azotized organic matter, absolutely void of saline or earthy substances, would probably produce no better effect; it is the admixture of these two classes of principles, of which the first is derived definitively from the atmosphere, while the second belongs to the solid part of the globe, which constitutes the normal manure that is indispensable to the improvement of soils.

Dead organic matter, subjected to the united influence of heat, of moisture, and of contact with the air, undergoes radical modifications, and passes by a regular course of transformation into a condition more and more simple. The tissues, so long as they form a part of the animated being, are protected against the destructive action of the atmospheric agents; in plants and animals this protection is not extended beyond the period of their existence; destruction commences with death, if the accessory circumstances are sufficiently intense; and then ensue all the phenomena of decomposition, of that putrid fermentation which, at the expense of the primitive elements of the organized being, generates bodies more stable and less complicated in their constitution, and which present themselves in the gaseous and crystalline conditions, forms which are affected by the inorganic bodies of nature in general.

The mineral substances which had been taken up in the organization become freed, and are thus again restored to the earth. The organized substances which change the most rapidly, are precisely those into which azote enters as a constituent principle. Left to themselves, whether in solution or merely moistened, these substances exhibit all the characteristic signs of putrefaction; they exhale an insupportable odor; and the result of their total and complete decomposition is finally the production of ammoniacal salts. The water wherein the phenomenon is accomplished facilitates it not only by weakening cohesion and enabling the molecules to move more freely, but it assists also by the very affinity which each of its own principles bears to the elements of the substance subjected to the putrescent fermentation. Proust observed that during the decomposition of gluten immersed in water, a mixture of carbonic acid and of pure hydrogen gas is disengaged, a phenomenon which he explains by the decomposition of the water; at the same time are produced ammoniacal salts, among which are acetates and lactates, whose acids are generated by the very act of fermentation. As a striking example of the agency of water in the transit of azote into the ammoniacal state in a quarternary compound, we may take the putrefaction of urea.

Urea is found in the urine of man and of quadrupeds; its composition, according to M. Dumas, is:

Carbon	20.0
Hydrogen	6.6
Oxygen	26.7
Azote	46.7
	100.0

The animal substances dissolved in urine, as the mucus of the bladder, &c., undergo, on contact with the air, a modification which causes them to act upon urea like ferments. By their influence the elements of water react upon this substance, and transform it into carbonate of ammonia.

Carbonate of ammonia is composed of:

Carbonic acid 56.41, containing	Carbon	15.39
	Oxygen	41.02
and Ammonia 43.59, containing	Hydrogen	7.69
	Azote	35.90

But 100 of urea have been found to produce by fermentation 130 of carbonate of ammonia.

	Carbon.	Hydrogen.	Oxygen.	Azote.
Previous to fermentation, 100 of urea contains	20.00	6.60	2.67	46.7
After fermentation, 130 of carbonate of ammonia contains	20.00	10.00	53.3	46.7
Difference	0.0	3.4	26.6	0.0

So that during its transformation, the urea has gained 3.4 of hydrogen, and 26.6 of oxygen.

In water the hydrogen is to the oxygen as 1 to 8. (: : 1 : 8.)

Now it is precisely in this proportion that hydrogen and oxygen are found to be acquired by the urea in passing into the state of carbonate of ammonia; whence it follows that the elements of water are fixed in the process.

The putrefaction of azotized substances is far from always presenting results equally precise; most frequently in decomposing they pass through a series of changes, still very obscure, before they attain their ultimate limit, viz. the production of ammoniacal salts. Thus from putrefying caseum diffused in water, M. Braconnot obtained, among other products and ammoniacal salts, a very remarkable substance which he calls aposepedine.

Aposepedine when purified is a white crystalline substance, soluble in water and in alcohol, capable of combination with the metallic oxides; azote is one of its elements.

This substance, although engendered by the act of putrefaction, is nevertheless itself capable of putrefying and giving birth to the last products of the spontaneous decomposition of azotized matter.

One of the most striking characteristics, at least that which is most readily remarked, is the fetid odor which animal substances exhale during putrefaction. It is not always the smell of ammonia which predominates; that of sulphuretted hydrogen gas is often very strong; yet that is not the emanation which is most repulsive: miasmata and nauseous principles are also developed which seem to be the changed matter itself carried away by the gases that are disengaged.

Sulphur, like phosphorus, is almost always a constituent of organic bodies; its minute proportion, however, would be insufficient to give out the hepatic odor so intense as we often find it during putrefaction. The production of sulphuretted hydrogen is connected with the very curious fact, first appreciated by M. O. Henri, that sulphates dissolved in a medium impregnated with azotized matter in decomposition, do themselves undergo an actual reduction, pass into the state of sulphurets, and immediately give out sulphuretted hydrogen, either by the action of the carbonic acid of the atmosphere, or by that which is formed during the putrefaction of the organic matter. It is by a similar action exerted upon sulphate of lime that M. Henri explains the origin of the sulphureous waters of Enghien, near Paris; and M. Fontan in his important work on mineral waters has generalized this explanation.

The causes of the destruction of sulphates under such circumstances is easily understood. During the decomposition of organized substances, the carbon belonging to them forms carbonic acid gas; by combining both with the oxygen of the substances themselves, and with the oxygen of the water, it is probable that the oxygen of the sulphuric acid contributes equally to this formation, and that the sulphur is liberated.

The hydrogen of the decomposed water, as well as that of the solid matter, in contact with sulphur in the nascent state, combines with it to form sulphuretted hydrogen, which straightway reacts upon the base of the sulphate, producing from it, as we know, water and a metallic sulphuret. This sulphuret being unable to exist when exposed to the continued disengagement of carbonic acid gas which takes place in the centre of the mass in putrefaction, yields, as a definite result, a carbonate on the one part, and sulphuretted hydrogen on the other.

The faculty which azotized organic bodies possess of undergoing spontaneous decomposition in presence of water, and under the influence of heat, seems to depend upon the tendency which azote has to unite with hydrogen in order to form ammonia.

This tendency is perhaps the determining cause of the phenomenon of fermentation taken in the most general acceptation of the term. Organic bodies void of azote decompose less easily, and the kind of alteration which they undergo from the action of water and air, differs in many respects from the putrefaction of azotized matter. Of this the difficulty experienced in effecting the fermentation of vegetable substances is a proof. Nevertheless, the vegetable refuse which goes to the dunghill, all without exception, contains azotized elements, often, it is true, in extremely minute proportions;

but we know that there is no example of a vegetable organic tissue from which they are completely excluded.

The refuse of plants, the most amply endowed with azote, such as cabbage, beet-root, beans, &c., are certainly those which are susceptible of the most rapid and complete putrid fermentation; straw, on the contrary, when alone, undergoes it slowly and imperfectly, the small quantity of the azotic principle which it contains is changed, and reacts upon the ligneous particles which surround it; but the effect is soon arrested, and even ceases entirely, unless substances richer in azote concur. The woody matter of the straw is exactly in the condition of sugar which has not had a dose of ferment sufficient for its total transformation into alcohol.

Most organized substances, whether they belong to the animal or vegetable kingdom when placed in certain conditions, undergo the profoundest changes from the action of hydrogen; these alterations ought to be studied with all the more care, because in practical agriculture we are interested successively in fostering or preventing the causes which produce them, according as our object is to accelerate the decomposition of vegetable refuse for manure, or to adopt the precautions which experience suggests, in order to preserve the produce of our harvests unchanged. Organic substances moistened and exposed to the air under a temperature, the minimum of which (I believe after several experiments) may be fixed at 48° or 50° F., seize upon the oxygen and absorb it, in part, in order to form water with their hydrogen, and carbonic acid with their carbon. When these substances are accumulated in a mass sufficiently great, the heat produced is not rapidly dissipated, the temperature rises, and promotes the reaction to such a degree as to produce active burning, a conflagration, in place of the slow combustion manifested at first. It is not very unusual to see hay, which had been gathered in too damp a condition, take fire in the stack; and the high temperature acquired by wet rags placed in the fermenting vats of paper mills, and the copious production of carbonic acid which results, show that we are right in assimilating this kind of action to the phenomenon of combustion. This sluggish combustion is not peculiar to azotized organic substances: it takes place equally in those destitute of azote. The alteration of organic matters, the combustion which goes on at a low temperature by the action of the air, differs in its results from the decomposition which is effected in the midst of a liquid mass. We have seen, for example, that gluten fermenting under water, disengages hydrogen gas. Now Berthollet has established, and Saussure has confirmed his observations, that an azotized body in putrefaction, the whole of whose parts are in contact with the air, never contributes either hydrogen gas or azote to the confined atmosphere in which it is placed; and on the other hand, Saussure has shown, that organic substances which do not emit hydrogen gas during their spontaneous decomposition in a medium void of oxygen, do not alter the volume of an atmosphere of which this gas forms a part; on the contrary, these substances condense oxygen as soon as they attain that stage of their alteration in which they give out hydrogen.

In pursuing with persevering sagacity the study of putrefaction, M. de Saussure discovered the cause of this condensation; it consists in the fact, that an organic substance in course of spontaneous decomposition, acts in some respects like the platina sponge placed in a mixture of oxygen and hydrogen gas; we know that platina recently heated and introduced into a mixture of these two gases, determines their union in the proportions required to constitute water. Now by substituting for the metal some moistened seeds, previously deprived of their germinating faculty, the same effect is produced: the two gases combine until one of the two entirely disappears. When this combustion of hydrogen, proceeding from the decomposition of organic substances, takes place in the body of the atmosphere which contains azote, it is possible that a minute quantity of ammonia may be produced together with water; nor is it going too far to suppose, that manures very slightly azotized, take up azote from the atmosphere during their fermentation; that during the act of vegetation itself, the hydrogen proceeding from the decomposed water, or yet more, that which makes part of the essential oils formed by plants, may, in oxidating afresh, introduce atmospheric azote into their composition.

Dead organized matter, such as wood, straw, or leaves, exposed to wet, and for a long time to the action of the air, ends by becoming transformed into a brown substance, which when damped is almost black, which falls to powder when dry, and which is commonly known by the name of humus or mould.

This is, so to speak, the last term of the decomposition of organic matter; mould appears to belong already to the mineral kingdom; and whatever may be the diversity of its origin, it presents a sufficient number of peculiar characteristics to be considered a distinct substance. In fact, the atmosphere continues to exert its action upon mould; its inflammable elements are dissipated by a slow and imperceptible combustion, giving place to water and carbonic acid. But in this ulterior decomposition, those fetid products which characterize putrid fermentation are no longer observed.

Wet saw-dust, placed for some weeks in an atmosphere of oxygen, forms a certain quantity of carbonic acid, and the volume of the gas does not perceptibly diminish. The wood at the surface acquires a deep-brown color. Several experiments made by Saussure, prove that dead wood does not absorb the oxygen gas of the atmosphere; it transforms it into carbonic acid, and the action takes place as if the carbon of the organic matter alone experienced the effects of the oxygen; for the gaseous volume remains the same. The loss, however, undergone by the ligneous fibre, during its exposure to the air, is greater than it ought to be if carbon alone were eliminated: whence Saussure concludes, that while the woody matter loses carbon, it also lets some of its constituent water escape.

As a consequence of these observations, the relative proportion of carbon ought to augment in the humid wood changed by the action of the atmosphere, since we have established that by this action the woody fibre suffers loss in the elements of water besides what it loses in carbon. This is confirmed by the following analyses. The first, made upon some oak wood, previously purified by washing in water and in alcohol, we owe to Messrs. Thénard and Gay-Lussac; the succeeding one to Messrs. Meyer and Will:

	Oak wood.	Id. rotten.	Id. rotten.
Carbon	52.5	53.6	56.2
Hydrogen and oxygen, water	47.5	46.4	43.8
	100.0	100.0	100.0

Wood decomposed under water, without being in direct contact with the air, undergoes a different modification; it is blanched instead of blackened, and the carbon far from increasing is diminished. Saussure thinks that this kind of alteration depends mainly on the loss of the soluble and coloring principles of the wood, principles containing more carbon than the ligneous matter itself: so that pure woody fibre exposed wet to the action of the air would yield products analogous to those which result from its decomposition under water.

The damp linen rags which are fermented in paper manufactories afford a product which is white, and but slightly coherent. The mass, which heats successively during the operation, loses about 20 per cent. of its original weight. This is exactly what takes place in wood decayed by the alternate action of water and air, namely, it becomes white and extremely friable.

Some oak arrived at this stage of decomposition contained, according to Liebig:

Carbon	47.6
Hydrogen	6.2
Oxygen	44.9
Ashes	1.3
	100.0

Compared with the composition of oak wood when sound, these numbers show that during its modification the wood has lost carbon, and that it has gained hydrogen. The elements of water must necessarily have intervened, and become fixed during the reaction. Ligneous fibre decaying under water is not thereby completely protected from the atmosphere. Water always holds some air in solution, and the oxygen of that air reacts exactly as if it were in the gaseous state.

Upon all the phenomena of decomposition connected with fermentation, with putrefaction, or with dilatory combustion, heat exerts an influence which has certainly not been sufficiently appreciated. Organic bodies sunk in a large mass of water are not exposed to changes of temperature so various and abrupt as when they are placed in the atmosphere; their decomposition is more gradual, more uniform, and the soluble materials which they contain, or which are the result of the alteration they are undergoing, are in a great measure dissolved. Temperature may also produce great differences in the final result of the decomposition. Peat, which is derived, as we know, from the slow decomposition of submerged plants, does not appear to be formed in the swamps of warm climates; it has, perhaps, never been encountered in the stagnant waters of the equinoctial regions; there the woody fibre appears to be totally dissipated in carbonic acid gas, and in marsh gas, the probable source of the insalubrity of those countries. Lakes with peat bottoms are not found except on the very high table-lands of the Andes, in localities where the mean temperature does not exceed 49° or 50° F.

The alkalies are powerful agents in the decomposition of certain organic substances, whether in determining or in accelerating it. There are some, indeed, which experience no change without their intervention, whatever may be the other conditions favorable to decomposition. Thus, according to M. Chevreul, many coloring substances may be preserved in solution almost indefinitely, without change, in contact with gallic acid; but the presence of a very small quantity of free alkali suffices for their immediately acquiring the power of absorbing oxygen, and at the same time of acquiring a brown tint. M. Chevreul observed that 3.087 grs. of hematine dissolved in potash will absorb 3.857 grs. of oxygen in forming 0.925 of carbonic acid. The oxygen which enters into the carbonic acid, therefore, represents nothing like the quantity which was fixed by the solution, and it is almost certain that this gas likewise reacted upon the hydrogen of the coloring matter. The use of the alkalies for accelerating the destruction of organized matter has been long known to agriculturists.

Straw, fern, and the ligneous parts of plants are sometimes stratified with quick-lime, in order to facilitate their disintegration, and consequently their decomposition. The utility of this old practice cannot be disputed while confined within certain limits; but it is often abused; for it is beyond doubt that alkalies mingled indiscriminately with manure become in reality more injurious than advantageous for the end proposed in their introduction.

The appearance of a certain brown substance, little soluble in water, but easily dissolving in alkalies, is a characteristic proper to all vegetable matter under decomposition; a characteristic which becomes more marked as the decomposition advances towards its last stage, namely, the production of humus. This substance is ulmine, which, on account of some acid properties which it possesses, is also named ulmic acid. It forms a part of mould, and M. P. Boullay constantly found it in the water of dunghills.

In 1797, Vauquelin discovered ulmine united with potash in the matter of the exudation from the ulcer of an elm-tree.

In 1804, Klaproth confirmed this observation. M. Braconnot succeeded in obtaining ulmine artificially by subjecting woody fibre to the action of alkalies. This substance is easily procured by carefully heating in a silver capsule, and continually stirring a mixture of equal parts of potash and of saw-dust slightly damped. At a certain time the woody matter softens and suddenly dissolves; the mass then begins to swell up, and the fire is slaked. The product obtained dissolves almost totally in water. The solution is of a very deep brown color, and contains as principal ingredient ulmine combined with potash; the ulmine is precipitated by the addition of a sufficient quantity of weak sulphuric acid. After having been washed and dried, ulmine is black and brittle, and resembles jet; while still wet, it reddens turnsole paper, and its solution in potash forms with several salts, and by the way of double decomposition, insoluble ulmates. M. Peligot assigns to ulmine the following composition:

Carbon	72.3
Hydrogen	6.2
Oxygen	21.5
	100.0

Dunghills, rotten wood, and mould, always contain a brown substance, which possesses properties very similar to those which characterize ulmine obtained by the action of alkalies upon ligneous fibre.

Mould which contains this ulmine in abundance, and in the condition most favorable to vegetation, ought on that account to be examined with attention. Its history has, indeed, been so ably traced by M. de Saussure, that science at the present day can add but little to the important deductions of the celebrated author of the *Recherches Chimiques*.

M. de Saussure defines vegetable mould (humus) to be the black substance which covers dead vegetables after they have been long exposed to the combined action of water and oxygen. His experiments refer to mould nearly pure; that is, separated by a fine sieve from the vegetable remains which are always mixed with it; to mould which had been gathered on high rocks, or from the trunks of trees, where it could not have been exposed to admixture or to any influence, other than that of the spontaneous decomposition by which it had been produced. All the varieties of mould collected in this way appeared fertile, especially when they were previously mixed with gravel, which supplies support to the roots of plants, and permits the access of the air. That variety, however, must be excepted which was obtained from the interior of trees, and had been formed in such a situation that the rain-water which entered found no free outlet; the humus then contained extractive principles, derived in part from the living plant, and which seemed to obstruct the pores of the vegetable to which it was applied as manure.

In making comparative calcinations in close vessels of different varieties of humus, and of plants similar to those from which they had proceeded, and collecting the charcoal on one hand, and the volatile and gaseous matters on the other, M. de Saussure discovered that they contained, for the same weight, a larger quantity of carbon and of azote than the vegetables whence they proceeded. The larger proportion of azote in the humus seems to imply that during their decomposition, vegetables do not throw off this element; but to this cause must be added that which might be connected with the spoils of insects which live in humus.

Weak acids have no other effect upon humus than to dissolve out the metallic, earthy, and alkaline elements which it contains. The more powerful acids, such as the sulphuric acid, frequently cause a disengagement of acetic acid. Alcohol scarcely acts upon humus, merely dissolving out of it a few hundredth parts of resinous matter, which probably pre-existed in the vegetable. Potash and soda dissolve humus almost completely, causing an evolution of ammonia. From this solution, acids throw down a brown, inflammable powder, possessing the characters which we have recognised in ulmine. The ulmine which is separated in this way, is far from corresponding with the weight of the matter treated with the alkalies, which is evidently due to the humus containing principles which are not precipitated from the alkaline solution.

A quantity of humus which yielded no more than one tenth of ashes by incineration, only lost one eleventh of its weight under repeated treatments with boiling water. The humus thus exhausted, was exposed in a moist state to the action of the air for three months, and gave a new quantity of soluble matter under renewed washing with water; and the same effect is constantly reproduced. By exposing moist insoluble humus to the air, therefore, a quantity of soluble extractive matter is formed. This matter, obtained by evaporating the water which is charged with it, is not deliquescent; it yields ammonia on distillation. The watery solution, brought to the consistence of sirup, is neutral to re-agents, and its taste is sensibly sweet.

It is familiarly known that the alkaline salts, which enter into the constitution of vegetable juices, but rarely exhibit the reactions that are proper to them; the plant or the sap must be dried and incinerated before their presence can be ascertained. It is the same with regard to the salts contained in humus.

Humus, as I have already observed, is the last term in the putrefaction of vegetable organic matter; its elements have acquired a stability which enables them to resist all fermentation. M. de Saussure preserved humus for a whole year in vessels filled with distilled water, and plunged in mercury, without remarking any emission of gas. Still it is unquestionable that the organic portion of humus is completely destructible when exposed moist to the action of the air; in the course of time it is dissipated, and by and by there remains nothing more than the fixed saline and earthy matters which it contained. This fact M. B. de Saussure had already perceived from his observations upon the vegetable soil that occurs in the country between San Germano and Turin. This destructibility of vegetable earth, says M. de Saussure, sen., is a fact without exception; and as often as agriculturists have proposed to supply the place of manure by repeated ploughings, they have had sad experience of its truth: the soil is gradually impoverished, and fertile fields ultimately become barren. I may add, that the nature of the climate has a vast influence upon the dissipation of the fertilizing principles of the soil, and that Europeans are certainly in error when they object to the superficial ploughings or hoeings which the land so commonly receives in tropical countries. It is there well known that too much stirring of the soil is often prejudicial even in irrigated lands, where consequently the bad effects cannot be attributed to too great a degree of dryness. The information which has lately reached the Academy of Sciences upon the agriculture of the French possessions in Africa, tend to make us perceive that the same cause produces the same effects in Algeria, and that it is not without reason that the Arabs only work their lands that are preparing for grain crops, very superficially.

Humus is, in fact, dissipated by a process of slow combustion in the air: in contact with oxygen, it produces carbonic acid, as is proved by the experiments of M. de Saussure. Pure humus, moistened with distilled water, confined in bell-glasses placed over mercury, formed carbonic acid, causing the disappearance of the oxygen of the air. The volume of the acid gas formed, corresponded in volume with that of the oxygen which had disappeared. Humus, therefore, in contact with air, gives off carbonic acid, and the phenomenon here still takes place as if carbon were not alone consumed. The loss experienced is greater than that which ought to occur from the quantity of carbon which unites with the oxygen; and Saussure concluded that there is, at the same time, a loss of the elements of water. The capital fact which results from these experiments of Saussure, the deduction directly applicable to the theory of manures is this: that humus is dissipated when it is exposed to the air, and that during the slow combustion which it undergoes, it is a constant source of carbonic acid gas.

To complete the views that may throw light on the part played by manures, I have still to speak of an important phenomenon which occasionally takes place under the same conditions as those that accompany the decomposition, the putrefaction of animal matters: I mean the spontaneous formation of nitric acid—the occurrence of nitrification as it is called. Nitric acid results from the union of azote with oxygen. Such at least is the constitution of this acid when it is combined in salts; but in its isolated state, it is always united with a certain quantity of water. It has not yet been obtained, and it appears indeed not to exist, in the perfectly dry or anhydrous state. The azote, therefore, does not combine directly with the oxygen; there must be, at all events, the intervention of water, and to effect the union of the two gases by means of the electric spark, the mixture, according to Cavendish, must be moist. Nevertheless, the combination of azote with oxygen appears to be singularly favored by the presence of earthy or alkaline bases, seeing that in nature the nitrates are met with in a certain abundance; but the circumstances which determine their formation are still involved in deep obscurity.

Three distinct origins may be assigned to the natural nitrates; 1st. certain soils, still indifferently studied, show an efflorescence of nitrate of potash on their surface, or by lixiviation yield large quantities of this salt. Such is the source of the saltpetre which is imported from India.

According to M. Proust, the soil of certain localities in the neighborhood of Saragossa is an inexhaustible mine of saltpetre. I have myself seen, near Latacunga, a short way from Quito, upon a soil formed of trachytic debris, a similar production of nitre taking place as it were under my eyes.

2d. On the coast of Peru, in the desert of Tarapaca, at a short distance from the port of Iquique, and in an argillaceous soil of extremely recent formation, there are numerous stratified deposites of nitrate of soda, analogous to, and perhaps contemporaneous with, the deposites of common salt which are worked upon the same coast, in the desert of Sechura, near the equator. This is, so far as I know, the only instance of a nitrate being dug out of the bowels of the earth as a mineral mass. The nitrate of soda of Tarapaca, reaches Europe at the present time in large quantities, and supplies the place of nitrate of potash in many chemical processes. Various experiments have also been made upon the value of the salt as a manure; but at present these experiments have been very contradictory, and further experience seems necessary before any definitive judgment can be come to on the matter.

3d. The greater number of the soils that are exposed to animal emanations—heaps of rubbish proceeding from buildings that have been long inhabited, the soil of stables, cow-houses, cellars, &c., almost always contain a quantity of nitrates. In countries where rain seldom falls, and where consequently these salts, which are extremely soluble, can accumulate in the soil, in Egypt, for example, the ruins of ancient cities are at the present time true nitre-beds. It is with the formation from nitre in such circumstances that we feel particularly interested. The presence of the salt is frequently proclaimed in our agricultural operations; it is formed during the preparation of our dunghills, in the midst of our cultivated fields, and we discover it in the plants which we gather. We are by so much the more interested in discovering its existence, and in ascertaining its mode of action, as in the actual state of our knowledge we are still unable to say whether or no nitre is an auxiliary in the phenomena of vegetation, and contributes to the production of the azotized principles which enter into the organization of plants.

To have nitrates formed, the presence of azotized organic matter is not sufficient; it is further necessary that this matter during its decomposition be in contact with alkaline, calcareous, or magnesian carbonates. It has been observed that rocks of a crystalline structure do not nitrify so readily when they are without the substances which have just been named. The calcareous and magnesian rocks which are most favorable to nitrification, under the influence of animal emanations and of vegetables in a state of decomposition, are those which are the least coherent, or which are most porous, such as chalk, tufa, &c. In those countries where the soil does not undergo spontaneous nitrification, certain arrangements of circumstances, known to favor the production of saltpetre, are made: artificial nitre-beds are prepared. In the north of Europe where the rocks are granitic, in a hut or shed built of wood, a mixture is made of common earth, of calcareous sand or marl, and of wood-ashes. This heap is watered with the urine of herbivorous animals, and the mixture is stirred or shifted from time to time to favor the access of the air; and with the same view, the workmen are very careful never to beat or press the heap, which is generally from two to two and a half feet in thickness, and usually of the whole length of the hut or shed. Experience has shown that the process of nitrification goes on best in the shade. In Prussia, the practice is to wet with the water of a dunghill a mixture composed of five parts of vegetable earth, and one part of wood-ashes and straw. With this kind of mortar, solid walls or masses from twenty to four and twenty feet in length, by about six feet and a half in thickness, are built, rods of wood being introduced during the construction in considerable numbers, and in such a way that they can be pulled out when the mass has acquired sufficient solidity; by this means it is obvious that a very free access of air is secured to the interior of these nitre walls, which are always built in damp places, and thatched over with straw to preserve them both from the sun and the rain. The mass is watered from time to time, and after the lapse of a year, the materials are held sufficiently impregnated with saltpetre to be worth lixiviating.

In these artificial nitre-beds we perceive the object to be, to combine the circumstances under which the nitrates are formed in the soil of stables, and in the cellars of human habitations. Organic matters, rich in azote, are, in fact, brought into contact with earthy alkaline carbonates. The necessity that is felt in the arrangement of nitre-beds for the introduction of substances of animal origin, leads us to presume that the greater part of the nitric acid which is produced, is derived from the azote of these substances. But whether this azote combines with the oxygen of the air, or with the oxygen of the organic principles, we do not know—we are still ignorant of the way in which the acidification is effected.

Professor Liebig, setting out from the fact that azotized organic substances always produce ammonia during their putrefaction, and next perceiving that during the combustion of ammoniacal gas, mixed with a large excess of hydrogen, there is always oxidation of the azote, concludes that nitrification is the result of the slow combustion of the ammonia which is the product of the azotized matters in progress of decomposition. The azote of ammonia is indeed oxidated under favor of divers conditions which it is easy to secure. In burning animal substances by means of oxide of copper, it is well known how many precautions must be taken to prevent the appearance of nitrous acid; and on the contrary, by taking measures to favor the production of this acid, for example, by passing a current of ammoniacal gas over peroxide of iron or manganese in a red hot tube, abundance of nitrate of ammonia is obtained. The same result follows exposure of a mixture of oxygen and ammoniacal gas to the action of incandescent spongy platinum. The determining cause of the acidification of the azote, which forms an element of the ammonia, is probably due to this, that during the combustion two bodies are formed which are capable of combining immediately: nitric acid, on one hand, and on the other water, without which this acid could not exist. The phenomenon, however, only takes place in this instance at a considerably elevated temperature. At ordinary temperatures, combustion of the elements of ammonia has not, as far as I know, yet been observed; and in a series of experiments which I undertook, proceeding all the while upon ideas completely in conformity with those advanced by Liebig, I did not succeed in forming any nitrates by enclosing mixtures of chalk, potash, &c., in an atmosphere composed of oxygen and ammoniacal vapor.

In a communication made to the Academy of Sciences, M. Kuhlman announces that he had ascertained the presence of nitrate of ammonia in the products of the putrefaction of animal matter. If this announcement be confirmed, if nitric acid be in reality one of the numerous products of the putrid fermentation, the nitrification of soils in contact with organic matters would be readily explicable. I must say, however, that I have sought in vain for nitrate of ammonia in the product of the putrid fermentation of caseum. And after all, we should still be at a loss to account for the formation of nitre in many places, where it appears to be produced in the absence of organic matter, as in the saltpetre soils of India, South America, and Spain. Dr. John Davy, who visited the nitre districts of Ceylon, and Proust, who long inhabited the Peninsula, have given it as their opinion that the nitre appears in soils which contain no vestiges of organic matter. The assertion of Proust, however, is open to suspicion, inasmuch as in his memoir he affirms that the lands close to those that produce the nitre are extremely fertile, so that they yield abundant crops without ever receiving manure. But at the present day, it is a law that every fertile soil must contain or receive dead organic matter. In Ceylon, according to Davy, the caverns, the walls of which become covered with an efflorescence of saltpetre with such rapidity, have a fertile and thickly wooded soil lying over them, the percolations from which may readily penetrate their interior. The observations which I had an opportunity of making upon the nitre soils near Latacunga, were not perhaps of sufficient precision; but I think I can affirm that the soil was not without humus: patches were perceived here and there that were covered with turf. It must still be admitted, however, that in the localities which have been particularly indicated there must exist some peculiar and permanent cause of nitrification; inasmuch as in other and fertile soils, saltpetre only appears as it were accidentally, and never in extraordinary quantity.

Whatever the value of the ingenious but still very imperfect theories of nitrification, it is still of importance to ascertain the existence or absence of nitrates in the soil. Wollaston recommended a process which enables us to do this very readily. It is founded on the property possessed by the aqua regia—a mixture of the nitric and hydrochloric acids—to dissolve pure gold, which, as is familiarly known, resists the action of either of these acids applied separately. The soil in which the presence of a nitrate is suspected is treated with boiling distilled water, and thrown upon a filter. The filtered fluid is reduced by evaporation to a very small quantity, which is then poured into a test tube, and a little pure hydrochloric acid is added; some particles of leaf gold are then introduced, and the fluid is stirred with a glass rod. If any nitrates have been present, the particles of gold are speedily dissolved.

Having now described the circumstances which determine, and the phenomena which accompany the decomposition of dead organic matter, I have next to treat of manures in particular, of their preparation, of their application, and of their relative values. Speaking generally, the manure which is derived from the dejections of animals, supplied in a farm-yard with abundance of food and of litter, used with the double object of cleanliness and health, is the best of all. The principal substances which contribute day by day to increase the mass of our dunghills are straw, and the excretions and urine of horned cattle, horses, hogs, &c. These various substances, besides the organic elements which enter into their composition, further contain the various mineral substances which are indispensable to the development of vegetables. Animal excrements of every kind, in fact, when they are burned, leave quantities of ashes which are frequently very considerable, and in which are encountered the same saline and earthy ingredients that pre-existed in the forage with which the animals were supplied. Excrements, therefore, necessarily vary in their composition according to the kind of food that is consumed, and the nature and the state of health of the animal which produced them. Those of the herbivora have never been sufficiently examined. Thaer and Einhof have merely ascertained that cow-dung contains an extractive principle, partly coagulable by heat, and that remains of the food may be separated from it. All excrementitious matters, in fact, contain a certain quantity of the alimentary matter which has escaped digestion, especially when animals are abundantly supplied with food. Some albuminous matter is also found there; but the substance after vegetable remains that appears to predominate is bilious.*

We know that after mastication, the food, mingled with saliva and the secretions of the mucous glands, passes into the gullet, and from thence into the stomach. There it imbibes gastric juice, turns sour, becomes modified, and is finally converted into a kind of pulp which is called chyme. Once formed, chyme passes into the small intestines, where it encounters the bile and pacreatic juice, which modify it, and cause it to separate into chyle, which is absorbed by the vessels of the bowels, and excrementitious residue, which descends into the large intestines, where it becomes a fetid mass that is expelled from time to time by the animal.

* The latest inquiries of the physiological chemists would lead us to suspect that this was not the case. Bile ought only to be an occasional, and even an unnatural constituent of animal excrement, if these views be well founded. It seems that the elements of bile added to the elements of starch supply the precise elements of fat; a substance so abundantly formed in the process of digestion. The bile that is poured into the upper part of the alimentary canal is probably all used up in forming fat.—Eng. Ed.

The bile which accompanied the fecal matter is secreted by the liver, and is familiarly known as a viscid, bitter fluid of a yellowish green color and a peculiarly nauseous odor. According to M. Thenard, the bile of the ox contains :—

Water	700
Picromel*	69
Fatty matter	15
Soda, phosphate of soda, chlorides of potassium and sodium, sulphate of soda	10
Phosphate of lime, oxide of iron	1
	795

Urine is a liquid secreted by the kidneys from arterial blood ; it passes into the bladder by the ureters. Its composition varies according to the animals which produce it. Urea is its most characteristic principle ; and in the water which it always contains in large proportion, various saline substances and animal matter, which is regarded as mucus of the bladder, are encountered. The urine of the horse, according to M. Chevreul, contains carbonate of soda, of lime, and of magnesia, sulphate of soda, chloride of sodium, hippurate of soda, urea, and a red-colored oil.

The urine of horned cattle has a similar compositon, with this difference, that it is much more watery. In the urine of our cow-houses which had undergone change, I have ascertained the presence of the alkaline carbonates, of common salt, and of the reddish oil mentioned above. Having at various times had occasion to evaporate considerable quantities of the urine of the horse, I always observed that on coming to the boiling point, a quantity of azotized matter which resembled albumen was coagulated. I also perceived in the urine of herbivorous animals a volatile acid, to which its odor is probably owing.

In the urine of the camel, M. Chevreul found the carbonates of lime and magnesia, silica, sulphate of lime, and oxide of iron, in very small quantities ; chloride of potassium, carbonate of potash, sulphate of soda, in small quantities ; sulphate of potash, in large quantity ; urea ; an alkaline hippurate ; and a reddish oil.

The urine of the rabbit, according to Vauquelin, contains carbonate of lime, of magnesia, and of potash, chloride of potassium, sulphate of potash, sulphur, urea, and mucus.

The urine of birds is distinguished by the large proportion of uric acid it contains. Food, however, has a great influence upon this proportion ; highly azotized aliments increasing it considerably. Wollaston observed that the excrements of a fowl which was fed

* Picromel, discovered in the bile of the ox by M. Thenard, is colorless, and of the consistence of sirup. It produces upon the tongue an acrid and bitter sensation, which rapidly changes to a flavor slightly sugary. The recent researches of Messrs. Tiedemann and Gmelin have discovered in ox bile substances which had escaped the first investigations. These chemists found : 1st. a substance having the smell of musk, and which is probably one of the causes of the odor peculiar to the excrement of kine ; 2d. fatty substances ; 3d. biliary resin ; 4th. a crystallized substance called taurine ; 5th. biliary sugar, of which azote forms one of the elements. According to Messrs. Tiedemann and Gmelin, the picromel of M. Thenard results from the union of sugar and biliary resin.

upon herbage contained only 2 per cent. of uric acid. That of a pheasant fed upon barley contained, on the contrary, 14 per cent.; and that of a falcon which fed upon flesh alone, yielded scarcely any thing but uric acid. The urine of an ostrich was found by Fourcroy and Vauquelin to contain uric acid in the proportion of about one sixteenth of its mass.

I have already given the composition of urea. Hippuric acid is an azotized acid which is readily obtained by adding a little hydrochloric acid to the fresh urine of the horse reduced by evaporation to about one tenth of its original volume, when a granular crystalline mass is precipitated. If the urine have been stale instead of fresh, benzoic acid and not hippuric acid is obtained; benzoic acid was, in fact, long admitted as one of the elements of the urine of herbivorous animals; but it is derived from the transformation of hippuric acid into benzoic acid and ammonia, the change being produced by contact with the organic matters which putrefy so quickly in urine. Liebig was the author of this observation; it was in operating upon unchanged urine that he discovered hippuric acid. The following is its composition :—

Carbon	60.7
Hydrogen	5.0
Oxygen	26.3
Azote	8.0
	100.0

Uric acid has not yet been met with in the urine of mammiferous herbivora; but it exists in that of man, having been first discovered in calculi from the bladder; whence it received the name of lithic acid. Liebig's analysis shows it to be composed of :—

Carbon	36.1
Hydrogen	2.4
Oxygen	28.2
Azote	33.4
	100.0

The litter most commonly used to absorb the urine of stall-kept animals is wheat straw, which consists in principal part of lignine or woody fibre: like all vegetable tissues, however, it contains an azotized principle, and substances that are soluble in caustic alkalies. In the ashes of straw, we have indicated silica as abundant, and various alkaline and earthy salts. The proportion of azote appears to vary in the ratio of from 3 to 6 per 1000. An analysis which I made of dry wheat straw gave the following elements :—

Carbon	48.4
Hydrogen	5.3
Oxygen	38.9
Azote	00.4 to 0.6
Ashes	07.0
	100

Agriculturists have, in all ages, admitted that the most powerful manures are derived from animal substances; an opinion or rather a fact, which, expressed in scientific language, amounts to this, that the most active manures are precisely those which contain the largest proportion of azotized principles. It is obvious indeed from every thing which precedes, that all the substances which contribute to form farm dung, contain azote; and that into many of them, such as uric acid, hippuric acid, and urea, this element enters very largely.

When we consider the immediate changes which all highly azotized substances undergo in the process of putrefaction, we can foresee that in their transformation into manure, they must give origin to ammoniacal salts; and well-established facts prove beyond doubt that salts, having ammonia for their base, must be ranked among the most powerful of all the agents in promoting vegetation. It is sufficient, for instance, to bear in mind that in the productive husbandry of Flanders, putrid urine is the manure that is employed with the greatest success; but we have seen that by putrefaction, the urea of the urine is entirely changed into carbonate of ammonia. The fields of Flanders are consequently fertilized with a solution of carbonate of ammonia in water.

Along a great extent of the coast of Peru, the soil, which consists of a quartzy sand mixed with clay, and is perfectly barren of itself, is rendered fertile, is made to yield abundant crops, by the application of guano; and this manure, which effects a change so prompt and so remarkable, consists almost exclusively of ammoniacal salts. It was with this fact before me that in 1832, when I was on the coasts of the Southern Ocean, I adopted the opinion which I now proclaim in regard to the utility of the salts having a basis of ammonia in the phenomena of vegetation. I have stated my views on this subject in a memoir published in 1837.* Previous to this publication, however, M. Schattenmann, one of the most ingenious manufacturers of Alsace, had already directed the attention of husbandmen to this important matter, by reminding them that it is the custom in Switzerland to add sulphate of iron or green vitriol to the urine-vats, for the purpose of changing the carbonate of ammonia into the sulphate, and thus obtaining a fixed instead of a highly volatile salt, liable to escape and to be lost. In a communication made in 1835 to the agricultural association of Bauchsweiler, M. Schattenmann announced positively that the drainings from dunghills thus prepared, applied upon meadow lands, produced very grea effects.

Such, to the best of my knowledge, are the practical facts which establish the useful influence of ammonia on the growth of plants far better than the experiments of the laboratory could have done. Nevertheless, it must be acknowledged that long before the dates above quoted, Davy had shown that water containing $\frac{1}{300}$th of carbonate of ammonia is singularly favorable to the growth of wheat, far more so, under circumstances exactly similar, than the hydrochlorate and the nitrate of the same base; and this influence, it is important to observe, Davy ascribed to the fact that carbonate of ammonia contains carbon, hydrogen, oxygen, and azote; in a word

* Annales de Chimie, t. lxv. 2e série, p. 301.

all the elements that are essential to the organization of plants. The illustrious English chemist concluded from his experiments, that the well-known efficacy of soot, as a manure, is due, in part, to the volatile alkali which it contains.

Professor Liebig, in adopting these opinions, has sought to generalize them; he has attempted to show, by very delicate experiments, that the air which lies immediately over the surface of the ground, always contains carbonate of ammonia, and that the same salt can be detected in rain and snow, and in spring water. The ammonia of the atmosphere, according to Liebig, concurs with that which is developed in manures, in the formation of the azotized principles proper to vegetables. These ingenious ideas correspond exactly with those which M. de Saussure made public in 1802, when he ascertained that the gaseous azote of the air is not directly absorbed by plants. "If azote be a simple substance, and not an element of water," says this celebrated observer, "we must admit that plants do not assimilate it, save in vegetable and animal extracts, and in the ammoniacal vapors or other compounds soluble in water which they absorb from the soil, or from the atmosphere. It is impossible," he continues, "to doubt the presence of ammoniacal vapors in the atmosphere when we see that the pure sulphate of alumina, exposed to the air, ends by becoming changed into the ammoniacal sulphate of alumina."*

In agricultural establishments, in which the importance of manure is duly appreciated, every precaution is taken both for its production and preservation. Any expense incurred in improving this vital department of the farm, is soon repaid beyond all proportion to the outlay. The industry and the intelligence possessed by the farmer, may indeed almost be judged of at a glance by the care he bestows on his dunghill. It is truly a deplorable thing to witness the neglect which causes the vast loss and destruction of manure over a great part of these countries. The dunghill is often arranged as if it were a matter of moment that it should be exposed to the water collected from every roof in the vicinity, as if the business were to take advantage of every shower of rain to wash and cleanse it from all it contains that is really valuable. The main secret of the admirable and successful husbandry of French Flanders may perhaps lie in the extreme care that is taken in that country to collect every thing that can contribute to the fertility of the soil. Our agricultural societies, which are now so universally established, would confer one of the greatest services on the community if they would encourage by every means at their command economy of manure; premiums awarded to those farmers who should preserve their dunghills in the most rational and advantageous manner, would prove of more real service than premiums in many other and more popular directions.

The place where the dung of a farm is laid ought to be rather near to the stables and cow-houses. The arrangements may be

* Recherches Chimiques, p. 207.

varied to infinity, but they ought all to combine the following conditions: 1st. That the drippings from the heap should not run away, but should be collected in a tank or cistern under ground; 2d. That no water, except the rain which falls on the dung-heap, or any water that may be thrown upon it on purpose, should be allowed to drain into this reservoir; 3d. That the place for the dunghill be of size enough to avoid the necessity of heaping the manure to too great a height. The ground upon which the dung is piled ought to slope gently one way or another—from each side towards the centre is best—so that the drippings may be collected in the tank or cistern. It is also desirable that the soil underneath should be clayey and impermeable; where it is not so, it becomes necessary to puddle, to cement, or to pave the bottom of the dunghill stance as well as the bottom and sides of the tank or cistern. The water which runs from the heap should be thrown back upon it occasionally, by means of a pump and hose, so as to preserve it in a state of constant moistness. The opening into the tank, which is best placed immediately under the centre of the dung-heap, is closed by means of a strong grating in wood or iron, the bars being sufficiently close to prevent the solid matters from passing through. One very important arrangement, one which, in fact, must on no account be overlooked, is that the drains from the stables and cow-houses be so contrived, that they all run to the dunghill. The litter, however abundant, never absorbs the whole of the urine, especially at the time when the cattle are upon green food; and it would be quite unpardonable in the husbandman did he not take measures to secure this, the most valuable portion of the manure at his disposal.

The litter mixed with the droppings of the animals, and soaked with their urine, ought to be carried from the stables to the dunghill upon a light barrow. The practice of dragging out the manure with dung-hooks, which is often permitted when the field upon which it is to be spread is at no great distance, ought on no account to be allowed; the loss from the practice is always considerable.

Materials ought not to be thrown on the dunghill at random or hap-hazard; they should be evenly spread and divided; an uneven heap gives rise to vacancies, which by and by become mouldy, to the great detriment of the manure. It is of much importance that the heap be pretty solid, in order to prevent too great a rise of temperature, and too rapid a fermentation, which are always injurious. Particular care must also be taken that the heap preserves a sufficient degree of moistness, not only of its surface but of its entire mass, which is effected by watering it frequently. At Bechelbronn, our dung-heap is so firmly trodden down, in the course of its accumulation, by the feet of the workmen, that a loaded wagon drawn by four horses can be taken across it without very great difficulty. The thickness of the heap is not a matter of indifference: besides the convenience of loading, which must not be forgotten, any great thickness may become injurious by causing the temperature to rise too high; circumstances occurring which should compel us to keep a mass in this state for any length of time, the decomposition would make such progress as to occasion very great loss. Experience has shown that the thickness of a dung-heap ought not to exceed from about four feet and a half to six feet and a half; it ought certainly never to exceed the latter amount.

With a view to prevent the drying of the dung-heap and its consequences, too great a rise in temperature and destruction of manure, it is the practice in some places to arrange the dung-heap on the north side of a building, which is undoubtedly advantageous, but not always to be realized, especially in connection with a farm of some magnitude, where the immediate vicinity of a large mass of matter in a state of putrid fermentation is not only unpleasant, but may be unwholesome. In the north of France, the dung-heap is sometimes shaded from the sun by means of a row of elms, and the shelter thus secured is vastly preferable to that which it has been proposed to obtain by means of a roof or shed, which, besides other inconveniences, would be found costly at first, liable to speedy decay, &c. If circumstances, such as the smallness of the farm, the permeable nature of the soil, &c., prevent the construction of a reservoir, there is risk of the dung-water being quite lost; but such waste may be prevented by covering the bottom of the pit or stance for the dung-heap with a bed of sand, peat marl, or any other dry and porous substance capable of absorbing liquids. This practice is often followed by the farmers of Alsace.

In some farms, the different kinds of dung are piled apart from one another in particular heaps; that of the stable being put by itself, as well as that of the cow-house, that of the hog-stye, and that of the sheep-pen. In great establishments, such a separation is often one of necessity; but the advantages which are ascribed to it are questionable at least, and the remarks that have been made upon it by writers do not appear founded on any accurate observation. Without denying that certain crops answer better when special manures are employed, it still seems to me more advantageous to pile every kind of manure together, when the difficulties of the situation are not such as to make this either particularly inconvenient or expensive. In this way, indeed, a dung-heap of medium constitution is obtained, which is regarded with reason as that, the application of which to the soil is attended with the greatest advantages in the majority of instances. The distinction which some have sought to make between the relative qualities of manures of different origins is far too absolute; and this is the reason, without doubt, which renders it so difficult to bring the observations of different agriculturists to agree. Thus, according to Sinclair, the dung of the hog-stye is the most active of all, the richest in fertilizing principles; according to Schwertz, on the contrary, it is the most indifferent manure of the farm-yard.

The fact is, that manures, which are the produce of the same animals, often present greater differences in regard to quality, than manures which proceed from different sources. I shall show by and by that the value of manure depends especially upon the feeding, the age, and the condition in which the animal is placed that produces it. It is well known that the dung of cattle, fed during winter upon straw, is greatly inferior to that which they yield when consuming food of a more nutritious quality.

When the litter mixed with animal excrements is accumulated in sufficient quantity in the pit or on the dung stance, fermentation speedily sets in, and abundance of vapor is disengaged. As carbonate of ammonia is among the volatile products of this decomposition, it is of importance to hold it under control; this is done by keeping the heap in a state of proper moistness, and in excluding as much as possible the access of air. The daily addition of fresh quantities of litter from the stables and stalls, contributes powerfully to impede the dispersion of the volatile elements, which it is so important to preserve in manure; duly spread upon the heap, each addition becomes, in fact, a fresh obstacle to evaporation; it forms a covering which plays the part of a condenser, at the same time that it protects the inferior layers from the direct contact of the air. So long as the dung-heap is kept up and attended to in this way, the fermentation is limited to the inferior layers of the mass. Thaer even satisfied himself that air collected from the surface of a dung-heap, undergoing moderate fermentation, does not contain much more carbonic acid than that which is taken from the mass of the atmosphere. Neither does a vessel containing nitric acid, when placed upon the fermenting mass, produce those dense white vapors which are a certain indication of the presence of ammonia. The slow decomposition which it is of so much importance to effect, is not readily secured save in masses sufficiently trodden down, and in which the litter of different kinds has been spread as evenly as possible.

It is an important point that the manure should be carried out to the field before the upper portions recently added begin to undergo change, otherwise the whole mass enters into full fermentation, and the volatile elements, being no longer arrested by the upper layer, escape and are lost. One means of preventing this loss in any case (which however can but rarely occur) in which there was a necessity for suffering the mass to become *made* through its whole thickness, would be to cover it with a layer of vegetable mould, in which the volatile principles would be condensed; the layer of earth would in fact thus be converted into a most powerful manure.

The loss of carbonate of ammonia, during the fermentation of farm-dung, is further prevented by the use of certain salts which have the power of changing the volatile carbonate into a fixed salt. It was with a view of bringing a re-action of this kind into play, that M. Schattenmann, the able director of the manufactories of Bauchsweiler, proposed to add to dung-heaps, in the course of their accumulation and preparation, a certain quantity of sulphate of iron or of sulphate of lime, either of which is decomposed by the carbonate of ammonia evolved, and a fixed ammoniacal salt (the sulphate) is produced.* The loss of ammonia from dung-heaps in the course

* Annales de Chimie, 3e série, t. iv. p. 116.

of regulated fermentation, must not however be estimated too highly; when the decomposition is carefully conducted, the mass having been well trodden and properly damped, the loss is really very small. The gentle fermentation, secured by these means, has characters which differ essentially from those that accompany the rapid putrefaction which never fails to take place when matters are not well managed. As an example of the rapid and injurious fermentation of which I speak, I may cite that which frequently takes place in piles of horse-dung: every one must have seen such dung-hills loosely thrown together, left to themselves, without any addition of water, acquiring a very intense heat in the course of a few days, and have even heard of their taking fire. I have seen piles of this kind reduced to their merely earthy constituents! Such are never the results of the moderate and gradual decomposition which farm-dung ought never to exceed. When the pit or stance is emptied, in which a slow and equal fermentation has taken place, the superior layer is seen to be very nearly in the same state in which it was when it was piled; the layer immediately beneath this one is changed in a greater degree, and sometimes exhales a slight ammoniacal odor. In the lower strata, the modification is yet greater in degree: the straw has lost its consistency, it is fibrous and breaks into pieces with the greatest ease; the mass is also progressively darker in color as we go deeper, and on the ground it is completely black; the smell which this part of the heap exhales, is that of sulphuretted hydrogen, and when it is tested, sulphate of iron is discovered; no doubt these sulphurous products are all the consequence of the decomposition, under the influence of the organic matter, of the sulphates which were contained in the manure. This is the sign by which I know that farm-dung is duly prepared; the presence of sulphurets and of the hydrosulphate of ammonia will have no ill effect upon vegetation; for scarcely is the manure spread upon the ground, than these products are changed into sulphates, and then the manure emits that musky smell which is peculiar to it. Further, there is no doubt but that the state in which a carefully tended dung-heap is found in the end, is due to the circumstances in which it has been placed and kept during the whole time of its preparation; its constituent elements would have gone through a totally different course in the progress of their modification had they been left exposed to the open air. To be satisfied of this, it is enough to remark the powerful and purely ammoniacal smell which meets us in a warm stable, especially during the summer season, upon the ground of which the urine of the animals it contains is left to decompose.

From what has now been said, it will be understood how destructive to good manure is the custom which obtains in certain countries of turning dung-heaps frequently, of airing them as it were, in order to hasten decomposition. Treated in this way, stable litter, &c., does in fact decompose much more rapidly; but it does so, and I own that I do not myself clearly perceive the object proposed by it, at the expense of the quality; for it is very evident that the volatile principles must be dissipated and lost in the same proportion as their points of contact with the air are multiplied.

The plan of collecting all the litter of the farm into one particular and appropriate place, is that which is generally adopted. Nevertheless, there are countries in which the dung is left to accumulate in the cattle-stalls, it being merely covered with fresh straw every day. The ground thus rises continually under the feet of the cattle, so that it is necessary to have moveable cribs which can also be raised by degrees. This method is so far convenient, that the necessity for cleaning out the stable continually is avoided ; but little is gained in the end in the matter of labor, for the same mass of manure has still ultimately to be removed. The fermentation of the manure would be greatly accelerated by the usual high temperature of the stables, did not the feet of the cattle tread the mass very closely, and this and the daily addition of straw together produce the same effect as I have indicated in treating of the management of the dung-heap out of doors : it condenses vapors and volatile particles, and prevents evaporation. The fact is, that in stalls and stables in which the dung is allowed to accumulate in this way, we are not sensible of any very offensive odor, and the animals which live in them breathe without inconvenience, it being always understood that all communication with the exterior is not interrupted, which in fact it ought never to be, even in cases where the stables and stalls are kept perfectly clean. This method of proceeding becomes almost impracticable when cattle are fed upon food that is not dry, but on the contrary that is extremely watery, such as roots, green clover, &c. ; the quantity of urine that is then passed is so considerable, and the excrements themselves are so copious and so liquid, that an enormous quantity of straw would be required to absorb the liquid parts ; in spite of any reasonable addition of litter, indeed, the animals would still be exposed to be kept in the mire, which would doubtless become a powerful cause of insalubrity among them.

In Belgium, according to Schwertz, manure is accumulated in the stables by guarding against the inconveniences which the last mode of proceeding generally implies. The cattle are placed upon a kind of platform raised above the pavement of the stable, and the droppings being withdrawn from under them, are trodden down and allowed to accumulate upon the floor.

One inconvenience attending the use of straw, is that it is frequently dear ; it is also scarce in some countries. In those parts of Switzerland, for instance, where all the available lands are meadows, they are obliged to economize litter as much as possible, so that they even wash it, and thus make it serve repeatedly. Although it would be difficult to give a reason for a practice which has the effect of increasing the bulk of the manure, adding to the expense of transport, and at the same time diminishing its quality ; it is, nevertheless, a fact that this mode of proceeding has been long in use in various Cantons. We probably only see here another means of securing even the last particle of the excrementitious matters passed by cattle, the process employed being in fact identical with that used by the chemist in his most delicate analyses. In Switzerland, the urine that is passed by the cattle flows along a gutter which communicates with a large reservoir containing water, in which not only are the solid excrements diffused, but in which the litter is washed, this being changed only twice a week. The reservoir is constructed under the floor of the cow-house itself, in order to be protected from the frost. The fermentation of a mass so diluted is scarcely perceptible, and, save from leakage, there is no loss of decomposing animal matter. The liquid manure is raised by means of a pump, and carried to the meadow in tubs placed upon carts. In Switzerland, it is also the usage to employ the urine of cattle separately as manure, under the name of *purin;* to this liquid manure, a quantity of sulphate of iron is frequently added with the view of bringing the volatile carbonate to the state of the fixed sulphate of ammonia, as I have already said.

Liquid manures have their advantages and their inconveniences. We shall immediately discuss their value comparatively with that of solid manures, and we shall be led to adopt the opinion of M. Crud in regard to them, viz., that the advantages ascribed to them in Switzerland are exaggerated. Whatever the form under which manures are applied, the question has been warmly discussed, whether it be to the interest or disadvantage of the agriculturist to employ them before or after they have undergone fermentation?

Organic substances, however, are in no condition to favor the growth of vegetables until they have undergone material changes which modify their nature. One of the results of this change, as we have seen, is the development of ammoniacal salts. Fresh manure, such as it comes from the stable, introduced immediately into the ground, there undergoes precisely the same changes, and gives rise to the same products as it does when subjected to preparation in a dung-heap in the manner already described; there is only this difference, that being scattered and mixed with a large quantity of inert matter, the decomposition takes place much more slowly than it does in the heap. The question which has been so actively discussed, therefore, reduces itself to this: is it advantageous to have the manure fermented in the soil it is intended to fertilize? We may be allowed to express surprise that such a question should have been raised in the present day, and still more that the affirmative answer should have been disputed by agriculturists of distinguished merit. Some have even gone so far as to maintain that fresh excrements were injurious to vegetation. Proofs to the contrary are readily obtained; it is enough to recollect that in the grazing and folding of sheep and kine, the dung and urine pass directly into the ground of our pastures and fields, and who shall say that the land is not benefited by what it thus receives? Unquestionably fresh manure in excess proves injurious to vegetables, but as much may be said in regard to the best-fermented dungs.

M. Gazzeri, an Italian chemist, has devoted himself with the most laudable perseverance to inquiries having for their object to show that the general custom of leaving manures to become decomposed before leading them out to the field is attended with a considerable loss of fertilizing principles, and that it is therefore advantageous to use them in the state in which they come from the stable. To remove all doubts which might yet be entertained upon the effects of unfermented manures, M. Gazzeri showed that wheat could be successfully grown in land which had received an extraordinary dose of pigeon's dung, which is regarded as one of the most active of all manures; and horse droppings, taken at the moment they were passed, mixed with earth, in the proportion of one-fourth of the whole bulk, had no injurious effect on the growth of the cereals. To ascertain the amount of loss which fresh manures suffered from fermentation, M. Gazzeri placed certain quantities, ascertained by weight, to putrefy under favorable circumstances; and the decomposition completed, he weighed them again. In this way, he ascertained that horse-dung, in the course of four months, lost more than the half of the dry matter which it contained before its putrefaction. Davy, indeed, had already shown that there is a loss of volatile principles, during the decomposition of fresh manures, that must be useful to vegetation. Davy's experiment consisted in introducing manure into a retort, the extremity of which communicated with the soil under turf, and he found that in the course of a few days the grass which was thus exposed to the emanations from the retort, grew with particular luxuriance. Although it appears certain, then, that in conducting the preparation of manure in the heap with prudence, the volatile and ammoniacal principles which appear in the course of the putrefaction may be retained, it is nevertheless unquestionable that the employment of manure directly and without previous fermentation, would most effectually prevent the loss of matters that must be valuable. Thaer, Schwertz, Mr. Coke, and others, have consequently admitted the advantages of the latter procedure. In agreeing with them completely, which I do, it still remains certain that on the greater number of farms, dung-heaps must be formed as matter of necessity. Manure is only available at certain determinate seasons of the year; it cannot be carried out and spread as it is produced. In Alsace it is carried out to the fields on which it is to be spread whenever circumstances will permit, and without regard to its more or less advanced state of decomposition. The circumstances which lead to its being kept in the pit for two or three months, also lead to the manure being half or more than half matured before it is led out; and this, after all, is perhaps the best state in which it can be put into the ground. It is then easily incorporated with the soil, and its fertilizing principles are already in that condition which enables them to act, within a limited time, with greater energy than they would do were the manure employed quite fresh. This is the condition in which our manures almost always are at Bechelbronn when we carry them out: it rarely happens that they have been three months on the stance before their removal. Speediness of action is a point which is not without importance. Fresh dung will always act more slowly than that which has reached a certain point of decomposition, and the advantage which mostly accrues to the farmer in forcing his crops, will often induce him to use manure that has ripened in the pit or stance.

In warm and moist countries, as may be conceived, it is almost matter of indifference whether the dung be put into the ground quite fresh, or in a state of decomposition further advanced; its decomposition, aided by the heat of the climate, is always effected rapidly enough. But it is otherwise in cold climates, where the temperature which excites and maintains vegetation is often of short duration, and must at once be taken advantage of. During a great part of the year, the ground is so cold that organic substances buried in it are preserved with comparatively little change. Under such climatic conditions, there is no doubt that manures in a state of forwardness are to be preferred. It is probably from such motives that the extensive use of liquid manures in Switzerland is derived, the action of these being, so to speak, immediate; and it is with such manure that in Flanders the cultivation of various plants that are of great value in manufacturing processes, is carried on.

When the fermentation of manure has been managed discreetly, and all the precautions requisite to prevent the dissipation of ammoniacal salts and the loss of soluble elements have been taken, there is the immense advantage attending it, besides obtaining immediate action, that a manure is produced of greater value under a smaller bulk and a less weight. The dung-heap often loses a third of its bulk in undergoing fermentation, a circumstance which occasions an important saving in carriage. A like saving may be effected with reference to fresh manures, by drying them in the sun, which I have sometimes seen done; they are thus reduced to one-third or one-fourth of their original weight, and when the distance to which they have to be carried is great, there may be real advantage in proceeding in this way.

An objection of some moment made to the use of fresh dung to corn lands is, that it usually contains the seeds of weeds and the eggs of insects which nothing but putrefaction will destroy. This objection of course loses all its weight when the land that is manured is to receive a crop which admits of hoeing; and the custom which obtains with us at Bechelbronn of using manure in every state of decomposition to the first crop in the rotation, is a guarantee that fresh manure is really productive of no inconvenience in practice. Another difficulty pointed out by Thaer, is that of covering in dung so long and full of straw as fresh stable or stall dung. This objection disappears when the manure is laid in furrows formed by the plough, as is done in Alsace, by which means the covering in is effected by a single operation.

If opinions are still divided upon the question whether dung ought to be employed before or after fermentation, they are no less so as to the mode of spreading it, and the best periods for laying it on the land. It may be imagined that the conclusion come to upon the first question necessarily influences the opinions held on the second. Those who believe that manure may be advantageously used in the state in which it comes from the stables, are altogether indifferent in regard to the times of carrying it out. They take advantage of every leisure moment that occurs for performing this necessary work, which is no trifling advantage; it is the practice which we follow at Bechelbronn—we carry out our manure as we find opportunity. The lands which are to be manured in the spring have frequently their supply carried out during winter when the frost enables us to get upon them. The dung first shot down in little heaps, at regular distances, is afterwards spread as equally as possible, frequently even upon the snow; and I have never seen any ill effect from the practice. The custom which some farmers have of keeping dung in large heaps in the field, in order that it may be all spread and worked under at the same time by the plough, is certainly objectionable; the places upon which these heaps have been laid are evidently too strongly manured; no manure, save that which is quite fresh and very long in the straw, or which it is proposed to spread immediately in furrows, ought ever to be laid down in large heaps upon the field.

The method which I have recommended, of leaving manure spread over the surface of the fields exposed to the weather for several weeks or months, has been severely criticised. By such exposure, it has been said the dung must lose its volatile elements, and the rain must wash out and carry off its more soluble parts. Influenced by such fears, some farmers do not spread their dung until the moment of ploughing it in. Such diversity of opinion among practical men, all personally interested in deriving the greatest possible amount of advantage from the manure they employ, must not be thought of lightly: when different modes of procedure in agriculture are the subjects of debate, we must not be in too great a hurry to come to general conclusions. Climate is not without its influence in the question which now engages us. In Alsace, experience has pronounced in favor of the practice followed; but in other countries there may be very good reasons for not proceeding in the same way. In Alsace, where the annual depth of rain amounts to 26.7 inches, no more than 4.3 inches fall during the three months of December, January, and February. In a district where a larger quantity of rain falls during the winter, the manure would probably suffer from the procedure followed in Alsace.

The quality of the manure must also be taken into consideration. A dunghill which contained a large proportion of carbonate of ammonia, which exhaled a strong smell of volatile alkali, would infallibly lose in value by any unnecessary or prolonged exposure to the air; but the loss becomes insignificant when the manure, by good management, is brought to contain but a small proportion of volatile ammoniacal salts, as happens with manures which have received additions of gypsum; or otherwise, when the dung-heap has been carried out fresh, and at a season so cold that it can be kept without material change until the period arrives for spreading it over or working it into the ground. When the rains are not excessive, the soluble parts of manure spread upon the ground penetrate and remain in its upper stratum, exactly as happens when, instead of being buried, it is spread upon the herbage in full growth. The plan of top-dressing is often of great use, and is another and a practical proof of how little detriment results from leaving manure exposed to atmospherical vicissitudes. The procedure by top-dressing has arisen from necessity: it was first resorted to with the view of giving the land an addition to the inadequate dose of manure which it had received before it was sown; but it has been found to answer so well in many districts, that it has been continued. We have employed it at Bechelbronn upon various occasions, even to hoed crops, and with decided advantage, the main one being, that time was gained for the production of manure.

In the district of Marck, the practice of top-dressing lands sowed with winter grain, is rapidly gaining ground; the dressing takes place when the blade is already above ground; and experience proves that the passage of the wagons over the field, and the feet of the horses and the men, cause no appreciable mischief; all traces of them very soon disappear. Nevertheless it is decidedly better to take advantage of a hard frost, when the land will bear carts, &c., for the performance of the process. This plan, according to Schwertz, is found to answer extremely well in Switzerland, for hemp, and indeed for almost every thing else. In my opinion, top-dressing ought to be viewed as a means of giving the soil, already under a crop, the manure which we had been compelled to refuse it at an earlier period. Still, Thaer assures us, and his authority is always of great weight, that he has too frequently seen the good effects of top-dressings to beans, peas, and leguminous crops in general, not to be satisfied of the general advantages of the method, in connection with light soils especially, in which the sowing may have been late.

The elementary composition of farm-dung is a point which is not undeserving of consideration. I have made repeated analyses of that of Bechelbronn, operating upon it in a medium state of decomposition. The animals which had produced this dung were thirty horses, thirty oxen, and from ten to twenty hogs. The absolute quantity of moisture was ascertained by first drying in the air a considerable weight of dung, and, after pounding, continuing and completing the drying of a given quantity in the oil-bath, in vacuo, at a temperature of 230° F.

The dung prepared in the winter of the year—

1837–8 contained	20.4	per cent. of dry matter.
1838–9	22.2	
Prepared in summer of 1839........	19.6	
Medium	20.7	
Water	79.3	

Analysis yielded the following results:

Times of preparation.	Carbon.	Hydrogen.	Oxygen.	Azote.	Ashes.
Winter of 1837–8	32.4	3.8	25.8	1.7	36.3
"	32.5	4.1	26.0	1.7	35.7
"	38.7	4.5	28.7	1.7	26.4
Spring of 1838	36.4	4.0	19.1	2.4	38.1
" 1839	40.0	4.3	27.6	2.4	25.7
" "	34.5	4.3	27.6	2.0	31.5

On the average, farm-dung dried at 238° contains:

Carbon	35.8
Hydrogen	4.2
Oxygen	25.8
Azote	2.0
Salts and earths	32.2
	100.0

When moist, its composition is represented by—

Carbon	7.41
Hydrogen	0.87
Oxygen	5.34
Azote	0.41
Salts and earths	6.67
Water	79.30
	100.0

The constitution of dung-heaps must of necessity vary; those, however, which have a common origin do not seem to present very great differences in the proportion of their elements. Thus, horse-dung, in the south of France, yielded, on analysis, in the dry state, 2.4 per cent. of azote. This manure contained only 61 per cent. of moisture.

Did we but know the composition and the quantity of the excretions passed, in the course of the twenty-four hours, by the various animals which contribute to the production of manure, we should be able to determine approximatively what the elements are which have been eliminated in the course of the fermentation. It would be sufficient to compare the elementary matter in the litter, or fresh dung, as it comes from the stable, with that which exists in the fermented or prepared manure. I have data which I think sufficient to enable me to institute this comparison. It must always be borne in mind, however, that the analyses which I shall now detail were made upon the excrements of a single individual of each kind. It would certainly have been preferable to have had average analyses of average qualities; but the object I had in view, when I undertook these experiments, was quite different from that which I have now before me.

EXCRETIONS OF THE HORSE.*

The horse was fed upon hay and oats. The urine and the excrements together contained 76.2 per cent. of moisture. In twenty-four hours the excretions weighed—moist, 34.2 lbs.; dry, 8.1 lbs.

Their composition was found to be—

	In the dry state.	Moist ditto.
Carbon	38.6	9.19
Hydrogen	5.0	1.20
Oxygen	36.4	8.66
Azote	2.7	4.13
Salts and earth	17.3	4.13
Water	"	76.17
	100.0	100.0

* The size of the horse was rather below the average usual size of farm horses.

EXCRETIONS OF THE COW.

The cow was fed upon hay and raw potatoes. The urine and the excrements together contained 86.4 of moisture. The weight of the excretions, in twenty-four hours, was—moist, 80.5 lbs.; dry, 10.9 lbs.

Their composition by analysis was:

	Dry.	Wet.
Carbon	39.8	5.39
Hydrogen	4.7	0.64
Oxygen	35.5	4.81
Azote	2.6	0.36
Salts and earth	17.4	2.36
Water	"	86.44
	100.0	100.00

EXCRETIONS OF THE PIG.

The pigs, upon which the observations were made, were from six to eight months old. They were fed upon steamed potatoes. The urine and the excrements lost by drying 82 per cent. of moisture. The average of the excretions yielded by one pig in twenty-four hours was: moist, 9.1 lbs.; dry, 1.6 lb.

Composition:

	Dry.	Moist.
Carbon	38.7	6.97
Hydrogen	4.8	0.86
Oxygen	32.5	5.85
Azote	3.4	0.61
Salts and earth	20.6	87.1
Water	"	82.00
	100.0	100.00

The litter that is generally employed is wheat-straw. This straw, in the condition in which it is used, contains 26 per cent. of moisture.

Its composition is:

	Dried.	Undried.
Carbon	48.4	35.8
Hydrogen	5.3	3.9
Oxygen	38.9	28.8
Azote	0.4	00.3
Salts and earth	7.0	5.2
Water	"	26.0
	100.0	100.0

At Bechelbronn each horse receives daily as litter 4.4 lbs.; each cow 6.6 lbs.; each pig 4.1 lbs. of straw.

To the stables and the cow-houses together are given every twenty-four hours 132.0 lbs. of straw for thirty horses; 198.0 lbs. for thirty horned cattle; 66.0 lbs. for sixteen pigs; making 396.0 lbs. of straw, estimated when dry at 292.6 lbs.

The composition of the materials which constitute the dung produced in one day are set forth in the following table:

Excretions yielded in 24 hours by	Weight when dry.	Weight in the wet state.	Elements of the dry matter.					Water constituting the wet matter
			Carb.	Hydrog	Oxygen	Azote.	Salts & earths.	
	lbs.	lbs.	lbs.	lbs.	lbs.	lbs.	lbs.	lbs.
Thirty horses	245.08	1028.28	94.60	12.32	89.10	6.60	42.46	783.20
Thirty horned cattle	326.36	2416.48	130.24	15.40	116.16	8.58	56.98	2089.12
Sixteen pigs........	26.40	146.74	10.12	1.32	8.58	0.88	5.50	120.34
Straw used in litter	292.60	396.00	41.68	15.62	113.74	1.10	20.46	103.40

The average or mean composition of this mixture may be taken as follows:

In the dry state.					In the wet state.					
Carbon.	Hydrog.	Oxygen.	Azote.	Salts.	Carbon.	Hydrog.	Oxygen.	Azote.	Salt.	Water.
42.3	5.0	36.7	1.9	14.1	9.4	1.2	8.2	0.4	3.2	77.6
That of the resulting Dung:										
35.8	4.2	25.8	2.0	32.2	7.4	0.9	5.3	0.4	6.7	79.3

On comparing the composition of the dung-heap with that of the different kinds of litter collected in a day, little difference is observed; the larger quantity of saline and earthy matters discovered in the fermented manure is readily explained from the additions of ashes incorporated with it, and also by the accidental admixture of earthy matters proceeding from the sweepings of the court, the earth adhering to the roots consumed as food, &c.—refuse of every kind, the residue after cleansing the various kinds of fodder for the stable and stall, &c., all goes to the dung-heap. Lastly, and with reference to the elements that are liable to be dissipated in the state of gas, or which may be changed into water, the azote is perceptibly in larger quantity in the prepared manure than in the unfermented litter and excretions. This is at once seen on comparing the composition of these two products after the saline and earthy matters have been deducted.

	Carbon.	Hydrogen.	Oxygen.	Azote.
The composition of fresh litter, is..........	49.3	5.8	42.7	2
That of dung...........................	52.8	6.1	33.1	3.0

Dung is, therefore, somewhat richer in carbon than litter, and it contains less oxygen. It is the property of lignine undergoing decomposition, that it yields a product which relatively abounds more in carbon than the original matter, in spite of the carbonic acid which is formed and thrown off during the alterations undergone; this is owing to the elements of water being thrown off in relatively still larger quantity at the same time.

Fermented dung contains less oxygen than that which comes from the stable; it ought also to contain less hydrogen: but this analysis does not proclaim. It must be observed, however, that the quantity of oxygen (4.6,) the loss of which appears would require no more than 0.57 of hydrogen to constitute water; and this is a quantity which it is impossible to answer for in experiments made upon such substances without excessive delicacy of manipulation. This much may be certainly concluded, viz., that manure which has undergone preparation contains a larger relative proportion of azote than the substances which have concurred in its production; and for this reason, it is very probable that upon the whole a very trifling loss of this element is experienced if the fermentation has been carefully managed, and the manure has been carried out and distributed upon the land before its decomposition is too far advanced. This conclusion, which I am particularly anxious to establish, is partly explained by the interesting researches of Mr. Hermann, which go to prove that woody fibre in rotting attracts and fixes a quantity of the atmospheric air.

Azote is in fact the element which it is of highest importance to augment and to preserve in dung. The organic substances which are the most advantageous in producing manures are precisely those which give origin by their decomposition to the largest proportion of azotized matters soluble or volatile. I say by their decomposition, because the mere presence of azote in matters of organic origin does not suffice to constitute them manure. Coal, for example, contains azote in very appreciable quantity; and yet its ameliorating influence upon the soil is absolutely null; this happens from coal resisting the action of those atmospheric agencies which determine that putrid fermentation, the ultimate result of which is always the production of ammoniacal salts, or other azotized compounds favorable to the growth of vegetables. While we admit the high importance, indeed the absolute necessity of azotic principles in manures, then, we must not therefore conclude that these principles are the only ones which contribute to fertilize the earth.

It is unquestionable that the alkaline and earthy salts are further indispensable to the accomplishment of the phenomena of vegetation; and it is far from being sufficiently shown that the organic principles void of azote play a merely passive part when added to the soil. But with few exceptions, the fixed salts, water or its elements, and carbon superabound in manure. The element which exists there in smallest proportion is azote, which is the one also that is most apt to be dissipated during the alteration of the bodies that contain it. For these reasons azote is really the element whose presence it is of highest moment to ascertain; its proportion is that in fact which fixes the comparative value of different manures.

Since it is by undergoing modification in the course of their decomposition by putrefaction that those azotized substances which are favorable to vegetation are developed in quaternary compounds, it will be readily understood that all things else being equal, a manure which is completely decompoundable into soluble or gaseous products in the course of a single season, will exert in virtue of this alone the whole of its useful influence upon the first crop. It is entirely different if the manure decomposes more slowly; its action upon the first crop will be less obvious, but its influence will continue longer. There are manures which act, it may be said, at the moment they are put into the ground; there are others, the action of which continues during several years. Nevertheless two manures, although acting within periods so different in point of extent, will produce the same final result if they severally contain the same dose of azotic elements, if they are of the same intrinsic value.

The durability of manures, the length of time during which they will continue to exert their influence, is a matter of great importance. It often depends on their state of cohesion, or on their insolubility, though climate and the nature of the soil have also a marked influence on their decomposition and consequent effects. It is not easy in the present state of knowledge to predict with certainty how long the beneficial effects of a given manure will continue to be felt; but we know well enough what will hasten the decomposition of manure, and what will retard this final result, and so apportion as it were the fertilizing principles among the different crops in the rotation. Aware of the importance of azote in manures, M. Payen and I undertook an extensive series of analyses, with a view to ascertain the proportion of this principle in the various matters and mixtures made use of in the improvement of the soil. This labor enabled us to class manures; and assuming farm-dung as the standard, to refer each to its place in a comparative scale, I shall give the conclusions to which we came in the tabular form; but before doing this, I think it necessary to premise a few observations upon the several manures, or substances usually employed in preparing manures.

Straw, woody stems, haum, leaves, and weeds. The straw of corn, the haum and stalks of various plants of farm growth, weeds of all kinds, and leaves collected in the woods, all contribute to increase the supply of manure.

Straw is the article that is generally employed for litter; its hollow tubular structure, which makes it apt to imbibe urine, renders it peculiarly valuable for this purpose; and it at the same time supplies a soft and warm bed for the cattle. The weight of the straw used as litter may be doubled by the absorption of urine and admixture with excrements; but it is by its very nature and of itself a manure which is not to be slighted, since it contains from 2 to 6 thousandths of azote.

The stems of leguminous plants—bean and pea straw—are much more highly azotized than the straw of corn; it is certainly best to consume this article as forage when it is not too woody and hard. As litter it is often unfit to form a good bed for cattle, and should therefore not be so employed alone; but it presents the twofold advantage of adding to the manure a large proportion of azotized principles, and at the same time of effecting a saving of straw. At Bechelbronn we have found it very advantageous to mix a certain quantity of the dried stems of the *Madia sativa* (gold of pleasure) with both our cowhouse and stable litter.

In forest districts, the leaves of trees are frequently used as litter; they perhaps absorb urine in smaller quantities than straw does. but as they are much more highly azotized, they greatly improve the quality of the dung. It is desirable that the materials used for litter should be capable of imbibing a large quantity of liquid; and these same materials are by so much the more advantageous as the proportion of azote which enters into their composition is high. The leaves of trees combine both of these conditions, and are therefore an immense resource in districts where they can be procured in abundance. Where the woods are strictly preserved, the removal of the leaves is generally prohibited; and it is doubtless injurious to deprive the soil of them in young plantations; but where the timber is further advanced, the objections to their removal are infinitely less, and it is therefore generally permitted to carry them away within certain limits. And when it is seen that from natural causes a great part of the leaves is actually lost to the soil of the forest, the wind sweeping them into the ravines, whence they are carried away by the rains, it is evidently far better to allow the poorer cultivators to profit by them. The benefit obtained appears the greater, as the time and labor bestowed in collecting the leaves is not taken into the reckoning.

Bean straw, and other stalks of a very hard and thready nature, make but indifferent litter, they are often so hard that they hurt cattle; and then their cuticle being impermeable, they absorb little or no urine. It has been proposed to crush them in the mill or to cut them in pieces, but either of these processes is attended with expense. The best thing to do would sometimes be to place them where they would get crushed under the wheels of the farm carts. The use of woody stems of every description would be attended with unquestionable saving in the useful article of straw, and it must never be forgotten that to economize straw as litter, is to increase the quantity of available forage. If, for example, it were possible to reduce to the state of litter the woody stems of the Jerusalem artichoke in places where this vegetable is grown to any extent, the advantages would be very decided; the quantity of these stalks collected from an acre may amount to from four to five tons; the pith of which they are almost entirely composed is of a very spongy nature and well fitted to absorb fluids. These stalks are light, and properly bruised, would probably replace an equal weight of straw, first as litter and then as an element of the dunghill, instead of being burned as at present to heat the oven or to boil the copper, which seems of all methods the worst to derive any advantage from the woody haum, whether of the Jerusalem artichoke, the potato, rape, &c. These substances contain about 4 per 1000 of azote, and are most profitably transformed into manure. We have found that by placing them at the bottom of the dung-heaps, they end by undergoing decomposition; even the most woody stems of vegetables, indeed, decompose pretty rapidly when they are impregnated with urine and mixed with the droppings of animals. Mere moisture without other addition does not suffice, they then rot with extreme slowness.

The green parts of vegetables buried in the ground with the water they contain, undergo decomposition rapidly; the best plan of using them as manure would therefore be to plough them in at once, were there not certain objections to this. In the first place it cannot always be done, on account of the season and the crops upon the ground; and then it might be imprudent to return to the earth the noxious weeds which had just been pulled up, frequently full of seeds, which would not fail to make their existence known before long. It is besides often impossible to bring loads of weeds to the farmstead; the best thing that can then be done is to change them rapidly into manure in a corner of one of the fields which has produced them. This is readily accomplished by means of lime; a bed is first made of the weeds about 14 inches thick, this is then covered with a thin layer of quick-lime, from half an inch to an inch in thickness; another layer of weeds is laid on, and then another layer of quick-lime, and so on in succession. After a few hours the action between the dry lime and the moist herbage begins, and it may be so intense as even to go the length of burning, to prevent which the pile must be covered with earth or with turf, and every means used to prevent the access of air. The process is generally complete within twenty-four hours, and the heap may then be spread as manure. Before proceeding to such an operation, however, it would be highly proper to calculate its cost. All depends on the price of the lime and the labor; and all things considered, I myself much doubt whether the plan could be followed with advantage.

Green manures. Under this title I include the green parts of vegetables which form part of our crops, such as the haum of potatoes, the outer leaves of carrots, cabbages, beet, turnips, &c. These articles are at once forage and manure, and it is for the husbandman to decide in conformity with his position and particular resources whether he ought to bury them at once, or to use them first as food for cattle.

From my own experience I should say that the leaves of beet and of turnips, and potato haum were articles which ought only to be given to cattle in cases of necessity. It is generally much better to bury them in the ground immediately after the crop is gathered; if they be very indifferent food, they are on the contrary excellent manure, superior in quality even to the best farm dung. From the experiments I have made on this subject, I find that the potato tops from an acre of ground may be equal to 6 or 7 hundred weight of that manure presumed to be dry; and the leaves of the beet, from the same extent of surface, are equal to more than 21 hundred weight of the same manure, also in a state of dryness. It is among green manures that we are to class the sea-weed or marine plants, which in many places are employed for improving the soil. These cryptogamic plants, which abound in azote, have a fertilizing power superior to that of common dung, a fact which explains the great store which is set in Brittany by the sea-weed that is collected on its coasts. Sea-weed is employed either fresh and as it comes from the sea, or half dried or macerated, or roasted, and even partially burned. It appears to act at once in virtue of the azotized organic matters which it contains, of the hygrometric properties which it possesses, and of the saline substances which enter into its composition. The agriculturists of Brittany have employed sea-weed as manure from time immemorial; and so have the people of Scotland and Ireland. In Brittany, the sea-weed is gathered at periods fixed by law. The first gathering, as well as that which has been cast up by the waves, is given up to the poor. The gatherings then take place at regular intervals by means of a kind of cutting rake. The sea-weed cut from the rocks is piled upon rafts or thrown into barges, and carried to the shore; and there is a trade carried on in the article all along the shores of the channel between Genest and Cape La Hogue, from the Chansey Isles, and from the coast of Calrados.

When sea-weed is employed in the fresh state, it is ploughed in as speedily as possible. For those kinds of crops which require made manures, the sea-weed is stratified with dung and so left to ferment. In some places the sea-weed is roasted or imperfectly burned, by which, while a large proportion of the vegetable tissue is destroyed, an azotized product is still left behind. Before burning the sea-weed, it is exposed for a time to the air and the rain, and it is then dried, being frequently turned. In this state, it is even used as fuel in places where wood is scarce. One great advantage in sea-weed which has been particularly indicated, is its total freedom from the seeds of noxious weeds.

Aquatic plants which grow in fresh water may also be employed as manure; the common reed cut and buried green, decomposes rapidly. And here I may mention that to destroy reeds which are often a cause of great annoyance in ponds, Schwertz recommends lowering the water about 16 inches, cutting the plant, and then raising the water to its old level; the water enters the interior of the stems and they all die in a very short space of time.

Crops which are buried green, for the improvement of the soil, are also to be ranked in the list of the manures which now engage us. The plan of burying green crops dates from the most remote antiquity; it was greatly recommended by the Romans, and is followed in Italy at the present day. The plants usually grown for the purpose of being burned green are colza or colewort, rape, buckwheat, tares, trefoil, &c. The preference, however, is given to one or other of the leguminous plants, such as tares, lupins, &c., plants which appear to have the highest power of extracting azotized principles from the atmosphere; and indeed the value of the whole process is founded upon this fact, for otherwise it would be impossible to give any reason for this long accredited mode of improving the soil. This, too, is one of the ways in which fallowing becomes useful; it is not merely the rest which the land thus obtains, it is also benefited by the vegetables which grow upon it spontaneously, which come to maturity and die, leaving in this way in the ground all they had attracted from the atmosphere, or fixed from the water with which they had been supplied.

Seeds, Oil-cake. It is in the seed that by far the largest propor tion of the azotized matter assimilated by vegetables during their growth is finally concentrated at the period of their maturity. Seeds are consequently very powerful manures, and great advantage is taken of them. In Tuscany, lupin seed is sold as manure; it contains 3½ per cent. of azote. It is employed after its germinating power has been destroyed by boiling or roasting. The cultivation of the lupin is carried on in districts, the situation of which is such that difficulty would be experienced in exporting more bulky crops. Grains from the brewery would also make excellent manure were it not generally found more advantageous to use them as food for cattle. In some places, however, where there is no adequate demand for them in this direction, they are dried upon a kiln, and are then equal to twice and a half their weight of farm dung; in some places they are actually sold at a proportionate price. The state of division of grains admits of their being regularly spread. In some parts of England, grains are used in the proportion of from 40 to 50 bushels per acre for wheat or barley.*

The refuse of the grape in wine countries contains a large quantity of azotized matter. The decomposition of the grape stones being slow, this refuse answers admirably as a manure for vines.

Oleaginous seeds after the extraction of the oil leave a residue which is an article of commerce, and is familiarly known under the name of *cake*. Oil contains no appreciable quantity of azote; this principle is contained entirely in the cake, which becomes through this alone most excellent manure. The proportion of azote which cake contains, varies from 3½ to 9 per cent. Oil-cake, from its mode of preparation, contains but very little moisture, and consequently offers great facilities in the way of carriage; it may be taken without difficulty to situations whither a load of dung could scarcely be carried.

Cake is applied in two modes: 1st. In powder, and by sowing upon the field, sometimes mixed with the seed. 2d. Mixed in water or in the drainings of the dung-hill, in which case the liquid containing the products of the decomposition of the cake is distributed over the land. By putrefaction under water, cake yields a matter of extreme fetor, comparable both in point of smell and of effects on vegetation to human excrement obtained from privies.

Although cake, from the large proportion of albumen and legumen which it contains, be an excellent food for cattle, it is still found more advantageous in many districts to use it as manure than for feeding. England imports oil-cake from all parts of the continent. France alone, from 1836 to 1840, exported more than 117,860 tons of the article. Oil-cake has been particularly recommended as manure for light sandy soils. When the soil is clayey and cold, Schwertz recommends a mixture of one part of lime and 6 parts of powdered cake. To me, however, the addition of lime has always appeared a questionable auxiliary in such manures as give rise readily to ammoniacal products, as is the case with oil-cake. For clayey lands, it would perhaps be advisable to employ oil-cake in a state of

* Sinclair, Agriculture, vol. i.

decomposition and diffused in water; its effects, I imagine, would not be doubtful.

Oil-cake, as a manure, is employed at very different seasons, according to the nature of the husbandry. It is always well to employ it in rainy weather. Its effect is always certain, if it comes on to rain two or three weeks after it has been put into the ground. Drought suspends its action; it frequently happens, indeed, that the first crop shows none of its good effects; but these never fail to appear in subsequent crops. Schwertz remarks very properly, that this circumstance has led many farmers to overlook the real advantages that belong to this manure. Cake, in fact, according to the dryness or moistness of the season, may act as a manure either of difficult or of easy decomposition, and so produce more immediate or more remote effects. In England about 800 weight of oil-cake per acre are commonly applied. Mr. Coke, of Holkham, ploughed in the powdered cake about six weeks before sowing turnips, but it is held more economical and more advantageous to strew it in fine powder along the furrow with the seed. The latter view, however, must not be too confidently acted on by farmers; the general recommendation to sow the fields with powdered cake, either some weeks before or some weeks after putting it in the seed, and when the plants have already sprung, appears to be the right one. We have various observations made by one of our most experienced practical farmers which prove that oil-cake used dry and without mixture often produces the most injurious effects upon germination. In September, 1824, M. Vilmorin, desiring to make a comparative trial of different pulverulent manures, strewed a quantity of powdered colewort-cake upon a piece of red clover. Upon all the parts of the field which had received other manures, applied in the same way, the clover sprung perfectly; but that which had received the oil-cake continued absolutely naked; the cake had been employed in the proportion of about 800 cwt. per acre. The same result was also obtained in a trial made with vetches and gray winter peas.* Duhamel, referring to similar facts, recommends the cake to be applied ten or twelve days before sowing. In Flanders, from 6 to 7 cwt. per acre is the quantity generally employed for wheat crops, and it is scattered over the surface before winter sets in, when the grain is already above the ground.

The *pulp of the beet-root* which has been employed in the sugar manufactories of France and Flanders, is an article which as food for cattle is known not to be inferior to the root before it has undergone expression, and it contains nearly the same proportions of sugar, albumen, &c. It is, therefore, always used as food to as great an extent as possible. But the article is kept with difficulty, and the production at times far exceeds the powers of consumption, so that it has to be made into manure, for which it answers excellently. The skimmings and dregs which are collected in the process of sugar making, are also available as manure. They contain about the same amount of azote or azotized matter as farm dung, and are therefore

* Vilmorin, in Maison Rustique, vol. i. p. 204.

of similar value. The animal charcoal of the sugar refinery, after it has served its office there, is an admirable manure. It is, in fact bone or ivory-black, mixed with the coagulated blood which has been employed to clarify the sirup by entangling impurities, and a very small quantity of sugar. This mixture, so rich in azotized principles, used actually to be turned into the sewers until the year 1824, when M. Payen showed its value as manure, since which time nearly 10,000 tons have been annually employed in ameliorating the soil, to the great advantage of practical agriculture. The importance of the trade in this residue of the sugar-house, and complaints of the occasional indifferent quality of the article, attracted the attention of the department of the Inferior-Loire in 1838, and led to the appointment of an inspector of the manure shipped from the port of Nantz. I may here observe, that in testing a manure it is by no means enough to limit attention to the quantity of organic matter which it contains. The only sure means is to determine the amount of azote; it is not organic matter, but the amount of azotized organic matter upon which almost alone depends the value of the manure.

The residue of the sugar refinery is another of those articles which presents an occasional anomaly in its application, and which must not be left unnoticed. Its effect upon the ground has not only been extremely variable, but it has sometimes happened that this manure, laid on very soon after coming from the manufactory, has been found decidedly injurious to vegetation. Kept for some time, for a month or two, in a heap before being applied, its effect has not only been found more certain, but also uniformly favorable.

It is not difficult to explain these divers and opposite influences: the sugar contained in the refuse undergoing fermentation yields first alcohol, and than acetic and lactic acids. Employed in this state, the substance must necessarily prove injurious to vegetation. It is only after it has lain for a sufficient length of time exposed to the air, to have had the animal matter it contains changed into ammonia, and the organic acids engendered saturated with this base, that it becomes truly useful to vegetation. The heap indeed then shows alkaline, not acid re-action.*

The *residue of the starch manufacturer*, the fetid water which is obtained in such quantity in the process of making starch from grain, is a powerful manure, and ought not to be suffered to run to waste.

The *pulp or residue of the potato* which is now produced in considerable quantity in the potato starch manufactories, is known to be an excellent article of food for hogs and cattle. Towards the end of the season, however, it is apt to be of very indifferent quality, and green food having by this time come in abundantly, it often goes to the dung-hill. In the dry state, it is worth its own weight of farm dung; wet, 100 of the pulp may be equal to about 131 of farm-yard dung. The water which has served for washing out the starch from the pulp, as in the case of wheat and other grain, contains an organic substance which when dried constitutes pulverulent

* Payen and Boussingault, Ann. de Chimie, v. iii. p. 95, 3e série

manure that is equal to about half its weight of the dry manure prepared from night soil, which the French call *poudrette*. M. Dailly made a comparative trial of these two kinds of manure, and from actual experiment found that 200 parts of the deposite from the starch manufactory might be used for 100 of poudrette. Even the water that is used in the manufacture, and from which the substance in question is deposited, is an excellent manure when thrown upon the ground, a circumstance which is by so much the more fortunate that this water by standing putrefies and throws off most offensive exhalations. By using the liquor to his fields, at once, M. Dailly prevents every kind of annoyance to himself and his neighbors, and moreover from his great starch manufactory he realizes in this way an additional profit which he estimates at upwards of £60 per annum. Analysis has shown that 100 of this water from the potato starch manufactory represents 17 of moist farm-yard dung.

In cider countries, the *pulp of the apples* that have been pressed is always thrown upon land as manure. At Bechelbronn we reserve it for our Jerusalem artichokes; in Normandy it is thought excellent for meadows and young orchards. Analysis of the pulp of apples grown in Alsace shows that when dry it contains a quantity of azote, which places it on the same footing as farm-yard dung. Sinclair informs us that in Herefordshire the pulp of the cider press is made into good manure by being mixed with quick-lime and turned two or three times in the course of the following summer. Doubtless the addition of lime will hasten the decomposition of the woody matter of the pulp; but it strikes me that this will take place rapidly enough of itself in the ground without contriving any means of accelerating the process.

Animal remains. The remains of dead animals and the animal matters obtained from the slaughter house are powerful manures, which are much sought after in places where their value is properly appreciated. Scraps and the refuse of skin, hair, horn, tendons, bones, feathers, &c., all form invaluable manure. The flesh of animals which die, and so much of that of horses that are slaughtered which cannot be used as food for animals, may be dried after having been previously boiled, and then reduced to powder and applied as manure. The blood of slaughtered animals is less proper as food for hogs, although it is often used in this way, than muscular flesh; it even occasionally gives rise to serious diseases among these animals. It is most easily prepared as manure, however, for which it answers admirably; it is enough to coagulate it by exposure to heat, and then having broken it down, to dry it in the air or in the stove. Liquid blood has been employed as manure, but decomposition then takes place so rapidly, that the products are exhaled without producing much effect. This objection may be remedied by two means, either by diluting the blood in a large quantity of water, with which the land is then irrigated, or by mixing it with a considerable mass of vegetable earth, which is then applied like ordinary manure. There is even a pulverulent manure of which blood forms the basis, prepared in special establishments in the vicinity of various large towns. The large quantity of azote contained in these manures shows how their value may be such as to permit of their being advantageously exported to great distances beyond seas.

Bones are employed in agriculture after having had the fat which they contained extracted from them by boiling. They are crushed by being passed between the teeth or grooves of a couple of cast-iron rollers. They must be regarded as a manure, the action of which is of long duration, because the animal matter contained in them decomposes slowly, protected as it is by the earthy casing which surrounds it. In England from 50 to 60 bushels of bruised bones per acre are usually put upon land prepared for turnips.

The employment of bones as manure has given rise to the most various and contradictory observations. In certain circumstances their effect upon vegetation has been almost null; in others their action has been decisive and most favorable. M. Payen has given a solution of these anomalies which is perfectly satisfactory. According to my learned colleague, bones in their interstices, contain a quantity of fat of various consistency, which may be removed by long boiling in water; the average quantity of grease obtained from fresh bones is about 10 per cent. It has been observed that this fatty matter diminishes gradually in bones that dry by long exposure; it even disappears almost entirely when they are dried at a high temperature. This happens from the water which is disengaged from the bony tissue by the effect of evaporation, being replaced by fat melted by the heat. The consequence of this is, that the organic tissue of bone, which was already sufficiently rebellious to decomposition, becomes still less alterable when it is impregnated with grease. The grease, in fact, by reacting upon the carbonate of lime of the bone, has formed an earthy soap which long resists atmospherical influences and change under ground.

It will readily be understood that bones in this condition can have little or no action upon vegetation, unless indeed they be reduced to very fine powder. This alone will explain how it may happen that some bones, after having remained four years in the ground, have been found to have lost no more than 8 per cent. of their weight, while those, the grease of which has been removed by boiling water, have lost in the same space of time from 25 to 30 per cent. of their weight.*

These observations of M. Payen show how completely Schwertz was mistaken when he ascribed the indifferent quality of the manure prepared from old bones, or from bones that had been boiled, to the absence of fat, which he regards, I know not on what authority, as a substance extremely favorable to vegetation. It is not very obvious how fatty substances should act as manures. I myself ascertained, from experiments made some years ago with a view to test the conclusions of an agriculturist who ascribed the good effects of cake to the fatty matters which it contained, that rape-oil had no kind of favorable influence upon the growth

* Payen, Maison Rustique, v. i. p. 194.

of wheat. I have said nothing here upon the importance of tne earthy matter of bones, particularly of the calcareous phosphate which they contain, but which is nevertheless acknowledged to be of great importance.

The *refuse from the glue-maker's*, washed and pressed, contains all the animal matters which have resisted the action of boiling water, such as portions of tendinous and skinny substance, hair, pieces of bone, of horn, and of flesh, a calcareous soap, and earthy matters. This mixture putrefies rapidly; but dried, it may be preserved for a great length of time. Analyzed dry, it yields about 4 per cent. of azote. From 4 to 5 cwt. per acre are employed, but it is necessary to manure every year.

The *refuse of the tallow-melter*, *graves*, as it is called, a residue consisting in great part of the membranes which have enveloped the fat of our domestic animals, mixed with a little blood, some flesh, and bony matter, and grease, has hitherto been employed almost exclusively as food for dogs. Of late, however, graves have been used as manure, and analysis shows that this substance must be estimated as equal to about 3½, farm-dung being fixed at 1. Used in this proportion, graves produce a marked effect. The action of graves, which may be thrown on in fragments and dry, or after having been steeped in hot water, and reduced to the state of a pulp, will continue for three or four years.

Shreds of *woollen rags* form a good manure for vines and olive-trees especially, though they are also available in husbandry of every description. The large proportion of azote, and the small quantity of water contained in woollen rags, constitute them not only one of the richest manures, but also one of those that is most easily transported; 25 cwts. per acre of woollen rags, the cost of which, in France, may be about £3, have been found sufficient as manure for three years. The slowness with which wool decomposes, indeed, causes its action to be continued during six or eight years. Twenty-five cwt. of woollen rags may be held equivalent to upwards of 40 tons of farm-dung, which, at the price of 5*s*. 10*d*. per ton, would cost £12 16*s*. At the end of three years, M. Delonchamps, an excellent practical farmer, gives his land a dressing of farm-dung for three years more, when he returns to the wool. Before spreading rags they must be cut into pieces, which is effected either by a machine, or by a piece of scythe-blade fixed in a block of wood. In England, the quantity of woollen rags allowed to the acre is generally about 13 cwt. Sinclair says that they are best suited for dry and sandy or chalky soils, and this because they attract moisture. I have not found the fact to be so. In the very dry soil of a vineyard manured with this article, I have found the pieces to decompose with extreme slowness, and, up to this time, the effect upon the vines has been scarcely perceptible.

The *raspings and shavings of horn* form a manure of great power, that seems applicable to every variety of soil. In England, about 40 bushels per acre are usually allowed.

Tendons, trimmings of hides, hair, feathers, &c., are manures very analogous to the last, and of which the value may be estimated from the quantity of azote which they severally contain. This value once determined, every farmer knows the quantity which he must lay upon his land; and he thus proceeds upon a much more rational foundation than when he takes for his guide one or other of those vague and arbitrary indications that have been given. Sinclair, for example, would have us lay on nine bushels of feather rubbish to the acre, and Schwertz recommends from four to five times as much more. Nothing, in fact, is more arbitrary and uncertain than to estimate such materials by the bulk; it must be obvious that the weight of a bushel of hide-trimmings, of horn-shavings, and of feather-rubbish, must differ very widely, not only with reference to one another, but also according to the state of division in which each is measured. As a general rule, it is by weight, and weight alone, that the quantity of manure must be estimated.

Shells and mud from the sea-shore and the bottoms of rivers, are matters that are not often very highly azotized; nevertheless they may contain an equivalent of the all-important element, azote, which may bring them near to wet farm-yard dung in point of value. The abundance of such matters in certain situations makes them extremely useful. The alkaline and earthy salts, which they generally contain in considerable quantity, also add to their fertilizing properties. The sea-sand which is employed in Brittany under the name of *marl*, (merl,) consists, in great part, of the remains of corallines, madrepores, and shells, mixed with a few hundredths of highly azotized organic manner. This marine marl is found in great abundance at the mouths of the river of Morlaix, where there is a considerable traffic carried on in the article. It is said to be reproduced, new banks of it being met with from time to time. It is obtained by dredging from barges, and the process is only allowed to go on from the 15th of May to the 15th of October, when the quays of the town of Morlaix are seen covered with the produce. It is carted to a distance of five leagues inland. A barge-load weighing seven tons, sells at from 6*s*. 6*d*. to 8*s*. This same species of marl is now obtained upon the coast of Plancourtrez and in the roads of Brest. It has also been discovered near the mouth of the river Quimpert. It appears, finally, that the shell sand so much employed by the farmers of Devonshire and Cornwall is of the same essential nature.

In the neighborhood of Morlaix, from five to six tons per acre of this calcareous sand are employed upon light dry soils; from eleven to twelve tons are given to clayey lands. This quantity would probably be too great for porous and damp soils, inasmuch as sea-marl belongs to the class of *warm* manures; that is to say, it under goes speedy decomposition. There can be no doubt that sea-marl acts further, in virtue of the calcareous matter which it contains, and also of its merely mechanical properties upon the strong argillaceous lands of Brittany, for which sand alone is an excellent improver. It is also to the carbonate of lime which it contains, that its good effects upon lands that show an inflorescence of iron pyrites must be ascribed. It is well to lay this shell-marl upon the land shortly after it is taken from the sea; by long exposure to the air, it suffers disaggregation and loses a portion of its good qualities.

There is another kind of sea-sand called *trez*, which forms banks in the neighborhood of Morlaix, and which is known under the name of *tanque* on the northern shores of France, which is favorable to vegetation, particularly after it has been washed and freed from the greater part of the salt which it contains. It is thrown upon the land in larger quantity than the marl. The small quantity of animal matter which it contains putrefies and is lost when it remains exposed to the air for any length of time, so that a distinction has been made between fresh or *live trez*, and old or *dead trez*, the one being the article as it comes from the sea, the other after it has been exposed some time on the shore; the article which has been exposed undoubtedly contains a smaller quantity of organic matter than that which is quite fresh. This variety of sea-sand is particularly available upon close and clayey lands, which sometimes receive as many as sixteen tons per acre with advantage; lighter lands, of course, require much less.

Shells, sand, slime, and sea-weed, are not the only useful materials supplied to agriculture by the sea; *fish*, or their *offal*, is frequently employed as manure. The practice of manuring with fish is very old, and is universal wherever it can be had recourse to. I have already had occasion to say, that at the period of the conquest of America, the Spaniards found it established among the Indians, on the shores of the Pacific ocean. The lands are occasionally manured with fish along the sea-board of Great Britain and Ireland, and the low lands of Lincolnshire, Cambridgeshire, and Norfolk, also receive occasional supplies of the same powerful manure. The offal of the herring fishery, of cod, of skate, and of the pilchard, in Cornwall, the dog-fish entire, and other kinds, that are either less esteemed, or that are caught in quantities greater than can be consumed as food, are all admirable manures. We have been recommended to mix the fish or fish-offal with quick-lime; but, unless in certain circumstances, the practice is very questionable; the addition is probably only proper where the materials are exceedingly oily, as is the case with pilchards, herrings, &c.: an earthy soap is then formed which prevents the injurious effects upon vegetation which wholly oleaginous matters scarcely fail to produce. One analysis of codfish, which I made along with M. Payen, gave us a proportion of azote of nearly seven per cent. This, of itself, is enough to explain wherefore the flesh, the cartilages, and the bones of fishes should be found such energetic manures.

The *slime* deposited by rivers also yields manure which may be employed to much advantage. The Nile, which periodically inundates the plains of Lower Egypt, owes its fertilizing action to the slime which it contains, and which it deposites before it again recedes into its bed. On the banks of the Durance, the mud or slime deposited by the river is carefully collected for distribution over the fields in its vicinity. The waters of this river are frequently turbid and improper for irrigation, until they have deposited the slime which they hold in suspension ; the waters are therefore turned into canals for the purpose of deposition before they are let upon the land ; and such is the quantity of slime that is precipitated, that two or three gatherings of it are made in the course of the year. It is dug out and thrown upon the banks to dry ; reduced to powder, it is fit to be laid upon the land ; and such is its fertilizing power, that a field which yielded but four for one, has been brought to yield twelve for one by its means.*

Wood and coal soot, and Picardy ashes. Soot has been known for a long period as a useful manure. M. Braconnot, in the soot of a chimney where wood had been the fuel, found the following ingredients :

Ulmic acid	30.0
Azotic matter, soluble in water	20.0
Insoluble carbonated matter	3.9
Silica	1.0
Carbonate of lime	14.7
Carbonate of magnesia (traces of)	
Sulphate of lime	0.5
Ferruginous phosphate of lime	1.5
Chloride of potassium	0.4
Acetate of potash	4.1
Acetate of lime	5.7
Acetate of magnesia	0.5
Acetate of iron (traces of)	
Acetate of ammonia	0.2
An acrid and bitter element	0.5
Water	12.5
	100.0

The analysis which M. Payen and I made of wood and coal soot, confirms the presence of the azotized principle detected by M. Braconnot. A considerable trade is carried on in soot for agricultural purposes in large towns ; it is thrown upon clovers and young wheats, in the proportion of about 20 bushels to the acre. Some have recommended that it should be mixed with lime ; but as soot always contains salts having a base of ammonia, the practice is evidently objectionable, unless indeed the object be to get rid of that which is most useful in the article, which will be effectually accomplished by adding lime to it. The proper procedure is to employ the soot without admixture during calm or wet weather. In Flanders, the colewort beds destined for transplanting are very generally manured with soot, which is believed to have the property of preserving the young plants from the attacks of insects. In the neighborhood of Lisle, they give from 55 to 60 bushels of soot per acre. Schwertz appeals to many facts which go far to satisfy us that the effects of soot upon clovers are particularly advantageous ; he says, moreover, that coal soot is preferable to wood soot. The superior properties of coal soot are evidently due to two causes : first, it is more dense

* Belleval, in Annals of French Agriculture, 2d series, vol. xiv. p. 261. The beds of many of the oozy-bottomed rivers in England near the sea are inexhaustible sources of the most valuable manure. The bed of the Thames, between London Bridge and Putney Bridge at low water, is a true gold mine if it were but rightly used.—Eng. Ed

than wood soot, and in a given bulk actually contains a larger quan tity of matter; secondly, I have found that, for equal weights, coal soot contains the larger quantity of azote.

Picardy ashes are prepared by the slow and imperfect combustion of the pyritic turf which is dug up in the department of the Aisne for the manufacture of sulphate of iron and of alum. This turf piled up, heats, and finally takes fire; the combustion continues for about a month, abundance of sulphureous vapors being disengaged. The residue is a gray ash, still containing a quantity of carbonaceous matter, which is found very advantageous in the way of top-dressing for meadows. It might be maintained that the utility of such ashes depends solely on the sulphate of lime which they contain; but it is ascertained that they are much more active as manure than this substance employed by itself; analysis, in fact, explains in some degree the fertilizing powers of these ashes, by showing that they contain more than $\frac{1}{2}$ per cent. of azote, to say nothing of the saline matters of which vegetables are so greedy. It is extremely probable that during the slow incineration of the turf, there is a quantity of sulphate of ammonia produced.

The ashes which remain after the lixiviation of the pyritic and aluminous lignites which are mined for the purpose of making green vitriol, are analogous to Picardy ashes, and are employed with equal success in agriculture. At Forges-les-Eaux, the pyritic earths after lixiviation are mixed with a quarter of their weight of turf ashes, and form an active manure which is employed very extensively in the country around the town of Bray in France: it is equally adapted to meadows and to land under roots, such as potatoes or turnips, green crops or corn. Analysis shows these ashes to have the folowing composition:

Soluble organic matter	2.7
Insoluble humus	49.8
Sulphate of protoxide and of peroxide of iron	1.8
Fine sand	39.0
Sulphuret of iron; Peroxide of iron	6.7
	100.0

The vitriolic ashes of Forges-les-Eaux are more highly azotized than those of Picardy; they contain 2.72 per cent. of azote.

The effect of the imperfect combustion of these pyritic turfs, the product which results from it, explains to a certain extent the beneficial effects of the practice of *paring and burning*, an important and widely spread practice, the utility of which it would be difficult to understand, were it not connected in some way with the production of ammoniacal ashes.

The useful effects of paring and burning, are, in all probability, connected with the destruction of organic matter, very poor in azotized principles; in the transformation of the surface of the soil into a porous, carbonaceous earth, made apt to condense and retain the ammoniacal vapors disengaged during the combustion; lastly, by the production of alkaline and earthy salts, which are familiarly known to exert a most beneficial influence upon vegetation. These conditions seem so entirely those, the object of which it is to realize by paring and burning, that in order to make the operation favorable to the soil which undergoes it, the vegetable matter which it has produced, must of necessity be transformed into black ashes; when it goes beyond this, as Mr. Hoblyn has well observed, when the incineration is complete, and the residue presents itself as a red ash, the soil may be struck with perfect barrenness for the future. The burning, therefore, that was not properly managed, that led to the complete incineration of all the organic matter, would, for the same reason, have a very bad effect in the preparation of the Picardy ashes; which might indeed act in the same way as turf ashes from the hearth and oven, but which, deprived of all azotized principles, would not ameliorate the ground in the manner of organic manures.

I have frequently seen the process of burning pertormed in the steppes of southern America. Fire is set to the pastures after the grass which covers them has become dry and woody; the flame spreads with inconceivable rapidity, and to immense distances. The earth becomes charred and black; the combustion of those parts that are nearest to the surface, however, is never complete; and a few days after the passage of the flame, a fresh and vigorous vegetation is seen sprouting through the blackened soil, so that in a few weeks the scene of the desolation by fire, becomes changed into a rich and verdant meadow.

ANIMAL EXCREMENTS.

Horse-dung. The composition of horse-dung would lead us to infer that its action must be more energetic than that of cow-dung. Nevertheless, agriculturists frequently consider it as of inferior quality. This opinion is, even to a certain extent, well founded. Thus although it be acknowledged that horse-dung covered in before it has fermented, yields a very powerful manure, it is known that in general the same substance, after its decomposition, affords a manure that is really less useful than that of the cow-house. This comes entirely from the fact that the droppings of the stable, by reason of the small quantity of moisture they contain, present greater difficulties in the way of proper treatment than those from the cow-house. Mixed with litter and thrown loosely upon the dung-hill, horse-dung heats rapidly, dries, and perishes: unless the mass be supplied with a sufficient quantity of water to keep down the fermentation, and the access of the air be prevented by proper treading, there is always, without the least doubt, a considerable loss of principles, which it is of the highest importance to preserve. I can give a striking instance of this fact in the changes that happen in the conversion of horse-dung into manure in the last stage of decomposition: fresh horse-dung in the dry state contains 2.7 per cent. of azote. The same dung laid in a thick stratum and left to undergo entire decomposition, gave a humus or mould, from which, reduced to dryness, no more than one per cent. of azote was obtained. I add, that by this fermentation or decomposition, the dung had lost nine tenths of its weight.

From these numbers every one may judge how great had been the loss of azotized principles. In practice, however, little care is bestowed on the preparation of horse-dung ; the fermentation is rarely, if ever, pushed to this extreme point indeed ; but it is not the less true that it is constantly approached in a greater or less degree ; and that the consequences, although not altogether so unfavorable as those which I have particularly signalized, are nevertheless extremely destructive. All enlightened agriculturists have, therefore, long been aware of the attention necessary to the management of horse-dung, which requires a degree of care, that may be perfectly well dispensed with when the business is to convert the dejections of horned cattle into manure. To obtain the best results in the management of horse-dung, it appears to be absolutely necessary to give it a much larger quantity of moisture than it can ever receive from the urine of the animal ; if it be not watered it necessarily heats, dries, and loses both in weight and quality ; while, by being kept properly moist, it produces a manure, which half rotted, is of quality superior, or at all events equal, to the same weight of cow-dung.

M. Schattenmann, who has the produce of stables containing two hundred horses to manage, follows a process of the most commendable description in the preparation of his manure, and which is attended with the very best results. His dunghill stance, of no great depth, is about 440 yards square in superficies, and divided into two equal portions. The bottom of this stance is so arranged as to present two inclined planes, which bring all the liquids that drain from it to the middle, where there is an ample tank for their reception, furnished with a pump for their redistribution to the dunghill. There is also another spring-water pump destined to supply the water that is necessary to preserve the dung-heap in an adequate state of moistness. The latter auxiliary is quite indispensable ; the quantity of water necessary is so considerable when masses of such magnitude are to be treated, that we cannot trust to any casual source of supply. The two portions of the area are alternately piled with the dung as it comes from the stables ; it is heaped to the height of 10, 12, or 14 feet ; it is trodden down carefully, as it is evenly spread, and plentifully watered from the spring-water pump. Due consolidation, and a state of constant humidity, are the two conditions that are the most indispensable to the successful preparation of horse-dung. M. Schattenmann is in the habit of adding to the liquid, saturated with the soluble matters of the dunghill, a quantity of sulphate of iron in solution, or of sulphate of lime (gypsum) in powder ; he also throws the same salts upon the surface of his heap : the object of this is evidently to transform into sulphate, the volatile carbonate of ammonia formed in the course of the decomposition, and so to prevent its escape and loss. By these means a pasty manure, as rich as that which is yielded by horned cattle, and of a quality, the excellence of which is proclaimed by the remarkable crops that cover the lands which receive it, is produced in the course of two or three months.*

* Schattenmann, Annales de Chimie, 3e série, vol. iv. p. 117.

It is almost useless to add, that great care must be taken not to introduce too large a quantity of sulphate of iron, which might have a prejudicial influence upon vegetation, into the dunghill or the drainings from it. In making use of sulphate of lime there is nothing to fear on this score ; this salt in excess would be rather favorable than hurtful ; in general, gypsum is certainly the preferable substance, both on account of its never doing mischief, and of its greatly inferior price.*

Farmers generally advise horse-dung to be reserved for argillaceous, deep, and moist soils ; this recommendation is given in connection with the manure that is obtained by the usual imperfect process of preparation. With regard to the horse-dung, prepared in the manner which I have just described, and as practised by M. Schattenmann, it is adapted to soils of all kinds ; and if it differs from the dung of the cow-house, it is only by its superior quality. This last fact is at once explained by the elementary analysis of the excrements of a horse fed upon hay and oats.

100 parts of the urine of the animal so fed, yielded 12.4 of dry extract, the composition of which was as follows :

	In the state of extract.	In the liquid state.
Carbon	36.0	4.46
Hydrogen	3.8	0.47
Oxygen	11.3	1.40
Azote	12.5	1.55
Salts	36.4	4.51
Water	"	87.61
	100.0	100.00

The droppings of the same horse after drying, gave 24.7 of fixed matter, the analysis of which indicated :

	Dry excrement.	Moist excrement.
Carbon	38.7	9.56
Hydrogen	5.1	1.26
Oxygen	37.7	9.31
Azote	2.2	0.54
Salts	16.3	4.02
Water	"	75.31
	100.0	100.00

The *dung of horned cattle* is often extremely watery ; it is especially so when furnished by animals kept upon green food ; this extreme humidity renders its preparation easy. Its equivalent number is higher than that of horse-dung ; it is, in fact, less highly azotized, and consequently less active. If the food have a great effect upon the quality of the manure, it is quite certain that the circumstances or states of the cattle have an effect which is scarcely less remarkable. Milch cows and cows in calf always furnish a manure that is less highly azotized than stall-fed and laboring oxen ; and this is readily understood : the azotized principles of the food are diverted to secretions, which concur in the development of a new being in the one case, in the production of milk in the other ; for the same

* Every farmer who should have something like a cart or wagon-load of gypsum brought to the farm every year would find his profit from the practice.—ENG. ED.

reason the dejections of young animals, all things else being equal furnish a manure of less power and value than those of adult animals. I shall have occasion to recur to this important subject, which has never yet been sufficiently studied.

The urine and excrements of a milch cow, which is giving about 12 pints of milk per diem, have shown upon analysis, the following quantities of elements: 100 of the urine contained 11.7 of dry extract, and had this composition:

	Urine dry.	Urine liquid.
Carbon	27.2	3.18
Hydrogen	2.6	0.30
Oxygen	26.4	3.09
Azote	3.8	0.44
Salts	40.0	4.68
Water	0.0	88.31
	100.0	100.00

100 of fresh excrement left on drying 9.4 of dry substance, and in each state contained:

	Excrement dry.	Excrement moist.
Carbon	42.8	4.02
Hydrogen	5.2	0.49
Oxygen	37.7	3.54
Azote	2.3	0.22
Salts	12.0	1.13
Water	0.0	90.60
	100.0	100.00

Hog's dung. From all I have seen, I conclude that hogs well kept and put up to fatten, yield dejections which are highly azotized, and which must consequently furnish a manure of excellent quality. Schwertz has, indeed, ascertained that this manure acts more powerfully than cow-dung.

Sheep-dung is one of the most active of manures, a fact which is confirmed by analysis; for it is by no means watery, and in the usual state contains upwards of one per cent. of azote. The mode of managing sheep generally implies that they manure the ground immediately. Schwertz calculates that in the course of a night, a sheep will manure something more than a square yard of surface; at Bechelbronn we have found the quantity manured to be about 4 square feet. The following are the details of one experiment:

Two hundred sheep were folded for a fortnight upon a rye-stubble, of an extent which gave as nearly as possible four square feet of surface per sheep. The manuring thus effected was found to produce a maximum effect upon the crop of turnips which followed the rye.

Pigeon's dung is known as a *hot* manure, and of such activity that it must be used with discretion. Pigeon's dung is available for crops of every description; Schwertz has made use of it for a long time, and always with the greatest success, mixed with coal ashes, upon clovers. The Flemish farmers procure pigeon's dung from the department of the Pas de Calais, where there are a great number of dove-cotes, one of which, containing from six hundred to six hundred and fifty pigeons, will let for the sum of about £4 per annum, merely for the sake of the dung; the quantity yielded in this time may be about a wagon-load. In the neighborhood of Lisle, this manure is applied particularly in the cultivation of flax and tobacco. According to M. Cordier, the dung of between seven hundred and eight hundred pigeons is sufficient to manure nearly $2\frac{1}{2}$ acres of ground. The dung of three hundred and twelve pigeons, therefore, would suffice for an acre. The value of pigeon's dung may be estimated from the large proportion of azote which it contains; that which I analyzed from Bechelbronn gave $8\frac{1}{3}$ per cent. of this principle, a result which ought not to excite surprise when it is known that the white matter that appears in the excrements of birds, consists of nearly pure uric acid. The manure of the hen-house is nearly or quite as good as pigeon's dung.

Guano is a manure of the same nature as pigeon's dung, and the use of which, long familiar on the coasts of Peru, has lately extended to these countries, the article being now imported in large quantities, both from the South American and African coasts. Guano appears to be the result of the accumulation for ages of the excrements of the sea-fowl, which live and nestle in the islets, in the neighborhood of the great southern continents of the new and old world. The mass in many places forms beds of between 60 and 70 feet in thickness. The principal places whence guano is obtained, are the Chinche islands near Pisco; but other deposites of the substance are known to exist more to the south,—in the islets of Iza and Ilo, at Arica, and in the neighborhood of Payta, as I had an opportunity of ascertaining during my stay in that port. The inhabitants of Chinche are the principal traders in guano; and a class of small vessels, called Guaneros, are constantly engaged in carrying the manure.*

Fourcroy and Vauquelin were the first who fixed attention on the nature of guano. The specimen which they examined was brought to Europe by M. de Humboldt, and contained: Uric acid (0.25,) oxalate of ammonia, chlorhydrate of ammonia, oxalate of potash, phosphates of potash and of lime, chloride of potassium, fatty matter, and sand.

Since this time Dr. Fownes has again analyzed guano. The sample upon which he operated was of a light brown color and extremely offensive smell; it yielded:

Oxalate of ammonia Uric acid Traces of carbonate of ammonia and organic matter	66.2
Phosphates of lime and of magnesia	29.2
Phosphates and alkaline chlorides, and traces of sulphates	4.6
	100.0

Another sample, deeper in color and without smell, contained: Pure oxalate of ammonia, 44.6; earthy phosphates, 41.2; alkaline phosphates, sulphates, and chlorides, 14.2=100.

The composition of guano would confirm, were there any occasion

* Humboldt, Annales de Chimie, vol. lvi. p. 258.

for confirmation, the opinion that has been formed as to its origin. The islets which supply it are still tenanted, especially during the night, by a multitude of sea-fowl. Nevertheless, from the calculations of M. de Humboldt, the excrements of these birds in the course of three centuries, would not form a layer of guano of more than one third of an inch in thickness;—imagination stops short, startled, in presence of the vast lapse of time which must have been necessary to accumulate such beds of the substance as now exist, or rather, as lately existed in many places; for it is rapidly disappearing since it has become a subject of the commercial enterprise of mankind.*

The average composition of guano must by no means be inferred from the preceding analyses of picked samples: earthy matters are usually present in much larger proportion than they are here stated. The guano generally imported into England and France yields a proportion of azote very far short of that which the 25 per cent. of uric acid which has sometimes been stated to exist in this substance would yield. In three trials the azote found was 0.14, 0.05, and 0.05; the mean would therefore be 0.08, which represents the quantity of azote in pigeon's dung.

The litter and excrement of the silkworm is used as manure in the south. Analysis indicates 3 per cent. of azote in its constitution.

Human excrements are regarded as one of the most active manures that can be employed. In countries where agriculture has made real progress, this article is highly prized, and no pains are spared to obtain so powerful a manure. In Flanders, feculent matters form the staple of an active traffic; and in the neighborhood of large towns, they form an invaluable material for the amelioration of the soil. The Chinese collect human excrements with the greatest solicitude, vessels being placed for the purpose at regular distances along the most frequented ways. Old men, women and children, are engaged in mixing them with water, which is applied in the neighborhood of the plants in cultivation.† The fresh excrement is occasionally worked up with clay, and formed into bricks, which are pulverized when dry, and the powder is applied as a top-dressing. One of the advantages resulting from the almost exclusive use of this manure in China is this, that the fields seem to grow nothing but the plant which is the object of solicitude with the farmer; it is there extremely difficult to meet with a weed. The quality of feculent matter as a manure depends much on the nature and abundance of the food consumed by those who furnish it. M. d'Arcet relates a curious anecdote in connection with this fact: a farmer had purchased the produce of the *cabinet* of one of the most

* Dr. John Davy, all whose scientific researches equal in accuracy the brilliant investigations of his illustrious brother, has lately turned his attention to this subject: he finds that we have collections of guano in Great Britain that are really not to be despised in some cases. The surface of the ground under old-established rookeries is a true guano bed; and removed and used as manure in the open field, produces most excellent effects. See Dr. Davy's paper in Ed. Lond. and Dub. Philos. Mag. Oct. 1, 1844.—Eng. Ed.

† Julien, Annales de Chimie, vol. iii. p. 65, 3d series.

celebrated restaurateurs or taverns of the Palais Royal; encouraged by the success he obtained in employing this manure, and desirous of obtaining a larger supply of the article, he rented the produce of several of the barracks of Paris. The manure which he now obtained, however, he found to produce an effect greatly less than he had anticipated, so that he lost money by his bargain. Berzelius found the following substances in human excrements:

Remains of food	7.0
Bile	0.9
Albumen	0.9
A peculiar extractive matter	2.7
Indeterminate animal matter, viscous matter, resin, and an insoluble residuum	14.0
Salts	1.2
Water	73.3
	100.0

The salts had the composition following:

Carbonate of soda	29.4
Chloride of sodium	23.5
Sulphate of soda	11.8
Ammoniaco-magnesian phosphate	11.8
Phosphate of lime	23.5
	100.0

Human urine is one of the most powerful of all manures. Left to itself it speedily undergoes putrefaction, and devolves an abundance of ammoniacal salts, as all the world knows. Its composition, according to Berzelius, is the following:

Urea	3.01
Uric acid	0.10
Indeterminate animal matter; Lactic acid, and lactate of ammonia	1.71
Mucus of the bladder	0.03
Sulphate of potash	0.37
Sulphate of soda	0.32
Phosphate of soda	0.29
Chloride of sodium	0.45
Phosphate of ammonia	0.17
Chlorhydrate of ammonia	0.15
Phosphate of lime and of magnesia	0.10
Silica	traces
Water	93.30
	100.00

The phosphates of lime and magnesia which it contains are extremely insoluble salts, and have been supposed to be held in solution by phosphoric acid, lactic acid, and very recently by Professor Liebig, by hippuric acid, which he now states to be a regular constituent of healthy human urine.

From the interesting inquiries upon urine made by M. Lecanu, it appears that a man passes nearly half an ounce of azote with his urine in the course of twenty-four hours. A quantity of urine taken from a public urine pail of Paris, yielded 7 per 1000 of azote. The dry extract of the same urine yielded nearly 17 per cent.

Human soil as commonly obtained consists of a mixture of feculent matters and urine. It may be applied immediately to the ground as it comes from the privy. In some parts of Tuscany it is mixed with three times its bulk of water, and so applied to the surface. I have myself seen night-soil as it was obtained, and without preparation, spread upon a field of wheat without any ill effect: so that the Tuscan preparation may be regarded as a simple means of spreading a limited quantity of manure over a given extent of ground.

It is in French Flanders, however, that human soil is collected with especial care; it ought to be so collected everywhere. The reservoir for its preservation ought to be one of the essential articles in every farming establishment, as it is in Flanders, where there is always a cistern or cess-pool in masonry, with an arch turned over it for the purpose of collecting this invaluable manure. The bottom is cemented and paved. Two openings are left: one in the middle of the turned arch for the introduction of the material; the other, smaller and made on the north side, is for the admission of the air, which is requisite for the fermentation.

The Flemish reservoir may be of the dimensions of about 35 cubical yards. Whenever the necessary operations of the farm will permit, the carts are sent off to the neighboring town to purchase night-soil, which is then discharged into the reservoir, where it usually remains for several months before being carried out upon the land.

This favorite Flemish manure is applied in the liquid state (mixed in water) before or after the seed is in the ground, or to transplanted crops after they have been dibbled in. Its action is prompt and energetic. The sowing completed, and the land dressed up with all the pains which the Flemish farmer appears to take a pleasure in bestowing upon it, a charge of the manure is carried out at night in tubs or barrels. At the side or corner of the field there is a vat that will hold from 50 to 60 gallons, into which the load is discharged, and from which a workman, armed with a scoop at the end of a handle a dozen feet in length or more, proceeds to lade it out all around him. The vat emptied in one place is removed further on, and the same process is repeated until the whole field is watered.*

The purchase, the carriage, and the application of this Flemish manure cannot be otherwise than costly; we therefore see it given particularly to crops which, when luxuriant and successful, are of the highest market value—such as flax, rape, and tobacco.

This manure, the sample of it, at least, which M. Payen and I examined, is of a yellowish green color, and with reference to smell cannot be compared to any thing better than a weak solution of hydrosulphate of ammonia. This salt is undoubtedly present; but exposure to the air converts it rapidly into the sulphate of the same base. According to M. Kuhlmann, the quality of the liquid Flemish manure is to be judged of by its smell, its viscidity, and its saline and sharp taste. By the fermentation which takes place in the cess-pools, which are never emptied completely, the feculent matter, kept for some time there, does in fact acquire a slight viscidity. When solid excrementitious matter predominates in the fermented mass,

* Cordier, Agriculture of French Flanders, p. 240.

its effect upon vegetation is of longer continuance ; but when it is derived entirely from urine, it acts almost immediately after its application. In either case, the effect of Flemish manure does not extend beyond the season ; like all the other organic substances which have undergone complete putrid fermentation, it is a true annual manure.

Occasionally, a quantity of powdered oil-cake is thrown into the reservoir. This is either when the manure is supposed to be too dilute, or when there is little night-soil at command. The following, according to Professor Kuhlmann, is an example of the employment of the Flemish manure in a rotation which is common in the neighborhood of Lisle, and in the course of which the crops are colza or colewort, wheat and oats.

First year. In October or November, the land is manured with farm-dung, which is ploughed in, in the usual way. At this time a dose of the liquid manure, amounting to about 5000 gallons per acre, is applied : a second ploughing is given, and the colewort is planted.

Second year. The colza is gathered, the ground is ploughed for autumn sowing; from 1000 to 1300 gallons or so of liquid manure are distributed, and the wheat is sown.

Third year. The wheat stubble is ploughed down at the end of the autumn, and about 1000 or 1100 gallons of the liquid manure per acre are distributed ; the oats are sown in the spring. If circumstances should prevent the application of the liquid manure in autumn, it is laid on in March, and then it has been found that one-fifth less will suffice ; but its application at this season is avoided as much as possible on account of the havoc that is made by the passage of horses, carts, and men over the surface of the soft ploughed land. It is with a view to avoid this disturbance of the surface that in many places oil-cake in powder is applied to the fields under colza when the manuring has to be performed after the crop is in the ground.

For beet, the dose of Flemish manure is carried the length of from 1300 to 1400 gallons per acre ; but when the root is intended for the manufacture of sugar, liquid manure is sedulously avoided, experience having shown that it has the very worst effect upon the production of sugar, a circumstance which is very easily explained upon grounds that have already been given.

The price of Flemish manure at Lisle is 2½*d.* for a measure containing 22 gallons. In Flanders, it is held that this quantity, which will weigh hard upon 2 cwt., is equal to about 5 cwt. of farm-yard dung. The liquid manure which I analyzed yielded 2 per 1000 of azote. Farm-yard dung, in its usual state, contains as much as 4 per 1000 ; it follows, therefore, that the real equivalent number of Flemish manure is 182, that of farm dung being 100 ; in other words, it would require 182 of Flemish manure to replace 100 of farm-yard manure ; a conclusion that differs widely from that which is usually acted upon. But it must be observed that from its nature, the Flemish manure produces its maximum influence in the course of the season in which it is applied. It seems to have no effect on the crop of the succeeding year. Farm-yard dung, on the contrary, only exerts a portion of the whole amount of its beneficial influence in the course of the year in which it is laid on ; it has still something, often much, in reserve for succeeding years. To compare liquid manure with farm-yard dung, with reference to an annual crop, is to compare this manure to the unknown fraction of the farm-yard dung which comes into play in the course of the first year, and from such a contrast no possible inference can be drawn in regard to the relative value of the two kinds of dung. I have insisted upon this circumstance, because it is often involved in the estimates that are made of the relative values of the different species of manure ; and because, from losing sight of it, unfavorable conclusions are frequently come to in regard to manures that undergo decomposition very slowly ; these manures, nevertheless, acting for a great length of time, produce both a greater amount and a more durable kind of amelioration of the soil. Rapidity of action, in a manure, is undoubtedly a quality that is highly valuable in many cases ; and Flemish manure possesses this quality in the highest degree. Nevertheless, it is also an advantage to possess a manure which elaborates gradually, and according to the exigencies of vegetables, those principles that contribute to their growth, and which suspend in a great measure this elaboration in the course of the winter—which remain during the cold and rainy season in an almost inert condition, when any fecundating matter produced would merely be washed away and lost. These advantages, to which must be added that of breaking up and lightening the soil, are all possessed by good farm-yard manure. They are such, in fact, that this manure, even in Flanders, is still indispensable ; the liquid manures of that country are nothing more than annual auxiliaries.

The method followed in Flanders of using night-soil is certainly highly rational ; it is the same as that which is adopted in Alsace, in the neighborhood of towns, with this difference, that our farmers collect no store of the material ; they go in quest of it at the moment it is wanted. It is applied as in Flanders, or it is incorporated with absorbent substances, such as straw, or with other more consistent manures. The night-soil of Paris, which in the course of a year amounts to an immense quantity, is treated in a totally different manner, which appears to be in opposition to the simplest notions of science, of economy, and of all that is conducive to health. I allude to the mode of preparing *poudrette.*

In the neighborhood of Paris there are places appropriated to the reception of the night-soil : it is thrown into reservoirs of no great depth, in comparison with their superficial extent, and of an aggregate capacity which is such that they will contain the whole of the products collected by the night-man in the course of six months. These reservoirs are arranged in stages, one above another. Into the upper one are discharged the matters collected in the course of the night. The upper reservoir full, a sluice is opened by being pushed partially down, which allows the more liquid matters to escape into the second reservoir placed under it. Repeated drainings are effected in this way, and when the second basin is also full, there is a deposition of solid matter as in the first; the more liquid particles are then let off from the second into the third reservoir, and so on in succession until the last and lowest is attained, from which the liquid used to be turned into a watercourse; but, of late, these contaminated liquids have been got rid of by means of what may be called absorbing artesian shafts—deep holes pierced in a dry and porous soil.

When the deposite in the first reservoir is held to be sufficiently consistent, it is drained by lowering the sluice more and more; no fresh matter is added, the new charges being deposited in another system of reservoirs. The deposite once drained is in the pasty condition; it is then taken out with the spade, and spread upon an earthen floor which slopes off on either side, and the mass is turned from time to time to favor the drying; this process, in fact, is continued until the material has become pulverulent. It is then stored under sheds, or thrown up into pyramidal heaps, the sides of which are well beaten in order to enable them to throw off the wet.

Poudrette is of a brown color, and weighs nearly 150 lbs. per sack. Put into a retort, and distilled with a heat of from 424° to 930° Fahr., it yields 52.6 of ammoniacal fluid, and 47.3 of dry matter, in which we encounter fixed ammoniacal salts, such as the sulphates, phosphates, hydrochlorates, &c. M. Jacquemart finds that in 100 parts of poudrette there is 1.26 of ammonia, the greater part in the state of carbonate; but it contains a quantity of animal matter besides, which, by dry distillation, yields a nearly equal amount of the same substance; whence it follows that poudrette contains nearly $2\frac{1}{2}$ per cent. of volatile alkali, or 2 of azote. By direct analysis, I obtained 1.6 of azote.

Poudrette is spread upon the land at the time of ploughing, from 26 to 34 bushels per acre being allowed. On meadow lands it produces very good effects in the dose of about 25 bushels per acre. The disgusting smell of night-soil is, to a certain extent, an obstacle to its general use. This obstacle, however, is only felt in places where agricultural industry, and the manufactures connected with it, are still in a backward state. One remarkable circumstance is, that the disgust which naturally arises from the manipulation of such articles, has been more especially got over in countries that are justly celebrated for their extreme attention to cleanliness and the easy position of their inhabitants. I quote Flanders and Alsace in proof of the fact. It has been said, moreover, that certain articles produced in soils manured with human excrement contract a smell and taste which give rather unpleasant information of the nature of the manure that has been employed to favor their growth. In the limited circle of my own experience on this subject, I can say that I have observed nothing which favors such a statement. However this may be, Mr. Salmon has succeeded in disinfecting night-soil completely, by mixing it with a kind of animal charcoal obtained by calcining in close vessels a porous earth impregnated with organic substances. This is the article which is sold under he name of *animalized black.* Its quality as a manure must depend especially, I might even say entirely, on the quantity of azotized organic matter which enters into its composition.

Composts. A great deal has been written, and much has been said on the advantages of composts, or mixtures, contrived with a view to the amelioration of the soil. The receipts for these composts are very numerous; they prove that the discovery of a compost is an easy matter, and requires but a small amount of ingenuity. To unite different matters in such a way as to obtain a compound that shall act advantageously, it is only necessary to make it up of substances which of themselves and isolatedly are good manures. But that it is possible to supply the scarcity of manure, to create it in some sort by means of composts, is a subject of dispute. In fact, when we look attentively at the numerous mixtures which have been indicated as leading to this end, we always perceive that the proposal amounts to an extension or dilution of some powerful manure with a substance that is either inert or has little activity. This mode of proceeding may have its advantages; it enables us to make a more equal distribution of the manure we have at our disposal, but it actually supplies us with none.

Earthy substances almost always figure in composts. Turf-ashes, wood-ashes, marl, and particularly lime, are constant ingredients. Marl may suit certain soils; lime is a substance of great activity, and which for this reason must be admitted into composts with caution; it may act in the disintegration of woody parts—of stalks, and stems, and leaves; but we must be very careful not to follow the recommendation of Schwertz, who would have us throw quick-lime into our privies with the view to bringing the matters there contained into a consistent and readily pulverizable state. By doing so we should infallibly lose the greater part of the principles that are truly useful in the soil. Much mischief and great destruction of manure, indeed, have been the consequence of the insensate and indiscriminate use of quick-lime under all circumstances; the business is much rather to preserve than to destroy the substances that are used as manures; the purpose is to fix, not to dissipate the volatile elements which they contain. One great objection to the extensive employment of composts is the amount of labor they require in the repeated turnings which are held necessary in their preparation, and in the large quantity of matter which has to be transported.

The following table will be found useful as giving a comprehensive view of the proportion of azote contained in the various kinds of manures which have been particularly examined, and of their equivalents, referred to farm-yard dung as the standard.

TABLE OF THE COMPARATIVE VALUE OF MANURES

DEDUCED FROM ANALYSES MADE BY MESSRS. PAYEN AND BOUSSINGAULT.

Kind of Dung.	Water per 100.	Azote in 100 of matter.		Quality according to state.		Equivalent according to state.		Remarks.
		Dry.	Wet.	Dry.	Wet.	Dry.	Wet.	
Farm-yard dung	79.3	1.95	0.41	100	100	100	100	Aver. of Bechelbronn.
Dung from an inn yard	60.6	2.08	0.79	107	107	94	51	From south of France.
Dung water	99.6	1.54	0.06	78	2	127	68	Washed by the rain.
Wheat straw	19.3	0.30	0.24	15	60	650	167	Fresh, of Alsace, 1838.
Idem	5.3	0.53	0.49	27	122.5	367	82	Old from environs of Paris.
Idem	5.3	0.43	0.41	22	102.5	453	98	Ditto stubble.
Idem	9.4	1.42	1.33	73	332.5	137	30	Ditto upper part, ear included.
Rye straw	12.2	0.20	0.17	10	42.5	975	235	Of Alsace.
Idem	12.6	0.50	0.42	26	105	390	95	Environs of Paris, 1841.
Oat straw	21.0	0.36	0.28	18	70	542	143	} Of Alsace.
Barley straw	11.0	0.26	0.23	13	57.5	750	174	
Wheat chaff	7.6	0.94	0.85	48	212.5	207	47	
Pea straw	8.5	1.95	1.79	100	447.5	100	22	
Millet straw	19.0	0.96	0.78	49	195	203	51	
Buckwheat straw	11.6	0.54	0.48	27	120	361	83	
Lentil straw	9.2	1.12	1.01	57	250	174	40	
Dried potato tops	12.9	0.43	0.37	22	92.5	453	108	
Withered madia stalks	14.3	0.66	0.57	33	142.5	295	70	After seeding.
Idem turned under while green	70.6	1.53	0.45	79	113	126	89	Before seeding.
Dried broom	10.4	1.37	1.22	70	305	142	33	Stalk and leaves.
Withered leaves of beet-root	88.9	4.50	0.50	230	125	43	80	Of mangel wurzel.
Ditto of potatoes	76.0	2.30	0.55	117	137.5	85	73	Wither'd top & leaves.
Ditto of carrots	70.9	2.94	0.85	150	212.5	66	47	
Leaves of heather	7.0	1.90	1.74	97	425	103	23	Dried in the air.
Ditto of pear-trees	14.5	1.59	1.36	81.5	340	127	29	
Ditto of oak	25.0	1.57	1.18	80	293	125	34	} Leaves fallen in autumn.
Ditto of poplar	51.1	1.17	0.54	66	134	167	74	
Ditto of beech	39.3	1.91	1.18	78	294	102	34	
Ditto of acacia	53.6	1.56	0.72	80	180	125	56	
Box-tree	59.3	2.89	1.17	147	293	68	34	Branches and leaves.
Burred clover-roots	9.7	1.77	1.61	90	402.5	110	25	Dried in the air.
Fucus digitatus	39.2	1.41	0.86	72	215	139	46	} Dried in the air.
Idem	40.0	1.58	0.95	81	237.5	123	42	
Fucus saccharinus	40.0	2.29	1.38	117	345	85	29	
Idem	75.5	..	0.54	..	135	..	74	Fresh.
Burnt sea-weed	3.8	0.40	0.38	20	95	488	105	
Oyster shells	17.9	0.40	0.32	20	80	488	125	
Sea shells		0.05	0.05	3	13	3750	769	Dried sea shells of Dunkirk.
Mud of the Morlaix river	3.7	0.42	0.40	21	100	464	100	} Sea sand.
Trez of Roscoff roads	0.5	0.14	0.13	7	32.5	1393	308	
Sea-side marl	1.0	0.52	0.51	26.5	128	377	78	
Salt cod-fish	38.0	10.86	6.70	557	1675	18	6	
Cod-fish washed and pressed	10.0	18.74	16.86	961	4215	10	2½	Dried in the air.
Fir saw-dust	24.0	0.22	0.16	11	40	886	250	} Dried in the air.
Idem	24.0	0.31	0.28	15	57.5	629	174	
Oak saw-dust	26.0	0.72	0.54	36	135	256	74	
White lupine seed	10.5	4.35	3.49	223	872.5	45	111½	Tuscan, boiled & dried
Malt grains	6.0	4.90	4.51	251	1127.5	40	9	
Grape husks	48.2	3.31	1.71	169	427.5	57	23	
Oil-cake of linseed	13.4	6.00	5.20	307	1300	33	8	
Ditto of colewort	10.5	5.50	4.92	282	1230	35	8	
Ditto of Arachis	6.6	8.89	8.33	655	2082.5	21	4½	
Ditto of madia	11.1	5.70	5.06	292	1265	34	8	
Ditto of sesame	6.5	5.93	5.52	304	1378	33	7¼	
Oil-cake of hempseed	5.0	4.78	4.21	245	1052	41	9½	
Ditto of poppy	6.0	5.70	5.36	292	1340	34	7½	
Ditto of beech mast	6.2	3.53	3.31	181	828	55	12	
Ditto of walnuts	6.0	5.59	5.24	287	1310	35	7½	
Ditto of cotton seed	11.0	4.52	4.02	231	1000	32	10	

Kind of Dung.	Water per 100.	Azote in 100 of matter.		Quality according to state.		Equivalent according to state.		Remarks.
		Dry.	Wet.	Dry.	Wet.	Dry.	Wet.	
Ditto from refiners .	10.0	3.92	3.54	201	885	50	11¼	Recent fat by means of poplar sawdust.
Ditto ditto	7.7	0.58	0.54	30	135	332	75	Fish oil by ditto ditto.
Cider-apple refuse .	6.4	0.63	0.59	32	147	309	68	Dried in the air.
Refuse of hops . .	73.0	2.23	0.56	114	140	88	67	
Beet-root refuse . .	9.3	1.26	1.14	64	285	155	35	Dried in the air.
Idem	70.0	..	0.38	64	85	..	106	Fresh from the press.
Squeezed beet-root .	94.5	1.76	0.01	90	2	111	4137	Process of Dombasle.
Potato refuse . . .	73.0	1.95	0.53	100	131.5	100	76	
Potato juice . . .	95.4	8.28	0.38	425	94	23	106	Settled and decanted.
Water of the starch manufactory . .	99.2	8.28	0.07	425	17.5	..	571	From washing in four volumes of water.
Deposite from the water of ditto . . .	80.	1.81	0.36	92	90.	108	111	Drainings from heap.
Idem	15.	1.81	1.54	92	384.5	..	24	Dried in the air.
Solid cow-dung . .	85.9	2.30	0.32	117	80	84	125	
Urine of cows . . .	88.3	3.80	0.44	194	110	51	91	
Mixed cow-dung . .	84.3	2.59	0.41	132	102.5	75	98	
Solid horse-dung . .	75.3	2.21	0.55	113	137.5	88	73	
Horse urine	79.1	12.50	2.61	641	652.5	15½	15⅓	The horse drank but little; the urine was thick.
Mixed horse-dung .	75.4	3.02	0.74	154	185	66	54	
Pig-dung	81.4	3.37	0.63	172	157.5	58	63	
Sheep-dung . . .	63.0	2.99	1.11	153	277.5	65	36	
Goat-dung	46.0	3.93	2.16	201	540	50	18½	
Liquid Flem. manure	..	..	0.19	..	47.5	..	210	In the normal state.
Idem	..	..	0.22	..	55	..	182	
Poudrette of Belloni .	12.5	4.40	3.85	225	962	44	10⅓	Dried in the air.
Ditto of Montfaucon	41.4	2.67	1.56	137	390	73	25½	
Urine of public vats .	96.	17.56	16.83	900	4213	11	2⅓	Dried in the stove.
Idem	96.9	23.11	0.72	1133	179	8½	56	Thin, ammoniacal.
Animalized black .	44.6	1.96	1.09	100.5	272	98	37	Prepared for 11 months
Idem from the neighborhood of Paris .	42.0	2.96	1.24	151.6	310.5	66	32	Recently made.
Idem, called Dutch manure	44.1	2.48	1.36	127	340	79	29½	Made at Lyons.
Animalized sea-weed	12.1	2.73	2.40	140	600	7	16½	Dried in stove, (from Marseilles.)
Pigeon's dung . .	9.6	9.02	8.30	462	2075	21½	5	Of Bechelbronn.
Guano imported into England	19.6	6.20	5.00	323	1247	31½	80	In the ordinary state.
Idem	23.4	7.05	5.40	361	1349	28	74	Sifted.
Do. imp. into France	11.3	15.73	13.95	807	3487	12½	28½	
Silk-worm litter . .	14.3	3.48	3.29	178.7	827	56	12	Fifth age.
Idem	11.4	3.71	3.29	190	822	53	12	Sixth age.
Chrysalis of silk-worm	78.5	8.99	1.95	461	485	21½	20½	
Cockchafers . . .	77.0	13.93	3.20	714	801	14	13	
Dried muscular flesh	8.5	14.25	13.04	730	3260	13½	3	Dried in the air.
Soluble dried blood .	21.4	15.50	12.18	795	3045	12½	3¼	As sold.
Liquid blood . . .	81.0	..	2.95	795	736	..	13⅓	From slaughterhouses.
Idem	82.5	..	2.71	795	580	..	15	From worn-out horses.
Blood coagulated and pressed	73.5	17.00	4.51	871	1128	11½	9	Just out of the press.
Insoluble dried blood	12.5	17.00	14.88	871	3719	11½	2¾	Dried in manufactory.
Dregs from Prussian blue manufactory .	53.4	2.80	1.31	144	526	7	30½	Animalized with blood
Melter's bones . . .	7.5	7.58	7.02	388	1754	26	6	Dried in the air.
Fresh bones	30.0	..	5.31	..	1326	..	7½	As sold by the melters.
Fat bones, not heated	8.0	..	6.22	..	1554	..	6½	Including 0.10 of fat.
Dregs of bone glue .	42.0	0.91	0.53	47	133	214	76	
Glue dregs	33.6	5.63	3.73	288.4	933.5	35	11	As sold by the makers.
Graves	8.2	12.93	11.88	663	296953	15	3⅓	
Animal black of the sugar refiners . .	47.7	2.04	1.06	104	265	96	38	As sent out.
Sugar refiner's black	27.7	19.01	13.75	974	3437	103	28	From Paris.
Scum from the sugar refinery	67.0	1.58	0.54	81	134	127	75	From the sugar bakery of Vigneux.
English black . . .	13.5	8.02	6.95	411.4	1738	24	6	Blood, lime, soot.
Feathers	12.9	17.61	15.34	903	3835	11	2½	
Cow-hair flock . .	8.9	15.12	13.78	775	3445	13	3	
Woollen rags . . .	11.3	20.26	17.98	1039	4495	9½	2¼	
Horn shavings . . .	9.0	15.78	14.36	809	3590	12⅓	3	
Coal soot	15.6	1.59	1.35	81	337.5	122	30	
Wood soot	5.6	1.31	1.15	67	287.5	149	35	
Picardy ashes . . .	9.2	0.71	0.65	36	162.5	275	62	
Veget. mould from humus dung (terreau)	..	1.03	..	53	..	1โ0	33	Dried in the stove.

It is almost unnecessary to give any explanation of the uses that may be made of the preceding table : I shall, however, give a few illustrations from instances which have actually occurred in my experience.

Oil-cake is cheap at this time, (1842 ;) and the question is, whether it could be advantageously employed in connection with the cultivation of wheat. The presumption is, that wheat obtains the whole of its azote in the soil, that it acquires none from the atmosphere ; and again, I assume that the whole of the azote put into the ground would be used up by the crop. Under the most favorable circumstances of heat and moisture, this would probably be the case ; were it not so to the letter, the active matter which remained in the ground would operate advantageously in succeeding years. The following, then, are the elements of the question :

1st. In the wheat grown at Bechelbronn there is on an average 0.025 of azote. 2d. In the straw of 1841, I have just found 0.003 of azote. 3d. The oil-cake which I propose to employ contains 0.055 of azote, and its actual price, crushing included, is 3*s*. 4*d*. per cwt. 4th. The relation in point of weight of the grain to the straw is as 47 : 100.

A sheaf or bundle of wheat, 220 lbs. in weight, consists of:

Wheat 70.4 lbs., containing	1.760 of azote, and is worth	4*s*. 8*d*.
Straw 149.6 lbs. "	0.415 " "	1*s*. 8*d*.
Total of azote	2.175 Total value	6*s*. 4*d*.
	Difference of value	5*s*. 4*d*.
To grow which, 39 lbs. of oil-cake would be required, of the value of		1*s*. 2*d*.

So that 39 lbs. of oil-cake, converted into a sheaf of wheat, would be increased in intrinsic value to the extent of 5*s*. 2*d*. Supposing that but one-half or one-third of this amount, as indicated by theory, is realized in practice, it is obvious that the addition of the oil-cake might be made with advantage ; and that no means should be neglected to ensure the success of its application as a manure.*

The production of oil-cake in France, the Netherlands, and other countries of Europe, is very considerable ; in round numbers, 100 of oleaginous seeds yield 60 of cake ; but it has been calculated, with rare ability, and from authentic documents, by M. Leroy de Béthune, that not only is the whole of the oil-cake, which is the produce of the soil of France, exported, but that likewise of the oleaginous seeds which she imports from other countries. This M. de Béthune looks upon as a very lamentable agricultural fact. I have shown, indeed, from the example which I have quoted, that every pound of cake represents a primary material, which, properly treated, may be transformed into nearly 6 pounds weight of wheat-grain and straw, having a value infinitely greater than that of the oil-cake originally employed.

* Our author has of course left many other elements very necessary to be included out of his calculation here, such as labor, seed, rent charge, interest on capital, &c.—Eng. Ed.

While I agree with M. de Béthune, that it is generally wise to encourage exportation, I also admit with him that there are substances in reference to which it would be prudent to discourage exportation ; oil-cake, this powerful means of giving fertility to the soil, might be placed in the foremost rank of such substances. I am far from adopting all the principles of economists, which appear to me to be frequently far too absolute. In my opinion, any exportation, the consequence of which is the impoverishment of the soil, ought to be prohibited. I should, for instance, oppose the exportation of arable soil ; and in the same way, to allow an active manure to pass into the hands of strangers, is, in my eyes, tantamount to exporting the vegetable soil of our fields, to lessening their productiveness, to raising the price of the food of the poor ; for as much labor is required, as much care and capital must be expended upon an ungrateful soil to obtain a little, as upon a fertile soil to procure an ample return. To permit the exportation of oil-cake is to hinder the husbandman from taking advantage of all the circumstances with which nature presents him ; it is as if a chill were to be brought over the genial climate of France.*

I have shown the advantages of the application of oil-cake in the growth of wheat. I shall now inquire whether or not it is equally useful in connection with hay and potato crops ; the price of the article being presumed to be the same as before.

Upland meadows, when they have not been soiled, yield miserable returns, and their situation renders them difficult of access to carts : oil-cake in such circumstances comes powerfully to our aid.

Taking the price of hay at 5*s.* per 220 lbs., which is about its present price in France, and taking into account the composition of the after-math, we may reckon the azote contained in the hay of natural meadows at 0.015.

220 lbs. of hay, containing 3 lbs. of azote, will be worth	5*s.* 0*d.*
To produce which 56 lbs. of cake (azote 3.3 lbs.) worth would be required.	1*s.* 8*d.*
Difference in value between the cost and the crop	3*s.* 4*d.*

Upon this showing, oil-cake may be advantageously employed in the amelioration of upland meadows. Besides the cost of the manure, however, there are the very necessary additions to be made of the price of labor and rent.

From the observations which I made at Bechelbronn in 1839, I

* I own I am surprised at this passage in my esteemed author. There is nothing parallel in the instances he quotes. Did not the French husbandmen and oil-pressers *profit* by the exportation of oil-cake they would keep it at home ; and the profit of the farmer and manufacturer is the profit of the whole community. To export the soil would indeed be madness : it would obviously be killing the goose that lays the golden eggs ; but to export that which the soil produces in abundance year after year, is a totally different affair. M. Boussingault's reasoning would lead the wine-growers of Bordeaux and Burgundy to refuse us a hogshead of their smallest growth : *they cannot send it to us without impoverishing their soil, any more than they can let us have a pound of their oil-cake.* But one half of the vegetables that grow, at least, are at work accumulating the materials from the atmosphere and water, out of which the other half are supplied, and so the process of waste and supply, of destruction and reproduction, goes on without limits, and without end.—ENG. ED.

find that the relation between the weight of potatoes as they come from the ground, and that of the tops or haum, supposed to be dry, is as 100 is to 6.4.

The tubers contain:
Azote 0.0036 per 10000 parts; and 220 lbs. contain 0.729 of a lb. of azote, and are worth .... 1*s*. 8*d*.
The tops or haum, dry, contain:
Azote 0.0230 per 10000 parts; and 14 lbs. contain 0.330 of a pound of azote.
Total of azote 1.122.
Now 20.4 lbs. of cake which would be required to produce 220 lbs. of potatoes, contain 1.1 lb. of azote, and are worth .... 0*s*. 7½*d*.
Difference .... 1*s*. 0½*d*.

The oil-cake at the price of 3*s*. 2*d*. per cwt. may therefore be advantageously used for the production of potatoes: rent, labor, seed, &c., considered as before. At the price of 7*s*. 6*d*. or 8*s*. 6*d*. per cwt., however, to which oil-cake occasionally rises, it would not be possible to employ it profitably in this way. The cost of the manure would then amount to nearly as much as the value of the crop.

The equivalent numbers in the table express the relative values of different manures; they proclaim the proportions in which one substance must be substituted for another, and when purchases are to be made, they will show at a glance which is the article that is really, and in fact, the cheapest. The equivalent number of one variety of oil-cake, for instance, is 7.25; that of farm-yard dung is 100; which is as much as to say that in reference to mere fertilizing elements, 100 parts—lbs. cwts. or tons, of farm-yard dung may be replaced by 7¼ parts—lbs. cwts. or tons of oil-cake;—2 cwt. of farm-dung, for instance, by 14½ lbs. of cake. The 2 cwt. of farm-dung is valued in the table at 6*d*., or about 5*s*. per ton; the 14½ lbs. of cake would cost 5¾*d*. It is obvious, therefore, that even at the above low price of oil-cake, there would be no real advantage in substituting it generally for farm-yard dung; in situations, however, remote from large towns, where it is almost impossible to procure dung, or where the carriage of large masses of dung would be both difficult and expensive, there would then be advantage in the substitution.

Woollen rags at the price of about 2*s*. 10*d*. per cwt. are more profitable than farm-yard dung at 3*d*. per cwt. The equivalent of the rags is 2.22, and this quantity (2.22 lbs. avoird.) of rags is worth about ¾*d*.; by the substitution of the rags for farm-yard manure, therefore, a saving is effected of about 2¼*d*. on every cwt. of the latter that must have been employed. In good farming, however, it is less with reference to the money advantage of substituting one manure for another, that calculations are made, than with reference to the possibility of procuring either one manure or another at a moderate price. The estimated value of the dung in one of the columns of the table gives us at once the price that may be paid for it; for this purpose it is enough to know the value of standard dung: let this be as it usually is, 3*d*. per cwt.; if we would now know what may be paid for a hundred weight of bones simply dried in the air, the number designating these being 1554, we have only to make a simple equation in the following terms :—100 : 3*d*. : : 1554 : *x*, to have the solution : =3*s*. 10½*d*.

The most careful consideration of the relative value of different manures under the guidance of the analytical elements which I have indicated, justifies the preference which is given in practice to one kind over another, which on simple examination appears to offer greater advantages. Thus, by diffusing oil-cake through water, and leaving the mixture to ferment, a manure is obtained which presents all the characters, which possesses all the properties of human soil that has undergone fermentation in privies or cess-pools. And it is to this mixture of putrid oil-cake that the husbandmen of French Flanders have recourse, as we have seen, when their supply of night-soil runs short. When oil-cake is low in price, say about 3*s*. 3*d*. or 3*s*. 4*d*. per cwt., it might seem advantageous to manufacture Flemish manure with it; expensive carriage and time would be saved; for night-soil has generally to be fetched from a distance, and containing but 0.002 ($\frac{2}{1000}$ths) of azote, it is bulky, and its equivalent is in the same proportion high. Cameline oil-cake contains 0.055 ($\frac{55}{1000}$ths) of azote ; to make Flemish manure that should contain 0.002 of azote, it would be requisite to add to every 100 parts of cake 2,650 parts of water ; the cwt. of this manure would then come to 1½*d*., while Flemish manure prepared with night-soil, would cost the farmer but 1¼*d*. I have here taken the cake at a low price; were it 7*s*. 6*d*. per cwt. instead of 3*s*. 9*d*., which is perhaps much nearer its usual average cost, it is obvious that the cwt. of manure prepared from it, would cost twice as much more.

The proportion of azote, the value, and the equivalents of the several manures are given in the table, both for the substances absolutely dry, and for the condition in which they are commonly employed. This distinction is one of great importance. The water, the quantity of which is indicated in the first column, is a most variable constituent ; its presence, of course, depreciates the manure in the precise ratio in which it occurs. The reference of all the elements of each particular manure to that manure in a state of absolute dryness, is a very important feature in the table. In purchasing manures, the precaution of drying them chemically must never be neglected, more especially in connection with articles, which by their nature are capable of absorbing water in considerable, and often in very different, quantities.

CHAPTER VI.

OF MINERAL MANURES OR STIMULANTS.

All the organic manures, when burned, leave ashes composed of earthy and saline substances. The action of these substances upon vegetation is quite unquestionable, and it is certain that an organic manure, were it ever so rich in azotized principles, and ever so assimilable, would still be imperfect did it not further contain the truly mineral matters which plants require to meet with in the soil, in order to complete their growth and bring their seeds to maturity. The most active organic manures are always abundantly provided with inorganic principles. Farm dung (dry) contains about one-fourth of its weight of such substances, and the water which is used for irrigation invariably holds saline matter in solution.

Nevertheless, repeated cropping will often end by depriving the soil of the mineral substances which plants require; the salts contained in the manure supplied are sometimes inadequate to meet the demands of successive crops, and then the return falls off. It is consequently necessary in certain cases to furnish the soil anew with saline matters, in order to supply the continued drain that is made upon it, or to meet the exigencies of particular crops which are known to require an unusually large quantity of salts for their successful cultivation. It is in this way that clover, lucern, and sainfoin require plaster, (gypsum;) the cereals, silica, and certain calcareous salts; the vine, potash, &c.

Practice got the start of science in the application of mineral manures or stimulants. If their useful influence cannot be denied, as it cannot, if the circumstances in which it is advantageous to administer them, if the conditions and the doses in which they ought to be given to the ground have been the subject of long and careful observation with farmers, it must still be admitted that we are far from understanding exactly in what way they act; this is another motive for continuing to study them with perseverance.

CALCAREOUS MANURES.

In certain soils we have said that the calcareous element is either wanting, or present in very small and inadequate quantity; other soils, again, abound in calcareous matter, and observation appears to prove that the presence of carbonate of lime in a soil adds unequivocally to its fertility. The majority of the good wheat lands hitherto examined have been found to contain a notable quantity of this earth or earthy salt.

It is usual to put lime into the ground in the state of caustic or quick-lime; this is *liming*, properly so called. But it is also applied in the state of carbonate, as when we make use of chalk or marl, or shell-sand from the sea-shore.

The limestone that is used for burning is seldom pure; it frequently contains clay, quartzy sand, metallic oxides, and occasionally carbonaceous matter; frequently too it is so largely mixed with magnesia that it acquires peculiar characters; this is the magnesian limestone or dolomite. The purest carbonate of lime, by exposure for some time to a white heat, loses 43.7 of carbonic acid, and consequently contains 56.3 of caustic lime. Limestone is one of the most common of rocks; in the crystalline and saccharoid state, or of closer and finer grain, it often constitutes mountain masses, and is met with in every part of the geological series; it meets us as chalk in beds of enormous thickness, filling up extensive basins in the tertiary series; such are the chalk beds of the south and west coasts of England, extending through the counties of Kent and Sussex, &c.

The only mineral substance with which chalk, limestone, or carbonate of lime is likely to be confounded, is gypsum or sulphate of lime. But it is easy to distinguish either of these salts from the other: carbonate of lime dissolves with effervescence in dilute hydrochloric acid; sulphate of lime is insoluble in this liquid. Carbonate of lime is quite insoluble in water; sulphate of lime is very sensibly soluble, and a copious precipitate falls on the addition of a solution of oxalic acid or of oxalate of ammonia. Gypsum is always so soft that it can be scratched with the nail; limestone, save in the state of chalk, is generally so hard that it resists the nail.

The burning of lime for agricultural uses is carried on in the same way as for building and other economical purposes. Burnt or quicklime is a very different article from chalk or limestone; it is powerfully caustic or destructive of the organic tissue, and instead of being altogether insoluble, it is now soluble in about 630 parts of cold water. All the world knows how lime from the kiln, when watered, rises in temperature, breaks first into larger and then into smaller pieces, and finally falls down into fine powder; but every one is not aware that there is a true chemical union of water with the earth, and that the resulting powder is in chemical language a hydrate of lime, a substance which is much less caustic than pure lime, but still distinctly alkaline in its reaction.

It is generally admitted that the soil which is without a certain, and that a considerable proportion of the calcareous element, never possesses a high degree of fertility. This in particular is the opinion of English agriculturists, who apply lime with a kind of profusion; and the great improvement it frequently produces on the crops of grain, leaves no doubt as to the advantages of the procedure. Still it is now generally recognized that liming ceases to be useful upon lands that are already sufficiently calcareous, or that rest on a sub-soil of chalk. It is, therefore, by supplying the calcareous element which land requires to constitute it a soil adapted to the growth of corn, that the application of lime becomes useful; liming, in fact, enables us to make this necessary addition at least cost. Like other mineral manures, lime of itself produces little or no effect;

it is in concurrence with organic manures that it becomes truly useful; it is nowise, and never can become, a substitute for these.

The geological constitution of a country is perhaps the best guide to the necessity or advantages of liming. Soils that are derived from plutonic or igneous rocks, in which felspar, mica, or quartz predominate, are on the face of things likely to be improved by the introduction of lime. Direct analysis would of course give more decisive information on the fact. In any case, the measure recommended by prudence is to make a few preliminary trials upon the small scale; the experimental method is the only safe one in agriculture, when the question is in regard to the adoption of new plans. In England it is customary in liming clayey lands to allow from 230 to 300 or 310 bushels of stimulant per acre; on lighter soils the dose may vary from about 150 to 200 bushels, according to their character. In France the quantity usually employed is greatly less, from about 60 to 70 bushels being all that is generally thought advisable, and this at intervals of seven or eight years. In the neighborhood of Lisle little use is made of lime, although there the land is generally any thing but calcareous; perhaps the want of lime is not felt in consequence of the universal practice of employing the Flemish manure, which, as we have seen, contains ammoniacal salts, (and both human urine and excrement contain a large quantity of phosphate of lime and phosphate of magnesia in addition, the very salts that the generality of vegetables crave.) In the vicinity of Dunkirk, however, lime is frequently applied in the dose of between 40 and 50 bushels per acre, and with effects that are said to continue for ten or twelve years.

The dose of lime introduced into the soil in different countries, is moreover in a certain relation with the time during which the action of the earth is believed to continue; as the quantity administered at once is small, the dose must be repeated more frequently. Near Dunkirk they use from 40 to 50 bushels per acre every 10 or 12 years; in the department of La Sarthe, according to M. Puvis, they scatter on some 9 or 10 bushels only; but they do so every three years. This would lead us to conclude that soils which really wanted lime should receive a dose in the proportion of about 3½ bushels per acre annually. But the crops gathered from the ground every year, certainly do not abstract any thing like this quantity of calcareous matter; which would induce us to infer, that after a certain time the land will contain such a quantity of lime as to make any further addition of it unnecessary, or at all events, unnecessary save at rare and distant intervals.

One of the great advantages which lime has over all the other forms or kinds of calcareous stimulants employed, is unquestionably the state of extreme subdivision which it acquires in the quenching. In the course of falling down into this extremely fine powder, lime, as has been said, combines with a large quantity of water. But the change experienced does not stop short here; the air always contains some 10,000ths of carbonic acid gas, for which the hydrate of lime has a powerful affinity, so that it absorbs this gas greedily, abandoning, at the same time, its constitutional water, by which, in due season, the hydrate of lime becomes changed into the anhydrous carbonate of lime. This process is always slow; more rapid at first, when the interchange between the carbonic acid and water takes place freely; it becomes gradually slower and slower as there is less and less water left in the particles: the affinity of the lime for the water seems to increase continually in the ratio of the smallness of the quantity which it still contains. It must, therefore, constantly happen that in incorporating lime, in powder and partially carbonated with the soil, we also introduce lime that has preserved its causticity in some measure; it must be observed, however, that, once intimately mixed with the soil, this lime must speedily pass into the state of carbonate, because the soil and the water with which it is moistened always contain a considerable quantity of carbonic acid. Though we commence operations with quick-lime, consequently, it is carbonate of lime that is definitively introduced into the ground. I have thought this a point of sufficient importance to engage our attention for a short time, inasmuch as it simplifies the view of the end that is to be sought in applying lime; this, as M. Puvis has most satisfactorily established, is neither more nor less than the introduction into the ground of that proportion of the calcareous element which it either wanted originally, or which it has lost in the course of repeated cropping, in order to enable it to produce abundantly. Quick-lime incorporated with the soil must pass, as I have shown, very rapidly into the state of carbonate; but, before attaining to this state, it may, unquestionably, react upon the organic substances it encounters, disorganize them, favor their decomposition, in a word, behave as it does when used in composts. On the other hand, in causing the destruction of organic particles already in a state of decomposition, it must produce an unfavorable influence.

Lime, previously quenched and cold, is generally spread by being raked out from the cart upon the field, in little heaps, from five to six or seven yards apart, each containing from half to two-thirds of a bushel. It is or ought then to be spread immediately as evenly as possible over the surface. There is only the disadvantage attending this mode of proceeding, that slaked lime is twice the bulk of lime in the shell or lump, and that, by slaking, it takes up at least one-fifth of its original weight of water. There is saving of labor, therefore, in distributing the lime unslaked, in heaps, and waiting the slow process of extinction and pulverization by the moisture of the atmosphere. The lime is often laid in a corner of the field, and covered lightly over with vegetable earth to undergo pulverization, and this plan answers very well. Sometimes the lime, before being laid on, is worked up into a kind of compost with vegetable mould and other matters; this is all matter of calculation as to cost. If our object be to supply the soil with the calcareous elements it wants, the proper procedure is quite obvious.

The mode of using lime with reference to other improvers of the soil varies in different places. In one place it is usual to lime and to dung alternately ; in others, the two operations are done together, or very close upon one another. There are some lands so fertile, that they produce abundantly under the influence of lime alone. In laying on lime, one general rule is, that the weather should be dry, and the ground well drained; the end of summer is probably the most favorable season. To say nothing of the difficulty of spreading the lime in wet weather, if it is at all fresh, its caustic qualities are brought into immediate play by the moisture, and it destroys the roots of living vegetables, and the organic elements of the soil; and, again, it is quite certain that lime produces very little effect upon undrained and wet lands. In England, lime is very commonly used upon fallows, in the course of the summer, and before sowing the wheat for which fallowing is always a preparation. When it is given to land destined for beet or potatoes, it is led out in the spring, and spread before the young beet is transplanted, in the one case, the seed-potatoes deposited, in the other.

It is always matter of great moment to have lime spread evenly ; a thorough harrowing and a double *superficial* ploughing incorporate it sufficiently. According to M. Puvis, who has made a particular study of the subject of liming, as practised in the department of the Ain, a quantity of lime, amounting to 8,250 bushels, spread upon seventy-seven acres of land, in the course of nine years, produced so decided an improvement that the returns from winter-grain crops became the double of what they had been before.

Marl. Marl, in a general way, may be regarded as a mixture of carbonate of lime and clay in very variable proportions. Occasionally the clay is replaced by sand ; whence the titles, sandy marl, argillaceous marl. The article, in short, contains from 15 to as many as 90 per cent. of carbonate of lime. It presents numerous shades of color. Geologically speaking, it is usually met with in fresh-water formations of the latest date—the upper strata of the Jura limestone are frequently covered with deposites of argillaceous marls, and we see its formation going on at the bottoms of lakes and ponds, at the present day.

The distinguishing property of a calcareous marl, whatever admixture of other matters it contains, is that of crumbling to pieces under exposure to atmospherical influences. Every limestone rock that has this property may be considered and employed as a marl. The grand purpose of putting marl upon land is to supply it with the calcareous element it wants. To *marl* land, is therefore tantamount to liming it: the effect is the same. The value of the article is, indeed, so well known, that considerable expense is constantly incurred to get at the beds of it that form strata in the crust of the earth, or that lie at the bottoms of lakes. It appears to have been employed from the remotest antiquity.

The reason why marl and marly limestones fall so completely into powder, is obvious. If the mass, when wet, form a pasty mass, it shrinks as it dries, and cracks in all directions ; if more consistent, it is still always porous, and having imbibed a large quantity of rain in the autumn, this congeals during the frosts of the succeeding winter, and the ice, expanding with almost irresistible force, separates the particles, which cohere, indeed, so long as the frost continues, but fall away from one another on the first thaw, by which the solid rock of the autumn and winter becomes a heap of dust in the spring. In the same way, we see chalk, exposed to the wet and the frost, fall down to powder, and, in virtue of this property, and its constitution as carbonate of lime, employed with perfect success in lieu of lime and marl. Wherever there is a bed of chalk at hand, it is needless to go further in search of marl and quick-lime, in so far at least as the calcareous principle is concerned.

Argillaceous marl and sandy marl must, of course, act in two different ways upon the soil: in virtue of the calcareous element in either case, and in virtue of the argillaceous principle in the one, of the sandy principle in the other; and the kind of soil for which they are severally adapted can be conceived beforehand. To a stiff, clayey soil, we would naturally add the sandy marl; to a light sandy soil we would supply the argillaceous product, and thus effect improvement by a kind of double tide. It is therefore very important to distinguish between these two effects produced by marl—one mechanical, connected with the presence of clay or sand; the other chemical, and depending on the presence of carbonate of lime. It is to these two effects, separately and combined, that all the influence of marl is usually ascribed by practical agriculturists. From certain inquiries common to M. Payen and me, however, it appears that marl must act in yet another way; our analyses show that it always contains a certain though variable proportion of azotized matter. And there is nothing extraordinary in the discovery of this fact; it is no more than might have been anticipated from the geological circumstances attending its production. Marls are, as has been said, always connected with the most recent formations of the tertiary series; they are constantly accompanied by remains, which attest the presence of organic beings, and frequently they consist of little else than shells, and the disintegrated dwellings and bodies of moluscas, and madrepores, and corallines, and other inferior forms of things that once had life. It is by no means astonishing, therefore, that deposites which have had such an original should still contain evidences of the presence of the softer and more decompoundable, as well as of the harder and more rebellious constituents of the beings to whose existence they are due. One sample of marl which we analyzed, gave 0.002 of azote; another, from the Lower Rhine, gave rather more than 0.001 of the same element. It were, therefore, very proper, in analyzing marls, chalks, &c., to have an eye to their organic or azotic, as well as to their mineral constituents; there can be very little question of the azotized elements being at the bottom of the really wonderful fertilizing influences of the marls of certain districts.

Marl ought, like lime, to be spread very evenly over the land; it is generally laid on in the same way as lime—in little heaps at regular distances, and then scattered abroad. It appears to be a very general opinion that it is not advisable to cover it immediately, or very shortly after it is dug from the bed that supplies it; the practice where its employment is most general, and probably best understood, is to let it lie exposed through the summer or winter, or even the whole year before laying it on the land. It is also held not to be proper to cover it in marl deeply. Marl is advantageously laid out in heaps upon stubbles in the autumn; and in the early spring when it has been pulverized by the frost, it is spread with the shovel. When it is to be used with winter wheat or rye, it is laid on in the summer, and spread at the time of ploughing; the latter plan of proceeding, however, as Schwertz observes, can only be followed with marl that pulverizes readily. In England it is also laid down as a kind of principle that marl ought to be exposed for as long a time as possible to the influences of the atmosphere; that it ought to have a summer's heat and a winter's cold before it is applied. And that, in fact, which is at all consistent, and has not been exposed to the frost, scarcely pulverizes sufficiently to be readily miscible with the soil even under the influence of repeated ploughings; moreover, it produces very little obvious effect upon the crop with which it is first used. After spreading, a rough harrow is passed over the surface of the ground, which is then ploughed superficially two or three times, the harrow being again had recourse to repeatedly to break lumps, and so bring out the effect of the marl.

The quantity of marl that may be advantageously given varies according to the circumstances of the district. Marl, it may fairly be said, is frequently abused. In an excellent paper on the subject, M. Puvis lays it down as a principle that the first element in the calculation of the proper dose of marl, is the quantity of calcareous matter that is wanting in the soil. He says that every soil which contains more than 9 or 10 per cent. of carbonate of lime can dispense with marl; and that soils in which the lime falls short of this quantity, may advantageously receive a dose or successive doses of the substance that will bring them up to the point. The proper dose, consequently, depends first on the proportion of carbonate of lime contained in the soil, and then on that which the marl itself includes.

Considered from the rational point of view which M. Puvis has taken, marling is no longer an arbitrary process, but one that may be conducted on determinate principles. The extravagant quantities that are often laid on without other assignable reason than blind custom, are shown to be, if not injurious, yet useless: the quantity of marl to be incorporated is determined by the quality of the substance which is at our disposal, and by the depth of the layer of vegetable earth taken in connection with its chemical constitution. To facilitate the calculation of the proper dose, M. Puvis has drawn up a table, which, as it may be found useful in practice, I append. It shows at a glance the quantity of marl in cubic feet that ought to be put upon an acre of ground, the depth of the arable soil being considered in connection with the composition of the marl at command:—

Table of the Number of Cubic Feet of Marl applicable upon an Acre of Land ploughed to the depth of:						When 100 parts of marl contain of carbonate of lime,
$3\frac{1}{10}$ inches.	$4\frac{3}{10}$ inches.	$5\frac{5}{10}$ inches.	$6\frac{3}{10}$ inches.	$7\frac{4}{10}$ inches.	$8\frac{6}{10}$ inches.	
333	444	554	666	776	888	10
$166\frac{1}{2}$	222	277	333	388	444	20
111	146	$184\frac{2}{10}$	222	$258\frac{2}{10}$	296	30
$83\frac{1}{10}$	111	$138\frac{2}{10}$	$166\frac{2}{10}$	194	222	40
$66\frac{3}{10}$	$88\frac{4}{10}$	$110\frac{4}{10}$	$133\frac{1}{10}$	$155\frac{1}{10}$	$177\frac{3}{10}$	50
$55\frac{3}{10}$	74	$92\frac{2}{10}$	111	$129\frac{2}{10}$	148	60
$47\frac{4}{10}$	$63\frac{3}{10}$	$79\frac{1}{10}$	$95\frac{1}{10}$	$110\frac{6}{10}$	$126\frac{6}{10}$	70
$41\frac{5}{10}$	$55\frac{4}{10}$	$69\frac{2}{10}$	$83\frac{2}{10}$	97	111	80
37	$49\frac{3}{10}$	$61\frac{1}{2}$	74	$86\frac{2}{10}$	$98\frac{6}{10}$	90
$33\frac{3}{10}$	$44\frac{4}{10}$	$55\frac{4}{10}$	$66\frac{6}{10}$	$77\frac{6}{10}$	$88\frac{8}{10}$	100

M. Puvis does not by any means give the doses in this table as those that should be invariably employed; the table is one of averages, deduced from practical results, and tested by experience as most truly useful. But special cases may occur that would make departure from these conclusions not only advisable, but advantageous.*

The use of marl produces an unquestionable effect on the productive properties of the soil. According to M. Puvis, the application of the proper dose of a sandy marl, containing from 30 to 60 per cent. of carbonate of lime, doubled the produce of a piece of parched land in the department of the Isère. Before the application of the marl nothing but dwarfish crops of rye were gathered, yielding at most three for one of the seed; at present, eight for one of seed, and that wheat, are obtained; and the good effects are found to continue for ten and even twelve years.

The action of marl is not unlimited any more than that of lime, as the last sentence will give the reader reason to conclude. With every harvest, a certain proportion of it is carried off, and the land is finally left with an inadequate quantity of the calcareous element, which then requires to be restored. The nature of the crop, however, has the most marked influence on the quantity of lime that is taken up and carried away from the soil; allowing the broadest margin, and judging from the composition of the ashes of the plants that form the subjects of our ordinary crops, we can see that the quantity of $3\frac{1}{2}$ bushels of marl of the usual composition per acre, which is assumed as the average quantity to be laid on, is vastly more than can be absolutely necessary.

Wood ashes contribute to improve the soil. They contain, besides silica, both phosphate and carbonate of lime and alkaline sulphates, phosphates, and carbonates. In a general way, every thing derived

* Puvis in Annals of French Agriculture, vol. xxviii. p. 328, 2d series.

from plants that have lived must be useful to plants that are about to live, or that are actually living. Although the utility of wood ashes, then, is generally admitted, the numerous purposes to which they are applied in the arts, and their high price, which is the consequence of this, enable the husbandman to employ them but rarely on his land ; they are almost always lixiviated in order to procure the carbonate of potash they contain. In countries which are thickly wooded, indeed, the trees are actually cut down and burned for the sake of their ashes, just as oxen are run down and slaughtered in the vast plains of South America for the sake of their hides.

The good effect of wood ashes upon vegetation is known to communities the least advanced in civilization. The Indians of South America burn the stems and leaves of the maize in order to improve the soil. The same practice occurs among the natives of Africa : on the banks of the river Zaire, according to Tuckey, the ground is prepared by having little piles of dried herbs placed on it, to which fire is set ; and upon the spots where the ashes are collected, they sow peas and Indian corn ; these ashes are in fact the only manure that is employed. In England, wood ashes are esteemed as particularly useful upon gravelly soils ; about 40 bushels per acre are applied in the spring, where the article can be obtained.

The *lye-ashes* from the soap-boiler contain a small quantity of soluble saline matter which has escaped the lixiviation, mixed with a large proportion of lime, partly in the state of carbonate, the lime having been added to bring the carbonate of potash employed in the manufacture of soap into the caustic state. This ash or refuse is much sought after, and is administered in quantities that vary from 45 to 70 bushels per acre, a dose in which its action is felt for ten years or more. In wooded districts, where there is a good deal of potash prepared, ash of this kind is obtained in large quantity ; it is there employed alternately with organic manures. Ashes are applied in the same way as lime, with this difference, that it is held better not to plough them in until they have received a little rain. There are places where the ashes that remain in the lixiviating tub are thrown on in the dose of 170 bushels per acre.

Turf or peat ashes. Peat is the result of a peculiar spontaneous change that takes place in vegetables. It is produced in bogs or swamps, and in connection with stagnant waters ; turfy deposites are also encountered on the banks of rivers, in valleys, at the bottoms of former lakes, and at the mouths of rivers. Peat is met with from the level of the sea to the elevated platforms of the Vosges and Alps ; it lies in horizontal beds, frequently divided by strata of gravel, sand, or clay. It is always a product of comparatively recent formation, a fact which is attested by the thin layers of vegetable soil that lie over it in many places, and the animal remains and products of human industry that are frequently encountered in it.

The state of decomposition of the vegetables that form turf or peat is seldom so far advanced as to make the remains of the plants which compose it doubtful. It is of different kinds : hard or woody, and soft or herbaceous peat. Some of it is extremely compact, black, and like vegetable mould in appearance; generally speaking it is light, spongy, and of a lighter or deeper shade of brown. When quite dry, it is often extremely light; a cubic metre, which is about one-eleventh more than a cubic yard, will weigh from 5 to 6 cwt.

The circumstances in which turf has been found lead us to infer that it must contain the elementary insoluble elements of the plants that produced it. It appears, however, to contain a somewhat larger proportion of azote than the average quantity met with in herbaceous vegetables, supposed dry; but we have seen that in the slow alteration of lignine, azote becomes concentrated, as it were, in the residue; and that, in fine, mould contains a larger quantity of azote than the wood from which it proceeds. It appears further from some experiments very lately performed by Mr. Hermann, that during the putrefaction of the woody principle, azote is actually taken from the air to concur in the formation of certain products that are perfectly definite. Mr. Hermann quotes the following experiment:

Twenty-eight parts of wood taken from a log already attacked with rot, and in which, indeed, there were several points already decayed, were moistened and enclosed in a jar containing atmospherical air over mercury. The bulk of the atmosphere contained in the bell-glass was 262 volumes. The wood was kept there for ten days at a temperature of 75.2° Fahr. The apparent volume of the air continued unaltered to the end of the experiment; but a large quantity of carbonic acid had been formed:

RESULTS ON THE INCLUDED AIR.

The air contained:	Before.	After.
Azote	207 vols.	194 vols.
Carbonic acid		40 "
Oxygen	55	28 "
	262	262

The moist wood in its decomposition during ten days had consequently caused thirteen volumes of azote and twenty-seven volumes of oxygen to disappear. And Mr. Hermann found that it now contained principles analogous to those of humus, one of which, nitrolin, is highly azotized, and by the ulterior action of air and moisture, gives rise to ulmate of ammonia. These experiments of Mr. Hermann are new, and the conclusions to which they lead are both interesting and important.*

Turf or peat is virtually the woody principle in the last stage of modification by atmospherical influences; but it appears still to contain, although modified, the usual principles which enter into the constitution of herbaceous vegetables also. M. Payen detected a quantity of fatty matter in it, analogous to that which exists in leaves, and M. Reinsch found it to contain tannin. One sample of

* Vide his paper in Journ. für prakt. Chemie, b. xxiii. s. 379.

turf (from the neighborhood of Moscow, by the way) examined by Mr. Hermann, yielded of carbonaceous matter, nitrolin and vegetable remains 77.5 ; of ulmic acid 17.0 ; extract of humus 4.0 ; ammonia 0.25 ; and ash 1.25=100.0 The elementary composition of these varieties of turf, analyzed by M. Regnault, gave from 57 to 58 of carbon ; 5.1 to 5.6 hydrogen ; 30.8 to 31.8 oxygen and azote ; and 4.6 to 5.6 ashes.

Turf or peat has consequently a certain resemblance to mould or humus ; it differs, however, in the absence of substances soluble in water ; and it is easy to imagine that, produced as it is in connection with water, continually soaked in moisture, soluble matters ought not to be expected in it in appreciable quantity. Peat might, in fact, be likened to the insoluble part of humus left after lixiviation. And there is this further resemblance, that peat, like the humus which has been thoroughly lixiviated, if exposed to the air, by and by acquires a quantity of soluble material, the evolution of which is also hastened by the contact of the alkalies. The employment of turf as manure, in some countries, confirms the propriety of this mode of viewing its nature and constitution ; and then it is well known that bogs consisting of pure turf, when drained and limed, become tolerably fertile lands, yielding magnificent crops of oats and turnips especially.

The ashes of turf we might expect to contain the mineral substances usually found in the ashes of plants, and further a certain quantity of additional earthy matter. But this is not the case : several alkaline salts, indeed, have been discovered in very small proportion ; but no chemist, to my knowledge, has ever even suspected the presence of any of the phosphates ; a special search which was made for them in my laboratory failed to discover them. This is a fact which, I own, amazed me ; some coal ash, and another ash produced from lignite, gave a result equally negative. We might imagine the disappearance of the soluble salts ; but how the earthy phosphates should disappear ; how the ashes of coal should come to be without a trace of phosphoric acid, when we see that the iron ore, in connection with the coal fields, is always more or less phosphorigerous,* is surprising.

Turf or peat ashes are valuable improvers of the soil, and are in great request among intelligent farmers. Analysis, in fact, indicates several substances in their composition as calculated to assist vegetation ; carbonate of lime, in a state of extreme subdivision ; occasionally sulphate of lime, (gypsum ;) calcined clay, whose action upon strong and retentive lands is always beneficial ; silica in a favorable state for assimilation ; finally, alkaline salts, chlorides, sulphates, carbonates, and, perhaps, in spite of the negative given by chemical analysis, traces of the phosphates.

The peat of the bogs of Sceaux, near Château-Landon, leaves 19 per cent of ashes, composed, according to M. Berthier, of :—

* Our author might have added the fact, that the common bog iron ore of this country is a phosphate of iron.—Eng. Ed.

Caustic and carbonated lime	63.0
Clay	7.5
Gelatinous silica	13.0
Alumina	7.0
Oxide of iron	9.0
Carbonate of potash	0.5
	100.0

The peat of Voitsumra, dug upon the frontiers of Bavaria and Bohemia, contains the remains of trees; it leaves 1.7 per cent. of ashes, composed, according to M. Fikenscher, of:

Silica	36.5
Alumina	17.3
Oxide of iron	33.0
Lime	2.0
Magnesia	3.5
Sulphate of lime	4.5
Chloride of calcium	0.5
Carbonaceous matter not incinerated	2.7
	100.0

The brown herbaceous peat of the neighborhood of Troyes, leaves 11 per cent. of residue; it contains:

Carbonic acid and sulphur	23.0
Lime	23.0
Magnesia	14.0
Alumina and oxide of iron	14.0
Clay and silica	26.0
	100.0

The peat of Vassy is compact, and of a brown color; it is mixed with fragments of chalk. On incineration, it leaves 7.2 of residue per cent., containing:

Clay	11.0
Carbonate of lime	51.4
Sulphate of lime	26.0
Oxide of iron	11.5
	100.0

The peat of Champ-du-Feu, near Framont, (Vosges,) leaves 3 per cent. of ashes, which consist of:

Silica	40.0
Alumina and oxide of iron	30.0
Lime	30.0
	100.0

The peat of the environs of Haguenau (Lower Rhine) produces 12.5 per cent. of ashes, which, according to the analysis made in my laboratory, contain:

Silica and sand	65.5
Alumina	16.2
Lime	6.0
Magnesia	0.6
Oxide of iron	3.7
Potash and soda	2.3
Sulphuric acid	5.4
Chlorine	0.3
	100.0

Supposing the whole of the sulphuric acid found to have been in -

combination with lime, this peat could only have contained 4.1 per cent. of gypsum.

These analyses will show that the composition of peat, or turf, is very various. The varying and dissimilar effects produced by turf-ashes, may probably be owing to this variety of composition. Turf-ashes, in a general way, may be used as a substitute for gypsum; but this is upon the presumption that they contain lime, either in the state of carbonate or of sulphate. The Vassy turf-ashes, for example, may be employed for gypsing meadows, inasmuch as they contain a quarter of their weight of sulphate of lime.

The ashes from pyritic turf ought not to be used without great circumspection; they usually contain a quantity of iron pyrites which has not been destroyed in the burning, and which, exposed to the action of the air, gives rise to the formation of green vitriol, or sulphate of iron, which may prove prejudicial to vegetation. These ashes are generally of a red color, and very heavy, in consequence of containing a quantity of the oxide of iron. Good turf-ashes ought to be white and light; the sack ought to weigh something less than a hundred-weight. Schwertz recommends us to keep them from the wet; but at Bechelbronn, where we use large quantities of peat-ashes, we find no ill effects from leaving them exposed to the rain; frequently, indeed, we moisten them with water from the dunghill, in order to add to their properties as a mineral manure, those that belong to organic manures. On the whole, however, it is certainly better, for many reasons, to keep them dry; they are more easily carried, and they are more easily spread.

Turf-ashes of a good quality, that is to say, which include in their composition a large proportion of calcareous and alkaline salts, are adapted to crops of every description; but it is upon clover especially that their influence is truly surprising. This fact is well established in Flanders; but one must have employed them one's self to have any adequate idea of the improvement they produce. There is no risk of giving too large a quantity. In winter, when we have peat-ashes at our disposal, we give as many as 60 bushels per acre to our clovers; we scatter them even upon the surface of the snow, and distribute them by means of the rake in the spring. The Dutch use these ashes in still larger quantity, applying, at two different times, from 100 to 160 bushels per acre to their clover fields. According to Sinclair, the Dutch also make use of an ash procured from a turf which during winter is in contact with brackish water, a circumstance which renders this ash particularly rich in alkaline salts. It is sowed by hand, in the spring, upon clover, and the following year an abundant crop of wheat is obtained. The same material is also used in the cultivation of the hop; and it is said that, administered in small quantity to the roots of the vine, they preserve the plant from the attacks of destructive insects.

Coal-ashes. Coal, like the two last combustible materials, is the product of vegetables, which, however, have undergone such a change as to have lost almost every trace of organization. Coal of different kinds contains from 1.4 to about 2.3 per cent. of ashes, and about 2 per cent. of azote. The ash of a variety of coal of very excellent quality gave of—

Argillaceous matter (silica ?) not soluble in acids	62
Alumina	5
Lime	6
Magnesia	8
Oxide of manganese	3
Oxide and sulphuret of iron	16
	100

Coal ash also contains very minute quantities of alkaline salts, which usually escape analysis when they are not especially inquired after. One specimen analyzed in my laboratory, gave nearly 00.1 of alkali. Coal-ash is particularly useful on clayey soils; it acts by lessening the tenacity of the soil; and further, doubtless, by the introduction of certain useful principles, such as lime and alkaline salts.

OF ALKALINE SALTS.

It is impossible to doubt that salts having potash and soda for their base are useful in agriculture. The influence of wood-ashes, and of paring and burning is unquestionable; and they are so, in some considerable degree at least, in consequence of the salts of these bases which they supply, and which always enter into the constitution of vegetables. There are even certain crops which, in order to thrive, require a particular alkali; the vine, for example, the fruit of which contains bitartrate of potash, and sorrel, which contains the binoxalate of the same base, must needs have supplies of potash. The plants which are grown for the production of soda, the *salsola, &c.*, from which barilla is made, must come in a soil that naturally contains a salt of soda, such as that of the sea-shore.

It would appear, however, that the salts of soda or potash, must not exceed a very small proportion in the soil. All the experiments that have yet been undertaken with a view to ascertain the action of different saline substances on growing vegetables, have led to no very certain conclusion but this, that they must be used very sparingly. M. Lecoq has published an account of some experiments, made apparently with great care, which go to prove that common salt, in the dose of from $1\frac{1}{4}$ to $2\frac{1}{2}$ cwts. per acre, favored the growth of barley, wheat, lucern, and flax. Chloride of calcium and sulphate of soda, he also found to have the same good effects. M. de Dombasle, however, came to conclusions totally opposed to them, with reference especially to common salt, which, applied in the doses advised by M. Lecoq, was not found to produce any sensible effect. M. Puvis also obtained results that were equally negative. It would perhaps have been well had M. Lecoq begun by determining the proportion of alkaline salts which existed previously in the soil on which he conducted his experiments. If he operated on a soil that was either destitute of these salts, or that contained them only in minimum proportion, very probably he did good by adding them.

Nitrate of potash has been repeatedly recommended as an agent useful in agriculture. The conclusions that have been come to, however, from its use, are far from accordant. In the processes or modes of using nitre to the soil, it is not uncommon to find it associated with soot, or with vegetable mould, substances which require no assistance of any kind to constitute them powerful manures, and the addition of which is therefore calculated to raise strong doubts of the advantageous qualities ascribed to nitre alone. Were the advantages of nitrate of potash much less questioned than they are, however, the high price of the salt would probably always oppose insuperable obstacles to its employment. This is the reason, in all likelihood that has turned the attention of English agriculturists, for several years past, to nitrate of soda, a salt that is imported in quantity from Peru, and of which the price per cwt. may be about forty shillings; a price which, were it found really useful, would permit of its being used. Admitting the accuracy of the experiments that have been made, indeed, we cannot doubt the efficacy of nitrate of soda on soil already furnished with organic manure. The quantity that has been recommended is about one cwt. per acre.

Mr. Barclay made a few experiments after having heard much of the nitrate of soda from his neighbors, of the results of which the following examples will suffice to give a comparative estimate:

	Without nitrate.	With nitrate.	Difference in favor of the nitrates.
Wheat........	31 bush. 2 pecks.	35 bush. 3 pecks.	5 bush. 3 pecks.
Straw..........	21 cwt. 0 qrs. 19 lbs.	23 cwt. 2 qrs. 26 lbs.	3 cwt. 2 qrs. 7 lbs.

The produce of the land treated with nitrate, however, did not fetch so high a price at market as that grown without it; and every item of expense taken into the reckoning, the use of the nitrate was attended with no commercial benefit. Still this does not militate against the fact, that the production of vegetable matter was increased upon land treated with the nitrate of soda. And indeed much of the information which M. de Gourcy collected in England, is of a kind that tends to confirm the favorable influence of this salt on vegetation. Wheat, clover, and Swedish turnips are particularly specified as benefiting from its use. These facts admitted, we may ask: how does the nitrate of soda act? The chemical constitution of the nitrates is such, that we might conceive their acting at once as mineral and as organic manures. The important point for solution was to ascertain whether the azote of the nitrate contributed in any way to the formation of the azotized principles of plants. Davy, in taking with much distrust the report of Sir Kenelm Digby's experiments on the influence of nitre in the cultivation of barley, shows no disinclination to believe that the azote of the salt may concur in the production of albumen and gluten.* This, however, is a point in physiology which may be put to the proof by experiment, and seems peculiarly worthy of being tested in this way. I have admitted it as extremely probable, that the azote of the azotized principles of plants has its source either in the ammonia, which is the special ultimate product of the organic manure we employ, or in the azote of the atmosphere, or in both simultaneously; but the opinion which should maintain that the ammonia derived from the organic constituents of the soil, passes into the state of nitric acid before penetrating the tissues of plants, would find support nearly in the same facts which I have quoted as favoring the former view. We have seen, moreover, in our general considerations on nitrification, with what facility the azote of ammonia undergoes acidification in certain circumstances, a fact from which an argument of much potency for the nitric acid theory naturally flows. I shall here add an observation to which I have, up to this time perhaps, attached too little importance. When M. Rivero and I examined the highly irritating and poisonous milky sap of the *hura crepitans*, we had occasion to leave a considerable quantity of the water derived from the sap, after separating the caseum, to itself; by the spontaneous evaporation of this water, we collected really a considerable quantity of nitrate of potash. Since this time I have had occasion to note the same salt in the sap of several trees of the tropics. In the leaves and fruit, however, I have never found more than very minute quantities.

Gypsum, sulphate of lime, or plaster of Paris, is a compound of 41.5 lime with 58.5 sulphuric acid; gypsum generally contains a quantity of constitutional water, in which case it consists of 79.2 sulphate of lime, and 20.8 water = 100. This hydrate of sulphate of lime is one of the abundant minerals on the surface of the earth; it is met with in the crystalline state, and in granular and fibrous masses in the strata of most recent formation. It has no sensible taste, but is slightly soluble in water, this fluid dissolving $\frac{1}{460}$ of its weight of the salt. Exposed for some time to a white heat, it loses its water of constitution, and passes into the state in which when ground it is known under the name of plaster of Paris.

Gypsum is one of the most commonly employed of the mineral manures. Its virtues appear not to have been unknown to the ancients; but until lately its employment was limited to a few circumscribed districts. It was only about the middle of the eighteenth century that the protestant pastor, Mayer, took up the study of gypsum in the principality of Hohenlohe, proceeding upon certain information which he had obtained from Hehlen of Hanover, in the neighborhood of which, it was employed as an improver.

By extending a knowledge of the virtues of gypsum, both by his example and his writings, Mayer did great service to agriculture. Experiments were soon instituted in all quarters. Tschiffeli in Switzerland, Schubart in Germany, and Franklin in America, wrote on its effects, or practically demonstrated them to the satisfaction of all. But it appears to be the fate of all useful discoveries, of all happy applications of principles, to be opposed at first, and only to be admitted after having been vainly disputed. The use of gypsum soon aroused formidable opposition; and there is a curious episode in the history of the paper war that was long carried on upon his subject, which I think worth noting. Among the most strenuous enemies of the use of gypsum, were the proprietors of the salt-pans.

They declared that gypsum was not only incompetent to replace *schlot* or the refuse of their pans, as had been proposed, but that it was injurious; *schlot* was the only real improver, the stimulant of stimulants, for which there was no substitute. But it turned out by and by, that the *schlot* of the salt-pan was found to be neither more nor less than sulphate of lime, than gypsum—the article that was not only inefficient, but injurious. These gentlemen were afraid that the use of gypsum extending, they would want a market for their refuse.

The use of gypsum once introduced, extended rapidly in France, particularly around Paris, whence it crossed the Atlantic, and the fields of North America were actually manured with the produce of the quarries of Montmartre. The lately cleared lands of America abound in humus, and the plants indigenous there were most beneficially acted on by gypsum, which really produced remarkable effects; in both the new and the old world, its power, as one of the most useful auxiliaries of vegetation, soon appeared to be established.

We must not blind ourselves to the fact, however, that the partisans of gypsum were guilty of exaggeration. They spoke of the substance as a universal manure, capable of supplying the place of every other, as advantageous for every description of crop, as applicable to every variety of soil. Experience soon set bounds to such indiscriminate laudation; it was found that gypsum alone was inadequate to produce fertility, that it always required the concurrence of organic manures, if the soil did not contain them of itself; that it only acted beneficially on a certain, and that a very small number of plants; lastly, that it was upon artificial meadows, constituted by clover, lucern, and sainfoin, that it produced its best effects; its action, on the contrary, being scarcely perceptible upon natural meadows, doubtful in connection with hoed crops, and null with the cereals. These negative results cannot be called in question; they were come to by parties who were every way interested in having the decision otherwise.

The best season for spreading gypsum is the spring, and when the clover, sainfoin, or lucern, has already made a certain degree of progress; calm and moist weather is the best for laying it on. Opinion was long divided as to whether it should be applied in its natural state, and simply ground, or first burned and then ground. But it is now generally admitted that burning adds nothing to the qualities of gypsum. Although the usual practice is to sow or powder the meadows with the ground gypsum, it is still acknowledged that good effects are obtained from incorporating the substance with the soil. The advantage of the practice of scattering it on in powder, so as to adhere to the wet leaves of the growing plants, I find explained in the equality of distribution which is by this means effected.

In some places, the number and extent of which are by no means inconsiderable, no good effect whatever has attended the application of gypsum, although it has been administered in favorable conditions, and in connection with crops that elsewhere derive the highest amount of advantage from its use. This anomaly has been explained by assuming, without proving experimentally, however, that the fact is so, that the soil in these districts naturally contains a sufficient dose of gypsum. It has also been said that gypsum produces no effect on low-lying and damp soils.

The quantity of gypsum employed in different places, varies greatly: from $1\frac{1}{2}$ to 16 cwts. per acre have been recommended. The quality of the article employed has a great influence on this question, to say nothing of the price, which in many places is high.

The opinions of practical men, with regard to the advantages and propriety of applying gypsum, although they agreed in certain determinate circumstances, were still far from being unanimous upon every point. A particular inquiry into the subject was therefore held worthy of its attention by the French government, and a comprehensive report on all the information collected, was made by M. Bosc to the Royal Central Agricultural Society of France. This report shows in a striking manner the advantage that may be derived from the lights of practical men; in a single line or sentence we frequently find a summary of twenty or thirty years of experience. It is, however, indispensable to go to these gentlemen for their information; the agriculturists who devote themselves to cultivation, it is notorious, write very little, and those who spend very little time in this way, on the contrary, write a great deal. It may be that the reason for the silence of the one, is that also for the eloquence of the other.

The following series of questions and answers I believe to embrace most of the points connected with the employment of gypsum, that are of interest.

1st. Does plaster act favorably on artificial meadows? Of 43 opinions given, 40 are in the affirmative; 3 in the negative.

2d. Does it act favorably on artificial meadows, the soil of which is very damp? Unanimously, no. Ten opinions given.

3d. Will it supply the place of organic manure, or of vegetable mould? *i. e.* will a barren soil be converted into a fertile one by the use of plaster? No, unanimously. Seven opinions given.

4th. Does gypsing sensibly increase the crops of the cereals? Of 32 opinions, 30 negative, 2 affirmative.

The information thus obtained, valuable as it is, cannot yet be held to embrace every thing that seems desirable. Happily, all that was wanting has been supplied by the individual inquiries of Mr. Smith in England, and of M. de Villèle in France.

The soil upon which Mr. Smith made his experiments was light, with a substrate of chalk; the vegetable earth was a yard in depth at the top of the field, and lessened gradually, in such a way that at bottom it was but three inches thick. Every precaution was taken that the respective breadths contrasted should be as nearly as possible in the same circumstances. The following table shows the results:

GROWTH OF SAINFOIN UPON SOILS GYPSED AND UNGYPSED IN 1792, 1793, AND 1794.

Number of experiment.	Remarks.	Dry herb per acre.	Seed per acre.	Weight of total crop.	Proportion of stalk to seed.
		lbs.	lbs.	lbs.	
1	Crop on the deeper ungypsed soil	3357	419	3776	100 : 12.5
	Crop upon the contiguous breadth, which had received about 15 bushels of gypsum in April, 1794.	5462	582	6044	100 : 10.7
	Difference in favor of the gypsed breadth	2105	163	2268	
2	Crop upon the same soil, of less depth, and not gypsed . .	2766	245	3011	100 : 8.9
	Crop on contiguous soil, dressed with about 15 bushels of gypsum in April, 1792 . . .	4381	379	4760	100 : 8.7
	Difference in favor of the gypsed breadth	1615	134	1749	
3	Crop on the same soil, 3 inches deep, and not gypsed . .	2068	66	2134	100 : 3.2
	Crop on contiguous soil, dressed with about 15 bushels of gypsum, 17th May, 1794 . .	4879	211	5090	100 : 4.3
	Difference in favor of the gypsed piece	2811	145	2956	
4	Crop on the contiguous soil of experiment, No. 3, gypsed with the same dose in May, 1792	4310	205	4515	100 : 4.8
	Difference in favor of the crop gypsed twice, at an interval of two years	2242	139	2381	

These results show to what extent gypsum is favorable to the production of sainfoin. The crop from the unplastered breadth being taken as 100, that upon the plastered breadth is 231; it is more than doubled. The influence of gypsum was also found by Smith to extend to grain; assuming the grain crops on the ungypsed land at 100, those on the gypsed soil were 192; they were nearly doubled.

On comparing the weight of the herbaceous portion of the sainfoin to that of the seed produced, widely different relations are apparent. These Mr. Smith attributed to the different depths of the vegetable soil in different parts of the field. In the first experiment, where the relative proportion of seed is highest, the arable soil was three feet in thickness; the other crops were taken from parts where the depth of vegetable mould was considerably less. Thus the gypsed soil produced at the rate per acre:

	cwts.	qrs.	lbs.	
In the first experiment of	5	0	22	the depth of soil being 3 feet.
In the second experiment of	3	1	15	" " 18 inches.
In the third experiment of	1	3	15	" " 3 inches.

With this interesting fact before him, Mr. Smith imagined that soils of little depth wanted some principle essential to fructification, which gypsum, in spite of the unquestionable assistance it gives, is yet incompetent to supply. This principle is in all probability organic matter, which is naturally more abundant in the layer of true vegetable mould which is deepest.

Mr. Smith's observations on white clover were quite as decisive in favor of gypsum as those on sainfoin, and are confirmatory of the conclusions of the generality of farmers on the subject. The gypsum in connection with this crop was applied in the dose of 6 bushels per acre, on the 22d of May, a date at which the clover looked pale, and seemed to want sap. A fortnight afterwards, the effects of the gypsum were obvious; although no rain had fallen in the interval, the clover had become vigorous, and soon formed a covering thick enough to protect the ground from the scorching rays of the sun, which burned up all the parts which had not been gypsed.

COMPARATIVE GROWTHS OF WHITE CLOVER, GYPSED AND UNGYPSED, BY MR. SMITH.

EXPERIMENTS.	Herb or stalk per acre.	Seed per acre.	Total weight of the crop.	Proportion of herb to seed.
	lbs.	lbs.	lbs.	
A. Gypsed	2226	316	2542	100: 14.3
A. Not gypsed	839	56	895	100: 6.7
Difference	1387	260	1647	
B. Gypsed	2270	174	2444	100: 7.6
B. Not gypsed	500	61	561	100: 7.0
Difference	1770	113	1883	

The mean of these two experiments shows that the crop of white clover on the ungypsed land being 100, that on the gypsed is 225—twice and a quarter more.

The experiments of M. de Villèle may be viewed as supplementary or complementary to those of Mr. Smith. They were performed in the south of France, in accordance with the routine that is generally followed, viz: clover-hay, or sainfoin, previous to grain, upon soils of considerably different nature, and with doses of gypsum that varied from 8 to 3 on the same extent of surface. His conclusions or crops are stated in the following table:

KIND OF SOIL.	Number of experiments.	Crop.	Gypsum per acre.	Dry crop on the gypsed ground per acre.	Dry crop on meadow not gypsed, per acre.	Excess of the crop gypsed over the crop not gypsed.	Money value of the excess of forage.	Cost of the gypsum employed.	Profit from the use of the gypsum.
			cwt. qr.	cwt. qr. lbs.	cwt. qr. lbs.	cwt. qr. lbs.	s. d.	s. d.	s. d.
Light, dry, exposed to the south, 6 to 9 inches deep, and on chalk.	1	Sainfoin	6 3	28 2 6	18 0 1	10 2 15	17 7	6 9	10 10
	2	Sainfoin	2 2	32 2 27	16 1 13	16 1 13	27 1	14 0	25 0
	3	Sainfoin	4 4	27 0 1	17 0 21	19 3 8	16 3	5 1	11 2
Stony clayey, moist, about 16 inch. deep on a stiff clay.	1	Clover	4 0	40 3 19	20 1 23	20 1 23	33 10	3 2	2 3
	2	Clover	5 3	32 2 27	19 2 16	13 0 11	21 8	12 5	14 2

The unquestionable fact of a mineral salt stimulating the growth of certain plants in so remarkable a manner as to double and even to triple the usual quantities grown per acre, naturally aroused the curiosity of mankind to inquire into and endeavor to discover the cause. Explanations in abundance have been proposed; but so little satisfactory in general, that I do not think myself bound to mention them all. I shall limit myself, indeed, to two; one proposed by Davy some time ago, and one advocated by Liebig very lately.

Davy assumes that the plants of artificial meadows simply absorb sulphate of lime. He assures us that he had found a large proportion of this salt in the ashes of vegetables grown in soil which had been treated with turf ashes abounding in the substance. He believed that the gypsum entered particularly into the constitution of the woody fibre. And it is not uninteresting to observe, that the plants which gypsum certainly favors in the highest degree, are of very rapid growth; and that in all probability they would find it difficult to obtain the whole of the sulphate of lime they require from ordinary or ungypsed soils within the period of their growth. Let it not be forgotten, however, that if it be true that saline substances are indispensable to the organization of plants, it is also true that these substances can only be absorbed within certain limits; a salt the best calculated by its nature to aid vegetation, would become injurious by its excessive proportion, did the water which moistened the general soil contain too large a proportion of it in solution: if a plant languishes when it has not enough of one or other of its natural saline constituents, it also dies when furnished with the same substance in excess.

Let us now remember that salts can only act on vegetables in the state of solution, and we shall understand how those only which are but sparingly soluble, can ever be advantageously employed in agriculture. Water, in fact, having the power to dissolve only a very limited quantity of the mineral manure, will present it to the growing plant nearly in a constant quantity, so long as the soil contains any fair proportion of the substance. It is in this way precisely that gypsum appears to gain its superiority over the generality of mineral or saline manures ; water does not take up more than $\frac{1}{460}$th part of its weight before it becomes saturated ; a certain proportion of the moisture of the earth being dissipated by evaporation, there is forthwith a precipitation of sulphate of lime ; but the moisture that remains is nevertheless charged as before, neither more nor less, and in the fittest state, as it seems, to administer to the wants of the growing plant. If instead of sulphate of lime we suppose some salt that is much more soluble, sulphate of soda for example, we have nothing of the same state of equilibrium between the quantity of moisture and its charge of saline ingredients maintained. Supposing the moisture of the ground to hold $\frac{1}{460}$th of sulphate of soda in solution, and this quantity calculated to produce good effects upon growing vegetables ; suppose now that a drought sets in, which by dissipating one-half of the moisture, increases the charge of saline matter to $\frac{1}{330}$th of its bulk, it may very well happen that this proportion, instead of proving beneficial, will be felt as injurious to vegetation.

The hypothesis of Davy, supported by these ingenious views of M. Chaptal, would therefore lead us to regard gypsum as behaving to plants in the same general way as the insoluble salts which usually form an element of the soil or of manures, the phosphate and carbonate of lime, in particular, salts which are made apt to enter the tissues of plants by the carbonic acid which is found in all the water that falls from the clouds and that moistens the soil, and which has the property of dissolving small quantities of them. But while the strength of these solutions, weak at all times, is liable through atmospherical vicissitudes to vary, when the mere traces of saline matter which at best they offer at any time are inadequate to meet the demands of a crop disposed to grow rapidly and luxuriantly, such as clover, sainfoin, and lucern, the solution of sulphate of lime, of the same strength at all times and under all circumstances, is ready to supply the plants with the mineral substance they require, however rapid and vigorous their growth.

The theory of the action of gypsum proposed by Professor Liebig is extremely ingenious. He admits, with M. de Saussure, the presence of carbonate of ammonia in the atmosphere, and consequently in rain-water. This fact established, and it appears undeniable, the influence of gypsum would consist in its faculty of fixing the infinitely small quantity of carbonate of ammonia which is brought down by the rain and the dew, and so preventing its dissipation on the return of drought and sunshine. Carbonate of ammonia, in fact, as we have already seen, when speaking of manures, in contact with the sulphate of lime decomposes this salt, carbonate of lime and sulphate of ammonia being formed. I shall by and by inquire whether the reaction that takes place is of the precise nature of that here stated; but admitting, for the present, that it is, it would still be competent for us to ask if the quantity of ammonia condensed in this way was likely to suffice for the production of such decided effects as we frequently witness in connection with the crops that are assisted by gypsum.

Professor Liebig observes that a pound of sulphate of lime once converted into sulphate of ammonia, would introduce into the soil a quantity of ammonia equivalent to that which would be afforded it by 6.250 lbs. of horse's urine; a showing upon which it would be easy to demonstrate, taking the composition of sainfoin to be as I have shown it, that a pound of plaster fertilizing the ground to this extent, would be adequate to increase one hundred-fold the quantity of dry fodder produced.

According to my manner of viewing this question, it must be examined on a totally different basis. It is certain, for instance, that gypsum has no effect upon natural meadows; positive experience has satisfied me of the absolute inutility of the substance here; so that upon my natural meadows at Bechelbronn, I now never employ a particle of it. But let us review Professor Liebig's theory in connection with the production of sainfoin and clover, which in a general way derive an advantage from gypsum, which no one disputes.

Our harvest of clover, taken as dry, amounts on an average from strongly gypsed land, to 2 tons 1 cwt. very nearly per acre; and this quantity agrees pretty well with that which appears common in Germany. It is generally allowed that by gypsing we double the produce. It would follow from this, that an acre which had not been gypsed, would yield no more than $20\frac{1}{2}$ cwts. of dry clover; in my opinion the reduction would be still greater. Dry clover hay, made from the plant cut when in flower, contains about 2 per cent. of azote. The $20\frac{1}{2}$ cwts. of forage gained by the intervention of the gypsum would consequently contain 110 lbs. of ammonia, equivalent to 134.2 lbs. carbonate of ammonia. This consequently is the quantity of carbonate of ammonia which the gypsum ought to have been the means of procuring from the rain which falls upon an acre of land during the time that clover is upon the ground, in order to furnish the azote contained in the increased quantity of the crop.

Now in Alsace, from the time of gypsing in April, to the time of mowing in July, there falls on an average 3.92, nearly 4 inches of rain, which would amount in round numbers to 982 tons per acre. Were the azote of what may be spoken of as the surplus produce, derived from the rain in fact, all the water that falls ought to contain $\frac{1}{17000}$ of its weight of carbonate of ammonia. It is very questionable, however, whether any such proportion of ammoniacal salts exist in rain-water; yet the proportion ought to be very much greater, inasmuch as we have supposed the whole of the rain that fell to penetrate the ground, none of it to run off; but the truth is, that a very considerable proportion of the rain that falls never sinks into the soil; once the surface is thoroughly soaked, much that falls drains off, passes away by the ditches, and is lost with all it may contain that would prove beneficial to vegetation. It is in fact altogether impossible to make any approximation, even of the roughest kind, in regard to the quantity of rain-water that soaks into and that runs off the ground; and thus no kind of estimate can be formed of the relation between the moisture absorbed by plants, and that which escapes direct by the evaporation, without passing through them at all.

But even in admitting that it was really the ammonia contained in the rain-water, to which the very considerable increase of the crop of clover, lucern, and sainfoin was owing, it would still be left for us to explain wherefore, meteorological and other circumstances remaining the same, the same relative effects were not produced upon natural meadows covered with grasses, upon hoed crops, such as beet and turnips, and upon wheat; finally, the most serious objection that can be urged against this theory is founded upon the fact, that gypsum has no truly beneficial effect upon artificial meadows, save and except when the soil to which it is applied contains an adequate proportion of azotized organic manure. In a moderately manured soil, gypsum, as all the world knows, produces no sensible improvement; and as M. Crud, one of those men whom long experience has placed at the head of practical farming, said: It is to throw away both money and trouble to put gypsum upon an unkindly and impoverished bottom. It would seem, however, that if gypsum really fixes ammonia in the soil, in consequence of its action upon the rain-water that falls, converting its carbonate into sulphate of ammonia, the ammoniacal salt once introduced into the soil, ought to act independently and without the concurrence of another manure. That it really does act isolatedly, and of its own proper force when it exists, has been proved by the experiments of M. Schattenmann, who demonstrated on the large scale the beneficial effects of the sulphate of ammonia directly applied to natural meadows. It is obvious, that if the theory which I discuss be true, the greater number of practical observations which I have quoted must necessarily be false; or, on the contrary, these observations being accurate, the theory must be erroneous.

I have given reasons for maintaining the accuracy of the practical results; nevertheless, the better to establish this conviction, I have thought it advisable to add a few facts to the many that are already extant. I was, therefore, induced to undertake a series of experiments with a view to study, independently of all hypothetical idea, the action of gypsum upon certain hoed crops and cereals.

These experiments were made upon patches of land of 440 square yards each. Every precaution was taken to render the experiments strictly comparable one with another. Thus the ground appropriated to each particular crop was divided into three equal contiguous zones. The first zone, A, always received gypsum in the ratio of $4\frac{1}{4}$ bushels per acre. The second zone, B, and the third zone, C, were not gypsed. Each zone was sowed with the same quantity of seed, or planted with an equal number of beet plants or potatoes

A and C were the surfaces which I proposed to myself to contrast; the intermediate zone, B, was a kind of neutral ground employed merely to prevent the immediate contact of the gypsed with the ungypsed zone. I may here remark, that it would be well always to take such a precaution in making experiments on the effects of different manures.

In 1842 I tried the effect of gypsum upon wheat coming after three different crops. 1st. After clover ploughed in. 2d. After beetroot. 3d. After potatoes.

The gypsum was applied the 19th of May, at which time the wheat looked extremely well. The crop was cut between the 21st and 26th of July, and the following are the results obtained:

Crops.	Weight of grain, corn, and straw. A. piece gypsed.	B. not gypsed.	C. not gypsed.
Wheat after clover	319 lbs.	323 lbs.	327 lbs.
Wheat after mangel-wurzel	195 "	176 "	158 "
Wheat after potatoes	235 "	158 "	264 "
Average of the three experiments	250 "	248 "	250 "

The year 1842 having been unfavorable to wheat in consequence of the long drought, the experiment required to be repeated. This was done in 1843; and it must be allowed, that an experiment could scarcely be conducted under circumstances of weather more favorable to the cultivation of grain; the results here are given for equal spaces of three French acres, equal to 385 square yards. The gypsed zones had been treated with 70 lbs. of sulphate of lime each:

Year 1843.	Sheaves. lbs.	Grain. lbs.	Straw, chaff, and waste. lbs.
Rye with gypsum	516	137	379
Rye without	472	127	345
Wheat with gypsum	462	147	315
Wheat without	510	156	254
Wheat without	453	143	310
Oats with gypsum	329	112	217
Oats without	368	113	255.

From these numbers it is obvious that gypsum produces no appreciable effect upon wheat, oats, and rye, conclusions that agree with those come to in the previous year.

EXPERIMENT WITH FIELD-BEET OR MANGEL-WURZEL, OPENING THE ROTATION WITH MANURED SOIL, 1842.

The plants were transplanted and watered, and the gypsum was applied at the time of earthing up; a good deal of rain fell, and shortly after having been laid on, he gypsum had become incorporated with the ground. The crop was gathered on the 8th of October, three months after the gypsing, and from two equal surfaces, each of 242 square yards in extent, weighed as follows:

	cwt.	qrs.	lbs.
From the gypsed ground	13 cwt.	2 qrs.	6 lbs.
From the ungypsed	12 "	2 "	3 "

The gypsum would therefore appear to have had no beneficial effect; for the difference in favor of the gypsed piece is so trifling that it cannot be reasonably ascribed to the mineral manure: in fact, the quantity obtained from the gypsed surface does not exceed that which we constantly take from fields in the ordinary course of cultivation, and which have received no gypsum.

The action of gypsum, limited as it is to certain crops, will not allow us to admit that it produces its effect by fixing in the ground the carbonate of ammonia contained in rain-water; were it connected with any fixation of ammonia, it would be manifested generally, and not in particular instances only. Davy's theory therefore appears the more plausible, and requires discussion. Did the ashes of the clover grown in gypsed soils actually contain a large proportion of sulphate of lime, as affirmed by the illustrious English chemist, the action of gypsum would be readily understood. The whole question, therefore, seems to turn upon the composition of the ashes.

I have analyzed the ashes of clover grown at Bechelbronn, without and with the concurrence of gypsum. I shall here give the conclusions come to in 1841, a year remarkable for the heavy crops of clover, and those also for the year 1842, when the clover crop was but indifferent. The first table contains the results in the order in which they were registered; the second contains those obtained after the deduction of the carbonic acid and carbon which had remained in the ashes examined:

Carbon and Carbonic acid included.	Extraordinary Crop of 1841. Ashes of Clover.		Unfavorable Crop of 1842. Ashes of Clover.	
	Ungypsed.	Gypsed.	Ungypsed.	Gypsed.
Carbonic acid	14.2	22.1	21.5	26.8
Chlorine	3.4	2.9	2.5	2.2
Phosphoric acid	8.0	6.9	5.4	5.8
Sulphuric acid	3.2	2.6	2.4	2.3
Lime	23.7	22.4	25.4	26.7
Magnesia	6.3	5.1	5.6	7.4
Oxide of iron, manganese; alumina	1.0	0.6	0.5	traces.
Potash	19.6	27.8	22.5	25.3
Soda	1.0	0.7	2.2	0.2
Silica	16.8	7.9	10.0	2.7
Loss and charcoal	2.8	1.0	2.0	0.6
	100.0	100.0	100.0	100.0
Carbonic acid and Loss deducted:				
Chlorine	4.1	3.8	3.3	3.0
Phosphoric acid	9.7	9.0	7.1	8.2
Sulphuric acid	3.9	3.4	3.1	3.2
Lime	28.5	29.4	33.2	36.7
Magnesia	7.6	6.7	7.3	10.2
Oxide of iron, manganese; alumina	1.2	1.0	0.6	traces.
Potash	23.6	35.4	29.4	34.7
Soda	1.2	0.9	2.9	0.3
Silica	20.2	10.4	13.1	3.7
	100.0	100.0	100.0	100.0

The analyses here do not indicate the modes in which the various substances found were combined in the ashes ; but supposing that the whole of the sulphuric acid existed in combination with lime, which it most probably did, the preceding results would meet us in the following shape : the ashes of the clover grown upon soil without gypsum contain 6.0 per cent. of sulphate of lime ; those of clover grown upon a soil with gypsum, 5.7 per cent.

As it is impossible to answer for so small a difference as $3\frac{1}{1000}$ parts in researches of this kind, we must presume that the two ashes contained the same proportions of sulphate of lime.

Here, however, as in all other agricultural questions, isolated analyses throw but little light on the subject of inquiry. In order that they may enable us to arrive at any definite conclusion, two new elements must be taken into the discussion : 1st. The proportion of ash furnished by a given weight of the forage gathered ; 2d. The quantity of forage yielded by a given surface before and after the use of gypsum.

I have taken from my own observations the quantity of dry forage yielded by the two cuttings of 2d year's clover after gypsum, as amounting to 41 cwt. per acre. The same surface in the 1st year, and before the use of gypsum, would have produced but 9 cwt. 100 of dry clover gave :

	Year.	Ashes.	Ashes freed from carbonic acid per cent.	Per acre.
Clover ungypsed	1841	12.0	10.3	103 lbs.
Idem	1842	11.2	8.8	89 "
Clover gypsed	1841	7.0	5.4	248 "
Idem	1842	7.7	5.6	257 "

MINERAL SUBSTANCES IN THE CROP FROM $2\frac{4}{10}$ ACRES.

	Chlorine.	Phosphoric acid.	Sulphuric acid.	Lime.	Magnesia.	Oxide of iron, manganese, alumina.	Potash.	Soda.	Silica.	Ashes freed from carbonic acid.
Year 1841.	lbs.	lbs.	lbs.	lbs.	lbs.	lbs.	lbs.	lbs.	lbs.	lbs.
Fallow ungypsed	10·1	24·2	9·6	70·8	18·9	3·0	58·7	3·0	48·9	248
Ditto gypsed	22·6	53·2	20·2	174·6	39·8	5·9	210·3	5·2	61·8	594
Year 1842.										
Fallow ungypsed	6·6	15.4	6·6	70·8	15·6	1·3	62·9	6·1	27·9	234
Fallow gypsed	18·4	50·3	19·8	226·1	62·7	—	213·8	1·7	22·8	616

It is therefore obvious, that in the course of the three months which followed the application of the gypsum, the soil must have supplied the plant with very considerable quantities of mineral substance ; the crops taken from the gypsed soils contained in fact two or even three times the quantity of these substances which those grown previously to the gypsing contained. Representing, for example, by unity the quantity of the several bases and acids of the crop grown without gypsum, we should have the quantities of the same principles contained in the crops produced upon the gypsed soils represented by the following numbers:

	Chlorine.	Phosphoric acid.	Sulphuric acid.	Lime.	Magnesia and metallic oxide.	Potash and soda.	Silica.
1841	2.2	2.2	2.1	2.5	2.1	3.5	1.0
1842	2.8	3.3	3.1	3.1	3.7	3.2	1.0

Silica appears to form the only exception here, which would lead us to conclude that this earth was only absorbed by clover in the first period of its growth. Potash and lime are the bases which enter in largest proportion into the mineral constitution of clover; and there is another fact made evident which deserves particularly to fix attention: it is that the lime assimilated subsequently to the gypsing, bears no kind of relation to the quantity of sulphuric acid fixed during the same space of time. The excess of acid and of lime obtained from the ash of the gypsed clover over that of the ungypsed, is for:

1841,	Sulphuric acid	4.8	Lime	47.2
1842,	"	6.0	"	70.6

Supposing further, that the sulphuric acid assimilated subsequently to the gypsing was taken up in the state of sulphate of lime, we find that:

In 1841 the gypsed crop absorbed 18 lbs. of this salt.
In 1842 " 22 lbs. "

These quantities are so small as to lead us to suppose that the utility of gypsing consists in furnishing the plant with the large proportion of lime which it seems to require. Gypsing would then be equivalent to the application of lime; and in fact, according to Schwertz, Paris plaster is replaced in Flanders by slaked lime, by the lye-washed ashes of wood, and by peat-ash, with decided advantage.* Some peat-ash contains sulphate of lime, others none at all. What is employed successfully, most likely presents sulphuric acid in the state of an alkaline sulphate.

Wood-ash, which is certainly the best manure for artificial meadows, may contain upon an average one per cent. of sulphuric acid, and when lye-washed, the proportion ought to be much less; if perfectly washed, it ought to be null; at all events, there is no sulphate of lime present to fix the ammonia of the rain-water. Independently of earthy phosphates, so useful to all plants, lye-washed ashes frequently yield more than 80 per cent. of chalk. We thus perceive, in a general way, that the manures which stimulate the vegetation of clover are always calcareous, the lime being either in the state of sulphate or carbonate, which exists abundantly in the crops, combined with organic acids, and freed consequently of nearly the whole of the inorganic acid with which it was originally associated. Assuming that gypsum acts like chalk, it may be conceived that when

* Schwertz, Culture des Plantes fourragères, p 72

he former is incorporated with the indispensable manure, it is de-omposed, and carbonate of lime, in a state of minute division, and or that reason easily absorbed, is the result. It is only upon this upposition that I can understand the elimination of the sulphuric cid of the gypsum; for if the lime really entered the vegetable in he state of sulphate, the ashes ought to be much richer in that acid han analysis shows. This same difficulty occurs in the hypothesis f Liebig. If the 56 lbs. of ammonia derived from the atmosphere enetrated the plant in the form of sulphate, there must enter at the ame time 130 lbs. of sulphuric acid, and which ought to be recover-d in the ashes of the crop from one acre. Now, the ashes of 41 wts. of gypsed clover, abstracting the carbonic acid, weigh 2 cwts. lbs., containing in the 100, $3\frac{1}{3}$ of sulphuric acid. But the amount f ash, were the acid of the ammoniacal sulphate fixed in the crop, vould rise to 3 cwt. 1 qr. 20 lbs., and the ash would then contain 0 per cent. of sulphuric acid.

Before promulgating this last objection against received theories, I hought it right to ascertain whether the ashes contain, in the state f sulphate, the whole of the sulphur pre-existing in the incinerated lant. For it was not impossible that at the high temperature em-loyed, the silica might react upon the sulphates so as to expel a ortion of the sulphuric acid. However improbable this expulsion, wing to the great excess of potash always present in clover ash, t seemed expedient to determine the fact.

After having made out as exactly as possible the quantity of ash eft by the hay, and also the sulphuric acid, I took a certain weight of the same hay, burned it in a platina crucible along with a mixture of chlorate and carbonate of potash, and then sought for the sulphu-ic acid in the product of the ignition.

1000 parts of the plant furnished directly 3 of sulphuric acid, and y analysis of the ash, 2.8.

Thus, the alkaline ashes retain all the sulphur pre-existing in the lant which produced them.

I have laid stress upon the small proportion of sulphuric acid in a crop of clover, because there yet remains for consideration a third theory of gypsing, which I have helped to propagate, although doubtful concerning the author. This is founded upon the assumption that the proportion of sulphur is much greater in the leguminous than in the cereal tribe. Now, as gypsum is generally adapted to the manurement of leguminous plants, the origin of the sulphur has been ascribed to sulphate of lime incorporated with the soil. This view appeared the more plausible to M. Dumas and myself, inas-much as in accordance with it plants operated as reducing agents. It is, besides, very probable that sulphur, as an immediate constituen principle of vegetables, is derived from sulphates; but do leguminous plants really contain more than the cereals? This seems doubtful since careful investigation of the azotized principles of plants has shown gluten, caseine, and legumine to be nearly identical in com-position. I moreover find upon analysis of the ashes, that clover haricots, and beans do not sensibly contain more sulphur than rye, wheat, oats, and potatoes. There appears to be no doubt, therefore that the sulphur required by plants is supplied abundantly by the soil enriched with ordinary manure, as happens in the culture of the cereals, roots, and tubers.

In a word, it may be presumed that Paris-plaster acts usefully on artificial meadows by introducing lime into the soil. This is consistent both with the analysis of the ashes of the crops produced and of the soil; for according to the researches of M. Rigaud de l'Isle, gypsum operates only upon soils which do not contain a sufficient dose of lime in the state of carbonate.*

OF AMMONIACAL SALTS.

The last products of the putrefaction of azotized matters being ammoniacal combinations, it necessarily follows that salts having ammonia for their base, must act usefully in vegetation. This is confirmed by the employment of guano, and by experiments in which ammoniacal compounds have been directly applied as manure. I have already pointed attention to the observations of Davy relative to the favorable effect of carbonate of ammonia upon the development of plants, and shall now detail some recent trials made by M. Schattenmann with the sulphate and muriate of the same base.

These salts were introduced into the soil as a solution marking one degree of Beaumé's areometer, and in the dose of 102 bushels per acre. In 1843, the effects produced upon wheat by muriate and sulphate of ammonia were most distinct; as was also the case with natural meadows, which yielded under the influence of this liquid manure 82 cwts. of hay per acre, precisely double the crop afforded by the same meadow land without the salts of ammonia; but another important fact, and which M. Schattenmann announces with confidence as having been proved by repeated trials, is this; that solution of sulphate of ammonia employed in the same dose, and at the same degree of concentration, causes no appreciable melioration upon trefoil and lucerne. The result of a solution of sal-ammoniac was equally negative.

These observations agree in certain points with those formerly made by Rigaud de l'Isle, and more lately by M. Lecoq. But they are in direct opposition with the experiments of several physiologists who have studied the action of ammoniacal salts presented separately to vegetables, a circumstance very different from that wherein ammoniacal solutions are incorporated with arable land. Thus, M. Bouchardat† has stated that young plants of *mentha aquatica* and *sylvestris*, and of *mimosa pudica* die very soon, when made to vegetate with the roots plunged in very weak solutions of muriate, nitrate, and sulphate of ammonia. In offering, some years back, divers considerations upon guano, I promulgated the opinion that ammoniacal salts, in order to serve as azotized manure, must always contain organic acids or carbonic acid. Perhaps the term *usefu-*

* Mémoires de la Société d'Agriculture, année 1844.
† Bouchardat, Comptes rendus de l'Académie des Sciences, tom. xvi.

salt may be now restricted to the carbonate alone. At least, having watered young plants of trefoil, grown in silicious sand, with solution of oxalate of ammonia at $\frac{1}{600}$th, I observed them die after the lapse of eight or ten days. Plants of the same size, sown under like conditions, but irrigated with distilled water, continued to grow, and flowered.

In treating of gypsing, I have assigned my reasons against admitting that the azote fixed during the culture of trefoil, proceeds from the sulphate or muriate of ammonia naturally absorbed. I again assert, that it is materially impossible that ammoniacal salts combined with inorganic acids, other than the carbonic, can be useful as manure to plants, when administered separately; they can only become advantageous when their composition has undergone modification.

Two cwt. (220 lbs.) of wheat-sheaves, (straw and grain,) contain upon an average 2 lbs. 1 oz. 14 dwts. of azote, and leave after ignition 11 lbs. 3 oz. 16 dwts. of ash, into which enter 1 oz. 7 dwts. of sulphuric acid, and 14 dwts. of chlorine.

In the ammoniacal salts:

100 of azote corresponds to 283 of sulphuric acid.
" " 257 of chlorine.

Let us now consider sulphate of ammonia, which, according to the experiments of M. Schattenmann, supplied an excellent manure for wheat. If the 2 lbs. 1 oz. 14 dwts. of azote contained in the 2 cwt. of sheaves be derived from the sulphate absorbed by the cereal, the sulphuric acid of the sulphate ought to be recovered in its ashes; and according to the above standard, these ashes ought to contain 6 lbs. 16 dwts. of sulphuric acid. They afforded by analysis only 1 oz. 7 dwts.

Applying the same reasoning to muriate of ammonia, we find, supposing the azote of the 2 cwt. of sheaves to emanate from this salt, that the ashes should contain 5 lbs. 6 oz. 2 dwts. of chlorine; whereas they really contain but 14 dwts.

Without doubt, the nitrogenous principles of the cereal cannot be referred solely to the ammoniacal salts in the trials of M. Schattenmann; the manure given to the soil, and the atmosphere must have contributed a share. The appreciation of the value of ammoniacal salts becomes more precise when the results obtained on meadow land are estimated. There the produce was doubled, and of every 2 cwt. of hay gathered, 1 cwt. may be ascribed to the action of the salt.

Two cwt. of hay, containing 4 lbs. 16 dwts. of azote, leave 16 lbs. 2 oz. of ash.

If the half (2 lbs. 8 dwts.) of the azote of the fodder comes from the ammoniacal salts, the ashes will contain:

lbs.	oz.	dwts.	
5	8	4	of sulphuric acid, if the sulphate has been used.
5	2	0	" " if the muriate has been used.

Now, the ashes of the hay yielded by analysis:

oz.	dwts.	
5	9	of sulphuric acid.
5	4	of chlorine.

It is then very probable that if the ammoniacal salts afford azote to the plants, they enter not as muriate, sulphate, or phosphate, because there is no reason to believe that acids united with an alkali are eliminated almost in totality during the act of vegetation. It necessarily follows, that the ammonia of these salts, in order to yield to vegetables its constituent azote, must reach their organs in the form of carbonate, inasmuch as that is the sole ammoniacal salt which seems to exercise a direct and favorable agency.

However, if such be the case, how comes it to pass, that ammoniacal salts, as the muriate, phosphate, and sulphate, are converted into carbonate when once incorporated in the soil? Good arable land almost always contains, it is true, carbonate of lime; but there is no ground for admitting an acid interchange betwixt the calcareous and ammoniacal salts. We know, on the contrary, that carbonate of ammonia reacts instantaneously upon muriate and sulphate of lime, the products of this reaction being on the one hand muriate and sulphate of ammonia, on the other, carbonate of lime. The gypsum theory of Liebig is based upon the fact of this double decomposition, whereby the ammoniacal carbonate of rain-water is fixed in the state of sulphate at the cost of the sulphate of lime put on the ground as manure.

This reaction of sulphate of lime upon carbonate of ammonia is incontestable as a laboratory experiment; but in well-tilled grounds containing just the proper quantity of moisture, the reaction takes place in the inverse sense. The carbonate of lime reacts upon the sulphate of ammonia, and there result carbonate of ammonia and sulphate of lime.

This fact, however singular at first sight, is explained upon principles established by Berthollet in his chemical statics:

When two saline solutions are mixed together, and from the mixture an insoluble salt results, the insoluble compound is formed and precipitated. This is what happens on pouring a solution of carbonate of ammonia into one of sulphate of lime. But if, instead of bringing the two salts together dissolved, they are mixed in a pulverulent state, and just sufficient water is added to promote the reaction without dissolving the products, a volatile compound forms and is evolved—namely, carbonate of ammonia.

The experimental proof is easy, and not without interest. If chalk previously washed be intimately mixed with crystallized sulphate of ammonia, no change ensues, provided the powders are very dry. Let moist sand be introduced so as to impart to the mixture the consistence of light arable land of the usual humidity; at the very instant vapors of carbonate of ammonia, cognizable by their action on vegetable colors and their odor, are developed. When water is added in excess, the disengagement of ammoniacal vapor immediately ceases. The carbonate of ammonia not yet volatilized, dissolves and acts upon the ready formed gypsum so as to constitute sulphate of ammonia and carbonate of lime. The ordinary reaction is restored. Finally, this watery mixture being exposed to the air, furnishes anew ammoniacal vapors in proportion as the water vapor izes, and the volatile salt is progressively evolved until the mass is quite dry. In maintaining a similar mixture in a fit and constant state of moisture we may in two or three days, under the influence of a temperature of from 68° to 80° F., dissipate the greater portion of the ammonia of the sulphate, and obtain a quantity of sulphate of lime indicating the progress of the reaction.

It is almost needless to remark that sulphate of lime, placed in a condition of moisture analogous to that of the chalk, undergoes no alteration from the carbonate of ammonia. Thus, in the ordinary circumstances of humidity of cultivated land, it is at least doubtful whether the plaster diffused through it definitively retains the ammonia of the rain-water in the state of sulphate.

To estimate the sulphate of lime produced by the reaction between the sulphate of ammonia and carbonate of lime, the mixture after having been entirely dried in the air is to be treated with cold water. The lime of the sulphate dissolved is next thrown down by oxalate of ammonia, and computed in the state of carbonate.

In order to ascertain the presence of sulphate of ammonia the mixture is to be digested in dilute alcohol, which dissolves this salt without taking up sulphate of lime.

Muriate, phosphate, and oxalate of ammonia comport themselves as the sulphate. Carbonate of lime decomposes them under the same circumstances, giving rise to equivalent products.*

The preceding facts may perhaps tend to reconcile the contradictory results obtained in the application of ammoniacal salts as manure. When muriate, phosphate, or sulphate of ammonia is presented to plants, these salts produce no useful effect; they are ab-

* 1. 15.4 grains of crystallized sulphate of ammonia incorporated with five or six times its weight of chalk, gave after two days' exposure of the moist mixture in the open air, 12.3 grs. of sulphate of lime. In another experiment 16.0 grs. of sulphate of ammonia were obtained from 13.1 grs. of calcareous sulphate. Now according to the proportions of ammoniacal salt there ought to have been obtained 13 grs. and 14 grs. of sulphate of lime. Thus about 9-10ths of the ammonia contained in the sulphate submitted to experiment had been converted into carbonate.

2. 16.9 grs. of sulphate of ammonia were mixed with 123.2 grs. of chalk, and from 924 to 1071 grs. of silicious sand. The mixture was kept moist and exposed to the air during four days, the temperature ranging from 68° to 80° F. The substance, after being dried by the action of the air, was set to digest in weak alcohol; the alcoholic liquor yielded only an insignificant trace of sulphate of ammonia. There was about 18.5 grs. of sulphate of lime.

3. A very simple means of determining the reaction in question consists in plunging a fragment of chalk into a solution of sulphate of ammonia; the fragment so imbued on being exposed to the air emits during several days fumes of carbonate of ammonia. Several bits of chalk moistened in the solution, and exposed by the aid of an aspirator to a continuous current of air, afterwards washed in muriatic acid, disengaged enough of ammoniacal vapor to form in the acid nearly 15 grs. of sal-ammoniac.

4. With a slightly moistened silicious sand were incorporated 30.8 grs. of gypsum; the mixture then watered with a solution containing 30.8 grs. of carbonate of ammonia, was allowed to remain in the air during eight days, having always the consistence and humidity of arable land. At the end of this time it was dried, then treated with weak alcohol; the alcoholic fluid evaporated left no sulphate of ammonia.

5. Phosphate of ammonia, mixed with chalk, and kept in a moist state for some days, comported itself exactly as the sulphate in the same circumstances; 9-10ths of the ammoniacal phosphate were converted into phosphate of lime.

6. The reaction of oxalate of ammonia upon chalk is visible even when the mixture is well watered; this is explained indeed by the powerful affinity of oxalic acid for lime. The totality of the alkaline oxalate is promptly changed into oxalate of lime and carbonate of ammonia.

sorbed in limited quantity, like the majority of soluble substances But if instead of administering them separately dissolved in water, as was done in the physiological experiments, they are incorporated with a loose and humid soil, these salts react upon the calcareous matter almost always existing in the ground, and are transformed into carbonate of ammonia, which exerts undeniably a favorable influence upon vegetation. From these facts it may be presumed that the introduction of lime and marl is not merely to supply the defective calcareous element, but likewise a principle, carbonate of lime, which produces a particular action upon the manure, changing, through double decomposition, the unassimilable ammoniacal salts there present into a carbonate capable of being assimilated, which transmits to the plant the azote of the organic matter of the dung and the carbon contained in the calcareous rocks.

These reactions which go on between soluble salts and one that is insoluble under the peculiar conditions united in arable land, show that we must not always conclude as to what passes in the ground from phenomena observed in the laboratory of the chemist; and it is probable that by extending the study of these singular reactions to alkaline salts generally, we shall better understand the mode of action and utility of saline substances in agriculture. Thus, for example, the operation of common salt as a fertilizer is still very obscure. Many skilful husbandmen question its efficacy; nevertheless, when moderately employed it seems to do good. In plants growing on the sea-coast soda is found in a great measure combined with organic acids, and the chlorine deduced by analysis from their ashes is nowise proportional to the alkali they contain. The whole sodium does not enter the vegetable as a chloride, but very likely as carbonate of soda, and that in virtue of a reaction analogous to the one which calcareous matter has upon ammoniacal salts.

It is quite certain that chloride of sodium in solution is not affected by carbonate of lime; but then it was proved by Clouet that if into sand moistened with this same solution powdered chalk be put, and the mixture left in contact with air, an efflorescence of sesquicarbonate of soda ere long makes its appearance. Thus by the conjoint effect of capillarity and the carbonic acid of the atmosphere, common salt in the conditions above mentioned undergoes by contact with chalk a partial decomposition, of which the result is carbonate of soda, a salt, like carbonate of potash, most favorable to the growth of plants. Accordingly, in furnishing sea-salt to a soil sufficiently calcareous, we really enrich it with carbonate of soda. We moreover perceive that the same salt diffused through land devoid of carbonate of lime may not produce any fertilizing effect.

OF WATER.

Water is not only indispensable to the life of plants, but likewise promotes vegetation after the manner of a manure, on account of the saline or organic substances it generally holds in solution. Rain is the source of the soft waters which flow in rivers, spring from the soil, or constitute lakes. Rain-water although nearly pure is not absolutely exempt from extraneous matters. The air, especially after continued drought, always holds dust in suspension; this yields to the rain by which it is precipitated, whatever soluble matter it may contain.

It is further ascertained by the experiments of Cavendish and Séguin, that whenever the electric spark traverses a humid mixture of oxygen and azote, nitric acid and nitrate of ammonia are produced. Now this frequently happens; and according to Professor Liebig storm-rain always contains nitric acid associated with lime or ammonia. Common rain seldom contains nitrates, merely faint traces of common salt.*

In river and spring-water there necessarily exists a larger amount of dissolved substances derived from the strata they pass through, varying in nature according to the geological structure of the locality. From old crystalline rocks, like granite, water issues sometimes so little impregnated with salts, as to be almost identical with distilled water; that, on the contrary, which rises from a calcareous or gypseous bed is always contaminated with salts of lime. Notwithstanding the minute quantity of saline or earthy ingredients in spring and river-waters, they are drinkable, and considered good when they are limpid, without odor, capable of dissolving soap, and fitted for vegetable cookery. These two last characters are essential, inasmuch as proving that the waters contain only infinitesimal quantities of soluble salts of lime.

The action of tests readily indicates the nature of the dissolved salts.

Water contains: *sulphates* or *carbonates*, if nitrate of barytes causes a precipitate; a sulphate, when the precipitate is not redissolved by the addition of nitric acid;

Chlorides, if it give with nitrate of silver a curdy precipitate, insoluble upon addition of nitric acid;

Lime, when rendered turbid by oxalate of ammonia;

Magnesia, if when mixed with pure ammonia, and preserved in a closely stopped vial, a white flocculent deposite ensues. This test, however, is only applicable to water that has been boiled sufficiently long to expel all the carbonic acid in solution, and which would tend to hold any carbonate of lime dissolved. Carbonate of lime is separated from water by ammonia, after some hours, in the form of granular crystals, which adhere to the sides of the vessel.

In order to render the operation of tests more sensible, the bulk of the water may be reduced to a half or a fourth by evaporation.

Besides fixed salts, river-water always contains those of ammonia, particularly the carbonate; this fact was first ascertained, relative to the Seine water, by M. Chevreul.† Subsequently, Professor Liebig has discovered the same ammoniacal salt in rain-water; and M. Hunefeld has proved, that spring-water likewise con-

* Annales de Chimie, t. xxxv. 2e série.
† Chevreul, Annales de Chimie, t. lxxxii. p. 56.

tains it.* Lastly, M. Hermann has even determined quantitatively, carbonate of ammonia in the ferruginous waters of a turf-pit. The water of the Nile is not exempt from it, judging at least from the analysis of its mud. According to Regnault, 100 parts of this mud dried in the air contain :†

Chloride of sodium, sulphate of soda, and *carbonate of ammonia*	1
Organic matter	9
Water	10
Oxide of iron	6
Silica	4
Alumina	48
Carbonate of lime	18
Carbonate of magnesia	4
	100

The beautiful synthetic experiments of M. Dumas demonstrate that water is formed of:

Oxygen	88.89
Hydrogen	11.11

When pure, it boils at a temperature of 212° F. under a barometric pressure of 30 (29.921) inches. It congeals at 32° F.

All natural bodies dilate, augment in volume, by the action of heat, and contract under diminution of temperature. Water is amenable to this law between rather wide limits; it deviates, however, and presents an anomaly as it approaches congelation. As with all liquids, the density of water gradually increases in proportion as it cools, until its temperature is 39°.38 F. Setting out from this point the density diminishes, the liquid dilates more and more, so that at 32° it occupies nearly the same volume that it did at 49°. From this remarkable property, it results that during the most intense cold the stagnant water which covers the meadows rarely attains a lower temperature than 39°, whereby the organs of plants suffer no damage.

Let us suppose, in fact, that at the beginning of winter a sheet of stagnant water has a temperature of about 54°; in proportion as the liquid at the surface cools, it becomes denser, descends, and is immediately replaced by inferior layers, which rise in the ratio of their less density; but these new superior layers, subjected to the same refrigerating cause, contract and descend alternately. There is then established in the fluid molecules, movements of ascension and descent, of which the result is the cooling of the entire mass. Let us now admit that in virtue of this continued mingling of the cooled superior layers with those below, the temperature of the sheet of water is lowered to 39°.38; at this degree of the thermometer, the water acquires its maximum density; in parting with its heat it not only contracts no more, but becomes lighter. If then a body of stagnant water at a temperature of 39° is exposed to the chilling action of the atmosphere, the superior layer, far colder than the inferior, will no longer descend, since it will become lighter as its

* Liebig, Traité de Chimie, Introduction, p. cii.
† Description de l'Egypte, t. ii. p. 406.

temperature diminishes. Thus it is that the water of a pond or lake freezes at the surface, while it preserves beneath a temperature some degrees above 32°. In a situation where the temperature of the air was 29°, Davy found the thermometer indicate 43° in the herbage of an inundated meadow completely covered with ice.*

Water is always impregnated with atmospheric air, and a minute quantity of carbonic acid. Deprived of air, it is not agreeable to drink; it is even said, when long continued, to prove unwholesome if the dissolved gases are expelled by ebullition. River-water usually contains $\frac{1}{30}$th in volume of air, and $\frac{1}{30}$th carbonic acid. In spring-water, the amount of the latter is sometimes far more considerable.

The quantity and nature of saline ingredients in drinkable water vary much: in an agricultural point of view, the study of the contained salts would certainly be useful. The waters which serve as drink to the cattle of a farm, introduce into the dung-heap all the matters which are dissolved or held in suspension. At Bechelbronn, for example, I find that more than 2 cwts. of alkaline salts get into the dung in this way every year. When a farmer has the choice of several waters for giving his cattle or irrigating his meadows, he will do well to select that which is richest in alkaline salts, and still good to drink. In the *steppes* of America, it is astonishing with what discernment the cattle choose waters for allaying their thirst, containing minute quantities of sulphate of soda or common salt.

I close these considerations with a tabular view of the most recent analyses. The quantities of salts put down have been deduced from 100,000 parts of water for drinking.

* Davy, Agricultural Chemistry, p. 352

SOURCE OF THE WATERS.	Carbonate of lime.	Carbonate of magnesia.	Silica.	Sulphate of lime.	Sulphate of magnesia.	Sulphate of Soda.	Chloride of calcium.	Chloride of magnesium.	Chloride of sodium (marine salt.)	Nitrates.	Organic matter.	Total weight of matter.	AUTHORS OF THE ANALYSES.
Of the Seine above Paris .	11.3	0.4	0.5	3.6	0.6	. .	1.0	0.8	. .	traces	traces	18.2	Bouchardat
Marne	10.5	0.9	0.6	3.1	1.2	. .	. .	1.7	. .	. .	idem	18.0	Ditto
Ourcq at St. Denis . .	17.5	2.0	2.0	15.3	7.0	. .	. .	4.0	traces	. .	idem	47.8	Ditto
Yonne at Avallon . .	4.3	. .	1.9	traces	. .	. .	1.5	. .	idem	. .	idem	7.7	Ditto
Benvronne	25.7	. .	. .	20.3	. .	. .	8.5		. .	. .	. .	54.5	Colin
Thérouenne	26.2	. .	. .	2.0	. .	. .	3.6		. .	. .	. .	31.8	Ditto
Gergogne	18.0	. .	. .	1.5	. .	. .	1.5		0.9	. .	. .	21.9	Ditto
Bièvre near Paris . .	13.6	. .	. .	25.1	. .	. .	10.9		1.2	. .	. .	50.8	Ditto
Arcueil	16.9	. .	. .	16.9	. .	. .	11.0		1.9	. .	. .	46.7	Ditto
Spring of Roye (Lyons) .	23.8	. .	traces	1.4	. .	. .	. .	. .	1.2	traces	traces	26.4	Boussingault
Fountain Spring (Lyons) .	23.4	. .	idem	1.7	. .	. .	1.3	traces	0.2	. .	idem	26.6	Dupasquier
Rhone at Lyons (July) .	10.0	. .	idem	0.6	traces	. .	traces	idem	traces	. .	idem	10.6	Boussingault
Ditto ditto (February) .	15.0	. .	. .	2.0	0.7	. .	0.7			of lime	idem	18.4	Dupasquier
Spring of the garden of plants at Lyons . . .	27.0	. .	. .	25.2	. .	. .	16.8	1.6	12.6	7.6	strong traces	90.8	Ditto
Of Lake Geneva . . .	7.2	0.7	0.1	2.6	3.1	. .	. .	0.9	. .	. .	0.6	15.2	Tingry
Of the Arve (in August) .	5.2	0.4	0.1	3.2	2.9	. .	. .	0.7	. .	. .	0.3	12.8	Ditto
Ditto in February . . .	8.3	1.2	0.2	6.5	6.2	. .	. .	1.5	. .	. .	0.4	24.3	Ditto
Loire near Orleans . . .	1.7	. .	. .	. .	. .	. .	5.1		traces	. .	. .	6.8	Guindant
Loiret	11.9	. .	. .	3.8	. .	. .	10.2		2.5	. .	. .	28.4	Ditto

The water of the Artesian well at Grenelle, near Paris, according to the analysis of M. Payen, contains, in 100,000 parts :

Carbonate of lime	6.80
Carbonate of magnesia	1.42
Bicarbonate of potash	2.96
Sulphate of potash	1.20
Chloride of potassium	1.09
Silica	0.57
Yellow matter, not defined	0.02
Organic azotized matter	0.24
	14.30

CHAPTER VII.

OF THE ROTATION OF CROPS.

§ 1. OF THE ORGANIC MATTER OF MANURE AND OF CROPS.

It is known that the atmosphere and the organic matters diffused through the earth concur simultaneously to maintain the life of plants ; but how far each contributes is undetermined. We shall now study the theory of the exhaustion of the soil by culture, and the rotation of crops.

When a succession of crops is grown upon fertile land without renewal of manure, the produce gradually diminishes ; and after a certain period, if it be grain, the quantity which at the outset was eight or nine times the amount of the seed, will be reduced to three times or even to twice the seed. Thus crops impair the fertility of the soil, and eventually exhaust it.

It has been long admitted that different species of plants manifest great diversity in their powers of exhaustion. Certain kinds, indeed, as trefoil and lucerne, far from exhausting it, communicate new vigor. As a general rule, however, every plant may be said to impoverish the soil in which it grows. This impoverishment is always manifest when the plant after maturity is completely removed, but is less sensible when much rubbish is left. Thus, for example, clover, after yielding two crops, which are generally cut as fodder, might still yield a third ; this last, however, is generally ploughed into the ground as manure, being buried along with a considerable quantity of roots. This plan of meliorating the soil by the cultivation of trefoil is what is called *manuring by smothering ;* a method practised from a remote period in the south of Europe, and which offers decided advantages in those districts where there is abundance of pasture land. Hence, in smothering trefoil, the soil is amended at the expense of the nutritive matter it contains.

Thaer, who endeavored to make theory and practice mutually agree, laid it down as a rule, that the exhaustion occasioned by cropping is proportioned to the amount of nutriment in the crops, estimating the nutritive value according to Einhof's determination. But the above deduction is founded upon error.

In fact, to adopt the above principle is tacitly admitting that the whole organic matter of plants originally comes from the soil. This, no doubt, contributes in a certain proportion to the development of plants, but so also do air and water. On the other hand, physiologists, in opposition to the ideas of the school of Thaer, have perhaps exaggerated the material withdrawn from the air. Thus, M. de Saussure reckons that a sun-flower derives from the ground during its growth not more than $\frac{1}{20}$th of its weight, supposing the plant dry. The reasoning upon which he formed his conclusion is based, on the one hand, upon a knowledge of the extractive matter of garden-mould; on the other, upon the quantity of water a plant like sun-flower may absorb in a given time, to return it again to the air by transpiration.*

Little objection could be urged against the above conclusion, did not the experiments of M. Gazzeri tend to prove that roots virtually exercise, by their contact with solid organic matter, an incontestable absorbent action in imparting solubility.† I might refer to an observation of M. de Saussure, in which he states that plants grown in garden-mould deprived of its soluble components by repeated washing, reached, nevertheless, perfect maturity, although the produce in seed was less abundant than it might have been.‡ It is most probable that both parties have promulgated extreme opinions. Plants possibly draw from the atmosphere more than agriculturists commonly suppose, and the soil furnishes, independently of saline and earthy substances, a proportion of organic matter larger than certain physiologists admit. There is every reason to believe, from what I could learn respecting guano during my sojourn on the coast of Peru, that the greater part of the azotized principles of plants originates in the ammoniacal salts which exist or are formed in manure.§

In discussing the advantage of one course of crops over another, the question always hinges upon that of exhaustion. Wherever an unlimited supply of dung and of handiwork can be procured, there is no absolute necessity for following any regular system of rotation. Under such favorable circumstances, it is expedient to ascertain what kind of cultivation is, commercially speaking, best suited to the climate and the soil. There is little to fear that by a continued succession of similar crops, the fields will get infested with noxious weeds, because this inconvenience may be obviated by labor. Nor is impoverishment of the soil to be dreaded, since that can be remedied by the purchase of manure. The whole craft of agriculture is reducible to comparison of the probable value of the crop with the cost of manure, labor, &c. Farming of this sort excludes the

* Saussure, Recherches Chimiques sur la Végétation, p. 268.
† Annales de l'Agriculture Française, No. iii. p. 57.
‡ Saussure, Recherches Chimiques, p. 171.
§ Annales de Chimie, t. lxv. année 1837.

keep and propagation of cattle, and may be strictly regarded more as gardening than as agriculture.

But where manure cannot be had from without, things must be reduced to a system; and the amount of produce which it is possible to export each year is fixed within bounds, which cannot be exceeded with impunity.

When by judicious cultivation land is rendered fertile, it is necessary, towards securing its fertility, to supply after every succession of crops equal quantities of manure. In considering this in a purely chemical point of view, it may be said that the produce which can be taken away without damaging the fertility of the land, is the organic matter contained in the crops, abstraction made of that present in the manure. Indeed, this latter substance must in some form or other return to the soil to fecundate it anew. It is capital placed in the ground, the interest of which is represented by the commercial value of the produce of all the other agricultural operations.

Where lands are extensive, population scattered, and means of communication difficult, there is less necessity for being tied down to systematic cultivation. There is always enough for a scanty population. A field yields grain, and after the harvest is converted for a series of years into meadow-land; such is the pastoral system in all its simplicity. To this primitive state of husbandry may be referred those plantations on cleared land in countries covered with forests. When the trees are felled and burned upon the spot, the soil yields for long and without manure, crops of maize and of wheat of surprising quality, at the cost of the fecundity acquired during ages of repose.

But when from increased population the land becomes more valuable, a larger amount of produce is demanded. Imperfect culture would prove inadequate. Accordingly a triennial rotation of crops was very anciently adopted in the north of Europe, consisting as is well known of fallow land frequently ploughed during summer, followed by two years of grain. The fallow land received a certain quantity of manure to repair the exhaustion occasioned by the two crops of grain; hence when this mode of rotation is adopted there should be always sufficient meadow-land to supply manure.

Leaving waste one third of the surface has always been held a grave objection against triennial rotation. Hence various attempts have been made to get rid of the summer fallow. Some encouragement was given to these attempts from what occurs in horticulture, where the ground is rendered continually productive.* In certain countries, moreover, tillage is only interrupted by severe weather.

On the other hand, it has been long remarked that it is not always beneficial to grow grain during several consecutive years in the same ground, even when it is fertile and manure is abundant, owing to the almost insurmountable difficulty of destroying weeds. The fallow was justly considered the most efficient and economic means of getting rid of these. For this purpose *fallow-crops*, as they were

* Thaer, Agriculture raisonnée.

called, were introduced. Peas, beans, vetches, were at first the only plants used as fallow-crops.

However, it was soon perceived that the fallow-crops occasioned a very sensible diminution in the produce of corn; to counteract this inconvenience recourse was had to a surcharge of manure; but as this cannot always be obtained, it was necessary either to reduce the cultivated surface or to appropriate a certain amount of meadow. Still the fallow-crops had this advantage, that they enabled the farmer to derive from land a greater amount of produce in a given time without prejudice to the raising of corn. Hence the plan of turning the fallow to account was soon generally adopted.

The introduction of clover so modified the system of fallow-crops as at one time to induce the belief that the point of perfection had been attained in agriculture.* This was when it was ascertained that trefoil, which had hitherto been only cultivated in small enclosures, might be sown in spring upon corn land, and occupy next year the place of the fallow in the triennial rotation. Trefoil, so far from exhausting the soil, was found to give it new fertility, and the succeeding corn crop yielded a plentiful harvest.

It may be easily conceived what advantages were expected in substituting for the unproductive fallow the cultivation of a plant which did not impoverish the land, and furnished a quantity of excellent fodder that served as food for an additional number of cattle. It was even alleged that this plant cleared the fields of weeds.

A few years' experience sufficed to show that trefoil did not possess all the advantages attributed to it. On renewing the clover every third year on the same piece of ground it sometimes failed. Schubarth, the most zealous and enlightened advocate for its use, limited the renewal of the artificial meadow at first to the sixth, and eventually to the ninth year; and finding that it did not completely destroy the weeds in corn, he had recourse to hoed-crops for that purpose.

The introduction of trefoil has gradually led to the system of alternate rotation of crops generally adopted at present; and moreover, contrary to the anticipations of Schubarth, it may be renewed every four or five years on the same parcel of land.

The impossibility of substituting trefoil for the fallow of the triennial rotation was offered as a fresh proof of the principle maintained from time immemorial by agriculturists, namely, that different species of plants should be cultivated in succession on the same land, and that the same species should not recur except at considerable intervals; the earth yielding much finer crops when the same species do not follow in immediate sequence.†

Attempts have been made at various times to explain this phenomenon. It was at first thought that different species of vegetables required a particular nutriment; but it was soon perceived to be otherwise, and that the organs of each plant derived the necessary juices from substances which concur in the nutrition of vegetables

* Thaer, Agriculture raisonnée. † Ibid.

generally. In effect, plants the most opposite in botanical character and properties, alimentary as well as poisonous, will live and flourish on the same mound of earth, and with the same manure. Moreover these plants reciprocally withdraw nourishment from one another, which could not occur did each species need different elements of nutrition.*

When it was taken for granted that the organs of plants elaborate a common nourishment derived from the manure, then vegetables of diverse organizations were supposed endued with the faculty of searching at different depths for the nutritive matter contained in the soil, by reason of a more or less considerable extension and development of their roots. This served to explain how a plant with long and perpendicular roots could, as a sequel to corn, derive benefit from manure situate in the undermost layers of ploughed land. It is possible that an action of this kind may take place under certain circumstances, but the explanation can never be generally received.

Another explanation of the necessity for alternate crops is based upon properties assigned to the excretions of the roots, as compared to animal excrements.

The excretion of roots, first observed by Brugman in the *Viola arvensis*,† has been confirmed by the recent observations of M. Macaire. This physiologist obtained the matter exuded from certain plants by keeping their roots in water; but, strange to say, could not discover it in silicious sand in which certain vegetables had been grown.‡ I myself likewise failed in detecting sensible traces of organic matter in sand which had served as soil during several months to wheat and clover; a result which renders the fact of radicular excretion doubtful. The excretion consequent upon immersion in water is perhaps the effect of disease.

Be that as it may, upon the assumption of the excretion from roots, Messrs. Von Humboldt and Plenck have explained the cause of the attractions and repulsions of certain plants.§ More recently M. de Candolle has reproduced this idea as the basis of a theory of rotation of crops. If it be supposed, in fact, that the excretion from the roots represents vegetable excrements, it may be easily imagined that these excretions once deposited in the soil may be as prejudicial to the plant which produced them as would be the excrement of an animal presented to it as food. On the other hand, by change of species, the plant newly implanted may profit by the excretions of the preceding crop, absorbing them as nourishment. This ingenious hypothesis is deficient in the groundwork, inasmuch as the fact of radicular excretion is not sufficiently established. Again, admitting the excretion, several facts concur to demonstrate that plants may thrive in soil charged with their own excrements.

The culture of corn, for example, may proceed uninterruptedly, as we find in the triennial rotation. I have seen in the table-lands

* De Candolle, t. i. p. 248.
† Ibid. t. ii. p. 1497.
‡ Ibid. t. iii. p. 1474.
§ De Candolle, t. iii. p. 1474

of the Andes wheat fields, which had yielded excellent crops annually for more than two centuries. Maize may likewise be continually reproduced upon the same ground without inconvenience; this fact is well known in the south of Europe; and the greater portion of the coast of Peru has produced nothing else, from a date long anterior to the discovery of America. Further, potatoes may come again and again upon the same soil; they are incessantly cultivated at Santa-Fé and Quito, and nowhere are they of better quality. Indigo and sugar-cane may be brought under the same category. In Europe the Jerusalem artichoke produces constantly in the same place.* It must be conceded, that if all these plants excrete from their roots, their excretions are not of such a nature as to interfere with the progress of vegetation of the species producing them.

But the capital objection to the hypothesis of De Candolle is this, that it would be very remarkable indeed did any soluble organic matter, like such secretions, not putrefy when lying in the ground. In a word, it is difficult to understand how it should resist for years, as is pretended, the decomposing influence of heat and moisture together.

That there is no absolute necessity for alternation of crops when dung and labor can be readily procured, is undeniable. Nevertheless, there are certain plants which cannot be reproduced upon the same soil advantageously except at intervals more or less remote. The cause of this exigence on the part of certain vegetables is still obscure, and the hypotheses propounded for clearing it up far from satisfactory.

One of the marked advantages of alternate cultures, is the periodic cultivation of plants which improve the soil. In this way a sort of compensation is made for exhaustion. The main thing to be secured in rotation of crops is such a system as shall enable the husbandman to obtain the greatest amount of vegetable produce with the least manure, and in the shortest possible time. This system can be alone realized by employing in the course of rotation those plants which draw largely upon the atmosphere.

The best plan of rotation in theory, is that in which the quantity of organic matter obtained most exceeds the quantity of organic matter introduced into the soil in the shape of manure. This does not hold quite in practice. It is less the surplus amount of organic matter over that contained in the manure, than the value of this same matter which concerns the agriculturist. The excess required, and the form in which it should be produced, must vary widely according to locality, commercial demand, and the habits of people, considerations wholly apart from theoretical provisions. One point

* To this list might be added, according to the recent researches of M. Braconnot, the bay-rose with double flowers, and *Papaver somniferum*. That distinguished chemist terminates his memoir as follows: "My experiments are unfavorable, as may be perceived, to the theory of rotation of crops based on the excretions of the roots. These excretions if really occurring in the normal state are so obscure and little known as to lead to the inference that the general system of rotations must be referred to some other source." (Recherches sur l'influence des plantes sur le sol, Annales de Chimie t lxxii. p. 27.)

in theory that should agree with practice is this, that in no case is it possible to export more organic matter, and particularly more azotized organic matter, than the excess of the same matter contained in the manure which is consumed in the course of the rotation. By acting upon another presumption the productiveness of the soil would be infallibly lessened.

This irrefragable condition as to the term of exportation from a farm suggests some critical remarks upon sundry notions lately promulgated. The manufacture of beet-root sugar is an instance. European agriculture may probably derive certain advantages from this modern branch of industry, although these have been much overrated by certain speculators, who contend that sugar may thus be obtained through rotation of crops without lessening the other produce of the domain; so that the sugar constitutes an additional source of income. This seems to me erroneous.

If an estate yields annually 100 tons of beet-root for the support of cattle, their number must be diminished if the root is to be used for making sugar. The organic matter of the sugar extracted therefrom, is just so much nourishment withheld from the cattle. To assert the contrary would be equivalent to saying that potatoes grown upon a couple of acres of land, and submitted to the process of distillation before being employed as fodder, would feed as many animals as if eaten directly: assuredly, the organic principles of the potato converted into alcohol are lost as regards nutrition.

This does not imply that the manufacture of indigenous sugar, and of potato spirit, is less productive than breeding and fattening cattle. My sole object is to show that only a limited quantity of organic matter can be advantageously exported from an agricultural establishment. It must depend upon local and commercial circumstances whether this is to be exported in the form of sugar, corn, spirit, or butcher-meat.

The above statement is in apparent contradiction with generally received notions. Many persons believe that the manufacture of sugar, instead of injuring, is favorable to the breeding of cattle. It appears, from a Parliamentary return on this subject, in 1836, that in certain estates where sugar was made, the number of animals was increased; the numerical results are no doubt exact, but this augmentation in cattle is rather to be ascribed to an improved mode of farming than to the manufacture of sugar. In establishments where the triennial rotation with fallow was pursued, a rotation of four or five years with clover and weed-destroying plants has been introduced; so that it is by no means to be wondered at, that independently of beet-root, there should have been a considerable increase in other things. The introduction of this root, where it was not formerly grown, is of itself an important melioration. But in highly cultivated countries, where the most productive rotations have been long followed, the extraction of sugar would not effect such advantageous changes as those announced in the above return. If at Bechelbronn a time should ever come, and at present it seems far distant, when it would be deemed expedient to make sugar from the beet there grown it would certainly be requisite to diminish the number of cattle, or else to annex more meadow land. It is only indirectly, therefo: e, that the manufacture of home-sugar can promote the breeding of cattle, and so prove serviceable to agriculture.

From the definition given by me of the most advantageous course of crops, theoretically considered, it may be inferred how closely the study of rotations is connected with that of the exhaustion of the soil. Hence, to discuss the value of divers rotations, we must, in consonance with theory, compare the quantity of organic matter in a sequence of crops, with that in the manure expended upon them.

From a well-managed farm, where for a series of years an invariable system of culture has been steadily pursued, we must look for data. This I have done, as regards Bechelbronn, determining by analysis the composition of the manures and crops, and also of the more ordinary kinds of fodder or food. For a long time, a five years' rotation has been there adopted in the following order:

1st year.—Potatoes or beet-root manured.
2d year.—Wheat sown the autumn of the first year; clover interposed in the spring.
3d year.—Trefoil (clover) two crops; the third crop ploughed in or smothered.
4th year.—Wheat on the clover-break, turnips after the wheat.
5th year.—Oats.

The crop of oats which ends the rotation is generally scanty. The soil is then brought back to the point of fertility which it had before being dunged; and it is known by experience that it will not now yield a crop of any value.

I now proceed to detail the analyses of the different substances which enter into the rotation, indicating at the same time the average produce per acre.

POTATOES.

In the rather strong soil of Bechelbronn one acre produces upon an average about 105 cwts. of potatoes. This is below the ordinary rate of Alsace, where the crop amounts to from 155 to 165 cwts. per acre. The leaves and stems are left upon the ground.

A potato was cut in two, in order to subject it to analysis with a proportional part of the peel. The half weighed 335.2 grs. Stove-dried and reduced to flour, it weighed 289.3 grs. By absolute desiccation in vacuo, at a temperature of 230° F. it was found that one of moist tuber became 0.241; 15.4 grs. left of ash 0.039.

The average quantity of azote is 1.2. In 1836, I found 1.8 of azote. This notable difference, perhaps, depends on the analysis not having been made immediately after the harvest; or it may be partly due to meteorological influences. To convince myself that it did not depend upon any error of analysis, I examined anew the potato of 1836, preserved in the farinaceous state: it yielded 1.8 of azote. I shall, therefore, reckon the azote at 1.5:

	I.	II.
Carbon	43.72	43.40
Hydrogen	6.00	5.60
Oxygen	44.88	45.60
Azote	1.50	1.50
Ash	3.90	3.90

WHEAT.

I analyzed the grain gathered in 1837: one of wheat, dried in vacuo at 230° F. was reduced to 0.885; one of dry wheat left of ash 0.0243:

Carbon	46.10
Hydrogen	5.80
Oxygen	43.40
Azote	2.29
Ash	2.43
	100.00

The mean produce in wheat at Bechelbronn varies from 20½ to 22 bushels per acre; this variation depends on the drill crop which commences the rotation. After potatoes the average crop is 19½ bushels; after beet-root, 17 bushels; on clover-breaks it is 24 bushels. The average weight of the grain is 63 lbs. per bushel.

WHEAT-STRAW.

I estimate the proportion of the produce in grain to that in straw, as 44 to 100.

One of straw completely dried in vacuo at 230° F. becomes 0.740; one of dry straw leaves 0.0697 of ash:

	I.	II.
Carbon	48.48	48.38
Hydrogen	5.41	5.21
Oxygen	38.79	39.09
Azote	0.35	0.35
Ash	6.97	69.7
	100.00	100.00

RED CLOVER.

Clover delights in clayey soils; it thrives generally in good wheat lands; in light and sandy ground it gets bare and frosted. During its growth, it always requires the shelter of some other plant. For this reason, in spring, it is generally sown among wheat, which is put in the preceding autumn, or barley sown the same spring. We generally give from 11 to 14 lbs. of seed per acre. Clover is mowed the second year, as it is coming into flower; but when it is not to be consumed as green fodder, the mowing may take place before the flowering; this is required from the difficulty of making it into hay. In fact, in the process of drying clover, there is great risk of losing part both of the leaves and flower; besides, the drying always requires a considerable time, during which the clover runs the chance of being damaged by rain, and clover hay-making is almost impracticable in wet weather. Schwertz proposed to dry the clover on a sort of parrot-perches stuck into the ground. These supports are but eight feet high, and capable of bearing a load of 2 cwt. of green fodder, mowed twenty-four hours, and already withered. This method, as I have seen it practised in the Duchy of Baden, answers well, but there is considerable cost for manual labor, and in the first instance for perches. Schwertz reckons that 2 cwts. of green clover yield 48 lbs. of hay. The relation of green to dry fodder varies with the age of the plant, and the meteorological circumstances under which it has grown. Subjoined is the result of some experiments which I performed on the making of clover hay:

1 ton of clover in flower, 2d year (1841) afforded in hay 7 cwts.
1 ton of clover 1st year (1842) " 4 cwts. 2 qrs. 24 lbs.

The average produce of this fodder reduced to hay at Bechelbronn is 41 cwts. 3 qrs. per acre.

One of clover hay, after complete desiccation, weighed 0.790; one of dry hay left 0.078 of ash:

	I.	II.
Carbon	47.53	47.19
Hydrogen	4.69	5.33
Oxygen	57.96	37.66
Azote	2.06	2.06
Ash	7.76	7.76
	100.00	100.00

TURNIPS.

When turnips are cultivated as a second crop, as after rye or wheat, the produce is very uncertain. Attempts are occasionally made to raise them after wheat which has followed clover.

When cultivated on fresh manured soil, the produce is considerable; in some places it amounts to from 28 to 33 tons per acre; but as a second crop, we only obtain upon an average 7½ tons per acre. This crop is only counted as a half-crop in the general produce of the rotation.

Turnip is the most watery root I have examined. A slice weighing 2 oz. 17 dwts. dried in the stove, was reduced to 4 dwts. After thorough desiccation, one of turnip weighed 0.075; consequently the root contains 92.5 per cent. of water; one of dried turnip incinerated, left 0.0758 of ash:

	I.	II.
Carbon	42.80	49.93
Hydrogen	5.54	5.61
Oxygen	42.40	42.20
Azote	1.68	1.68
Ash	7.58	7.58
	100.00	100.00

OATS.

As this grain closes the rotation, the produce is not great. The average crop is 37 bushels per acre, at the weight of 33½ lbs. per bushel;* one of oats completely dried weighs 0.792; one of dried oats leaves 0.0398 of ash:

	I.	II.
Carbon	50.32	51.09
Hydrogen	6.32	6.44
Oxygen	37.14	36.25
Azote	2.24	2.24
Ash	3.98	3.98
	100.00	100.00

* This is but a light weight for a bushel of oats.—Eng. Ed.

OAT STRAW.

Oat straw is estimated at about 15 cwts. per acre; one part becomes, when dried in vacuo, 0.713; one part burned leaves 0.0509 of ash:

Carbon	49.93	50.25
Hydrogen	5.32	5.48
Oxygen	39.28	38.80
Azote	0.38	0.38
Ash	5.09	5.09
	100.00	100.00

FIELD BEET OR MANGEL-WURZEL.

On a freshly manured soil, the average produce of beet at Bechelbronn is 10 tons, 15 cwts. 1 qr. per acre. The worst crops do not fall below 5 tons, 2 cwts. 1 qr. 14 lbs., and the best do not exceed 16 tons, 7 cwts. 1 qr., results which I took occasion to observe varied sensibly from those obtained in different places. I stated that Schwertz and Thaer make the average 14 tons, 14 cwts. 2 qrs. 16 lbs. Mœlinger, after taking the mean of ten years, adopts 11 tons, 1 cwt. 3 qrs. 6 lbs. At Roville, M. de Dombasle speaks of 7 tons, 3 cwts. 26 lbs. as the mean of seven years.

At Bechelbronn, the leaves of the beet are not given to cattle; they are left upon the ground. A piece of beet-root weighing 1 oz. 16 dwts. was reduced to $4\frac{7}{10}$, say 5 dwts. after being stove dried. After complete desiccation, at 230° F. one part of root became 0.122; one part of root left upon incineration 0.0624 of residuum:

Carbon	42.75	42.93
Hydrogen	5.77	5.94
Oxygen	43.58	43.33
Azote	1.66	1.66
Ash	6.24	6.24
	100.00	190.00

RYE.

Rye is seldom introduced into the rotation at Bechelbronn. They reckon its produce at 26 bushels per acre, when it has had a supplementary dose of manure. The bushel weighs fully 58 lbs. I have taken the proportion of grain to straw as 45 is to 100. One part of rye, dried at 230° F. weighed 0.834; one part incinerated left 0.0237 of ash:

	I.	II.	III.
Carbon	46.35	45.72	46.38
Hydrogen	5.38	5.70	5.74
Oxygen	44.21	44.52	43.82
Azote	1.69	1.69	1.69
Ash	2.37	2.37	2.37
	100.00	100.00	100.00

RYE-STRAW.

One part of straw, completely dried, weighed 0.813; one part of which, incinerated, left 0.0368 of ash:

Carbon	49.88
Hydrogen	5.58
Oxygen	40.56
Azote	0.30
Ash	3.68
	100.00

WHITE PEAS

Raised on manured land yielded 16 bushels per acre, weighing fully 62 lbs. per bushel. One part of peas, after complete desiccation, weighed 0.914; one part of dried peas left of ash 0.0314:

	I.	II.
Carbon	46.06	46.94
Hydrogen	6.09	6.24
Oxygen	40.53	39.50
Azote	4.18	4.18
Ash	3.14	3.14
	100.00	100.00

PEA STRAW.

One acre of peas produces about 22 or 23 cwts. of straw; one part of the straw after desiccation weighed 0.802; one part after incineration 0.1132:

Carbon	45.80
Hydrogen	5.00
Oxygen	35.57
Azote	2.31
Ash	11.32
	100.00

JERUSALEM POTATO OR ARTICHOKE.

In Alsace, Jerusalem artichokes are always grown on one and the same piece of land, which is manured every two years. At Bechelbronn, on a somewhat shallow soil, the produce per acre amounts to:

Tubers	10 tons
Dry stems	11½ cwts.

A tuber which weighed on being taken from the ground 1 oz. $15\frac{6}{10}$ dwts., weighed $\frac{7}{10}$ dwts. after it was dried in the stove. After absolute desiccation, one part was reduced to 0.208; one portion of the dry tuber left 0.0594 after incineration:

		II.
Carbon	43.02	43.62
Hydrogen	5.91	5.80
Oxygen	43.56	43.07
Azote	1.57	1.57
Ash	5.94	5.94
	100.00	100.00

DRIED STEMS OF JERUSALEM ARTICHOKES.

These stems had stood through the winter where they grew, and were almost wholly composed of pith. One part after desiccation weighed 0.871: one part left of ash 0.0276.

Carbon	45.66
Hydrogen	5.43
Oxygen	45.72
Azote	0.43
Ash	2.76
	100.00

I fear that in this analysis the carbon and azote are rated too low.

I have collected in two tables the results of the analyses as detailed above. The first exhibits the quantity of dry matter and moisture contained in each specimen; the other, the elementary composition. On careful examination of the numbers given in the second table, certain substances will be found very analogous in composition. If the ashes be deducted, the analogy becomes complete; for many substances differing widely both in character and properties, nevertheless appear to possess the same composition; a fact which I do not undertake to explain.

TABLE OF THE PROPORTIONS OF WATER CONTAINED IN DIFFERENT SUBSTANCES.

Substances.	Dry matter.	Water.
Wheat	0.855	0.145
Rye	0.834	0.166
Oats	0.792	0.208
Wheat-straw	0.740	0.260
Rye-straw	0.813	0.187
Oat-straw	0.713	0.287
Potato	0.241	0.759
Field-beet	0.122	0.878
Turnip	0.075	0.925
Jerusalem potato	0.208	0.792
Peas	0.914	0.086
Pea-straw	0.882	0.118
Clover-hay	0.790	0.210
Jerusalem potato-stems	0.871	0.129

COMPOSITION OF THE SAME SUBSTANCES DRIED IN VACUO AT 230° FAHR.

SUBSTANCES.	Ashes included.					Ashes deducted.			
	Carbon.	Hydrogen.	Oxygen.	Azote.	Ash.	Carbon.	Hydrogen.	Oxygen.	Azote.
Wheat	46.1	05.8	43.4	02.3	02.4	47.2	06.0	44.4	02.4
Rye	46.2	05.6	44.2	01.7	02.3	47.3	05.7	45.3	01.7
Oats	50.7	06.4	36.7	02.2	04.0	52.9	06.6	38.2	02.3
Wheat-straw	48.4	05.3	38.9	00.4	07.0	52.1	05.7	41.8	00.4
Rye-straw	49.9	05.6	40.6	00.3	03.6	51.8	05.8	42.1	00.3
Oat-straw	50.1	05.4	39.0	00.4	05.1	52.8	05.7	41.1	00.4
Potato	44.0	05.8	44.7	01.5	04.0	45.9	06.1	46.4	01.6
Field-beet	42.8	05.8	43.4	01.7	06.3	45.7	06.2	46.3	01.8
Turnip	42.9	05.5	42.3	01.7	07.6	46.3	06.0	45.9	01.8
Jerusalem potato	43.3	05.8	43.3	01.6	06.0	46.0	06.2	46.1	01.7
Peas	46.5	06.2	40.0	04.2	03.1	48.0	06.4	41.3	04.3
Pea-straw	45.8	05.0	35.6	02.3	11.3	51.5	05.6	40.3	02.6
Clover-hay	47.4	05.0	37.8	02.1	07.7	51.3	05.4	41.1	02.2
Jerusalem potato-stems	45.7	05.4	45.7	00.4	02.8	47.0	05.6	47.0	00.4

RELATIONS OF MANURES TO CROPS.

The manure employed at Bechelbronn is what is commonly called farm-yard dung, a compost made up of the excrements of horses, oxen, and straw litter impregnated with urine. The dung of fowls and pigeons, and the sweepings of the yard, are sometimes applied to special purposes. The animals whose excrements form the dung which I have examined were horses, oxen, and swine.

The manure is put upon the ground when it has undergone fermentation in the heap : it is manure half-made : the straw litter is not entirely decomposed, but is soft and filamentous ; in this state manure retains a great deal of moisture.

DESICCATION OF HALF-MADE OR HALF-DECAYED MANURE.

EXPERIMENT I.

A quantity of manure prepared during the winter of 1837–1838, which in the state in which it was being put on the ground, weighed 257 lbs after it had been dried so as to be easily reduced to powder, weighed 57 lbs. The loss of water was therefore about 77.3 in 100 This number comes very near the estimate of several German agriculturists, who reckon the moisture in farm-yard dung at 75 per cent. Still this loss does not represent the whole of the water ; for after desiccation at 212° F. the 57 lbs. weighed 54 lbs. In fine, after desiccation in vacuo, at 230° F. it was found that one part of stove-dried manure lost 0.039. Thus the manure parted in totality with 79.62 per cent. of water, and contained in consequence 20.4 of dry substance.

EXPERIMENT II.

Of the manure prepared in the winter of 1838–1839, 220 lbs. after being chopped and dried weighed 56 lbs. One part of this manure was reduced in dry vacuo at a temperature of 230° F. to 0.872. The 220 lbs. would therefore have weighed when dry 48 lbs.

EXPERIMENT III.

Of the manure prepared during the summer of 1839, 660 lbs. weighed after desiccation 151 lbs. ; of this dry manure reduced to powder, one part lost by desiccation in vacuo at 230° F. 0.1461.

The 151 lbs. would therefore have lost 22 lbs. ; consequently the 660 lbs. of manure contained 129 lbs. of dry matter, that is, 19.64 per cent.

Subjoined is a summary of the per centage of dry matter :

First experiment	20.4
Second "	22.2
Third '	19.6
Average	20.7
Moisture (average)	79.3

ANALYSES OF HALF-MADE MANURES.

I. Manure prepared during the winter of 1837–1838 :

Matter 0.5595, gave carbonic acid 0.528, water 0.157 : C. 32.4, H. 3.8, Azote 1.7.—1.0 gave ashes 0.462.

II. Manure prepared during the winter of 1837–1838:

Matter 0.575, gave carbonic acid 0.676, water 0.212 : C 32.5, H. 4.1, Azote 1.69.—1.000 gave ashes 0.357.

III. Manure prepared during the winter of 1837–1838:

Matter 0.567, gave carbonic acid 0.791, water 0.232 : C. 38.7, H. 4.5, Azote 1.73.—1.000 gave ashes 0.264.

IV. Manure prepared during the spring of 1838:

Matter 0.586, gave carbonic acid 0.759, water 0.208 : C. 36.4, H. 4.0, Azote 2.4.—1.000 gave ashes 0.381.

V. Manure prepared during the spring of 1839:

Matter 0.445, gave carbonic acid 0.643, water 0.171 : C. 40.0, H. 4.3, Azote 2.4.—1.000 gave ashes 0.257.

VI.

Matter 0.427, gave carbonic acid 0.543, water 0.150 : C. 34.7, H, 3.9.
" 0.427, " " 0.530, " 0.127 : " 34.3, " 4.8,
Azote 2.0.—1.000 gave ashes 0.315.

COMPOSITION OF THE MANURES ANALYZED.

	Carbon.	Hydrogen.	Oxygen.	Azote.	Salts and earths.
I.	32.4	3.8	25.8	1.7	36.5
II.	32.5	4.1	26.0	1.7	35.7
III.	38.7	4.5	28.7	1.7	26.4
IV.	36.4	4.0	19.1	2.4	38.1
V.	40.0	4.3	27.6	2.4	25.7
VI.	34.5	4.3	27.7	2.0	31.5
Mean . . .	35.8	4.2	25.8	2.0	32.2

In all these analyses, the combustion was promoted by the addition of chlorate of potash; some oxide of antimony was likewise added. The carbonic acid of the ash was determined and struck off.

The measure of dung in use at Bechelbronn is the wagon drawn by four horses. After repeated weighings it was found that this measure contains nearly 1 ton, 15 cwts. 2 qrs. 23 lbs. of moist material, or 7 cwt. 1 qr. 15 lbs. if that be computed dry. The first course of the rotation receives 27 loads of this manure, weighing about 48 tons, 14 qrs. 5 lbs., equivalent to 9 tons, 19 cwts. 0 qr. 2 lbs. of dry manure.*

The preceding analyses show that this charge of manure, which is to fertilize the soil during the course of the rotation (five years) contains:

Carbon	8027 lbs.
Hydrogen	925
Oxygen	5767
Azote	447
Salts and earth	7188
	22355

Such are the principles which together form the organic matter that is to be consumed and in major part assimilated by the crops

* I presume that the quantity above specified is that which is laid on the French hectare, equal to 2.4 acres English. To get at the quantity laid on per acre, it would therefore be necessary to divide by 2 4-10 : Thus 48 tons, 14 cwts. 5 lbs. per hectare will be equal to 20 tons, 1 cwt. 3 qrs. per English acre.—Eng. Ed.

grown. I say partly, because I do not believe that the whole organic matter necessarily enters into the constitution of the plants which spring up during the rotation; no doubt a considerable portion of the manure is lost through spontaneous decomposition, or is carried away by the rain; and another portion may remain a long time dormant in the soil, to act as a fertilizer at a more or less distant period; just as in the present rotation the manure formerly introduced co-operates with that recently added. One thing is certain, viz., that the proportion of manure indicated is essential for average crops; by diminishing it the produce is necessarily lessened. Lastly, it is proved that after the rotation the crops have consumed the manure, and the earth will not yield its increase unless a fresh quantity be added.

I now proceed to consider the relation subsisting between the quantity of organic matter buried in the soil as manure, and what is recovered in the crops. In this way the respective proportions of elementary matter which various crops derived from the air and the soil, may be determined approximately, and a knowledge obtained of those rotations which least exhaust the land, or in other words, which obtain from the atmosphere the largest amount of organic matter.

The rotations set down in Tables I. and II. are those definitively adopted at Bechelbronn, and throughout the greater part of Alsace. These two rotations, which differ only in the hoed crop introduced, potatoes in one, beet-root in the other, are almost identical; nearly the same quantity of dry matter being produced per acre, and nearly the same quantity of organic material withdrawn from the atmosphere.

The rotation No. 3 was introduced by Schwertz at Hohenheim; theoretically, it is one of the most advantageous; it was tried at Bechelbronn but abandoned, because, from meteorological causes, peas and vetches fail frequently.

Table No. 4, shows the triennial rotation with manured fallow; this is disadvantageous in point of theory. The organic constituents of the crop exceed but little those of the manure. Supposing that even the whole of the straw were converted into manure, the farmer would still be compelled to procure manure from abroad, in compensation for the out-going of wheat. It is thus obvious why triennial rotation always requires a great deal of meadow land.

In table No. 5, the result of the continuous cultivation of Jerusalem artichokes is given. At Bechelbronn these are dressed every two years with about ten loads of dung per acre. Upon an average 20 tons of tubers and about 2 tons of woody stems are gathered in the course of two years. It will be perceived from perusal of this table, that the culture of Jerusalem artichokes presents, theoretically, considerable advantages. The organic matter of the crop greatly exceeds that of the manure. Moreover, in Alsace, where it is very common, it is held to be most productive. Still, the organic matter of the stems must be taken into account, which, practically speaking, are nearly worthless.

Table No. 6 comprises the data relative to a quadrennial rotation, adopted by M. Crud, and in which are grown successively: 1st. Potatoes or beet-root. 2d. Wheat. 3d. Red clover. 4th. Wheat. The first sowing is dressed with about 18 tons of half-wasted farm-yard dung. The gain in organic matter obtained by this rotation surpasses that of the preceding; but as the clover crops are not very sure when repeated every four years, M. Crud, for reasons which may be called in question, follows this rotation with one of lucern, which gets a fresh supply of manure. It cannot be denied that lucern furnishes a great mass of fodder, and in this respect the fertility of the land ought to be vastly enhanced, were this consumed on the spot; but I can discover no objection to the renewal of clover, if the lucern succeeds so well as M. Crud says it does. From too frequent repetition, farmers have gone into the opposite extreme of cultivating clover only every five or six years. This subject offers an important field for research. It is not impossible that the ill-success depends often on premature mowing of the clover during the first year, and before its roots have acquired sufficient vigor. This practice has been abandoned with us for some years, and there is now every thing to assure us that the second year's crop is thereby secured.

ROTATION COURSE No. 1.

Years.	Substances.	Crops per acre.	Crops dry.	Carbon.	Hydrogen.	Oxygen.	Azote.	Salts and earths.
		lbs.	lbs.	lbs.	lbs.	lbs.	lbs.	lbs.
1st	Potatoes	11733	2828	1244	164	1264	42	113
2d	Wheat	1231	1052	485	61	457	24	25
	Wheat-straw . .	2798	2070	1002	110	805	8	145
3d	Clover-hay . . .	4675	3693	1750	185	1396	78	284
4th	Wheat	1521	1300	599	75	564	30	31
	Wheat-straw . .	3456	2557	1237	135	995	10	179
	Turnips (2d crop) .	8754	656	2832	36	278	11	50
5th	Oats	1232	975	494	62	358	21	39
	Oat-straw . . .	1650	1176	593	63	458	5	60
	Total	37050	16307	10236	891	6575	229	926
	Manure employed .	4495	9314	3426	391	2403	185	2999
	Difference . . .		6993	6810	500	4172	44	2073

ROTATION COURSE No. 2.

Years.	Substances.	Crops per acre.	Crops dry.	Carbon.	Hydrogen.	Oxygen.	Azote.	Salts and earths.
		lbs.	lbs.	lbs.	lbs.	lbs.	lbs.	lbs.
1st	Mangel wurzel .	2383	2907	1244	157	1262	49	182
2d	Wheat	1086	928	428	53	403	21	22
	Wheat-straw . .	2468	1827	883	98	710	7	128
3d	Clover-hay . . .	11675	3693	1749	185	1396	77	284
4th	Wheat	1520	1300	599	75	564	30	31
	Wheat-straw . .	3456	2557	1237	135	995	10	179
	Turnips	8754	655	281	36	277	11	50
5th	Oats	1232	975	495	62	358	21	39
	Oat-straw . . .	1650	1176	589	63	458	5	60
	Total	27224	16018	7505	864	6423	231	975
	Manure employed	4495	9314	3426	391	2403	185	2999
	Difference . . .		6704	4079	473	4020	46	2024

ROTATION COURSE No. 3.

Years.	Substances.	Crops per acre.	Crops dry.	Carbon.	Hydrogen.	Oxygen.	Azote.	Salts and earths.
		lbs.	lbs.	lbs.	lbs.	lbs.	lbs.	lbs.
1st	Potatoes	11733	2828	1244	164	1264	42	113
2d	Wheat	1231	1054	485	61	457	24	25
	Wheat-straw . .	2798	2070	1002	110	805	8	145
3d	Clover-hay . . .	4675	3693	1750	185	1396	78	284
4th	Wheat	1515	1300	599	75	564	30	31
	Wheat-straw . .	3456	2558	1238	135	995	10	179
	Turnips	8754	656	282	36	278	11	50
5th	Peas (dunged) . .	1001	915	425	56	366	38	28
	Pea-straw . . .	2558	2256	1033	112	803	52	255
6th	Rye	1539	1278	590	71	565	22	30
	Rye-straw . . .	3420	2780	1387	155	1129	8	100
	Total	148280	21388	10035	1160	8622	323	1240
	Manure employed	148285	11176	4000	470	2883	223	3599
	Difference . . .		10212	6035	690	5739	100	2359

ROTATION COURSE No. 4.

Years.	Substances.	Crops per acre.	Crops dry.	Carbon.	Hydrogen.	Oxygen	Azote.	Salts and earths.
		lbs.	lbs.	lbs.	lbs.	lbs.	lbs.	lbs.
1st	Dunged fallow . .	. . .	. . .	. . .	. . .	. . .	. .	. .
2d & 3d	Wheat	3041	2600	951	150	1128	60	62
	Straw	6875	5080	2462	270	1979	20	356
	Total	9916	7680	3413	420	3107	80	418
	Manure employed	18330	3795	1358	159	979	76	1222
	Difference . . .	8414	3885	2055	261	2128	4	804

No. 5, CONTINUOUS POTATO CROPS.

Years.	Substances.	Crops per acre.	Crops dry.	Carbon.	Hydrogen.	Oxygen.	Azote.	Salts and earths.
		lbs.	lbs.	lbs.	lbs.	lbs.	lbs.	lbs.
1st & 2d	Potatoes	48473	10083	4366	585	4366	161	605
	Stalks	25850	22497	10289	1215	10289	90	630
	Total	74323	32580	14655	1800	14655	251	1235
	Manure employed .	41663	8624	3087	362	2225	172	2777
	Difference . . .		23956	11568	1438	12430	79	1542

No. 6, QUATRENNIAL ROTATION, ADOPTED BY M. CRUD.

Years.	Crops grown.	Crops per acre.	Crops dry.	ELEMENTARY INGREDIENTS OF THE CROP.				
				Carbon.	Hydrogen.	Oxygen.	Azote.	Salts and earths.
		lbs.	lbs.	lbs.	lbs.	lbs.	lbs.	lbs.
1st	Half acre of potatoes	9167	2209	972	128	987	33	88
	Ditto of beet-roots .	18333	2237	957	130	970	38	141
2d &4th	Wheat, 153 bushels .	3331	2847	1312	165	1235	65	68
	Wheat-straw . . .	7333	5243	2537	278	2040	21	347
3d	Clover three cuttings	7333	5793	2746	290	2190	121	446
	Total	45497	18329	8524	991	7422	278	1110
	Manure consumed .	40333	8349	2989	350	2154	167	2688
	Difference		9980	5535	641	5268	111	1578

SUMMARY.

Rotations.	Dry manure expended upon one acre in one year.	Azote contained in the manure.	Dry produce obtained in one year upon one acre.	Azote contained in the produce.	Gain in organic matter in one year upon one acre.	Gain in azote in one year upon one acre.
	lbs.	lbs.	lbs.	lbs.	lbs.	lbs.
No. 1	1862	37	3261	46	1996	9
No. 2	1862	37	3204	46	1746	9
No. 3	1862	37	3564	54	2513	17
No. 4	1247	23	2567	26	1565	3
No. 5	3710	86	16299	125	12758	39
No. 6	2087	42	4582	70	2889	28

From all that precedes, it is obvious that rotations which include trefoils, red clover, lucern, and sainfoin, are those that afford considerably the largest proportion of organic matter; a fact, indeed, which if not legitimately established, has still been long acted on in that system of cropping which embraces forage plants as an element. Lucerns, too, when they have taken kindly, yield an extraordinary quantity of forage, as every one may see by turning to the produce of the piece under that crop which in the system of M. Crud succeeds the quatrennial rotation. At the end of his rotation, M. Crud always lays on manure in the ratio of 18 tons per acre, which lasts for six years, and may be said to suffice for the succession of crops in the appended table:

Crops.		Produce per acre.	Contents in azote.
Lucern dry,	1st year	3080 lbs.	72 lbs.
"	2d year	9240	215
"	3d year	11458	269
"	4th year	9240	213
"	5th year	7333	172
Wheat,	6th year	1448	28
Straw		3645	11
			980
Dung employed		40233	205
Total gain in azote			775
Gain in azote per annum and per acre			130

In glancing at these tables, it is obvious that the azote of the crop always exceeds the azote of the manure. Generally speaking, I admit that this excess of azote is derived from the atmosphere: but I do not pretend to say in what precise manner the assimilation takes place. I shall only quote the conclusion of a paper which I published on the subject in the year 1837.* Azote may enter immediately into the constitution of vegetables, provided their green parts have the power of fixing it; azote may also enter vegetables dissolved in the water which bathes their roots, and which always contains it in a certain proportion. Lastly, it is possible that the air may contain an infinitely minute quantity of ammoniacal vapor, as some natural

* Annales de Chimie, t. lxix. p. 366.

philosophers* have maintained, and that this assimilated, decomposed, and recomposed anew by the plant, is the source of its azotized constituents.

§ 2. OF THE RESIDUES OF DIFFERENT CROPS.

The vegetable matter which is produced in the course of a season is never found entirely in the crop. A certain quantity of it, for instance, always remains in the ground. It is, therefore, a point of interest to ascertain what quantity of elementary matter is left in the soil after each kind of crop in the rotation; precise knowledge of this description may even be important in calculating rotations, for it is obvious that the remains of the crop now on the ground must influence that which is to follow, and in the course of a rotation the sum of the residuary matters must be regarded as a supplement or addition to the manure put into the ground at its commencement.

In the systems of rotation very generally followed at the present time, the influence of these residuary matters is manifest, and it is partly by their means that we can explain how a quantity of manure, frequently very moderate, should suffice for the whole of the crops in a productive rotation. The remarkable effect of clover has not failed to arrest attention even from the most unobserving. The wheat crop which comes after our drill crop in Alsace, beet or potatoes, averages from 18 to 20 bushels per acre; but the wheat crop that succeeds our clover averages from 23 to 24 bushels per acre.

The improvement of the soil, so obvious in connection with clover, in all probability also occurs in connection with the residues of other crops; but as in most instances the residue merely compensates the loss, or lessens its extent, the effect produced is less remarkable, and is less, indeed, in amount. All the world acknowledge, then, that the residues of the crops that enter into a rotation compensate in greater or less degree for what is carried away in the shape of harvest, and that in some cases they even add to the fertility of the soil; for in growing crops that leave a large quantity of residue, it is precisely as if a smaller quantity were taken from a given extent of surface. But what is the amount of residue or refuse which is returned to the soil by such and such a crop? What, in a word, is the value of this residuary matter considered as manure? This is a point upon which only the most vague and indefinite ideas are generally entertained; and it was with the purpose of substituting positive facts for mere guesses, that I determined on weighing and analyzing the vegetable residue of the several crops that enter as elements into our more usual rotations.

My experiments were made upon breadths of land which varied from 120 to 500 square yards in extent. The clover roots and stubble were taken up with the spade, and before being dried, were freed from adhering earth by washing. The beet-leaves and potato-tops were dried at once in the oven; and it was from each of the

* Saussure Recherches Chimiques, and Liebig, Agricultural Chemistry.

general masses reduced to powder that samples were taken for ultimate analysis, before proceeding to which, they were carefully dried in vacuo at 230° F.

It is not likely that any accurate mean result should have been come to from an examination of the produce of a single season. I should even say that the year in which these inquiries were undertaken was little favorable to them, inasmuch as the crops were generally bad; but it is obvious that they form a nucleus, around or by the side of which the results of other seasons may be arranged, and an average from larger premises come to.

POTATO TOPS OR HAUM.

A piece of land measuring 120 square yards, marked off from a spot that had suffered from drought, yielded 47.0 lbs. of green tops, which were reduced by drying to 18.4 lbs. A similar extent of surface, selected from a part of the field that looked well, gave green tops 79 lbs., which dried in the air were reduced to 16 lbs. We should thus have 23½ cwts. of green, and 6½ cwts. of dry tops per acre. The crop of potatoes in 1839 yielded at the rate of 101½ cwts. per acre. One hundred grammes, or 3 oz. 4 dwts. 8 grs. troy, of the top dried in the air, lost 12 grammes, or 7 dwts. 17 grs. by thorough drying at 230° F. The weight of the tops yielded per acre, taken as dry, consequently amounts to 5 cwts. 2 qrs. 14 lbs., and by elementary analysis they were found to have the following composition:

Carbon	44.8
Hydrogen	5.1
Oxygen	30.5
Azote	2.3
Salts and earths	17.8
	100.0

LEAVES OF FIELD-BEET, OR MANGEL-WURZEL.

Upon a surface of 500 square yards, 976 lbs. of leaves were gathered, the weight being taken two days after the roots were pulled up.

55 lbs. of leaves reducible to powder by drying in an oven, were brought to 6.6 lbs.

3 oz. 4 dwts. of leaves dried and pulverized, lost by desiccation at 230° F. 3¾ dwts. of moisture. The 6.6 lbs. brought to that state of dryness would have weighed 6.1 lbs. With these data it is found that the 9 3 lbs. of green leaves gathered upon 500 square yards would have weighed when dry 108.9 lbs.; and that the acre produced 85¾ cwts. of green and 9¾ cwts. of dry leaves. The crop of roots which answers to that quantity of leaves, was in 1839 but 6 tons, 2 cwts., that is to say, little more than half a crop; for our average is about 10½ tons.

COMPOSITION OF DRY LEAVES.

Carbon	38.1
Hydrogen	5.1
Oxygen	30.8
Azote	4.5
Salts and earths	21.5
	100.0

WHEAT STUBBLE.

From 120 square yards of ground we have obtained 13 lbs. of stubble dried in the air. The same surface in another field produced 17½ lbs.

Thus we have 5¾ cwts. of stubble per acre; but as wheat recurs twice in the rotation, the residues must be doubled; say, 11½ cwts. Stubble loses 0.26 of moisture when dried completely at 230°.

In 1839, the wheat after the drilled crop, or after clover, was only 17 bushels per acre.

I have assigned to stubble the same composition as that of straw.

CLOVER ROOTS.

A surface of 120 square yards gave 44 lbs. of roots, weighed after being thoroughly dried in the sun; when pulverized after drying in the stove the weight was reduced to 37 lbs.

3 oz. 4 dwts. of powdered roots lost by drying, at a temperature of 230° F., 5 dwts. of moisture. Thus the 44 lbs. of roots dried in the sun would have weighed 34 lbs., and one acre would have furnished 12¾ cwts. of residue perfectly dry.

In 1839, the clover crop when reduced to hay was far below the average.

COMPOSITION OF THE ROOTS.

Carbon	43.4
Hydrogen	5.3
Oxygen	36.9
Azote	1.8
Salts and earth	12.6
	100.0

OAT STUBBLE.

The residue of the oat crop, which concludes the rotation course, does not act upon the present, but on the next rotation; in the same way as the organic remains left in the ground by the oats which terminated the antecedent course, exerted their influence upon the present one. In 1839, the oat crop was above the average; it was as high as 16 cwts. 2 qrs. 18 lbs. per acre.

One French *are* of the land, equal to 120 square yards English, yielded 20 lbs. of stubble dried in the air, or at the rate in round numbers of 8 cwts. per acre.

In the following table I have given a summary of the results above stated, combining therewith the quantity and the composition of the manure expended in the rotation.

SUMMARY OF THE FOREGOING RESULTS.

Nature of the crop.	Produce per acre in 1839.	Produce dried at 212 deg. F.	Nature of the residues buried in the soil.	Residues obtained upon one acre.	Residues dried at 110 deg.	Elementary matter of the residues.				
						Carbon.	Hydrogen.	Oxygen.	Azote.	Salts and earths.
	lbs.	lbs.		lbs.	lbs.	lbs.	lbs.	lbs.	lbs.	lbs.
Potatoes .	11367	2739	Potatoe tops .	2632	630	282	32	189	14	112
Beetroots .	13678	1668	Beetroot leaves	9599	1070	410	55	330	48	236
Wheat . .	2150	1837	Stubble. . . .	1283	950	460	50	369	4	67
Clover-hay.	2292	1810	Roots dried in the sun . . .	1833	1418	615	75	523	26	178
Oats	1862	1474	Stubble	836	596	299	32	232	2	30
Total . . .	31349	9527		16182	4664	2066	244	1643	94	617
Manure employed	44995				9314	3335	391	2403	186	2995

It therefore appears that the refuse or residue of the several crops of a rotation represent, both in quantity and nature, somewhat less than one half of the manure originally put into the ground; I say somewhat less, because it must be remembered that in the sum of these residuary matters, the beetroot leaves and potato tops must not be allowed to stand together, the one crop naturally excluding the other, or at all events the two hoed or drilled crops not entering in this proportion into the same rotation.

The large quantity of organic matter restored to the soil by several of the crops in the series, consequently explains how the rotation may be closed without its being found indispensable to supply any additional manure in its course. It seems indubitable that without this addition of elementary matter, the fertility of the soil would decline much more rapidly than it does; the residue of each crop is nothing more than a portion of the crop itself restored to the ground; it is as if we only carried off one portion, the larger portion of the crop, and buried another portion green.

In the five years' rotation, it may be observed that there are two crops, the hoed crop and the forage crop, which yield substances to the ground that are both abundant in quantity and rich in azotized matter, and it is unquestionable that these crops act favorably on the cereals which succeed them. But data are wanting for the appreciation of their specific utility in the general rotation. We see, for instance, that despite the large proportion of residuary matter left by the beet or mangel-wurzel, this plant lessens considerably the produce of the wheat crop that comes after it. The potato, though it leaves much less refuse than the beet, seems nevertheless to act less unfavorably than this vegetable. Clover leaves more residue than the potato, and on this ground, alone, ought to favor the cereal that follows it; but it has a favorable influence out of all proportion with its quantity, contrasting this with the residue of either of the hoed crops; a fact from which we learn that the visible appreciable influence of the residuary matters of preceding crops, upon the luxuriance of succeeding crops, does not result solely from their mass, even supposing each to be possessed of equal qualities; this other, this additional effect, depends especially on an influence exerted on the soil by the crops which leave them. Had these crops been powerfully exhausting, we should expect that their refuse or residue, however considerable in quantity, could do no more than lessen the amount of exhaustion produced; in which case, its useful influence, however real, would pass unnoticed, were it estimated by the produce of the succeeding crop. If, on the contrary, a crop has been but slightly scourging, whether in consequence of the smallness of its quantity, or because it may have derived from the air the major part of its constituent elements, the useful influence of the residue will not fail to be conspicuous. When the relative value of different systems of rotation is discussed in the way we have done, we in fact estimate the value of the elementary matter derived from the atmosphere by an aggregate of crops; but the procedure generally followed is silent when the question is to assign to each crop in particular the special share which it has had in the total profit. To reply to this question, of which a knowledge of the various residues is one of the elements, we must first ascertain the quantity of elementary matter supplied by the soil and the atmosphere, with reference to each of the crops which enter into the rotation; in other words, the same investigations must be undertaken in reference to each plant considered by itself, that have been made in reference to the series collectively. There is unquestionable room, in this direction, for an important series of experiments.

§ 3. OF THE INORGANIC SUBSTANCES OF MANURES AND CROPS.

We have but just considered the organic matter developed in a series of successive harvests. To complete the study of rotations, to the extent at least that this can be done in the present state of science, we have still to examine the relations that may exist between the mineral substances which enter into the constitution of the produce, and those that make part of the manure given.

We have already shown in a general way that certain mineral salts, certain saline matters or salifiable bases, are essential to the constitution of vegetables. To the best of my knowledge, no seed has yet been met with that is without a phosphate; and it is now known that the alkaline salts powerfully promote vegetation.

Such is their ascertained influence, indeed, that tobacco, barley, and buckwheat sown in soils absolutely without organic matter, but containing saline substances, and only moistened with distilled water, produced perfect plants, which flowered and fruited, and yielded ripe seeds.* Whence it follows, that the presence of saline matter favors remarkably the assimilation of the azote of the atmosphere during the act of vegetation.

The importance of considering rotations in connection with the inorganic substances that are assimilated by plants was perfectly well known to Davy. "The exportation of grain from a country which receives nothing in exchange that can be turned into manure, must exhaust the soil in the long run," says the illustrious chemist;

* Liebig, in Journ. de Pharmacie, vol. iv., 3d series, p. 94.

who ascribed to this cause the present sterility of various parts of Northern Africa and of Asia Minor, as well as of Sicily, which for a long succession of years was the granary of Italy. Rome unquestionably contains in its catacombs quantities of phosphorus from all the countries of the earth.

Professor Liebig, in insisting with the greatest propriety on the useful part played by alkaline bases and saline matters in vegetation, has shown the necessity of taking inorganic substances into serious consideration in discussing rotations. It is long since I came to the same conclusion myself; but it strikes me, that to be truly profitable, such a discussion must necessarily rest on analyses of the ashes of plants which have grown in the same soil, and been manured with the same dung, the contents of which in mineral elements were already known. There is in fact a kind of account current to be established between the inorganic matter of the crop and that of the manure. Although I give every credit to the fidelity of the analyses of vegetable ashes that have been published up to the present time, I have not felt myself at liberty to make use of any of them in the direction which I now indicate. I have not thought that it would be fair or reasonable to contrast such heterogeneous compounds, as the ashes of plants grown at Geneva and Paris, under such dissimilar circumstances, with those of vegetables produced on a farm of Alsace, where the point to be explained, through the results of this contrast, had reference to a particular series of agricultural phenomena. And then my business was not merely with the scientific question; the manufacturing or commercial element in the consideration also touched me. I had to ascertain how I was likely to stand at some future time, did I presume to act upon the conclusions to which I came. There was nothing for me therefore but to analyze the ashes of the several vegetables which entered as elements into the rotation followed at Bechelbronn, but confining my inquiries to that portion of the vegetable which is looked upon particularly as the crop, so much of the plant as remains on the ground and is turned in again, of course taking nothing from it.

The ashes examined were almost all from the crops of 1841, two analyses having generally been made of each substance: and here I ought to say, that in this long and tedious labor, in which I spent nearly a whole year, I was most ably seconded by Mr. Letellier. By way of preface, I should say that in these analyses, losses will frequently be apparent, which for the most part exceed the limits that in the present day are tolerated in the more careful operations of the laboratory. These deficiencies, which puzzled me a good deal at first, I by and by discovered to proceed from the difficulty of incinerating certain vegetable substances completely. When they abound in alkaline salts, they leave ashes that melt so readily, that it becomes difficult to prevent their agglutination, and the charcoal that is not consumed is then effectually protected against any further action of the fire. There is nothing for it in such cases but to incinerate at the lowest temperature possible, and then a little moisture is apt to be left; the charcoal, however, is the substance that occasions the main difficulty, and the more important loss. To quote one of the instances, that of wheat, where the loss or deficiency is as high as 2.4 per cent. I may say that a direct inquiry after charcoal brought it out equal to 2, by which the actual deficiency is reduced to 0.4. I have not, however, introduced any correction for carbon, but present the reader with the results as they actually presented themselves to me. Among the number of the products of the analyses, alumina figures beside the oxide of iron. Alumina is an earth which I have always met with in minimum quantity in the ashes of plants, and is perhaps accidental; it may proceed from the earth which adheres to all herbaceous plants, and from which it is so difficult to free them completely.

COMPOSITION OF THE ASHES PROCEEDING FROM THE PLANTS GROWN AT BECHELBRONN.

Substances which yielded the ashes.	Acids.			Chlorine.	Lime.	Magnesia.	Potash.	Soda.	Silica.	Oxide of iron, alumina, &c.	Charcoal, moisture and loss.
	Carbonic.	Sulphuric.	Phosphoric.								
Potatoes	13.4	7.1	11.3	2.7	1.8	5.4	51.5	traces	5.6	0.5	0.7
Mangel-wurzel	16.1	1.6	6.1	5.2	7.0	4.4	39.0	6.0	8.0	2.5	4.2
Turnips	14.0	10.9	6.0	2.9	10.9	4.3	33.7	4.1	6.4	1.2	5.5
Potato tops	11.0	2.2	10.8	1.6	2.3	1.8	44.5	traces	13.0	5.2	7.6
Wheat	0.0	1.0	47.0	traces	2.9	15.9	29.5	traces	1.3	0.0	2.4
Wheat-straw	0.0	1.0	3.1	0.6	8.5	5.0	9.2	0.3	67.6	1.0	3.7
Oats	1.7	1.0	14.9	0.5	3.7	7.7	12.9	0.0	53.3	1.3	3.0
Oat-straw	3.2	4.1	3.0	4.7	8.3	2.8	24.5	4.4	40.0	2.1	2.9
Clover	25.0	2.5	6.3	2.6	24.6	6.3	26.6	0.5	5.3	0.3	0.0
Peas	0.5	4.7	30.1	1.1	10.1	11.9	35.3	2.5	1.5	traces	2.3
French beans	3.3	1.3	26.8	0.1	5.8	11.5	49.1	0.0	1.0	traces	1.1
Horse beans	1.0	1.6	34.2	0.7	5.1	8.6	45.2	0.0	0.5	traces	3.1

If these analytical results be now applied to the produce of an acre of ground, we should have the precise quantities of mineral substances abstracted from the soil by each of the several crops that enter into the rotation. Here they are in a table:

MINERAL SUBSTANCES TAKEN UP FROM THE SOIL BY THE VARIOUS CROPS GROWN AT BECHELBRONN UPON ONE ACRE.

Crop.	Dry crop.	Ashes per cent.	Quantity of ashes per acre.	Acids.		Chlorine.	Lime.	Magnesia.	Potash & Soda.	Silica.	Oxide of iron, alumnia, &c.
				Phosphoric.	Sulphuric.						
	lbs.	lbs.	lbs.	lbs.	lbs.	lbs.	lbs.	lbs.	lbs.	lbs.	lbs.
Potatoes	2828	4.0	113	13	8	3	2	6	58	6	1¾
Beet-roots	2908	6.3	183	11	3	9	13	8	82	15	4¾
Half crop of turnips, consumed off the ground,	656	7.6	50	3	5	1¼	5	2	19	3	0.7
Potato tops	5042	6.0	303	33	7	4	[illegible]	5	135	39	16
Wheat	1052	2.4	25	12	0.3	. .	8	4	7	0.4	..
Wheat-straw	2558	7.0	179	5	1.5	1	1	9	17	121	1¾
Oats	975	4.0	39	6	0.4	0.2	1½	3	5	21	0.6
Oat-straw	1176	5.1	60	1½	2.5	3	5	15	17	24	1
Clover	3693	7.7	284	18	7	7	70	18	77	15	0.9
Manured peas	915	3.1	28	8	1.2	0.3	3	3	10	0.5	traces.
French beans	1448	3.5	51	13	0.7	0.1	3	6	25	0.6	traces.
Horse beans	1944	3.0	58	20	0.75	0.5	3	5	26	0.3	traces.

On looking at this table we perceive that a medium crop of wheat takes from one acre of ground about 12 lbs., and a crop of beans about 20 lbs. of phosphoric acid; a crop of beet takes 11 lbs. of the same acid, and further, a very large quantity of potash and soda. It is obvious that such a process tends continually to exhaust arable land of the mineral substances useful to vegetation which it contains, and that a term must come when, without supplies of such mineral matters, the land would become unproductive from their ab straction. In bottoms of great fertility, such as those that are brought under tillage amidst the virgin forests of the New World at the present day, it may be imagined that any exhaustion of saline matters will remain unperceived for a long succession of years; for a succession of ages almost. And in South America, where the usual preliminary to cultivation is to burn the forest that stands on the ground, by which the saline and earthy constituents of millions of cubic feet of timber are added to the quantities that were already contained in the soil, I have already had occasion to speak of the ample returns which the husbandman receives for very small pains.* Under circumstances, in the neighborhood of large and populous towns, for instance, where the interest of the farmer and market-gardener is to send the largest possible quantity of produce to market, consuming the least possible quantity on the spot, the want of saline principles in the soil would very soon be felt, were it not that for every wagon-load of greens and carrots, fruit and potatoes, corn and straw, that finds its way into the city, a wagon-load of dung, containing each and every one of the principles locked up in the several crops, is returned to the land, and proves enough, and often more than enough, to replace all that has been carried away from it. The same principle holds good in regard to inorganic matters, which we have already established with reference to organic substances.

The most interesting case for consideration is that of an isolated farming establishment—a rural domain, so situated that it can obtain nothing from without, but exporting a certain proportion of its produce every year, has still to depend on itself for all it requires in the shape of manure. I have already shown, with sufficient clearness, I apprehend, how it is that lands in cultivation derive from the atmosphere the azotized principles necessary to replace the azotized products of the farm, which are continually carried away in the shape of grain, cattle, &c. I have now to show how the various saline substances, the alkalies, the phosphates, &c., which are also exported incessantly, are replaced. I believe that I shall be able, with the assistance of chemical analysis, to throw light on one of the most interesting points in the nature and history of cropping, and succeed in practically illustrating the theory of rotations. In what is to follow immediately, I shall always reason on the practical data collected at Bechelbronn, and which have already served for the

* The first breaks of the early English settlers in North America are now either very indifferent soils, or they have only been restored to some portion of their original fertility by manuring; so that the supply of fertilizing elements is not inexhaustible — Eng Ed

illustration of other particulars. My farm, I may say, by way of preliminary, is an ordinary establishment; the lands which have been brought by a system of rational treatment to a very satisfactory state of fertility, are not rich at bottom and originally, and they fall off rapidly if they have not the dose of manure at regular intervals, which is requisite to maintain them in their state of productiveness.

My first business was to determine the nature and the quantity of the mineral substances contained in my manure; and with a view to arrive at this information, I burned considerable quantities of dung at different periods of the year, mixed the ashes of the several incinerations, and from the mixture took a sample for ultimate analysis. The mean results are represented by:

Acids	Carbonic	2.0
	Phosphoric	3.0
	Sulphuric	1.9
Chlorine		0.6
Silica, sand		66.4
Lime		8.6
Magnesia		3.6
Oxide of iron, alumina		6.1
Potash and soda		7.8
		100.0

But our farm-yard dung is not the only article we are in the habit of giving to our land; it further receives a good dose of peat-ashes and gypsum. I here recall to the reader's mind that the mean composition of peat-ashes is this:

Silica	65.5
Alumina	16.2
Lime	6.0
Magnesia	0.6
Oxide of iron	3.7
Potash and soda	2.3
Sulphuric acid	5.4
Chlorine	0.3
	100.0

In the system followed at Bechelbronn, the farm-yard dung laid upon an acre contains 26 cwts. 3 qrs. of ashes. On our clover leas we spread the first year 7 cubic feet of turf-ashes; and in the beginning of spring of the second year, we lay on as much more, say 14 cubic feet, in all weighing about 2 tons. I do not take the 8 cwts. of gypsum which, in conformity with usage, the second year's clover generally receives, because I believe this addition to be perfectly useless after the very sufficient dose of peat-ash which we employ.

The whole of the mineral substances given to the land in the course of five years per acre is as follows, viz.: Ashes contained in the manure and in the peat-ashes, 7624 lbs.; consisting of phosphoric acid 90 lbs., sulphuric acid 304 lbs., chlorine 4.5 lbs., lime 532.5 lbs., magnesia 135.6 lbs., potash and soda 339 lbs., silica and sand 4630 lbs., oxide of iron, &c., 353 lbs.

It is therefore easy to perceive, from the preceding data, that what with the manure and the ashes it receives, the land is more than supplied with all the mineral substances required by the several crops it produces in the course of the rotation. Let us cast a glance over these with reference to their mineral or inorganic constituents, as we have already done in so far as the organic matters are concerned; let us compare, in a word, the quantity and the nature of the mineral substances removed in the course of five successive years, in contrast with the quantity and the nature of the same substances supplied at the commencement of the series, and we shall find that the sums of the phosphoric acid, sulphuric acid, and chlorine, and of the alkaline and earthy bases of the crops, are always smaller than the quantities of the same substances which exist in and are supplied to the arable soil.

I shall institute the comparison with the rotation No. 1, which begins with potatoes; and further, with a continuous crop which, as the one that is most common and convenient, shall be Jerusalem artichokes. I have not thought it advisable to discuss the rotation No. 2, in which beet replaces the potato, because the ashes of these two crops are so much alike, that it may be assumed to be matter of indifference which of the two enters as the drill-crop element into the series. With reference to the Jerusalem artichoke, I shall only remind the reader that the piece of land where it grows receives a dose of manure every two years, in the proportion of 41245 lbs. per acre, which manure contains 2776 lbs. of mineral constituent. Further, in the course of each winter peat-ashes, in the ratio of 2700 lbs. per acre, are laid on the land; and that the stems are generally incinerated on the spot, and the ashes they contain returned directly to the soil.

TABLE OF THE MINERAL MATTERS OF THE CROPS AND MANURES IN THE COURSE OF A ROTATION.

Average crop per acre.	Mineral substances in the crops.	ACIDS. Phos-phoric	ACIDS. Sul-Phuric	Chloride.	Lime.	Magnesia.	Potash and Soda.	Silica.
ROTATION NO. I.	lbs.	lbs.	lbs.	lbs.	lbs.	lbs.	lbs.	lbs.
Potatoes	113	13	8	3	2	6	58	6
2d and 4th years: wheat	50	24			1	8	15	
Ditto: wheat-straw	358	11	4	2	30	18	34	242
3d year: clover	284	18	7	7	70	18	77	15
5th year: oats	39	6			1	3	5	20
Ditto: oat-straw	60	1⅔	2½	3	5	1½	17	24
2d crop turnips; half crop	50	3	5	1	5	2	19	3
Sum of mineral substances	927	76⅔	27	16	114	56	225	310
Mineral substances of the manure	7582	90	304	32	533	136	339	5049
Excess over the mineral matters of the crops		14	277	15	417	79	114	4736
INCESSANT PRODUCTION OF THE JERUSALEM POTATO.								
1st and 2d years: mineral matters of the tubers	605	65	13	9	14	11	270	79
Mineral matters of dung	2777	65	53	17	239	100	219	1843
Ditto of turf ashes	4583		248	14	275	27	105	3002
Whole mineral matters of manures		83	301	31	514	127	324	4845
Difference in favor of the manures		18	287	20.5	500	117	52	4767

It was at one time asserted, that in order to ensure to a crop of wheat the necessary quantity of phosphates, its cultivation was preceded by one of roots or tubers, or leguminous plants, which were supposed to contain a much less proportion of these salts. By reference, however, to the table of mineral substances, removed from the soil by different crops, the absurdity of such reasoning becomes evident. For example, beans and haricots take 20 and 13.7 lbs. of phosphoric acid from every acre of land; potatoes and beet-root from the same surface take but 11 and 12.8 lbs. of that acid, exactly what is found in a crop of wheat. Trefoil is equally rich in phosphates with the sheaves of corn which have gone before it, and this large dose of phosphoric acid withdrawn from the soil, will nowise diminish the amount which will enter into the wheat that will by and by succeed the artificial meadow. It may be readily understood, that if the ground contains more than the quantity of mineral substances necessary for the total series of crops in a rotation, it is a matter of indifference whether the crops draw upon the soil in any particular order, and these succeed according to rules generally adopted for quite different reasons. It suits well, for instance, to begin a rotation with a drill crop sown in spring, and which, consequently, follows in our system the oats which closed the preceding rotation; it is a great advantage to be able to collect and cart out the manure during winter. Besides, the order is quite at the farmer's discretion, and there are places where, from particular reasons, quite another course is pursued. One part of the produce returns, as has been shown, to manure, after having served as fodder for the animals belonging to the farm. The inorganic matters are restored to the earth from which they came, deducting the fraction assimilated in the bodies of the cattle. Lastly, the whole of the wheat, and a certain amount of flesh will be exported, and with these a notable quantity of inorganic matter. Thus, in the above described rotation of five years, the minimum exportation of saline substances which must be removed from every acre of land, may be represented by $27\frac{1}{2}$ lbs. of phosphoric acid, and from 36 to 45 lbs. of alkali; this is just so much lost for the manure, and as there is definitively found at the end of the rotation a quantity of manure equal and nearly similar to that disposed of at the commencement, it is essential that the loss of mineral substance be made up from without, unless it be naturally contained in the soil.

In my first researches on the rotation of crops,* I stated that wherever there are exportable products, it becomes indispensable to keep a large proportion of meadow land, quoting, as an extreme case, the triennial rotation with manured summer-fallow. It is, in fact, the meadow which restores to the arable land the principles which have been carried off. This point, advanced upon analogy, is amply confirmed by the results of analysis.

I have examined, in reference to this question, the ashes of the hay of our meadows of Durrenbach, irrigated by the Sauer. The

* Memoir communicated to the Académie des Sciences, in 1838

analyses were made with ashes furnished by the crops of 1841 and 1842.

	I.	II.	III.	Average
Acids: Carbonic	9.0	5.5	"	7 3
Acids: Phosphoric	5.3	5.3	5.5	5.4
Acids: Sulphuric	2.4	2.9	"	2.7
Chlorine	2.3	2.8	"	2.6
Lime	20.4	15.4	"	17.9
Magnesia	6.0	8.3	"	7.2
Potash	16.1	27.3	"	21.7
Soda	1.2	2.3	"	1.8
Silica	33.7	29.2	"	31.5
Oxide of iron, &c.	1.5	0.6	0.5	0.9
Loss	2.1	0.4	"	1.0
	100.0	100.0		100.0

No. 1 yielded 6.0 per cent. of ash.
No. 2 " 6.2 idem.

In admitting as the average yearly return of our irrigated meadows, 3666 lbs. of hay and after-grass for the acre, it appears that we obtain, from a corresponding surface of land, 223.6 lbs. of ash, containing:

Acids: Carbonic	16.3
Acids: Phosphoric	12.1
Acids: Sulphuric	6.0
Chlorine	5.7
Lime	39.1
Magnesia	16.1
Potash and soda	52.0
Silica	70.4
Oxide of iron, and loss	4.2
	221.9*

In reckoning, as I have done, the lowest annual exportation of mineral substance from one acre of arable land at 5.5 lbs. of phosphoric acid and 8.2 lbs. of alkali, (potash and soda,) there must, in order to make up for loss, arrive each year at the farm a quantity of hay corresponding to about 1800 lbs. for every acre of ploughed land, which would establish between the arable and meadow land, a relation somewhat less than 1 to $\frac{1}{2}$.

In practice, the relation in question is sensibly less than that deduced from analysis; in some farms the meadow-land only occupies a fourth or fifth of the whole surface. When rye replaces wheat, the extent in meadow-land may be still more limited. It deserves notice, that I have supposed the arable land as destitute of proper inorganic matter, and that all came from the manure ashes and lime laid on, which is not rigorously true. There are soils containing traces of phosphates, and it is difficult to find clay or marl exempt from potash. Nevertheless, many clear-headed practical men begin to suspect that meadow has been too much sacrificed to arable land. In localities placed in similar conditions to those in which we are, removed from every source of organic manures, which, as I have shown in concert with M. Payen, are always furnished with saline

* The sum is only too small here from the number of places of decimals not having been carried out far enough.—Eng. Ed.

principles, an attempt has been made to imitate what is done in more favored districts, where it is possible, for example, to add animal remains to the manure. The corn crops felt this new procedure; nor could it be otherwise. But now there is a reaction in the opposite sense, and I could name most thriving establishments, where one-half of the farm is in meadow. The ever-increasing demand for butcher-meat will further this movement to the great advantage of the soil. In consequence of our peculiar position at Bechelbronn, nearly half our land is meadow, which allows of a large exportation of the produce of the arable land. In applying the results of the preceding analyses, I find that each year, provided there is no loss the hay ought to bring at least:

1254 lbs. of phosphoric acid,
627 " sulphuric acid,
602 " chlorine,
4155 " lime,
1672 " magnesia,
5456 " potash and soda,
7312 " silica.

This large amount of mineral substances is supplied by the meadows, which have no other manure than the water and mud thereby deposited, after flowing over the Vosges' freestone; they receive no manure from the farm, but are merely earthed with the sludge and mire borne down by the stream; these are real sources of saline impregnation. Meadows without running water ought not to be ranged in the same category, they only give the principles naturally contained in them; hence, they must be always manured every three or four years, and indeed, if not situate upon a naturally rich soil, are, according to my experience, very far from profitable.

The excess of mineral matters introduced into the ground over those that issue with the crops, an excess that ought always to be secured by judicious management, enriches the soil in saline and alkaline principles, which accumulate in the lapse of years, just as vegetable remains and azotized organic principles accumulate under a good system of rotation. By this, even in localities the most disadvantageously situate for the purchase of manure, temporary recurrence may be had to the introduction of such crops as flax, rape, &c., which being almost wholly exported, leave little organic residuum in the earth, and at the same time carry off a considerable quantity of mineral substance; circumstances which determine, as may be easily conceived, the maximum of exhaustion, and for that reason tend to reduce a soil becoming over-rich to what may be called the standard fertility.

In reviewing the chief points examined it will be seen, that as far as regards organic matter, the systems of culture which in borrowing most from the atmosphere, leave the most abundant residues in the land, are those that constitute the most productive rotations. In respect to inorganic matter, the rotation, to be advantageous, to have an enduring success, ought to be so managed that the crops exported should not leave the dung-hill with less than that constant quantity of mineral substance which it ought to contain. A crop which abstracts from the ground a notable proportion of one of its mineral elements, should not be repeatedly introduced in the course of a rotation, which depends on a given dose of manure, unless by the effect of time mineral element has been accumulated in the land. A clover crop takes up, for example, 77 lbs. of alkali per acre. If the fodder is consumed on the spot, the greater portion of the potash and soda will return to the manure after passing through the cattle, and the land eventually recover nearly the whole of the alkali. It will be quite otherwise if the fodder is taken to market; and it is to these repeated exportations of the produce of artificial meadows that the failure of trefoil, now observed in soils which have long yielded abundantly, is undoubtedly due. Accordingly, a means has been proposed of restoring to these lands their reproductive power, by applying alkaline manure.* If under such circumstances carbonate of soda would act as favorably as carbonate of potash or wood-ashes, the soda salt, in spite of its commercial value, might prove serviceable, and deserves a trial.

The lime manures naturally promote the growth of plants of which calcareous salts form a constituent; but here a capital distinction must be made. A soil may contain from 15 to 20 in the 100 of lime, and still be unable to dispense with calcareous manure; because the lime is in some other state than as it exists in chalk, as in the rubbish of pyroxene, mica, serpentine, and the like. A soil of this kind, although replete with lime, might still require gypsum for artificial meadow, and chalk for wheat and oats. It is from the carbonate that plants of rapid growth derive the lime essential to them, as was established by the researches of Rigaud de Lille, researches which have been censured by agricultural writers to whom they were unintelligible. I advocate the opinion of Rigaud, because in the Andes of Riobamba I have seen lucern growing in augitic rubbish, very rich in calcareous matter, and yet greatly benefited by liming.

The operation of gypsum is to introduce calcareous matter into plants. This I have endeavored to demonstrate from the analysis of the ash on the one hand, and on the other, from the consideration that finely divided carbonate of lime, as it exists in wood-ashes, acts with equal efficacy upon artificial meadows. By what means gypsum, if it does not enter the vegetable as a sulphate, parts with its sulphuric acid, is at present conjectural. It appears highly probable that calcareous matter is chiefly beneficial from the particular action it exercises on the fixed ammoniacal salts of the manure, transforming these successively, slowly, and as they may be wanted, into carbonate of ammonia. In the most favorable condition, the earth is only moist, not soaked with water, but permeable to the air. New researches will perhaps illustrate the utility of ammoniacal vapors thus developed in a confined atmosphere, where the roots are

* Information communicated by M. Schattenmann.

in operation. At least, it would be difficult to assign any other office to chalk in the marling or liming of land intended for corn, when we know how little lime corn absorbs. If, indeed, gypsum promotes the vegetation of trefoil, lucerne, sainfoin, &c., by furnishing the needful calcareous element, it could not fail to exercise an equally favorable agency upon wheat and oats, did they require it. The experiments adduced prove it not to be so, and their results are in some measure corroborated by analysis. Thus, if we compare the different quantities of lime withdrawn from the soil by trefoil and corn, we find them as follows:

The clover crop takes from 1 acre of ground nearly	70 lbs.	of lime.
Wheat " " "	16	"
Oat " " "	6.4	"

With this comparison before us, it seems evident that if the marling and liming of corn lands had no other object than the introduction of the minute portion of lime which is encountered in the crops, it would be difficult to justify the enormous expenditure of calcareous carbonate which is proved by daily experience to be advantageous.

It may be inferred from the foregoing, that in the most frequent case, namely, that of arable lands not sufficiently rich to do without manure, there can be no continuous cultivation without annexation of meadow; in a word, one part of the farm must yield crops without consuming manure, so as to replace the alkaline and earthy salts that are constantly withdrawn by successive harvests from another part. Lands enriched by rivers alone permit of a total and continued export of their produce without exhaustion. Such are the fields fertilized by the inundations of the Nile; and it is difficult to form an idea of the prodigious quantities of phosphoric acid, magnesia, and potash, which in a succession of ages have passed out of Egypt with her incessant exports of corn.

Irrigation is, without doubt, the most economical and efficient means of increasing the fertility of the soil, out of the abundant forage which it produces, and the resulting manure. Plants take up and concentrate in their organs the mineral and organic elements contained in the water, sometimes in proportions so minute as to escape analysis; just as they absorb and condense, in modified forms, the aeriform principles which constitute but some 10,000th parts in the composition of the atmosphere. It is thus that vegetables collect and organize the elements which are dissolved in water, and disseminated through the earth and the air, as a preparative to their being assimilated by animals.

CHAPTER VIII.

OF THE FEEDING OF THE ANIMALS BELONGING TO A FARM; AND OF THE IMMEDIATE PRINCIPLES OF ANIMAL ORIGIN.

§ 1. ORIGIN OF ANIMAL PRINCIPLES.

It is now generally admitted that the food of animals must necessarily contain azote; and this circumstance has led to the inference, that the herbivorous tribes obtain from their food the azote which enters into the constitution of their bodies.

In a general way, the individual consuming a certain portion of food every day, nevertheless does not increase in his average weight. This is what occurs with animals upon the quantity of food which is known to be sufficient for their keep; and it has been found that the human subject, living very regularly, returns at a certain hour, or at certain hours of the day, to a certain mean weight. Grooms, farm servants, &c., are perfectly well aware of the fact, that with a certain allowance of hay and corn, a horse will be kept in the condition necessary to do the work required of him without either gaining or losing in flesh.

Under such circumstances, the whole of the elementary matter contained in the food consumed, ought to be found in the dejections, the excretions, and the products of the act of respiration. And assuming that this is so, it might then be maintained that none of the elements is assimilated, assimilation being taken in the sense of an addition of principles introduced with the food to the principles already present in the body. Yet is there unquestionably assimilation, in the sense that the alimentary matters of the food become fixed in the system, having there undergone modification or change; and that they replace, or come instead of other elements of the same kind, which are daily thrown off by the vital acts of the economy.

During the nutrition of a young animal, and also in the process of fattening an adult, things go on differently; here there is unquestionably definitive fixation of a portion of the matter contained in the food: there is no longer balance between the waste and the supply; an animal then increases in weight notably and rapidly.

Looking at the question of feeding in the most general way, then, I admit that an adult animal, upon the daily allowance, voids a quantity of matter in its various excretions precisely equal to the quantity which it receives in its food:* all the elements, the same in nature and in quantity, which are contained in the food, are also contained in the excrements, vapors, and gases, which pass off from the living body; carbon and azote, hydrogen and oxygen, phospho-

* Boussingault, Annales de Chimie, 2e série, t. lxxxi, p. 113.

rus, sulphur, and chlorine, calcium, magnesium, sodium, potassium and iron, as they are all encountered in the food so are they all encountered in the body, and also in the excretions of an animal; and it seems certain, that no one of these primary or simple substances can be wanting in the nutriment without the body very speedily feeling the ill effects of its absence. Iron, for example, is a constant principle in the coloring matter of the blood; it also exists in large quantity in the hair; and he who should live on food that contained no trace of it would certainly, and before long, become disordered in his health.

In what has just been said, I take it for granted that animals do not absorb or assimilate any of the azote which forms so large a constituent in the air they breathe; and I am warranted in this by the researches of every physiologist of any name or distinction. Not only do animals obtain no azote from the atmosphere, but they actually exhale it incessantly, as was proved by M. Despretz in the course of his numerous experiments, and as I myself also demonstrated in the inquiries I undertook to ascertain whether herbivorous animals obtained azote from the air or not. The azote exhaled, it was discovered, proceeded entirely from the food consumed by the animal; a fact which, already of great importance in a physiological point of view and in reference to general physics, bears at the same time so immediately upon one of the most important questions of agriculture, that I think it well to give the particulars of one of the procedures by which it has been established.

The experiments in this case were performed on a milch-cow and a full-grown horse, which were placed in stalls so contrived that the droppings and the urine could be collected without loss. Before being made the subjects of experiment, the animals were ballasted or fed for a month with the same ration that was furnished to them during the three days and three nights which they passed in the experimental stalls. During the month, the weight of the animals did not vary sensibly, a circumstance which happily enables us to assume that neither did the weight vary during the seventy-two hours when they were under especial observation.

The cow was foddered with after-math hay and potatoes; the horse with the same hay and oats. The quantities of forage were accurately weighed, and their precise degree of moistness and their composition were determined from average samples. The water drunk was measured, its saline and earthy constituents having been previously ascertained. The excrementitious matters passed were of course collected with the greatest care; the excrements, the urine, and the milk were weighed, and the constitution of the whole estimated from elementary analyses of average specimens of each. The results of the two experiments are given in this table:

FOOD CONSUMED BY THE HORSE IN 24 HOURS.

Forage.	Weight in the wet state.	Weight in the dry state.	Elementary matter in the food. Carbon.	Hydrogen.	Oxygen.	Azote.	Salts and earths.
	lbs.	lbs. oz.	lbs. oz.	lb. oz. dwt.	lb. oz. dwt.	lb. oz. dwt.	lb. oz. dwt.
Hay, - - -	20	17 4	7 11	0 10 7	6 8 8	0 3 2	1 6 14
Oats, - -	6	5 2	2 7	0 3 18	1 10 14	0 1 7	0 2 10
Water, - -	43	..	..	..	..	..	0 0 8
Total, - -	69	22 6	10 6	1 2 5	8 7 2	0 4 9	1 9 12

PRODUCTS VOIDED BY THE HORSE IN 24 HOURS.

Products.	Weight in the wet state.	Weight in the dry state.	Elementary matter in the products. Carbon.	Hydrogen.	Oxygen.	Azote.	Salts and earths.
	lb. oz. dwt.	lb. oz. dwt.	lb. oz. dwt.	lb. oz. dwt.	lb. oz. dwt.	lb. oz. dwt.	lb. oz. dwt.
Urine, - -	3 6 15	0 9 14	0 3 10	0 0 7	0 1 2	0 1 4	0 3 10
Excrements, -	38 2 2	9 5 6	3 7 17	0 5 15	3 6 14	0 2 10	1 6 10
Total, - -	41 8 17	10 3 0	3 11 7	0 6 2	3 7 16	0 3 14	1 10 0
Total matter of the food, -	69 0 0	22 6 0	10 6 0	1 2 5	8 7 2	0 4 9	1 9 12
Difference, -	27 3 3	12 3 0	6 6 13	0 8 3	4 11 6	0 0 15	0 0 8

WATER CONSUMED BY THE HORSE IN 24 HOURS.

	lbs. oz.
With the hay, - - - - -	2 3
With the oats, - - - - -	0 14
Taken as drink, - - - - -	35 3
Total consumed, - - -	38 4

WATER VOIDED BY THE HORSE IN 24 HOURS.

	lbs. oz.
With the urine, - - - -	2 6
With the excrements, - -	23 8
Total voided, - - - - - -	25 14
Water consumed, - - -	38 4

Water exhaled by pulmonary and cutaneous transpiration, - - - - - -	12 6

FOOD CONSUMED BY THE COW IN 24 HOURS.

Fodder.	Weight in the wet state.	Weight in the dry state.	Elementary matter of the food. Carbon.	Hydrogen.	Oxygen.	Azote.	Salts and earths.
	lb. oz. dwt.	lb. oz. dwt.	lb. oz. dwt.	lb. oz. dwt.	lb. oz. dwt.	lb. oz. dwt.	lb. oz. dwt.
Potatoes, - -	40 2 5	11 2 1	4 11 2	0 7 15	4 10 17	0 1 12	0 6 13
After-math hay,	20 1 2	16 11 0	7 11 11	0 11 7	5 10 17	0 4 17	1 8 6
Water, - -	160 0 0	..	..	..	..	..	0 1 12
Total, - -	220 3 7	28 1 1	12 10 13	1 7 2	10 9 14	0 6 9	2 4 11

PRODUCTS VOIDED BY THE COW IN 24 HOURS.

Products.	Weight in the wet state.	Weight in the dry state.	Elementary matter in the products. Carbon.	Hydrogen.	Oxygen.	Azote.	Salts and earths.
	lb. oz. dwt.	lb. oz. dwt.	lb. oz. dwt.	lb. oz. dwt.	lb. oz. dwt.	lb. oz. dwt.	lb. oz. dwt.
Excrements, -	76 1 9	10 8 12	4 7 0	0 6 13	4 0 9	0 2 19	1 3 8
Urine, - -	21 11 12	2 6 17	0 8 7	0 0 16	0 8 3	0 1 3	1 0 6
Milk, - -	22 10 10	3 1 0	1 8 3	0 3 3	0 10 6	0 1 9	0 1 16
Total, - -	120 11 11	16 4 9	6 11 10	0 10 12	5 6 18	0 5 11	2 5 10
" matter of food,	220 3 7	28 1 1	12 10 13	1 7 2	10 9 14	0 6 9	2 4 11
Difference, -	99 3 16	11 8 12	5 11 3	0 8 10	5 2 16	0 0 18	0 0 19

WATER CONSUMED BY THE COW IN 24 HOURS.

	lbs. oz.
With the potatoes, - - - -	23 12
With the hay, - - - - - -	2 9
Taken as drink, - - - - -	132 0
Total consumed, - - -	158 5

WATER VOIDED BY THE COW IN 24 HOURS.

	lbs. oz.
With the potatoes, - - - -	53 10
With the urine, - - - - -	15 14
With the milk, - - -	16 3
Total voided, - - - - -	85 11
Water consume', - - - - -	158 5

Water passed off by pulmonary and cutaneous transpiration, - - - - -	72 10

From these sums it appears that the azote of the excrements is less by from 339.6 to 455.0 grains than that of the forage consumed. It appears also that the whole quantity of elementary matter contained in the excrements is less than that which had been taken as food; the difference is of course due to the quantities which were lost by respiration and the cutaneous exhalation.

The oxygen and hydrogen that are not accounted for in the sum of the products have not disappeared in the precise proportions requisite to form water; the excess of hydrogen amounts to as many as from 13 to 15 dwts. It is probable that this hydrogen of the food became changed into water by combining during respiration with the oxygen of the air.

The loss of carbon, which is very considerable, seeing that in the two experiments it amounts to nearly 12½ lbs., must have gone to form the carbonic acid, which is known to be so large and important a constituent in the expired air, and which is also exhaled from the general surface of the body. Neglecting the latter, it appears that each of the animals produced in the course of twenty-four hours upwards of 13 cubic feet of carbonic acid gas, the thermometer supposed at 32° F., the barometer at 30 inches.*

During respiration, then, or as a consequence of respiration, the carbon and hydrogen of the food have disappeared and given rise, by the concurrence of the oxygen of the air, to carbonic acid and water, precisely as if they had been burned. And an animal may, in fact, be regarded as an apparatus or system, in which a slow combustion is incessantly going on; there is perpetual disengagement of carbonic acid gas and of the vapor of water, just as there is from a stove in which any organic substance, wood, for example, is burning. In either case there is evolution of heat; all animals have a temperature above that of the medium which surrounds them, and the excess of the elevation is in some sort relative to the activity of the respiratory process, or, in other words, to the intensity of the combustion.

Under the influence of the oxygen that is taken into the body, the soluble principles of the blood pass through a series of modifications, the last of which is carbonic acid, which is exhaled and dissipated in the air; and it is in this way that a portion of the carbon of the food is returned to the atmosphere, after having accomplished the important function of supplying the animal with the heat that is necessary to its existence. Far from deriving any thing from the air, consequently, animals, on the contrary, are continually pouring carbon into it. The food is, therefore, the only source whence animals derive the matter that enters into their constitution; and, as the primary food of animals is obtained from vegetables, herbivorous creatures must necessarily find in the plants they consume all the

* The large quantity of carbonic acid shows the necessity for large and well-ventilated stables and cow-houses. A cow, it appears, will vitiate 66 cubic feet of air in a day. It will be observed in the table that the saline and earthy matters of the ejecta exceed those of the ingesta in both instances. This is from error in observation, and is owing to the difficulty of determining exactly the quantities of these substances. The error is less in the case of the horse than in that of the cow.

elements they assimilate. It might be expected from this, that the material constitution of animals should approach, and sometimes even be identical with that of vegetables; and it is found, in fact, that a considerable number of ternary or quarternary organic compounds, of either kingdom, present the greatest analogy to one another; their identity, in some cases, is even complete. Some fatty substances of animal origin do not differ in any way from vegetable fats; the margaric acid which is obtained from hog's lard has the precise characters of the margaric acid which is furnished by olive oil, and the same identity is preserved through the entire series of quarternary azotized principles, as a glance at the following table, which contains the results of the analyses performed by Messrs Dumas and Cahours, will show.

	FIBRINE.		ALBUMEN.		CASEINE.	
	Animal.	Vegetable.	Animal.	Vegetable.	Animal	Vegetable.
Carbon	52.8	53.2	53.5	53.7	53.5	53.5
Hydrogen........	7.0	7.0	7.1	7.1	7.0	7.1
Oxygen..........	23.7	23.4	23.6	23.5	23.7	23.4
Azote...........	16.5	16.4	15.8	15.7	15.8	16.0
	100.0	100.0	100.0	100.0	100.0	100.0

These principles, to which must be added gelatine, the fats and several earthy and alkaline salts, constitute the frame-work of the animal tissues, or the fluids which penetrate them; it is therefore necessary for us to examine each of them shortly.

Gelatine is met with in almost all the solid parts, in the bones, tendons, cartilages, skin, cellular tissue, muscular flesh—all contain it. It is readily soluble in boiling water; cold water only takes up a small quantity of it. Two or three parts of gelatine dissolved in 100 parts of hot water, suffice to turn the fluid into a tremulous jelly when it has become cold. Tannin, or infusion of gall-nuts, precipitates gelatine completely from its solution, the precipitate being very bulky and perfectly insoluble in water; and it is this chemical combination or principle which lies at the bottom of the art of tanning.

Gelatine is extensively used in the arts, under the familiar name of glue. Isinglass consists of gelatine nearly pure, and, according to Mulder, contains:

Carbon......................	50.8
Hydrogen	6.6
Azote.......................	18.3
Oxygen......................	24.3
	100.0

Fibrine occurs in a state of solution in the blood, and forms the principal ingredient in muscular flesh. It is readily obtained by whipping a quantity of blood just taken from the veins of a living animal; the white stringy masses that adhere to the rod are fibrine, which, by gentle kneading under water, become colorless. Fibrine, when moist, is a highly elastic and flexible substance; dried, it loses about 30 per cent. of water, and becomes brittle, horny, semi-transparent. Thrown into water, it gradually imbibes all it had lost by drying, and regains its former properties. Burned and incinerated, fibrine leaves a quantity of ash, which consists, for the major part, of phosphate of lime, with which is mixed a small quantity of phosphate of magnesia and of oxide of iron.

Albumen exists in large quantity dissolved in the water or serum of the blood, and in the white of the egg; it is also found in almost all the animal fluids that are not excretions, or destined to be thrown off as useless to the system. Albumen, as familiarly known, has the remarkable property of coagulating or setting into a soft fluid, at a certain temperature—158° F.

Caseum, or *caseine*, is the distinguishing principle of milk. By combining with acids it forms an insoluble compound; and it undergoes a remarkable coagulation, as all the world knows, in contact with a piece of the inner membrane of the stomach of a young animal: from a fluid it sets into a soft solid, which by degrees separates into two portions—whey and curd. The curd, or caseum, always contains fat, and, when burned, leaves a considerable quantity of ash.

Physiologists distinguish three principal tissues in the bodies of animals; the muscular, the nervous, and the cellular.

The *muscular tissue* consists of an assemblage of contractile fibres, here disseminated through the masses of organs, there collected into bundles and constituting the flesh. This is the instrument by which animals perform all their voluntary motions, and it is that also by which all the active but involuntary movements of the body are excited. Muscular flesh is always a compound substance, however; it consists of fibrine, the contractile or proper element, albumen, fat, gelatine, an odorous extractive matter, lactic acid, different salts and the coloring principle of the blood.

Put into cold water, so long as the temperature is below from 130° to 140° F., little effect is produced beyond the solution of the soluble salts which it may contain, and of a portion of its extractive matter and albumen. At from 175° to 195°, the albumen which had been dissolved, coagulates and rises to the top as scum, and the fat melts and floats on the surface. The fibrinous element of the meat, however, preserves its characters even after the action of boiling water continued for some time.

The *nervous tissue* constitutes the brain, spinal marrow, and nerves, distributed to all parts of the body. Brain in its composition contains a large quantity of water,—80 per cent.—certain fatty matters, albumen, osmazone, phosphorus in combination with fat, sulphur, and phosphates of potash, lime, and magnesia. The composition of the brain of animals, the dog, the sheep, the ox, appears to be very analogous to that of the human subject.

Cellular tissue is the general connecting medium throughout the animal body, and is not only met with, it may be said, everywhere, but forms a main element in many of the textures of the body, such as the serous and mucous membranes, the cartilages, the bones themselves, which are in fact only cellular tissue impregnated with calcareous salts. Tendons may be viewed as condensed ropes of cellular tissue ; by long boiling in water they melt entirely into gelatine.

Bones consist of cellular tissue, as stated, resolvable into gelatine, and of a large proportion of saline earthy matter, consisting principally of phosphate of lime. The presence of this phosphate is not extraordinary, inasmuch as we have found that it forms an element in all the vegetables upon which animals are supported. By boiling bones even reduced to powder under the usual pressure of the atmosphere, but a small quantity of their gelatine is obtained ; but by putting them into a Papin's digester, and subjecting them to a considerably higher temperature than that of boiling water, we can dissolve the whole, or nearly the whole of the animal matter, and leave the earthy parts unchanged ; or by proceeding in another way, by soaking bones for a time in dilute muriatic acid, we can dissolve out the earthy matter, and leave the bone, having its original form indeed, but as an elastic, pliant gristle.

The relations between the earthy and organic matter of bone, vary with the species, but especially with the age of the animal. In early life the cellular element predominates ; in adult age the salts predominate. We have three analyses of bone, which I shall here present :

	Man.	Ox.	Ox.
Cartilage susceptible of change into gelatine	33.3	33.3	50.0
Sub-phosphate of lime	53.0	57.4	37.0
Carbonate of lime	11.8	3.9	10.0
Phosphate of magnesia	1.2	2.0	1.2
Soda, and a trace of common salt	1.2	3.4	"
	100.0	100.0	98.3

Hair has a very complex composition, no fewer than nine different principles or substances having been detected in its constitution ; among the number, mucus, various oily matters, sulphur, and iron ; wool, fur, and horn, are all similar in their composition to hair.

Blood, in all the higher animals, is a sluggish fluid, of a deep red color ; in many of the inferior tribes, however, such as insects, crustaceans, and shell-fish, it is limpid, and generally colorless. Under the microscope, red blood is seen to consist of two distinct portions, a serum or whey, in which float a multitude of minute, solid, opaque corpuscles—the globules of the blood of physiologists, particles which have different characters in different classes of animals.

Blood is a very heterogeneous compound. Left to itself, after being drawn from a vein, it sets or coagulates into a soft gelatinous solid, which by and by begins to separate into two portions, one watery, of a yellowish color, and opalescent, the water, whey, or serum ; another solid, of a deep red or reddish brown color, the clot or coagulum. The watery portion contains a large quantity of albumen in solution. M. Lecanu, in his analysis of the blood, speaks of as many as twenty-five different substances as entering into its composition :

Water	790 4
Oxygen, azote, free carbonic acid Iron Hydrochlorates of soda, potash, ammonia Sulphates of potash and of soda Subcarbonate of lime and magnesia Phosphates of soda, lime, and magnesia Lactate of soda A soap, having soda and fixed fat acids for its elements An odorous, volatile salt, a fat acid A fatty substance, containing phosphorus Cholesterine Seroline	11.0
Albumen dissolved in the water	67.8
Globules and fibrine	130.8
	1000.0

The blood globules consist principally of albumen combined with a little fibrine and red coloring matter. Any difference observed between one sample of blood and another, is connected especially, almost exclusively, with the relative proportions of the liquid part or serum, and the solid part or clot. The solids are in larger proportion in males than females, in grown-up persons than in aged individuals and children, in subjects well and abundantly fed than in those indifferently supplied with food. No analysis that has yet been made has thrown any true light on the cause of the difference of color perceived between arterial and venous blood; nevertheless, it is positively known that it is by the concurrence of the oxygen of the atmosphere that the arterial blood in the living body acquires the characters which distinguish it, and that carbonic acid gas is evolved or thrown off in the course of the action that takes place.

Ox blood, thoroughly dried, has been found to consist of:

Carbon	52.0
Hydrogen	7.2
Azote	15.1
Oxygen	21.3
Ash	4.4
	100.0

Milk. This well-known fluid may be said to combine in itself all the organic principles and mineral substances which enter into the constitution of organized beings. Caseum, identical with fibrine and albumen, fatty matters, sugar of milk, and different salts, among the number of which the phosphates stand distinguished.

The caseum, the sugar, and a portion of the salts, are in solution; the fatty matters are held in suspension in the milk in the form of globules. The following table will be found useful, as giving a comprehensive survey of the composition of different kinds of milk.

Milk.	Caseum, albumen and insoluble salts.	Fatty matter.	Sugar of milk and soluble salts.	Water.	Dry matter in 100 of milk.	Remarks.	Authors of the analyses
Of the cow..	3.6	4.0	5.0	87.4	12.6	Average of 12 analyses at Bechelbronn.	Le Bel and Boussingault.
Of the cow..	3.8	3.5	6.1	86.6	13.4	Average of 6 analyses in the environs of Paris.	Quevenne.
Of the cow..	4.5	3.1	5.4	87.0	13.0	Idem.	Henri and Chevalier.
Of the cow..	5.6	3.6	4.0	86.8	13.2	Idem.	Lecanu.
Of the cow..	5.1	3.0	4.6	87.3	12.7	An analysis, Giessen.	Haidlen.
Of the ass...	1.7	1.4	6.4	90.5	9.5	Average of 5 analyses.	Péligot.
Of woman...	3.1	3.4	4.3	89.2	10.8	Of good quality.	Haidlen.
Of woman...	2.7	1.3	3.2	92.8	7.2	Of middling quality.	Haidlen.

Cow's milk always shows slight alkaline reaction; its density is about 1.03. According to M. Haidlen, it contains no salt formed by an organic acid, no lactates, and the alkali is in combination with caseum, the solution of which it assists. It may contain about a half per cent. of ash, the several constituents of which appear to be very stable, though their proportions vary greatly. In 100 parts of milk, taken from two different cows, Haidlen found the following salts:

Phosphate of lime	0.231	0.344
Phosphate of magnesia	0.042	0.064
Phosphate of iron	0.007	0.007
Chloride of potassium	0.144	0.183
Chloride of sodium	0.024	0.034
Soda	0.042	0.045
	0.490	0.677

As cow's milk is that which is by far the most directly interesting to agriculture, I shall enter somewhat particularly into its history; having, however, already spoken of caseum, its distinguishing constituent, and albumen, I shall here confine myself to the subject of the sugar and the oil or butter.

Sugar of milk is prepared for commercial purposes, in countries or districts where cheese-making is carried on to a great extent, and the quantity of whey at command is very large. In some Cantons of Switzerland, sugar of milk is obtained by simply evaporating whey properly clarified, to the consistence of sirup, which deposites the sugar in the crystalline form as it cools. This first produce is brown, and contaminated with various impurities, from which it is freed by repeated solutions and crystallizations. It then becomes colorless, transparent, and nearly tasteless, feeling gritty between the teeth, and having only an obscure sweet taste. It requires from 8 to 9 parts of cold water to dissolve it; in hot water it is more soluble. According to Proust, it consists of:

Carbon	40.0
Hydrogen	6.7
Oxygen	53.3
	100.0

Butter. To understand the preparation of butter thoroughly, it is absolutely necessary to know the physical constitution of the milk from which it is obtained. Now the microscope shows us that milk holds in suspension an infinity of globules of different dimensions, which, by reason of their less specific gravity, tend to rise to the surface of the liquid in which they float, where they collect, and by and by form a film or layer of a different character from the fluid beneath; the superficial layer is the *cream*, and this removed, the subjacent liquid constitutes the *skim-milk*. This separation appears to take place most completely in a cool temperature from 54° to 60° F.

Allowed to stand for a time, which varies with the temperature, milk becomes sour, and by and by separates into three strata or parts: cream, whey, and curd, or coagulated caseum. By suffering the milk to become acid before removing the cream, it has been thought that a larger quantity of this, the most valuable constituent of the milk, was obtained; and the fact is probably so; but in districts where the subject of the dairy has been most carefully studied, it has been found that it is better to cream before the appearance of any signs of acidity have appeared. When a knife can be pushed through the cream, and withdrawn without any milk appearing, the cream ought to be removed.*

Butter is obtained from cream by churning, as all the world knows; by the agitation, the fatty particles cohere and separate from the watery portion, at first in smaller and then in larger masses. The remaining fluid is buttermilk, a fluid slightly acid, and of a very agreeable flavor, containing the larger portion of the caseous element of the cream coagulated, and also a certain portion of the fatty principle which has not been separated.

The globules of milk appear, from the latest microscopical observations,† to be formed essentially of fatty matter, surrounded with a delicate, elastic, transparent pellicle. In the course of the agitation or trituration of churning, these delicate pellicles give way, and then the globules of oil or fatty matter are left free to cohere, which they were prevented from doing previously, by the interposition of the delicate film or covering of the several globules. Were the butter simply suspended in the state of emulsion in the milk, we should certainly expect that it would separate on the application of heat; but this it does not: cream or milk may be brought to the boiling point, and even boiled for some time, without a particle of oil appearing. Could M. Romanet show any of these pellicles, apart from the oil-globules they enclose, it would be very satisfactory, and would certainly enable us to explain the effect of churning.

Churning is a longer or shorter process, according to a variety of

* Thaer, Principes, &c., t. iv. p. 341.
† M. Romanet, MSS.

circumstances; it succeeds best between 55° and 60° F. So that, in summer, a cool place, and in winter a warm place, is chosen for the operation. There is no absorption of oxygen during the process of churning, as was once supposed; the operation succeeds performed in vacuo, and with the churn filled with carbonic acid or hydrogen gas.

On being taken out of the churn, the butter is kneaded and pressed, and even washed under fair water, to free it as much as possible from the buttermilk and curd which it always contains, and to the presence of which must be ascribed the speedy alteration which butter undergoes in warm weather. To preserve fresh butter it is absolutely necessary to melt it, in order to get rid of all moisture, and at the same time to separate the caseous portion. This is the process employed to keep fresh butter in all the warmer countries of the world. In some districts of the continent, it is also had recourse to with the same view. The butter is thrown into a clean cast-iron pot, and fire is applied. By and by the melted mass enters into violent ebullition, which is owing to the disengagement of watery vapor; it is stirred continually to favor the escape of the steam, and the fire is moderated. When all ebullition has ceased, the fire is withdrawn, and the melted butter is run upon a strainer, by which all the curd is retained. M. Clouet has proposed to clarify butter by melting it at a temperature between 120° and 140° F., and keeping it so long melted as to dissipate the water and secure the deposition of the cheesy matter, after which the clear melted butter would be decanted. I doubt whether by this means the water could be sufficiently got rid of, a very important condition in connection with the keeping of butter, though certainly all the caseum would be deposited.

The moisture and curd contained in fresh butter may amount together to about 18 per cent.; at least we find that we lose about 18 lbs. upon every 100 lbs. weight of butter which we melt at Bechelbronn.

The information which we have on the produce in butter and cheese, from different samples of milk, is very discordant, so that I prefer giving the results of a single experiment made under my own eyes. From 100 lbs. weight of milk, we obtained:

Cream	15.60 lbs.
White curd cheese	8.93 "
Whey	75.47 "
	100.00

The 15.60 lbs. of cream yielded by churning:

3.3 lbs. butter, or 21.2 per cent., and
12.27 " buttermilk.

The reckoning with reference to 100 lbs. of milk consequently stands thus:

Cheese	8.93
Butter	3.33
Buttermilk	12.27
Whey	75.47
	100.0

Taking the whole of the milk obtained and treated at different seasons of the year, I find that 36,000 lbs. of milk yielded 1080 lbs. of fresh butter, which is at the rate of 3 per cent. From the statement of M. Baude, it appears that near Geneva a proportion of butter so high as 3 per cent. is never obtained, probably because there a larger proportion of fatty matter is left in the cheese. In the dairy of Cartigny, 2200 gallons of milk gave:

Butter	363 lbs.	or about	1.6	per cent
Gruyère cheese	1515	"	6.9	"
Clot from the whey, obtained by boiling	1140	"	5.2	"

In the same neighborhood, another dairy, that of Lullin, gave from the same quantity of milk:

Butter	418 lbs.	or 1.9	per cent.
Cheese	1485	67.5	"
Clot from whey	968	4.4	" *

OF THE FOOD OF ANIMALS AND FEEDING.

The identity, in point of composition and properties, which appears to obtain between certain substances derived from either kingdom of nature, naturally led to the conclusion that animals do not form or originate the substances which enter into their organization, but that they find these ready formed in their food, and merely appropriate them; whence we must conclude, that herbivorous animals assimilate several of the proximate principles of plants immediately, causing them to undergo but slight modifications, and that the elements of the animal tissues and fluids pre-exist in vegetables, which further contain the earthy phosphate that forms the distinguishing characteristic in bone.†

The food of herbivorous animals must, therefore, always contain, and in fact always contains, four essential principles, which, by their combination or reunion, constitute nutritious matter, properly so called:—1st. An azotized matter, such as albumen, caseine, gluten, substances which are probably the original of flesh. 2d. An oily or fatty matter, which approaches more or less closely to fatty bodies in general. 3d. A substance having a ternary composition, sugar, gum, fecula. 4th. Certain salts, particularly phosphates of lime, magnesia, and iron. This mixed constitution, which a forage plant must needs offer, justifies the general ideas propounded by Dr. Prout on nutrition. This able chemist has said that milk was to be viewed as the standard food, and that all alimentary matters must resemble it in composition, in greater or less degree: that is to say, besides phosphates, food must contain an azotized principle, a non-azotized principle, and a fatty body, to stand in lieu of caseum, sugar, and butter.

The fundamental principle that animals find the several substances which make up their bodies, ready formed in the substances they

* In all the dairy counties of England, the milk is never required, like the ground, to give a double crop; it yields either butter or cheese, not both. Hence the greater richness of English cheese in general.—Eng. Ed.

† Dumas and Boussingault. The Chemical and Physiological Balance of Organic Nature, post 8vo. London. Baillière, 1843. (A very useful little work.—Eng. Ed.)

consume, seems very well calculated to assist the practical farmer in managing the food of the animals upon his land; for if flesh, fat, and bone exist all but ready formed in the food, it is obvious that the best kind will be that precisely which, under the same weight, contains the largest quantity of the various matters of the organization.

It is by no means easy to ascertain precisely the amount of the azotized constituents, gluten, and albumen, contained in plants; to do so requires both time and pains. But let it be once admitted that the nutritive properties of forage increase in the precise ratio of these matters, this is clearly as much as to say that the value is in proportion to the quantity of azote contained in the food, and that it becomes a matter of the highest moment to have at hand a ready mode of determining the point. I believe it infinitely better to get at the quantity of azote immediately, which is easily done, than by any roundabout and laborious process to ascertain the amount of albumen and gluten: the quantity of azote ascertained, it is most easy to deduce the quantity of albumen and gluten—in other words, of *flesh*—contained in each particular species of food examined; for, as a general rule, vegetable food does not contain any other azotized principle. It is true, indeed, that all the azotized principles of vegetable origin cannot be considered as nutritious; some of them, on the contrary, are virulent poisons or active medicines, according to the dose in which they are administered. But these poisonous substances are not met with in appreciable quantity in the plants which are commonly grown for the food either of man or beast. Still, all the truly nutritious articles of food contain an azotized principle. The experiments of M. Magendie have shown, that substances which contain no azote, such as sugar, starch, oil, will not support life; and, on the other hand, it is ascertained that the quality of alimentary matter, flour for example, increases with the amount of gluten which it contains. It is because the seeds of the leguminous vegetables are richer in azotized principles—that is, in *flesh*—that they are also more highly nutritious than the seeds of the cereals.

These several considerations, therefore, induce me to conclude, that *the nutritious principles of plants and their products reside in their azotized principles, and consequently that their nutritious powers are in proportion to the quantity of azote they contain.* From what precedes, however, it is obvious that I am far from regarding azotized principles alone as sufficient for the nutrition of animals; but it is a fact, that every highly azotized vegetable nutritive substance is generally accompanied by the other organic and inorganic substances which concur in nutrition.

In seeking to learn the precise quantity of azote contained in a great number of articles used as food for cattle, I have had it in view particularly to find a standard or fixed point for estimating their comparative nutritive properties. It is long since more than one of the most distinguished farmers, both of England and Germany, essayed to resolve this important problem in rural economy. Thus Thaer and many others have given tables of the quantities by weight in which one article of alimentation might be substituted for another These tables are in fact tables of equivalents with reference to food. But it is unfortunate that there should be considerable diversity of statement among their authors. Yet, even up to the present time, it could not well have been otherwise, and these discrepancies will only surprise those who are unacquainted with the difficulties of the subject. One grand cause of difference probably exists in the degree of dryness of the article subjected to experiment. The nature of the soil, a very dry or very rainy season, the climate, &c., must all be regarded as so many causes influencing the quantity of water contained in plants, and in consequence their actual nutritive qualities. The only sure mode of proceeding, in short, appears to be, to reduce the several articles to a state of complete dryness, and to make their quantity in this condition the first element in the reckoning. I may state, that the theoretical data obtained by proceeding in this way have already been approved by practical applications.

Hay may be assumed as the most common or universally used of all kinds of fodder: it is in some sort the staple food of the animals that are particularly attached to an agricultural concern, and may therefore be appropriately made the standard of comparison for all other kinds of food or forage. Hay itself, however, varies greatly in point of quality: in assuming it as the standard, I have therefore to state, that meadow hay of good quality is to be understood. The analyses which I have made of this article at different times, satisfy me that in the state in which it is commonly used, it contains from 1.0 to 1.5 of azote per cent. In choosing a specimen for analysis, it is, of course, highly necessary that it be an average specimen; that it consist of equal or rather relative proportions of the several elements which enter into its constitution, such as stalks, leaves, flowers, and seeds. Taking a sample of hay, for instance, weighing exactly 5 lbs. avoird., I found that it was made up of—

Hard woody stems	2.393 lbs.
Bottoms of leaves and very fine stems	0.847
Flowers, leaves, and a few seeds	1.760
	5.000

The ultimate analysis of which gave:

Of azote per cent.			1.19
Military contract hay of 1840 gave of azote per cent.			1.21
Hay made in Alsace in 1835	"	"	1.04
Hay made in Alsace in 1837	"	"	1.15
Average of azote per 100			1.15

Hay, as it is generally used, contains from 11 to 12 per cent. of water, which is got rid of by thorough drying. And as albumen, caseum, and vegetable gluten contain 16 per cent. of azote, we perceive that the azotized matter which is the representative of *flesh*, in hay may be represented by the number 7.2 per cent. Hay does not, indeed, always contain so much azote; that which is won from marshy lands contains much less; and again there are samples that contain more. After-math, or second-crop hay, is certainly more nutritious than first-crop hay, a fact which we have ascertained epeatedly at Bechelbronn ; but this hay is nevertheless held less uitable for horses, probably because, being made late in the season, t is commonly stacked more or less damp, and suffers change in onsequence :

After-math hay gave	2.0	per cent. of azote
A choice sample of the best hay	1.29	"
The flower or ear, containing little woody stem	2.1	"

These examples suffice to show, that when an animal is to be put pon another kind of food than hay, it is very necessary to take the uality of the latter article, which has been employed, into the acount. In the table which I shall immediately present, I have asumed good meadow-hay, containing 1.15 of azote and 11 of water er cent. for my standard.. The importance of a table of equivalents or forage has long been felt by farmers ; and they who have given heir attention to the accumulation of data for its construction, deerve our best thanks. The use of a table of equivalents is extremey simple : the numbers placed underneath the value of hay inicate the weights of the several kinds of forage named in the first olumn, which may respectively be substituted for 100 parts of hay y weight. Thus, according to Block, 366 lbs. of carrots may be ubstituted for 100 lbs. of meadow-hay. Pabst holds 60 lbs. of oats o be equivalent to 100 lbs. of hay. If the question be to replace .26 or 7¼ lbs. of oats in the ration of a horse by Jerusalem artihokes, we find in the table that 60 of oats are equivalent to 274 erusalem potatoes, and we therefore infer that 35.2, say 35¼ lbs. s the weight of the root to be substituted for that of the oats.

Certain information on the nutritive value of the various articles onsumed by cattle as food, is really of high importance in rural conomy ; it is obviously the only guide for the feeder in the use or urchase of forage. Let us suppose, for example, that a measure of otatoes (22 gallons) weighing 165 lbs. is worth 10*d.* at market, nd that hay is worth 2*s.* 6*d.* the cwt. ; 2 cwts. or rather 220 lbs. ould cost 5*s.* Let us now admit, on theoretical grounds, that this uantity of hay is equivalent to 693 lbs. of potatoes; it plainly apears, on looking at the cost of these equivalents, that there would e a positive advantage in using potatoes, inasmuch as they are orth no more than 3*s.* 6½*d.* There would indeed be money to be ade by selling hay, and purchasing its equivalent in potatoes.

The equivalents which I have deduced from my elementary anases, agree on many occasions with the conclusions of practical en ; in others, they differ notably from them ; at the same time it ust be observed, that the practical equivalents differ from one nother in at least an equal degree. We see, for instance, that chnee and Thaer think 220 lbs. of hay will be replaced by 1465 s. of wheat straw, while Flottow gives 429 lbs. as the equivalent umber. According to Mayer, 630 lbs. of turnip are equivalent to 20 lbs. of hay, while Middleton gives 1760 as the equivalent numer of turnips, a number which coincides remarkably with that infered from theory. Block assigns 66 as the equivalent number of eas. Thaer makes it more than twice as high, viz. 145. Mangel wurzel, according to Thaer, is represented by 1012; while Mayer and Pabst call it but 550, and M. de Dombasle states it a little higher, viz. 574. However highly we estimate the difficulties of coming to accurate conclusions on the subject of alimentation or feeding, it is not easy to account for such discrepancies among practical men; and then, as to the astonishing similarity which their conclusions bear to one another upon many heads, it is impossible to overlook the fact, that the resemblance is far more in appearance than in fact; for it is notorious, that the generality of those who have committed themselves to writing have generally copied each other. Indeed, it is not always very obvious whether the equivalent number which we find assumed, has been determined by the farmer from his own observation or experience, or has been adopted from some other observer. No one who is not a total stranger to the art of making experiments will ever be brought to believe that eleven experimenters, operating separately, could have fallen plump upon the number 90 as the equivalent for lucern, or even that any five of them could have lighted upon 600, neither more nor less, as the equivalent number for cabbage!

The method which I have myself pursued, that namely of inferring the nutritious quality from the contents in azote, is far from being free from objection; on the whole, it may be said to place the equivalents somewhat too low, inasmuch as by the process of elementary analysis, the quantity of azote is apt to come out a little too high, some portion of it being derived from the nitrates present in vegetables, which are certainly of no avail in nutrition. This is the source to which I ascribe the anomaly presented by the leaves of mangel-wurzel. And, then, it is not to be forgotten that in dosing the azote we have regard but to the *flesh* contained in the article of food, which although unquestionably the principle that is of highest value, and the one which is apt to be most deficient, is still not all. The neutral non-azotized substances, starch, sugar, gum, oil, are indispensable as auxiliaries in the alimentation of cattle; the three first undergo changes in the course of the digestive process which fit them to be absorbed immediately, and the oil is brought to the state of an emulsion, and so is taken up and adds to the fat. The woody fibre alone of vegetables appears to have no direct share in the nutrition of animals; it is discovered almost or altogether unchanged in the dejections.

It is therefore every thing but matter of indifference whether a particular article of forage contains a larger or a smaller proportion of starch, sugar, &c., associated with a given quantity of azotized or truly animalized matter. The potato and meadow-hay brought to the same state of dryness, contain as nearly as possible the same proportions of azote—from 1.3 to 1.5 per cent.; in other words, about $8\frac{1}{2}$ per cent. of albumen and gluten, *i. e.*, of flesh. But in the potato, almost the whole of the $91\frac{1}{2}$ per cent. of the remainder consists of starch; while in hay it is woody fibre, inert matter as we must presume it, that is present in by far the largest proportion. And this explains the higher value of the same weight of dry potato as an article of sustenance. To give our theoretical equivalents all the precision that is really desirable, it would be necessary to ascertain the quantity of organic matter which escaped digestion with reference to each particular species of food. This is an inquiry which it is my purpose to enter upon by and by. The labor completed, we should then be in possession of tables in regard to the proportion of the non-azotized as well as the azotized principles; and further, to the quantity of inert matter which it would be proper to deduct from the weight of the ration allowed in each case.

To have determined the azote in an article of food, then, is not to have done all that is strictly necessary: still azote is the scarce element in all kinds of vegetable food; starch, gum, sugar, pectine, oil, are universally present, and generally in adequate quantity. As articles, as unlike one another as possible, I have mentioned potatoes and meadow hay. Now the theory indicates 300 of the root for 100 of the dried grass; and I can state positively, from long and repeated observation, that it is not advisable in practice to substitute less than 280 of potatoes for 100 of meadow-hay.

The state of dryness of certain kinds of forage may have a marked influence on their nutritious qualities. They may even decline in nutritive value by the process of drying, so that analysis of itself may lead us into error in regard to the nutritive value of dry articles of food. Breeders have in fact long suspected that green fodder is more nutritious than dry fodder; that grass, clover, &c., lose nutritious matter by being made into hay. That the thing is so in fact, appears to have been demonstrated by a skilful agriculturist, well acquainted with the art of experimenting,* who found that 9 lbs. of green lucern were quite equal in foddering sheep to $3\frac{3}{10}$ lbs. of the same forage made into hay, while he at the same time ascertained that 9 lbs. of green lucern would not on an average yield more than 2.02 lbs. of hay. In allowing each sheep $3\frac{3}{10}$ lbs. of lucern hay as its ration, consequently, it was as if the animal had had 14.34 or more than $14\frac{1}{4}$ lbs. of the green vegetable for its allowance.

These practical facts are obviously of great importance; they prove beyond a shadow of doubt that the belief of agriculturists in general as to the immense advantages of consuming clover and lucern as green meat is well founded. Nor is this all; it is not merely the absolutely greater feeding value of the crop green than of the crop dried and made into hay; there is further, the saving of expense in making the hay, and still further, the escape of all risk from loss through bad weather during the process, by which that which was valuable fodder but a few days before, may become fit only for the dung-hill. Still, because 100 of green clover or lucern represent 23 of the same articles dried, it does not follow that the feeding properties of the fodder in each of the two states can be truly represented by the ratios of these numbers to one another. Messrs. Perrault find from their experiments that the true relation is 8 to 3. By assuming 71.5 lbs. as the quantity of dry forage obtained from

* M. Perrault de Jotemps, in Journ. d'Agricult. v. iii. p. 97.

220 lbs. of green clover or lucern, the quantity which is actually obtained on an average, the ratio comes out 8 to 2.6, a number which falls somewhat short of that which is assumed, but not much. With regard to the difference in the feeding or nutritive value of green and dried fodder, the loss may in a general way be ascribed to loss of the more substantial parts of the plants especially experienced in the process of drying. This is the conclusion, at all events, to which M. Crud came; I have myself, however, found that clover-hay, made in the field and ricked in the usual way, had not the same nutritive value as a quantity of the same crop carefully dried in the laboratory.

By way of pendant to the conclusions of Messrs. Perrault, from their valuable observations, I shall here add the average of some experiments that were made at Bechelbronn, in 1841, on the conversion of clover into clover-hay. The clover crops of this season were magnificent; the plant in its second year growing to more than a yard in height. Green clover on the average may be considered as consisting of:

Clover-hay	29.85
Water	70.15
	100.00

As extremes in our experiments of 1841, we add:

Clover-hay	35.7	25.0
Water	64.3	76.0
	100.0	100.0

Analysis gave the number 75 as the nutritive equivalent number of clover-hay. Assuming 76 to represent the moisture lost during the drying, the equivalent becomes 311 for the same fodder in the green state, meadow-hay, the standard, being 100.

But practice is not here in harmony with theory; the value of clover-hay, in point of nutritive power, is found not to differ essentially from that of meadow-hay; and the equivalent of green clover is generally placed between 425 and 500. And I may say, that daily experience in the stable tends to show that the theoretical equivalent of clover-hay is too high, that its nutritious properties are not so great as they are inferred to be. From a mean of four weighings, I find that four cows upon green clover consume 2499 lbs., or 624¼ lbs. each per diem. The usual allowance to one of our cows, however, is 33 lbs. of hay of good quality; from which it would follow, that the equivalent of green clover would be 445. But the animals on the green fodder fattened apace, and every thing showed that they were very differently nourished than they would have been with their 33 lbs. of meadow-hay. According to theoretical data, each cow in its 624½ lbs. of green food per day received an equivalent of 47.3 lbs. of hay; and if it be considered, that during the season of green forage they have it almost at will, it must be conceded that during this period the quantity of food consumed is actually greater than when it is regularly doled out. Additional experiments are therefore necessary to decide the question as to whether forage eaten green is really more nutritious than the same forage consumed when converted into hay. For my own part, I should not be surprised, from what I have seen, were it found that dry fodder, previously moistened and carefully portioned out, was actually more nourishing than the same food would have been had it been eaten green. Green forage, of a very soft or watery nature, is notoriously possessed of purgative properties, which must lessen its value as food; but my observation leads me to say, on the other hand, that animals kept upon dry fodder require more care with regard to watering than is generally bestowed upon them. The absolute necessity of a sufficient degree of moistness in the food, in order to secure its due and easy digestion, greatly countenances the practice which is beginning to be introduced in some places of steeping hay for some time in water before giving it to cattle. This necessity further explains the great advantages in associating with dried fodder other very watery articles, such as roots and tubers, turnips and field-beet, potatoes and Jerusalem artichokes.

The oleaginous seeds contain a considerable proportion of animalized matter, similar in composition and qualities to the caseum of milk; and the cake which comes from the oil-mill retains almost the whole of this substance. The proportion of from 0.05 to 0.06 of azote, indicates nearly 42 per cent. of the representative of flesh in oil-cake. Theory, in fact, rates the nutritious power of this substance so high, that 100 of hay may be replaced by from 22 to 27 of cake.

The almost universal use of oil-cake in the feeding and fattening of cattle, is of itself sufficient evidence of its highly nutritive qualities. It has even been found possible to keep sheep and oxen upon this food almost exclusively. M. Bouscaren finding considerable difficulty in getting rid of his oil-cake, thought of associating with his oil-mill an establishment for feeding cattle; and he found that oxen put up to fatten throve perfectly upon a mixture of the refuse of the wine-press and oil-cake. Cows, upon a diet of this kind, give on an average 12⅓ pints of milk per diem. The allowance per head is about 15 lbs. of oil-cake in three meals, given each time immediately after the animals have been watered, and in the interval, each is allowed about 12 lbs. of straw or chaff. The cake broken in pieces is steeped in water, and worked up into a paste of the consistency of dough. If the animals show any disinclination to this food at first, they are brought to like it by having a ball of it, the size of the fist, administered to them two or three times.

Supposing that the cows fed in this way would be adequately maintained upon 33 lbs. of hay, and that 13 lbs. of straw are equivalent to 3 lbs. of hay, it appears that in the allowance given, 15 lbs. of oil-cake will supply the place of 30 lbs. of hay; the equivalent of the cake, therefore, is 51.5, a number very different from the 22 deduced from analysis. The equivalents of those who have sought to appreciate the alimentary value of oil-cake are, however, sufficiently at variance with one another. It will be seen in the table, that the numbers assigned by different authorities are 42, 57, and 108; and M. Perrault, from direct experiment, found the equivalent number of colza-cake to be 36, analysis giving 23 as the theoretical number. On the whole it may be said, that in practice, the results, although sufficiently different, still agree in ascribing to oil-cake a nutritive value inferior to that indicated by theory.

I have thought it important to insist upon the discrepancy which is here so conspicuous between the inferences from chemical analysis and those arrived at by experience, because it appears to me to depend upon a particular circumstance which frequently intervenes in the feeding of cattle, and which it is very important to be aware of: I allude to the influence of the *bulk* of the allowance of food.

Vegetable food of every description has nearly the same specific gravity; it is but little above that of water; the bulk of the allowance therefore depends upon its weight. Every one will conceive that a ration of highly nutritious food, which for this reason would occupy but little space, would be open to many objections. A cart-horse, of the ordinary size, from what I have myself repeatedly observed, requires from 26 to 33 lbs. of solid food, and about the same quantity of water in the twenty-four hours. The bulk of this allowance, when masticated and brought to the state in which it is swallowed, will be upwards of 9½ cubic feet. Now, if for the ordinary forage, one that is five times more nutritious were substituted, oil-cake, for example, the dry ration, according to the rule of equivalents, would be reduced to 6.6, or a little more than 4½ lbs., and its bulk would not surpass 5¼ cubic feet. The animal would not feel satisfied with this allowance, it would still feel hungry, or the food given in such a concentrated shape would disagree with it. If, on the contrary, a forage that is very little nutritious were substituted, such as wheat-straw, the equivalent of which is 500, the ration would then become too bulky to be eaten in the course of a day, it would amount to as many as 165 lbs. It is therefore absolutely necessary to take into consideration the bulk of the food allowed: the belly must of necessity be filled; whatever the nutritive value of any article, it must be given in a certain quantity; and in the case of such a substance as oil-cake, the consumption to fill the stomach would cease to be in any kind of proportion to the nutritive equivalent.

It is extremely difficult to appreciate the precise limits beyond which an article of forage or a given ration ceases to be nutritious. When any addition is made to an allowance known and admitted to be sufficient, the effect of the extra quantity is scarcely perceptible; so that, in practice, we are apt to fall into the error of estimating at too low a rate the nutritious powers of food given in too large quantities. I have had proof of this in a series of experiments on the maintenance of a number of milch-kine. To a cow which was receiving the equivalent of 33 lbs. of meadow-hay in dry fodder and Jerusalem potatoes, an addition was made of 6½ lbs. of oil-cake, by which the allowance of nourishment was doubled theoretically; the animal only ate the half of the cake, however: still, the quality of the milk was not improved. Experience here would compel us to set down the $3\frac{1}{4}$ lbs. of cake consumed as nil; yet it is positively ascertained that the article is one of the most substantial known.

The hard and husky grain which is given to cattle frequently escapes digestion, because it has escaped the teeth—a circumstance which leads to the formation of an estimate of its nutritious qualities inferior to those it actually possesses. To prevent this loss, oats are now often bruised, as are beans and peas also; or they are mixed with chopped hay or straw, which the animals are compelled to chew thoroughly before they can swallow it; or the corn is steamed or steeped in boiling water before it is put into the manger. Some experiments that were instituted by order of the French veterinary commission, however, seemed to show, that the loss of corn from passing through the stomach and bowels unchanged was really so trifling, that it might be safely left out of the account.

Tubers and roots are invaluable fodder for horned cattle, and in the course of the winter, come instead of hay to a considerable extent. Our experience at Bechelbronn also enables us to say, that horses are readily brought to a regimen of the same description, which, judiciously instituted, becomes the means of great economy in the maintenance of these animals.

Roots, turnips, and mangel-wurzel, are frequently thrown down whole before the animals. It is vastly better; nay, it is so much better that it ought to be made an invariable rule never to give them save cut into slices and mixed with cut straw or chaff. There is always a great advantage in combining any very soft and watery article of food with one that is dry and hard, to say nothing of the chaff absorbing and rendering useful the juices that would escape and be lost.

Mangel-wurzel, turnips, carrots, and Jerusalem potatoes, are always given raw. The potato is frequently steamed or boiled first; yet I can say positively that horned cattle do extremely well upon raw potatoes; and at Bechelbronn, our cows never have them otherwise than raw: they are never boiled, save for horses and hogs. The best mode of dealing with them is to steam them; they need never be thoroughly boiled as when they are to serve for the food of man. The steamed or boiled potatoes are crushed between two rollers, or simply broken with a wooden spade or dolly, and mixed with cut hay or straw or chaff before being served out. It may not be unnecessary to observe, that by steaming, potatoes lose no weight; whence we conclude that the nutritive equivalent for the boiled is the same as that for the raw tuber. Nevertheless, it is possible that the amylaceous principle is rendered more readily assimilable by boiling, and that by this means the tubers actually become more nutritious. Some have proposed to roast potatoes in the oven; and there can be little question but that, treated in this way, they answer admirably for fattening hogs or even oxen. Done in the oven, potatoes may be brought into a state in which they may perfectly supply the place of corn in the foddering of horses and other cattle. There is but the expense of the firing to be taken into the account.

The only mode of ascertaining the favorable or unfavorable influence of any particular system of diet or regimen upon animals, is by weighing them. In regard to full-grown animals performing regular work, such as cart and plough horses, and to milch-kine, the allowance ought to be such as will maintain them at the same or nearly the same weight. Any thing like stinting is immediately followed by loss of flesh and of weight, of strength and spirit, in the animal. The allowance being continued the same, similar effects will follow any increase of work, any exaction of unusual effort on the part of the animal. An essential condition, therefore, in all experiments on the due dieting or feeding of animals, is, that they be performed under precisely similar conditions of labor. Young animals receiving a sufficiency of wholesome food, increase from day to day by a quantity which we shall have occasion immediately to mention; and all changes of regimen are followed at once by notable variations in the ratio of the growth; if the new regimen be less nutritious than that which went before it, the balance immediately proclaims the fact.

Cattle put up to fatten are always supplied with a superfluity of fodder; the excess may be regarded as an addition to the quantity requisite to maintain them in health and strength. The increase in the weight of an animal is often so great within a given time, as to be very appreciable by weighings made even at very close intervals; the balance also shows us that the rate of increase varies at different periods of the interval during which the fattening is going on. An animal put up to fatten for the butcher, is not the best subject for coming to conclusions upon in regard to the nutritive value of different articles of sustenance; still it is useful, in a practical point of view, to determine the influence of this and of that course of regimen on the production of fat. Any misapplication of nutritive equivalents is speedily proclaimed by the animal's losing weight, instead of maintaining or gaining upon the amount to which it had attained.

When the quantity of fodder has been ascertained which an animal ought to have in the twenty-four hours to maintain it in full health and vigor, or that may be necessary to enable it to lay on additional flesh and fat, it is to be weighed, and the article or mixture of articles which it is the business of the experimenter to try, is to be given in part or in whole. After the lapse of a certain time the animal is weighed again, and the weight upon this occasion enables us to say whether the new or amended ration is superior, equal, or inferior, to that which had preceded it. Such is the procedure generally followed: but in putting it in practice myself, I saw that it was liable to lead to rather serious mistakes, which I then used every effort to diminish or to nullify in the experiments which I undertook on the keep of horses—experiments which I think interesting enough to deserve being particularly related.

In a considerable number of observations with which I had become familiar, I saw that the course had not always been continued for a sufficient length of time; so that changes which were the effect of mere accident must frequently have been ascribed to the effects of regimen. In a general way, it is acknowledged that an adult animal, upon the ration that is known to be adequate for its maintenance, returns at the same hour every day to the yesterday's weight: this, however, is only strictly true in reference to a series of weighings continued through a number of days, to make any irregularity between one weighing and another disappear.

With a view to discovering the amount of variation which an animal experiences in point of weight when it is fed in the same uniform manner, is foddered precisely at the same hours, &c., I weighed a horse and a mare, which were leading the most regular and unvaried life possible, for they were both employed in working an exhausting machine for several days in succession, the weighings being performed at noon each day before they were watered, and from four to five hours after their breakfast. Here are the results in a tabular form:

Date of the weighings.	Weight of the horse.		Weight of the mare.	
	kil.	lbs. avoird.	kil.	lbs. avoird.
16 December, 1841	453.0	996.6	494.0	1086.8
17 " "	455.0		497.0	
18 " "	456.0		497.0	
19 " "	454.0		497.5	1092.7
20 " "	449.0	988.9	487.0	
21 " "	449.5		487.5	
22 " "	449.0		492.0	
23 " "	454.0		496.5	
24 " "	454.0	998.8	484.5	1065.9
25 " "	459.5	1010.9	..	
27 " "	448.0	985.6	490.5	
28 " "	452.0	994.4	496.0	
29 " "	454.0		491.0	
30 " "	448.0		484.0	1064.8
31 " "	452.5		491.0	
Mean weights	452.0	994.4	491.8	1081.9
Maxima	459.5	1010.9	497.5	1092.7
Minima	448.0	985.6	484.0	1064.8
Greatest difference above the mean		16.5		10.8
Greatest difference below the mean		8.8		17.1
Difference between the extreme weights		7.7		6.3

Another horse (Old Fox) 12 years old, taken fasting, at four o'clock in the morning of the 28th of April, 1842, weighed 1051 lbs.; at the same hour of the 29th, he weighed 1060 lbs.; ditto on the 29th, 1038 lbs.

It is obvious, therefore, that a horse foddered most regularly and weighed at the same hour, nevertheless presents differences in his weight that may amount to nearly 30 lbs.; and which, without assurance of this fact, we should be disposed to ascribe to the effect of our regimen. This is enough to satisfy us that in all experiments upon feeding, it is absolutely necessary to carry them on for some considerable time, in order to escape, or at all events to lessen the errors that would be introduced into the conclusions by these accidental differences of weight. They may vary with reference to different animals; they are necessarily smaller in amount among those that are young and small, such as calves and sheep, than in adult oxen and horses; bu; they do not occur the less on that account, and must, therefore, occasion errors of the same description. What, then, shall we say of those small variations in the weight in a ewe or a ram, amounting perhaps to 1½ or 2 lbs., ascertained in the course of an experiment carried over two or three days, though conducted with the most scrupulous attention to accuracy in the world? That they may very possibly have been purely accidental.

The first in every series of experiments on the maintenance of animals, ought in fact to have it in view to ascertain the amount of accidental variation in the weight of the creatures which are their subjects; as this variation is now on this side now on that, there is an obvious advantage in having a certain number upon trial at a time; any error that occurs will thus be more apt to be corrected; and the results may be held more worthy of confidence in proportion as the numbers have been large from which they have been deduced Another cause of error, which I had occasion to discover in the course of my experiments, appears to be connected with the weight of the allowance. Equal in nutritious value, different allowances may still have very different weights: it is obvious, that a ration of hay and corn will weigh much less than its equivalent in roots, tubers, or green meat. Animals that have been kept for some time upon a dry diet, if put on one that is very bulky and watery, will immediately increase very considerably in weight; and their increase is both so sudden and so great, that it is impossible to ascribe it to augmented nutrition, to flesh and fat laid on. The animals are simply distended, their paunch and bowels are filled with a larger quantity of food than they were before; and the state of distension continues, though it suffers accidental variations, so long as the new course of feeding is persisted in. In opposite circumstances, as when animals that have been long upon soft and watery food, are suddenly put upon hard diet, they always drop very considerably in weight. These sudden changes throw disorder and contradiction into the conclusions, and puzzled me greatly until I discovered their cause. It is obvious that no kind of reliance can be placed upon the conclusions which have been come to from single weighings made at the end of each particular course of alimentation. To get at results which shall be worthy of any credit, the animals that are to be made the subjects of experiment must be fed for several days upon the particular ration that is to be approved, in order to be brought to the state of body which may be said to belong in particular to each system of dieting, before being weighed; it is only when this is attained, indeed, that the experiment can be held to be properly begun; and then it is to be continued for a sufficient length of time to lessen the influence of those accidental variations of weight, of which I have spoken so particularly. It is perhaps needless to observe, that any increase in weight and the maintenance of that increase, are not always of themselves sufficient signs for affirming that the course then followed is superior or equal to the one which preceded it. Various other circumstances of divers character must be taken into the reckoning, and in particular the state of the animals. It is very necessary to have an eye to the state of the coat, to the spirit or liveliness of the animal, to the nature of the dejections, the size of the belly, the disposition of draught animals for their work, the quantity of milk given by milch-kine, &c. Nevertheless, and as a general proposition, it may be said that a stationary condition, or a slight increase of weight, is almost always in favor of the course along with which it is gained or maintained, while any loss is almost always an indication of an inadequate allowance or of deficient nutritive qualities in the ration, taken in connection with the work required or the milk obtained.

The experiments which I am about to detail were undertaken to determine the nutritive value of a variety of forages associated with the ordinary articles in keeping the horse. The great dearth of forage that was felt in Alsace, in consequence of the extraordinary droughts of 1840, led us to feel the full importance of researches in this direction; for then we were compelled to replace by potatoes a very large proportion of the hay usually consumed in the stable. And, indeed, by assuming the theoretical equivalent as the basis of this substitution, I found that I saved money by the course, at the same time that the health and strength of my draught cattle were maintained unimpaired. Still, as every question that bears upon the keep of the animals attached to a farm is too important to be left to the decision of theory alone, I thought it imperative on me to control the inferences of chemical analysis by the results of experience.

The best food for horses has long been admitted to be hay and oats in combination; neither article alone would have the same happy effect that the two together produce. A ration of hay alone would be too bulky; one of oats alone would not be bulky enough. But the horse is not particular in his food. Barley in southern countries replaces oats, and answers equally well. I have myself kept horses and mules for long periods of time on maize and the tops of sugar canes exclusively; and on the elevated tablelands of the Andes, and in the steppes of South America, the horses, though they do much hard work, are kept wholly on green meat. Much of course depends on the way in which the animal has been brought up.

In the circumstances in which we are generally placed in this country, I do not imagine that there would be any actual advantage in replacing the ordinary food of our horses by roots and tubers; I doubt even whether the substitution would have good effects. I know, indeed, that horses have been kept through the winter upon potatoes and mangel-wurzel; but it is a different matter to feed an animal and keep him standing quiet in the stable without work, and to feed him at the same time that a certain quantity of labor is required of him every day. A horse in full work would scarcely get through the bulky ration, which should consist of beet-root alone; his meal-times are restricted; if he has certain hours for his work, so has he certain hours for his breakfast, dinner, and supper also.

This is one reason why carriers' horses and post-horses, horses, in a word, which have long and severe work to perform, receive the larger portion of their allowance in corn. The inconveniences of bulky rations are much less felt in the cow-house than in the stable; not to speak of their particular organization, which actually enables them to take in a much larger quantity of food than the horse, the steer and the cow have always a longer time allowed them for their meals than are regularly given to the horse.

The experience of nearly a whole year having satisfied me that a cart-horse may have half his ration in roots or tubers, I set out from this fact in the experiments which I instituted.

EXPERIMENTS ON THE MAINTENANCE OF HORSES WITH MIXED FOOD.

The usual allowance to a horse at Bechelbronn for the twenty-four hours consists of:

Hay	22 lbs.
Straw	5½
Oats	7¼

With this ration the teams are kept in excellent condition. Two teams were selected as subjects of experiment, each consisting of four horses; these I shall distinguish by the titles, Team No. 1 and Team No. 2. Each remained under the care of the same servant throughout. Team No. 1 was composed of:

Braun, a mare,	7	years old.
Schimmel, a horse,	7	"
Hans, do.,	16	"
Gaty, do.,	8	"

Team No. 2 was composed of:

Old Fox, a mare,	16	years old,
Braun, do.,	5	"
Nickel, do.,	14	"
Hengst, a horse,	5	"

EXPERIMENT I.

One half the allowance of hay was replaced by potatoes lightly steamed; 280 of the tubers being assumed, according to theory, as equivalent to 100 of hay. The ration, therefore, consisted of:

Hay	11 lbs.
Straw	5½
Oats	7¼
Potatoes	30 8-10

The potatoes were broken down and mixed with chopped straw, and never put into the mangers until cold.

From accidental circumstances, particularly bad weather during the course of the autumnal labors, the teams were exposed to very hard work, an event which of course throws uncertainty over the results of this trial. After having been upon the course of food indicated for a few days, the teams were weighed once, and again after an interval of twenty-four hours:

	Team No. 1.	No. 2.	Both teams.	Mean per horse.
First weighing	4617.8	4461	9079.4	1134.9
Second weighing	4554.0	4334	8888.0	1111.0
In 24 hours .....loss	63.8	127	191.4	23.9

The loss experienced here authorized me to conclude, that the allowance under the circumstances was not sufficient. The 30.8 lbs. of steamed potatoes could not have adequately replaced the 11 lbs. of hay; it would have been highly interesting to have ascertained how horses kept on the standard and usual allowance would have stood the same amount of fatigue. Unfortunately this comparison could not be made, all the horses in the stable having been put on the potato regimen at the same time. There is this much to be said for the particular course tried, however, that the animals did their work with great spirit, and continued in excellent health.

EXPERIMENT II.

INTRODUCTION OF JERUSALEM POTATOES INTO THE RATION.

Jerusalem potatoes are held excellent food for the horse; they are eaten greedily, and he thrives on them. In this second experiment, $30\frac{3}{10}$ths lbs. of Jerusalems cut into slices were substituted for 11 lbs. of hay, the same theoretical equivalents being assumed for them as for the common potato. The ration now consisted of:

Hay	11 lbs.
Straw	$5\frac{1}{2}$
Oats	$7\frac{1}{2}$
Jerusalem potatoes .	30.8

Having been accustomed to this regimen for some days, the teams were weighed, and having gone on for eleven days they were weighed again:

	Team No. 1.	No. 2.	Both teams.	Means per horse.
First weighing	4556	3245	8901	1112.7
Second weighing......	4611	3412	8923	1113.6
In 11 days.......	gain 55	loss 33	gain 22	gain 0.9

A result which leads to the conclusion, that the equivalent assumed for the Jerusalem potato was correct; the animals had done their work, and gained, one with another, $\frac{9}{10}$ths of a pound in weight.

EXPERIMENT III.

RATION OF HAY AND POTATOES.

Eleven pounds of hay, in the usual allowance, were replaced by 30.8 lbs. of potatoes; the whole of the oats and straw, by 15.4 lbs. of hay. These substitutions were made upon the supposition, that 100 of hay was equivalent to 280 of potatoes, to 50 of oats, and to 520 of straw. The ration, then, was composed as follows:

Hay	26.6 lbs.
Potatoes	30.8 "

This was a ration which it was the more interesting to try, from the circumstance of Professor Liebig* having come to the conclusion, from certain theoretical views, that it must be impossible to keep horses in health and strength upon hay and potatoes exclusively. The experiment was continued for a fortnight:

	Team of No. 1.	No. 2.	Both teams.	Mean weight per horse.
First weighing	4620	4312	8932	1116.5
Second weighing	4675	4697	9372	1171.5
In 14 days gain	55	385	440	55.0

In one fortnight, consequently, the weight of eight horses had increased by an aggregate sum of 440 lbs., or 55 lbs. per head—an increase at the rate of, as nearly as possible, 3.9, say 4 lbs. per diem; and allowing the greatest latitude for error, it seems that we cannot estimate the increase per head at less than 1.76, say $1\frac{3}{4}$ lbs. per diem. The condition of the horses was most satisfactory; the dejections were healthy in appearance; the only inconvenience observed was, the considerable bulk of the allowance, and the additional time which had to be given the teams to their meals. This inconvenience was particularly obvious in the case of the older horses. Besides the two experimental lots, other twelve horses were put upon the same regimen, and with the same good effects. The equivalents adopted in the composition of the ration, in this third experiment, may therefore be regarded with perfect confidence as suitable. Experience, indeed, would rather lead us to conclude, that the nutritive power of the potato had been estimated at somewhat too low a rate.

EXPERIMENT IV.

SUBSTITUTION OF OATS AND STRAW FOR A PORTION OF THE HAY.

The ration here consisted of:

Hay	11 lbs.
Straw	11 "
Oats	12.1 "

The horses, having been two days on this diet, were weighed. The experiment was continued for eleven days:

	Team No. 1.	No. 2.	Both teams.	Average per horse.
First weighing	4584.8	4348.3	8933.1	1116.7
Second weighing	4593.6	4352.7	8946.3	1118.2
In 11 days gain	8.8	4.4	13.2	1.5

Under this regimen, consequently, the weight of the teams remained very nearly the same as it was before beginning the experiment; still there was something gained.

In conducting this experiment, we had an opportunity of observing how important it is to habituate the animals to their new regimen before weighing for the first time. Had this precaution been neglected, the result would have come out against the ration, for the animals were found, when first entered on it, to weigh together as many as 9372 lbs., and two days afterwards no more than 8933 lbs.,

* Agricultural chemistry.

which would have indicated a loss of 449 lbs.; the difference being due, however, in great part, or entirely, to the less bulky or weighty food employed.

EXPERIMENT V.

POTATOES SUBSTITUTED FOR A PORTION OF THE HAY.

The ration made use of in the first experiment looks so well, in reference to economy of hay, and, indeed, answered so well under the peculiar circumstances in which it was tried, that I thought it would be advisable to try it again when the horses were doing ordinary work. The ration consisted of:

Hay	11	lbs.
Straw	5.5	"
Oats	7.23	"
Steamed potatoes	30.8	"

The first weighing took place after the horses had been over a week on the ration, and the experiment was continued for 63 days. In team No. 1, Braun, from indisposition, had been replaced by Rapp, a horse nine years old, and weighing 1157 lbs.:

	Team No. 1.	No. 2.	Both teams.	Average weight per horse.
First weighing	4425	4362	8848	1106.1
Second weighing	4501	4428	8929	1116.2
In 63 days gain	76	66	81	10.1

In the course of two months, consequently, on a ration in which 11 lbs. of hay were replaced by 30.8 lbs. of dressed potatoes, the weight of the horses may be said to have been more than maintained. This experiment seems to show satisfactorily, that the equivalent of the potato cannot be far from the number 280.

EXPERIMENT VI.

JERUSALEM POTATO FOR A PORTION OF THE HAY.

The horses were brought back to the same conditions as in the second experiment, 30.8 lbs. of Jerusalems being substituted for 11 lbs. of hay. The team No. 2 was alone subjected to this experiment, being kept on it for 16 days, and first weighed after having had it for some time:

First weighing	No. 2.	4395 lbs.	Average weight per horse	1098.9
Second weighing	"	4396.7 "	" "	1099.1
In 16 days	gain	1.7		0.2

This result confirms that which was elicited by the second experiment.

EXPERIMENT VII.

INTRODUCTION OF FIELD-BEET, OR MANGEL-WURZEL, INTO THE RATION.

Horses readily get accustomed to field-beet. The root is sliced, and mixed with chaff, (cut straw.) For 11 lbs. of hay, which I retrenched, I allowed 44 lbs. of beet; i. e. I took 400 as the equivalent number of the root. The ration consisted as under:

Hay	11 lbs.
Straw	5.5 "
Oats	7.2 "
Beet	44.0 "

A horse, after having been kept on this diet for some time, was weighed; and the regimen having been continued for a fortnight, he was weighed again:

First weighing		1014.0 lbs.
Second weighing		1023.0 "
In a fortnight	gain	9.4

This horse was all the while doing rather hard but very regular work; for eight hours every day he was in the shafts of a grinding mill. He did not alter in condition; the dejections were healthy.

During the winter of 1841–2, our cows ate a considerable proportion of our beet; and, as a substitute for the 33 lbs. of meadow-hay, which is their usual allowance, we gave $72\frac{1}{2}$ lbs. of beet. The ration then stood thus:

Hay	22 lbs.
Beet	72.6 "
Straw	4.4 "

Upon this regimen, the weight of the inmates of one of our stables was:

On the 29th January	24615 lbs.
On the 21st April	26488
Increase due to births and to growth	1837

It thus appears that, in foddering kine, the quantity of beet allowed with advantage may be large; but it is also obvious, that the nutritive value of the root is not great. At Bechelbronn, at all events, we found it requisite to replace 9 or 10 of hay by 40 of root. Our beet, it is true, contains but 12 per cent. of dry matter; in other places, where the proportion of dry substance to the water is larger, it is possible that a smaller proportion would be found to answer the end.

EXPERIMENT VIII.

INTRODUCTION OF THE SWEDISH TURNIP INTO THE RATION AND REPLACING A PORTION OF THE HAY.

Swedish turnip, combined with some dry forage, answers excellently with the horse. Analysis, indicating 280 as the equivalent of this article, two horses were put upon the following ration, in which 11 lbs. of the usual allowance of hay were replaced by Swedish turnip:

Hay	11 lbs.
Straw	5.5
Oats	7.2
Swedes	30.8

It was obvious before the lapse of but a few days, that the horses were falling off upon this regimen, that they were not fed; and on weighing them, this plainly appeared:

First weighing	2283.6	Aver. of each horse	1141.8
Second weighing, 9 days afterwards	2178.0	"	1089.0
Loss in 9 days	105.6		52.8

The equivalent for the Swedish turnip adopted, had therefore been too high; the allowance was not sufficient. This led me to analyze the article again; and I discovered that the true equivalent of the sample with which I was operating, was at least 676, and not 280 as I had presumed before. Indeed, in another experiment with the same pair of horses where the equivalent of Swedish turnip was assumed at 400, I found that though the animals kept up their weight at the point to which it had fallen, they gained nothing; whence it may be safely inferred that the No. 400 was still too low, and that the new equivalent 676 is nearer the truth.

EXPERIMENT IX.

INTRODUCTION OF CARROTS INTO THE RATION.

Horses are extremely fond of carrots; and there is no root perhaps, the nutritious qualities of which have been more vaunted or exaggerated. Yet, analysis appears to indicate that 350 of carrot are required to replace 100 of good meadow-hay. On one occasion, in the stable at Bechelbronn, when the potato in one of our rations was replaced by an equal weight of carrots, the effect was highly disadvantageous; and even in following the theoretical equivalent of the carrot (350) we had still no reason to be perfectly satisfied. I now believe, in fact, that as many as 400 of carrots may be found requisite to replace 100 of good meadow-hay.

The carrot crop of 1841 having been a failure, I had to limit myself to observations made on a single horse, which was put upon a ration in which 11 lbs. of hay were replaced by 38.5 lbs. of carrots. The horse, habituated to this diet,

Weighed	1025.2 lbs.
A fortnight after	1014.2
Loss in a fortnight	11.0

Nevertheless he remained in good condition, so that the equivalent 350 is probably not far from the truth. I ought to say, however, that the men think this number too low; an opinion in which they would be borne out, could we but be certain that the loss of weight of the horse just indicated was not accidental.

EXPERIMENT X.

BOILED RYE AS A SUBSTITUTE FOR OATS.

It has been stated, that rye boiled till the grain bursts may be used as a substitute for an equal bulk of oats in the keep of a horse. The experiment which I made on the point is very far from bearing out any thing of the kind. By preliminary trials I had ascertained that rye of good quality swells to twice its former bulk by boiling.

The two horses that were made the subjects of experiment now, had been kept for some time on a ration formed of:

Hay	2.2 lbs.	
Oats	5.5	=8.8 pints.

For the oats, the same quantity by measure, 8.8 pints of boiled rye were substituted, containing 4.4 pints of raw grain, weighing 4.15 lbs. On the 11th day it was deemed prudent to interrupt the experiment, of which the following are the results:

First weighing:	Both horses	2010 lbs.	Average of each	1004.5
Second "	"	1927	"	963.0
Loss in 11 days		83		41.5

In fact, with such a ration as this, in which *water* was made to replace solid corn, no other result could reasonably be expected. In continuing it, the health of the horses would very certainly have soon been seriously compromised. There is no objection to rye in itself as an element in the food of a horse; but then it must be substituted in the quantity indicated by the table of equivalents, by adopting which, Mr. Dailly found that he could keep the post-horses of Paris in good heart, at a time when the difference between the price of oats and rye made it advantageous to substitute the latter for the former. The experiments of Mr. Dailly on the subject were so decisive and so ably conducted, that I felt myself relieved from the necessity of inquiring further into it myself.

From these experiments, the particulars of which have now been given, it may be concluded that the nutritive equivalents of the potato, beet, Jerusalem potato, and carrot, as they come out upon analysis, or as they are inferred from the amount of azote they contain, may be adopted without detriment to the health of horses. If they err at all, it is that they assign equivalents somewhat too high, which is the same as saying that their actual nutritive power is rather less than these numbers give it; so that a portion of the hay of the standard ration being substituted for its equivalent of tuber or root, the diet will be improved.

Thus, 100 of good meadow-hay may be taken, as ascertained by experiment, to be equivalent to:

280	potatoes—by analysis, equal to	315
280	Jerusalems	311
400	beet	548
400	Swede (too little)	676
400	carrot	382

In the following table of nutritive equivalents, to the numbers assigned by the theory, I have added those of the whole which I find in the entire series of observations that have come to my knowledge. I have also given the standard quantity of water, and the quantity of azote, contained in each species of food. When the theoretical equivalents do not differ too widely from those supplied by direct observation, I believe that they ought to be preferred. The details of my experiments, and the precautions needful in entering on and carrying them through, must have satisfied every one of the difficulties attending their conduct; yet all allow how little these have been attentively contemplated, and what slender measures of precaution against error have been taken. Our equivalent for field-beet is 400, a number come to by introducing 44 lbs. of the root into the ration, in lieu of 11 lbs. of hay; had we introduced 56 lbs., the equivalent number would have come out 500; and it is questionable whether the final result would have been affected by this substitution. In my opinion, direct observation or experiment is indispensable, but mainly, solely as a means of checking within rather wide limits the results of chemical analysis.

TABLE OF THE NUTRITIVE EQUIVALENTS OF DIFFERENT KINDS OF FORAGE.

TITLE.	Standard water per cent.	Azote per cent.	Azote per cent. in the article not dried.	Theory.	Block.	Petri.	Meyer.	Thaer.	Pabst.	Flottow.	Pohl.	Rieder.	Gemerhausen.	Crud.	Weber.	Dombasle.	Krantz.	Schwertz.	Schnee.	Midleton.	Murre.	André.	Boussingault.
Ordinary natural meadow-hay	11.0	1.34	1.15	100	100	100	100	100	100	100	100	100	100	100	100	100	100	100	100	100	100	100	0
Ditto, of fine quality	14.0	1.50	1.30	98	--	--	--	--	--	--	--	--	--	--	--	--	--	--	--	--	--	--	--
Ditto, select	18.8	2.40	2.00	58	108	100	--	--	--	100	--	--	--	--	--	--	--	--	--	--	--	--	--
Ditto, freed from woody stems	14.0	2.44	2.10	55	--	--	--	--	--	--	--	--	--	--	--	--	--	--	--	--	--	--	--
Lucern hay	16.6	1.66	1.38	83	--	90	--	90	100	--	--	90	90	90	90	90	90	100	90	--	--	90	--
Red clover-hay, 2d year's growth	10.1	1.70	1.54	75	100	90	--	90	100	--	--	--	--	90	--	--	--	100	--	--	--	--	--
Red clover cut in flower, green, ditto	76.0	..	0.64	311	430	--	--	450	425	500	450	--	--	--	--	--	--	--	--	--	--	--	--
New wheat-straw, crop 1841	26.0	0.36	0.27	426	200	360	150	450	300	175	--	500	500	--	--	--	--	--	500	--	--	--	--
Old wheat-straw	8.5	0.53	0.49	235	--	--	--	--	--	--	--	--	--	--	--	--	--	--	--	--	--	--	--
Ditto, ditto, lower parts of the stalk	5.3	0.43	0.41	280	--	--	--	--	--	--	--	--	--	--	--	--	--	--	--	--	--	--	--
Ditto, ditto, upper part of ditto and ear	9.4	1.42	1.33	86	--	--	--	--	--	--	--	--	--	--	--	--	--	--	--	--	--	--	--
New rye-straw	18.7	0.30	0.24	479	200	500	150	666	35	175	--	--	660	--	--	--	--	--	666	--	--	--	--
Old ditto	12.6	0.50	0.42	250	--	--	--	--	--	--	--	--	--	--	--	--	--	--	--	--	--	--	--
Oat-straw	21.0	0.36	0.30	383	200	200	150	190	200	175	--	--	190	--	150	--	--	400	182	--	--	-	--
Barley ditto	11.0	0.30	0.25	460	193	180	150	150	200	175	--	--	150	--	--	--	--	400	154	--	--	--	--
Pea ditto	8.5	1.95	1.79	64	165	200	150	130	150	200	90	--	--	--	--	--	--	--	143	--	--	--	--
Millet ditto	19.0	0.96	0.78	147	--	250	--	--	--	--	--	--	--	--	--	--	--	--	--	--	--	--	--
Buckwheat ditto	11.6	0.54	0.48	240	--	200	--	--	--	--	--	--	--	--	--	--	--	--	--	--	--	--	--
Lentil ditto	9.2	1.18	1.01	114	160	200	--	130	150	--	--	--	--	--	--	--	--	--	191	--	--	--	--
Vetches cut in flower and dried into hay	11.0	1.16	1.14	101	--	125	--	--	100	--	--	--	--	--	--	--	--	--	--	--	--	--	--
Potato tops	76.0	2.30	0.55	209	--	300	--	--	--	--	--	--	90	90	--	--	--	--	90	--	--	--	--
Field-beet leaves	88.9	4.50	0.50	230	600	--	--	--	600	--	--	--	--	--	--	--	--	--	--	--	--	--	--
Carrot ditto	70.9	2.94	0.85	135	--	--	--	--	--	--	--	--	--	600	--	--	--	--	--	--	--	--	--
Jerusalem potato stems	86.4	2.70	0.37	311	--	--	--	--	325	--	--	--	--	--	--	--	--	--	--	--	--	--	--
Lime-tree young shoots	55.0	3 25	1.45	79	73	--	--	--	--	--	--	--	--	--	--	--	--	--	--	--	--	--	--
Canada-poplar ditto	62.5	2.29	0.86	134	67	--	--	--	--	--	--	--	--	--	--	--	--	--	--	--	--	--	--
Oak ditto	57.4	2.16	0.92	125	83	--	--	--	--	--	--	--	--	--	--	--	--	--	--	--	--	--	--
Acacia ditto (autumn)	53.6	1.56	0.72	160	--	--	--	--	--	--	--	--	--	--	--	--	--	--	--	--	--	--	--
Drum cabbage	92.3	3.70	0.28	411	556	500	250	429	600	600	--	600	600	500	600	--	--	--	--	--	--	--	--
Swedish turnip	91.0	1.83	0.17	676	--	300	--	300	250	--	350	370	350	--	--	--	--	200	--	--	--	350	--
Turnip	92.5	1.70	0.13	885	533	600	290	526	450	500	525	525	525	525	500	--	--	450	--	800	667	--	--
Field-beet (1838)	87.8	1.70	0.21	548	366	400	250	460	250	300	--	--	460	255	--	261	--	333	--	--	--	--	400
Ditto, white Silesian	85.6	1.43	0.18	669	366	--	--	--	--	--	--	--	--	--	--	220	--	--	--	--	--	--	--
Carrots	87.6	2.40	0.30	382	205	250	225	300	250	--	266	270	266	260	266	307	266	270	--	338	--	--	380
Jerusalem potatoes (1839)	79.2	1.60	0.33	348	--	--	--	--	--	--	--	--	--	--	--	--	--	--	--	--	--	--	280
Ditto (1836)	75.5	2.20	0.42	274	--	--	--	--	--	--	--	--	--	--	--	--	--	--	--	--	--	--	--

TITLE.	Standard water per cent.	Azote per cent	Azote per cent. in the article not dried.	Theory.	Block.	Petri.	Meyer.	Thaer.	Pabst.	Flottow.	Pohl.	Rieder.	Gemerhausen.	Crud.	Weber.	Dombasle.	Krantz.	Schwertz.	Schnee.	Midleton.	Murre.	André.	Boussingault
Potatoes (1838)	65.9	1.50	0.36	319	216	200	150	200	200	250	200	--	--	210	--	187	227	200	--	--	--	--	280
Ditto (1836)	79.4	1.80	0.37	311	--	--	--	--	--	--	--	--	--	--	--	--	--	--	--	--	--	--	--
Ditto after keeping in the pit	76.8	1.18	0.30	383	400	--	--	--	--	--	--	--	--	--	--	--	--	--	--	--	--	--	--
Cider apple pulp dried in the air	6.4	0.63	0.59	195	--	--	--	--	--	--	--	--	--	--	--	--	--	--	--	--	--	--	--
Beet-root magma from the sugar mill	70.0	..	0.38	303	--	--	--	--	--	--	--	--	--	--	--	--	--	--	--	--	--	--	--
Vetches in seed	14.6	5.13	4.37	26	30	54	--	66	40	--	--	--	--	--	--	--	--	--	--	--	--	--	--
Field beans	7.9	5.50	5.11	23	30	54	50	73	40	--	50	--	--	--	--	--	--	--	--	--	--	--	--
White peas (dry)	8.6	4.20	3.84	27	30	54	48	66	40	--	47	--	--	--	--	--	--	--	--	--	--	--	--
White haricots	5.0	4.30	4.58	25	--	39	--	--	--	--	--	--	--	--	--	--	--	--	--	--	--	--	--
Lentils	9.0	4.40	4.00	29	--	--	--	--	--	--	--	--	--	--	--	--	--	--	--	--	--	--	--
New maize	18.0	2.00	1.64	70	--	52	--	--	--	--	--	--	--	--	--	--	--	--	--	--	--	--	59
Buckwheat	12.5	2.40	2.10	55	--	64	--	--	--	--	--	--	--	--	--	--	--	--	--	--	--	--	--
Barley (1836)	13.2	2.02	1.76	65	33	61	53	76	50	--	--	52	--	--	--	47	--	--	--	--	--	--	--
Barley-meal	13.0	2.46	2.14	54	--	--	--	--	--	--	--	--	--	--	--	--	--	--	--	--	--	--	--
Ditto	13.0	2.20	1.90	61	--	--	--	--	--	--	--	--	--	--	--	--	--	--	--	--	--	--	--
Oats (1838)	20.8	2.20	1.74	68	--	71	--	86	60	--	50	55	--	--	--	--	--	--	--	--	--	40	50
Ditto (1836)	12.4	2.22	1.92	60	--	--	--	--	--	--	--	--	--	--	--	--	--	--	--	--	--	--	--
Ditto (Parisian)	14.0	1.93	1.70	68	--	--	--	--	--	--	--	--	--	--	--	--	--	--	--	--	--	--	--
Rye (1836)	11.5	1.70	1.50	77	33	55	51	71	50	44	38	52	--	--	--	--	65	--	52	--	--	--	--
Ditto (1838)	11.5	2.27	2.00	58	--	--	--	--	--	--	--	--	--	--	--	--	--	--	--	--	--	--	--
Wheat (1836, Alsace)	10.5	2.33	2.09	55	27	52	46	64	40	--	--	--	--	--	--	--	44	--	--	--	--	--	--
Ditto (1838)	14.5	2.30	2.00	57	--	--	--	--	--	--	--	--	--	--	--	--	--	--	--	--	--	--	--
Ditto from highly manured soil	16.6	3.18	2.65	43	--	--	--	--	--	--	--	--	--	--	--	--	--	--	--	--	--	--	--
	12.1	2.60	2.80	41	--	--	--	--	--	--	--	--	--	--	--	--	--	--	--	--	--	--	--
Recent bran	37.1	2.18	1.36	85	105	--	--	--	--	--	--	--	--	--	--	--	--	--	--	--	--	--	--
	13.8	2.77	2.30	50	--	--	--	--	--	--	--	--	--	--	--	--	--	--	--	--	--	--	--
Wheat husks or chaff	7.6	0.94	0.85	135	160	--	--	--	--	--	175	--	--	--	--	--	--	--	--	--	--	--	--
Rice (Piedmont)	13.4	1.39	1.20	96	--	--	--	--	--	--	--	--	--	--	--	--	--	--	--	--	--	--	--
Gold of pleasure seed (Madia)	8.0	4.00	3.67	31	--	--	--	--	--	--	--	--	--	--	--	--	--	--	--	--	--	--	--
Ditto, cake	11.2	5.70	5.06	23	--	--	--	--	--	--	--	--	--	--	--	--	--	--	--	--	--	--	--
Linseed cake	13.4	6.00	5.20	22	42	180	--	--	--	--	--	--	--	--	57	--	--	--	--	--	--	--	--
Calza ditto	10.5	5.50	4.92	23	--	--	--	--	--	--	--	--	--	--	--	--	--	--	--	--	--	--	--
Madia ditto	6.5	5.93	5.51	21	--	--	--	--	--	--	--	--	--	--	--	--	--	--	--	--	--	--	--
Hemp ditto	5.0	4.78	4.21	27	--	--	--	--	--	--	--	--	--	--	--	--	--	--	--	--	--	--	--
Poppy ditto	6.8	5.70	5.36	21	--	--	--	--	--	--	--	--	--	--	--	--	--	--	--	--	--	--	--
Nut ditto	6.0	5.59	5.24	22	--	--	--	--	--	--	--	--	--	--	--	--	--	--	--	--	--	--	--
Beech mast ditto	6.2	3.53	3.31	35	--	--	--	--	--	--	--	--	--	--	--	--	--	--	--	--	--	--	--
Arachis (?) ditto	6.6	8.89	8.33	14	--	--	--	--	--	--	--	--	--	--	--	--	--	--	--	--	--	--	--
Dry acorns	--	..	0.80	143	--	--	--	--	--	--	--	--	--	--	--	--	--	--	--	--	--	--	--
Refuse of the wine press, air-dried	48.2	3.31	1.71	68	--	62	--	--	75	--	--	--	--	--	--	--	--	--	--	--	--	--	--

To complete the preceding ample table, I shall still add the equivalents of a few articles of forage that have not yet been examined chemically.

100 of meadow-hay are replaced by :

From 85 to 90 of sainfoin hay,	according to	Petri and Meyer.
By 90 of spurry hay	"	Petri.
325 to 500 of green spurry	"	Pabst and Flottow
42 to 50 of chestnuts	"	Block and Petri.
By 50 of Indian chestnuts	"	Petri.
62 of turnsole seeds	"	Petri.
109 of rye-bran	"	Block.

In the list of substances there are some which are used almost exclusively for the food of man, and I have thought it not uninteresting to contrast these different articles with reference particularly to the quantity of azote they contain. I have composed the following table or list of equivalents on this basis; having assumed wheaten flour as the standard and called it 100. As all herbs, roots, leaves, &c., may be pulverized after drying, I have spoken of these articles dry under the name of meal.

Wheat flour (good quality)	100	White-heart cabbage	810
Wheat	107	Cabbage meal	83
Barley-meal	119	Potatoes	613
Barley	130	Potato meal	126
Rye	111	Carrots	757
Buckwheat	108	Carrot meal	95
Maize	138	Turnips	1335
Yellow peas	67	Mealy bananas (Ficus Indica)	700
Horse-beans	44	Manihot (casava plant)	700
White French beans	56	Name ? (discorea sativa)	300
Rice	171	Apio ? (arracacha)	1050
Lentils	57		

Judging from the equivalents, leguminous vegetables must be possessed of a much higher nutritive value than wheat; and it is known, indeed, that haricots, peas, and beans, form in some sort substitutes for animal food. The difference indicated is so great, however, that it may surprise those who have never thought of the subject that engages us. In a general way we are all perhaps disposed to regard the articles that enter habitually into our food as highly nutritious. The fact, however, is, that tubers, roots, and even the seeds of the cereal grasses are but very moderately nutritious. If we see herbivorous animals getting fat upon such things, it is only because their organization enables them to consume them in large quantities. I doubt very much whether a man doing hard work could support himself on bread exclusively. I am aware that countries are quoted where the potato and where rice form the sole articles of food of the inhabitants; but I believe also that these instances are incomplete. In Alsace, for example, the peasantry always associate their potato diet with a large quantity of sour or curdled milk; in Ireland with buttermilk. The Indians of the Upper Andes do not by any means live on potatoes alone, as some travellers have said they do; at Quito, the daily food of the inhabitants is *lorco*, a compound of potatoes, and a large quantity of cheese. Rice is often cited as one of the most nourishing articles of diet; I am satisfied, however, after having lived long in countries where rice is largely consumed,

35

that it is any thing but a substantial, or, for its bulk, nutritious article of sustenance. This is an important question, inasmuch as in some departments of the public service rice is sometimes served out as a substitute for other articles of diet. In the French navy, for example, 60 grammes, or about 20 dwts. of rice may be substituted for 60 dwts. of split peas or haricots; but I cannot hold such a substitution to be either fair or reasonable. At a period when I had myself the charge of the rations for a detachment of men, I found that the experience of the country where I was, assigned 3 lbs. of rice as the equivalent of 1 lb. of haricot beans; and analysis confirms this practical conclusion.

Haricots, in fact, contain about 0.046 of azote; rice no more than 0.014. And if the nutritious properties be really in proportion to the amount of azote, it is obvious that 3½ of rice will be required in lieu of 1 of the leguminous seed.

We hear it constantly repeated that rice is the sole nutriment of the nations of the whole of India. But the fact would appear not to be precisely so; and I may here quote M. Lequerri, who, during a long residence in India, paid particular attention to the manners and customs of the inhabitants of Pondicherri. "The food," says M. L., "is almost entirely vegetable, and rice is the staple; the inferior castes only ever eat meat. But all eat kari, an article prepared with meat, fish, or vegetables, which is mixed with the rice boiled in very little water. It is requisite to have seen the Indians at their meals to have any idea of the enormous quantity of rice they will put into their stomachs. No European could cram so much at a time; and they very commonly allow that rice alone will not nourish them. They very generally still eat a quantity of bread."*

§ II. OF THE INORGANIC CONSTITUENTS OF FOOD.

We discover in the bodies of animals the several mineral substances, the existence of which we have ascertained in vegetables. The bones, as we have seen, contain a large quantity of phosphate of lime; it is requisite therefore that the elements of this salt, phosphoric acid and lime, should form part of the ration or diet-roll; this is a point upon which all physiologists are agreed; but the point upon which there is nothing like uniformity yet attained has reference to the precise quantity of mineral matter which must enter into the constitution of the food. The analyses of ashes which I have given show that if vegetable aliments all contain nearly the same inorganic principles, they still contain them in very different proportions: thus potatoes, wheat, oats, and beans, contain much less lime than clover, straw, and peas. The phosphoric and sulphuric acids and the alkalies do not vary less; so that we are led to ask whether a ration compounded of such and such an article, or of such and such articles, will furnish the animals to which it is supplied with the neces-

* The Irish peasantry, who live so much on potatoes, have buttermilk with them at least, often salt herring; and a laboring man, it is said, will consume 12 or 14 lbs per diem!—ENG. ED.

sary dose of inorganic principles, which must be assimilated daily, and which is quite indispensable to maintain them in health and vigor.

It is easy to arrive at a knowledge of the mineral principles which are necessary as elements of the diet, by ascertaining their quantity in the ration, which long experience has shown to be sufficient. Yet as there is reason to believe that in many cases mineral substances are present in excess, I have thought that it might be useful to determine by means of analysis the nature and the proportion of the inorganic elements which are actually assimilated by an individual, in order to have a minimum which might serve as a basis for any reasonings or inferences on the subject. My experiments were performed in two opposite circumstances in which I regard assimilation as most rapid and most complete: videlicit, a calf in full growth, and a milch-cow in calf.

The calf was six months old, and weighed 369 lbs. Some days before being made the subject of experiment it was fed with hay. During the two days when it had this fodder ad libitum, it ate 19 lbs.

In the course of the 1st day the calf passed 21.49 lbs. of excrements.
2d day " 20.39 "

41.88

Which, dried, was reduced to 7.41 lbs.

In the course of the two days 5.584 lbs. of urine were collected, which, evaporated, yielded 2933.2 grains of extract, the animal having in the same interval drunk 45.7 pints of water.

Analysis discovered in 100:

	Azote	Ashes
Of the hay	1.6	7.6
Of the dry excrements	2.1	12.7
Of the urinous extract	4.0	40.0

Now if we inquire from these data in regard to the quantity of azote and of mineral matters which were assumed with the food in the course of two days, we have:

	Azote (half-drachms.)	Mineral substances (half-drachms.)
In the food, discarding fractions,	69	328
In the excrements	50.5	214
In the urine	3.8	38
Together	54.3	252

Therefore, azote fixed or exhaled in 2 days 14.7 half-drachms.
Mineral substances fixed in 2 days....... 76 "

The composition of the ashes obtained from the hay and from the excrements, shows us approximatively both the quantity and the nature of the several inorganic substances which had been assimilated. The composition of these ashes is as follows:

	Of the hay.	Of the excrements.	Of the urine.
Acids: Carbonic	9.0	2.0	17.3
Acids: Phosphoric	5.3	5.1	0.2
Acids: Sulphuric	2.4	2.3	7.0
Chlorine	2.3	1.9	9.9
Lime	20.4	16.0	0.9
Magnesia	6.0	6.5	6.0
Potash and soda	17.3	12.5	57.3
Oxide of iron, alumina	1.5	1.0	"
Silesia	33.7	51.0	1.2
Loss	2.1	1.7	"
	100.0	100.0	100.0

If the hay consumed contained 328 half-drachms of ash or mineral matter, the excrements and urine 252 half-drachms of the same matters; the difference between the two sums, 76 half-drachms,* is the quantity of mineral matter fixed in the course of two days, of which 200.6 grains were phosphoric acid, and 494.0 grains were lime. This quantity of lime, however, is more than four times as much as is necessary to constitute a subphosphate of lime such as exists in the bones. It is true, indeed, that there is always a quantity of carbonate of lime associated with the subphosphate in bones; 10 of carbonate for 38 of phosphate, according to Fourcroy and Vauquelin, in those of the ox. Still the quantity of lime assimilated was vastly more than it ought to have been, had it only gone to assist in the formation of bone. If there was no error in the observations, it is probable that the base in question enters into the constitution of the salts with organic acids which are encountered in all parts of the animal body.

By a series of weighings, I ascertained that my calf, fed simply upon hay, increased every day by a quantity equal to 9725.9 grains troy, in which were included 858.35 grains of mineral substances, the calcareous phosphate and carbonate of the bones in this quantity being represented by 262.4 grains, or nearly 3 per cent. of the entire weight acquired in the course of twenty-four hours.

In the experiment with the milch-cow in calf, I limited my inquiries to the phosphoric acid and the lime taken in and given out. The animal, four years old, was 2½ months gone with calf, and weighed 1452.6 lbs. She had the same allowance during the experiment as she had had for several days before, and which for twenty four hours consisted of—

Hay	16.5 lbs.
Cut wheat-straw	9.9
Beet	59.4

The experiment was continued for four days, during which the excrements, the urine, and the milk, were carefully collected and weighed, and the ashes, both of the food consumed and of the products rendered, were determined by chemical analysis. Suffice it to say, that, representing the quantity of mineral matters assumed into the body in the course of the experiment by 849.9 half-drachms, the quantity voided amounted to no more than 556 half-drachms. In the quantity assumed, there were 100.2 half-drachms of phosphoric acid, and 203.8 half-drachms of lime; in the quantity voided, there were but 68.2 half-drachms of phosphoric acid, and 116.8 half-drachms of lime: this is at the rate of about 8 half-drachms of phosphoric acid, and 22 half-drachms of lime assimilated in the course of twenty-four hours. Here, as in the case of the calf, the quantity of lime assimilated is greatly superior to what it ought to be, in order, by combining with the phosphoric acid, to constitute the phosphate of lime of the bones.

From these inquiries into the nutrition of a calf and of a cow in

* The exact quantity is 2392.8 grains troy.—Eng. Ed.

calf, it follows that there is a portion of the mineral substance taken in with the food, which remains definitively fixed to concur in the growth or in the evolution of the individual. In an adult animal it is to be presumed that no such definitive fixation of inorganic principles takes place, or that it is much less considerable; that in the dejections and several secretions ought to be found the whole of the phosphoric acid, of the lime, &c., taken in with the food. And this presumption is confirmed by experience; for on instituting an inquiry into the matter upon a horse, it was found that the mineral matters assumed were almost exactly balanced by those discharged. Nevertheless, and granting this to be quite true, which it is, it would be a grave mistake to suppose that an adult animal could go on for even a very short period of time upon food that contained no mineral matter. Precisely as in the case of organic matter, it appears that a portion of inorganic matter is also fixed in the living frame, where for a time it forms an integral element in the wonderful structure; and a supply of the latter kind is undoubtedly no less necessary than is the supply of the former description recognised by all the world. Were there an inadequate quantity of phosphoric acid, of lime, &c., in the food, no question but that the body would speedily feel the effects of the deficiency, and that disease and death would by and by put an end to life. So much, indeed, seems demonstrated by the very interesting experiments of M. Chossat, in which he kept granivorous animals upon a diet rich in azotized principles and in starch, but deficient in lime. From some previous inquiries, M. Chossat had observed that pigeons even require to add a certain proportion of lime to their ordinary food, the quantity naturally contained in which does not suffice them. Wheat, as we have seen, though it contains a large proportion of phosphate of magnesia, contains very little phosphate of lime; and pigeons put on this grain, though they do perfectly well at first, and even get fat, begin by and by to fall off. In from two to three months, the birds appeared to suffer from constant thirst; they drank frequently; the fœces became soft and liquid, and the flesh wasted, and in from eight to ten months the creatures died under the effects of a diarrhœa, which M. Chossat attributed to *deficiency of the calcareous element* in the food. And it is neither uninteresting nor unimportant to observe, that the same thing occasionally occurs in the human subject during the period when the process of ossification is usually most active. But one of the most remarkable features of M. Chossat's experiments was observed in the state of the bones of the pigeons; they became so thin and weak that they broke during the life of the birds with the slightest force.* The conclusion from this fact is obvious. Supplies of all the elements of all the parts of the body are indispensable to the maintenance of health, to the continuance of life.

A pigeon will eat about 463.140 grains of wheat per diem, containing 9.725 grains of ash, in which analysis discovers 4.569 grains of phosphoric acid, and 0.277 of a grain of lime. But this small

* Chossat, in Comptes Rendus, t. xiv., p. 451.

quantity of lime is incompetent to maintain the bones in their standard condition. I have thought it of moment to insist upon these facts, because I see that they may sometimes come into play in practical rural economy. No breeder or feeder ought to be ignorant of the influence of mineral substances on nutrition. It is not only indispensable that the allowance of an animal in full growth be sufficient to support, and even to add to the soft textures; it must further contain the elements requisite for the nutrition of the osseous system: and it is not impossible but that, in managing the feeding of young cattle or young horses in such a way as to reduce to a minimum, or to give in excess, certain of the inorganic elements of the food, we may succeed in impressing one character or another upon a race. It is even possible that the empirical rules which are acted upon with a view to increase or diminish the quantity of bone, the weight of flesh or of fat, &c., are all connected with various proportions of phosphoric acid, of lime, magnesia, &c., in the food. It will probably be discovered, some day, that Bakewell's art is to be explained through the composition of the ashes of the food.

Wheat is not the only alimentary matter that contains an insufficient quantity of lime; maize or Indian corn contains still less: and if that which is grown in the tropics contains as little as that which is produced in Europe, it would be difficult to explain how the grain should answer so well as it unquestionably does for food.* It is true that it is seldom or never consumed alone and without addition; and in South America, where the animals have it largely, I have observed that they frequently eat earth. The habit which certain tribes of the natives have of eating earth, too, which has been particularly remarked upon by travellers and missionaries as an instance of depravation of taste, presents itself to me in quite another light, since I became acquainted with the composition of the ashes of the ordinary article of diet in the countries where it occurs.†

The calcareous and other salts necessary to nutrition, however, are not derived from the food exclusively; the water that is generally consumed contains a quantity which is by no means to be neglected. A horse or a cow, for instance, which drinks from 15 to 45 quarts of water per diem, will even, if the water be as pure as that of the Artesian well of Grenelle, take in from 35 to 108 grains of saline matter in which carbonate of lime predominates; water that is less free from saline impregnation would of course introduce a much larger proportion; some waters in the quantities above specified will contain from 138 to upwards of 400 grains of saline matter, one half of which may be carbonate of lime. And I am here speaking of clear or filtered water; that which is muddy or turbid contains a still larger quantity of earthy matter in suspension than in solution. In an experiment made for the purpose of getting at the amount of earthy matter taken by a milch-cow from the watering-

* An ash of maize, analyzed in my laboratory by M. Letellier, contained but 1.3 per cent. of lime to 50.1 of phosphoric acid and 17.0 of magnesia.

† I several times saw children chastised in Indian villages who had been caught eating earth.

trough in the course of the day, I found that it amounted to about 770 grains troy.

Notwithstanding these facts, it is still doubtful whether the lime contained in ordinary well-water would prove sufficient to supply a growing animal with the material requisite to the formation of its bones; in adults, indeed, changes in the elements of the bones appear to proceed so slowly that a very small quantity of calcareous matter probably suffices to repair losses; but it is otherwise with young and growing animals. I have shown that a calf six months old receives with its forage a quantity of phosphoric acid which corresponds to 555.7 grains of phosphate of lime. A calf a few weeks old, when it has 17 or 18 pints of milk per diem, receives 802.7 grains of mineral substances, into which subphosphate of lime or bone earth enters in the proportion ot 370.5 grains. It would be interesting to ascertain what quantities of these substances were assimilated by so young an animal, and at a period when the growth is so rapid that the increase from day to day sometimes exceeds 2 pounds.

The importance of the inorganic principles of the food once recognised, it concerns us to take note of their nature and quantity in the ratio we allow to our domestic animals. It is in fact this consideration which has led me to determine the quantities of phosphoric acid and lime contained in the various articles of food the ashes of which have been analyzed. With these data the proportion of bone earth contained in a given ration is forthwith perceived.

One thousand parts of the forage gathered at Bechelbronn in its ordinary state contained:

Forage.	Mineral Substances.	Azote.	Phosphoric acid.	Lime.	Bone earth.
Hay	62.33	11.50	3.37	10.04	6.96
Potatoes	9.64	3.70	1.09	0.17	0.33
Beet	7.70	2.10	0.46	0.54	0.95
Turnip	5.70	1.30	0.35	0.62	0.72
Jerusalem Potato	12.47	3.75	1.35	0 29	0.56
Wheat	20.51	20.50	9.64	0.60	1.16
Maize	11.00	16.40	5.51	0.14	0.27
Oats	31.74	17.87	4.73	1.17	2.27
Wheat-straw	51.90	3.00	1.61	4.41	3.32
Oat-straw	35.70	3.00	1.07	2.97	2.21
Clover-hay	73.50	21.00	4.63	18.08	9.85
Peas	30.00	38.40	9.03	3.03	5.83
Haricots	35.00	45.80	9.38	2.03	5.94
Beans	30.00	51.10	10.26	1.53	9.27

We seem here to observe a certain relation between the proportion of azote and that of the phosphoric acid contained in the food; the most highly azotized are also those that generally contain the largest quantity of the acid, a circumstance which seems to indicate that in the vegetable kingdom the phosphates are connected more especially with the azotized principles, and that they accompany them in passing into the textures of animals. With the assistance of the above table it is easy to ascertain the quantity of phosphate of lime which enters into a given ration. Let us take that given to the horses in experiment 3d, in which the half of the hay was replaced by potatoes, one of the articles that contains the smallest proportion of lime, and we find in the

26.6 lbs. of hay	632.9	grs. phosphoric acid and	1867.9	grs. of lime.
30.8 lbs. potatoes	387.7	"	37.0	"
	1020.6		1904.9	

numbers which correspond with 1798.5 grains of bone earth, and 978.7 grains of uncombined lime.

In his usual allowance a work-horse at Bechelbronn receives :

Hay	22 lbs.	containing	524.8	grs. phosphoric acid, and	1543.8 grs. of lime.
Straw	5.5	"	60.7	"	
Oats	7.2	"	230.2	"	
			81.57		

In other words, 1735 grains of bone earth, and 864 grains of free lime.

I have found that very young foals, growing rapidly, and weighing about 374 lbs., consume per diem :

Hay....	19.8 lbs. containing of phosphoric acid	463 grs. ;	lime	1389 grs.
Oats ...	7.2 " "	231	"	58.6

which represent 95 of bone earth or subphosphate of lime. As a consequence of the relation which appears to exist between the azote and phosphoric acid of an article of sustenance, it comes to pass that like nutritive equivalents also indicate like proportions of phosphoric acid ; so that by introducing a suitable quantity of hay or clover, articles that abound in lime, into the ration, we are always certain of having food favorable to the development of the osseous system, whatever the nature and quality of the other articles that enter into the constitution of the allowance.

The relation of the phosphoric acid to the azote approaches the ratio of 3 to 10, in the more ordinary articles of forage ; but the same relation is no longer apparent in the cereals and leguminous vegetables ; in grain and in peas, beans, &c., the phosphoric acid amounts to about a fourth of the azote contained. Thus we have :

	Theoretical equivalent.	Phosphoric acid in the equivalent.		Theoretical equivalent.	Phosphoric acid in the equivalent.
Hay	100	0.34	Oat-straw	380	0.40
Potatoes	320	0.35	Oats	68	0.32
Beet	548	0.28	Maize	70	0.38
Turnip	885	0.31	Wheat	43	0.41
Jerusalems	273	0.37	Peas	25	0.23
Dry clover	75	0.34	Haricots	27	0.25
Wheat-straw	235	0.37	Beans	23	0.24

§ III. OF THE FATTY CONSTITUENTS OF FORAGE : CONSIDERATIONS ON FATTENING.

When fat was observed accumulating in the tissues of the animal body, and it was unknown that the presence of fatty matters in plants is what may be termed a general fact, men naturally conceived that the fat was produced from the food in the act of digestion, that it was composed in the animal body much in the same way as it is formed in the seed and leaf of the living vegetable.

The inquiries which I am about to present, however, all tend to make us conclude that fatty substances are only produced in vegetables, and that they pass ready formed into the bodies of animals, where they may either undergo combustion immediately, so as to evolve the heat which the animal requires, or be stored up in the tissues in order to serve as a magazine of combustible matter.

This latter view appears the most simple ; but before discussing the experiments which bear it out, it seems necessary to pass in brief review the notions that have been entertained at different times on the formation of fat. When the great burying-place of the Innocents was emptied, for example, it was commonly imagined that one of the effects of the putrefactive process was to convert the flesh, the brain, the viscera, &c., into fat—adipocire, as it was called ; it was not, indeed, till after the researches of M. Chevreul had been undertaken, and that it was discovered adipocire contained the same acids as human fat, which had, in fact, only been partially saponified by ammonia,—until the inquiries of M. Gay-Lussac were made public—that it was acknowledged that muscular flesh or fibrine subjected to putrefaction leaves no larger a quantity of fat than can be obtained from it by proper solvents before it has undergone any change : the effect of putrefaction is to destroy the fibrine, and so to expose the fatty substance which it contained.

It may therefore be said, that all these fortuitous opinions upon the supposed formation of fat by chemical processes, have vanished as they have been successively subjected to careful examination.

Let us now turn to the inferences come to by physiology. The bodies of carnivorous animals are often loaded with fat ; and none can be detected in any of their excretions. It is therefore in these animals that it must be most easy to ascertain the source or origin, and mode of disappearance of fatty matter.

When the progress of digestion is watched in a dog, it is soon discovered that the chyle is far from being a fluid having uniformly the same characters and qualities. That which is produced under the influence of a vegetable diet, abounding in the starchy principle and in sugar, or after a meal of perfectly lean meat, is always and alike poor in molecules or globules. The chyle is then nearly transparent, extremely serous, and yields very little fat when washed with ether. But if the animal have a meal of fat food, the chyle that results from it is opaque like cream, very rich in particles, and, digested with ether, yields a large quantity of fatty or oily matter to that solvent.

These facts, observed by M. Magendie, and confirmed with more ample details by Messrs. Sandras and Bouchardat, show that the fatty principles of our food minutely subdivided or made into an emulsion by the act of digestion, pass without undergoing any essential change into the chyle, and from that into the blood, whither they can in fact be followed, and in which they can be shown to remain for a longer or a shorter time unaltered, at the disposal of the economy, as it were. Such observations have naturally led physiologists to conclude that the fatty principles of the food were the principal, if not the only sources whence animals derive the fat which is met with stored up in their tissues, or which appears in the butter of their milk. And this view, so long as the carnivorous tribes alone are considered, has not a single feature which makes it objectionable. But when we would extend it to the herbivorous tribes, two difficulties meet us on the threshold of the inquiry.

1st. Do vegetables actually contain such a quantity of fatty matter in their structure as will explain the fattening of cattle and the formation of milk?

2d. Is it not more simple to suppose fat and butter the product of certain transformations undergone by starch and sugar in the animal economy?

It appears at first sight most opposite to nature to suppose that the feeding ox finds the whole of the fat he lays on ready formed in the food he eats; it is only, in fact, after having made repeated analyses of plants, and discovered fatty matters almost everywhere, and in quantities generally superior to any that had been suspected in the composition of plants, that the idea begins to acquire likelihood; finally, the chemist becomes convinced that it is so when he finds a regular association of neutral azotized substances and fatty principles in all the articles usually employed as food for cattle,—in the grasses and cereals, in the leaves and stems and seeds of plants.

Fatty substances appear to be principally formed in the leaves, where they frequently show themselves under the form and with the properties of wax. Taken into the bodies of animals, mingled with the blood, and exposed to the influence of the oxygen of the inspired air, they will undergo an incipient oxidation, whence will result the stearic or oleic acid that is found as a constituent of suet. By undergoing a second elaboration in the bodies of the carnivora, the same fatty substances, oxidated anew, would produce the margaric acid which characterizes their fat. These divers principles, by a still further degree of oxidation, would give rise to the fat volatile acids which make their appearance in the blood and in the perspiration. Finally, did they suffer complete oxidation, *i. e.* combustion, they would be changed into carbonic acid and water, and be in this shape eliminated from the economy.

Among the various properties possessed by fatty substances, there is one which may play an important part in the phenomena of fattening; this is the solvent power which they severally possess in regard to one another; the property of mixing in all imaginable proportions, still preserving the general features which severally distinguish them. In the stomach, in the intestines, in the chyle, and in the blood, fatty substances of various kinds may form homogeneous matters by their intimate admixture, and become divided into globules of complex composition, but everywhere the same.

Another property of fatty matters of every kind, which deserves particular attention, is that of insolubility in water. We find, in fact, that when an animal eats a soluble substance, that in general it is consumed by a true process of combustion, which converts its carbon into carbonic acid, and its hydrogen into water; or otherwise, it is simply eliminated without change in the urine.

Fatty matters may, indeed, disappear under the first form; but so long as they escape remarkable modification, it is certain that they do not pass off by the urine, and that the quantity eliminated by the perspiration is insignificant. Their insolubility, therefore, retains them in the economy once they have entered the blood or the tissues; and it is in virtue of this quality that they constitute a kind of magazine of combustible matter in the animal body. This is the principal reason wherefore individuals supplied with food in excess get fat, and that those insufficiently fed fall lean; the fatty matter being deposited in the interstices of the tissues in the former case, being taken up from them and burned in the second.

This explanation is attractively simple; but in our attachment to it we must not forget that other explanations have also been given; and in particular it must be contrasted with a view which has been formed upon certain inquiries undertaken by M. Dumas. It is known, for instance, that sugar may be regarded as a compound of carbonic acid, water, and olefiant gas. Now there is nothing to prevent olefiant gas becoming detached and taking different states of condensation, to give rise to bodies which by undergoing oxidation would produce fat acids and consequently fats. Since it has been known that the *oil of potato spirit* is also met with in the spirit obtained from the refuse of the grape, and in the spirit procured from malt, and from the molasses of beet-root sugar, the assurance that the oil is a product of the fermentation of sugar appears to be complete.

We ought even to be prepared to admit a phenomenon of the same kind as taking place in plants, when we see the sugar of their stems disappearing in the same ratio as their seeds or fruits become charged with oleaginous matter: all the palms elaborate sugar before producing oil.

It is upon chemical views of this kind that the second opinion as to the source of fat in animals has been formed, and which may be said to stand in direct contrast with that which assumes this substance as pre-existing in the food, which regards it as produced in the blood itself, under the influences of the most intimate forces of animal life. For my own part, I adopt the view which supposes an animal to be supplied with fat already formed, mainly because it presents itself to me as more in harmony with the facts which I observe in our stables. Still I do not deny that it may be possible for a certain quantity of fat to be elaborated in the bodies of herbivorous animals, under the influence of a special fermentation of the sugar which forms an element in their food; although I feel satisfied, from practical facts, that sugar plays no essential part in the fattening of cattle.

The formation theory, nevertheless, is not without data of a very curious and important kind, which require notice. Huber had found that bees fed upon honey, and even upon sugar, did not lose the power of producing wax for a long period. And Messrs. Milne Edwards and Dumas have lately confirmed the occasional accuracy at least of this conclusion. Their first experiments were unfavorable to the conclusions of the celebrated bee-master of Geneva. A swarm fed with lump-sugar yielded very insignificant quantities of wax; three other swarms, placed in glazed hives, and fed on honey and water, failed to produce any wax; but a fourth swarm gave a totally different result.

The procedure by analysis was now instituted. The absolute quantity of wax contained in the body of a bee, or in the bodies of a certain number of bees, was first ascertained; second, the quantity of wax in the combs constructed by the laborers was determined; third, the quantity of wax contained in the honey consumed was discovered. The final result of the inquiry was, that a swarm of 2005 laborers, after having in the course of a month consumed 12889.43 grs. of honey, had produced 2407 grs. of wax.*

Granting the accuracy of this conclusion, admitting that the bee fed upon honey has the power of producing wax, I might still ask whether it was therefore legitimate to conclude that the ox was endowed with any faculty of the same kind? Still, to the interesting physiological fact above quoted, may be associated the remarkable fact of the conversion of sugar into butyric acid, observed by Messrs. Pelouze and Gélis, a conversion effected by mixing a small quantity of caseum with a solution of sugar, and adding a sufficiency of chalk to neutralize acid as it was formed. This mixture, placed in a temperature of from 77° to 86° F., by and by passes through a series of changes, the last term of which is the butyric fermentation.

This butyric acid is a volatile, colorless, and very limpid fluid, having an odor that brings to mind at once that of vinegar and rancid butter. Although it has a resemblance to the acids which prevail in different kinds of fat, it nevertheless, by uniting with glycerine, constitutes a fatty body, butyrine, which forms about one-hundredth part in the constitution of butter, and must therefore be received as one of the elements of the fats; and the observations of Arthur Young would even incline us to presume that butyric acid exerts a favorable influence in the fattening of animals. Comparative experiments satisfied Young that hogs fattened more quickly on food that had become sour, than on the same food before it had turned. Now it is very probable that there was production of butyric acid in the course of the fermentation.

The question in reference to the formation of fat is of much more limited interest to the farmer than to the physiologist. The agriculturist, in fact, cares little whether a couple of pounds of honey consumed by a hive of bees will give origin to some 10 dwts. or so of wax; the matter that concerns him is, as to the degree in which the roots or tubers that he grows are fattening; and whether or not he can advantageously substitute a cheaper for a more costly article in his piggery or stalls? And here, as in so many other places,

* Vide Comptes rendus de l'Académie des Sciences, t. xvii. p. 131.

practice got the start of theory; and I own, with perfect humility, that I think its conclusions are in general greatly to be preferred; the universal custom of giving oil-cake and oleaginous seeds to our milch-kine and fatting oxen and sheep, appears to me to supply an argument of much greater force than any that can be obtained from chemical researches pursued in the laboratory.

Such articles as the potato and the banana, which answer admirably for the keep of hogs after they are weaned, are not adequate to fatten them for the butcher; they contain but very small quantities of fatty matter, and though the animals will grow rapidly upon them, and even attain to and maintain a certain condition, all that is required can only be secured by adding some other article to the ration that is richer in oleaginous or fatty principles. At Bechelbronn, whatever others have said on the subject, we find that our hogs will not fatten on potatoes alone; so that I agree with Schwertz when he says, that while hogs will get into good flesh upon potatoes, and even seem to fatten for a time, they soon cease from improving, and only begin to advance again when they receive in addition an allowance of barley or of split peas or beans.

A young pig will consume about **13** lbs. of potatoes per diem, into which, as analysis of the ashes informs us, there enters but about 17.2 grs. of lime. But this quantity is probably too small to meet the demand for bone earth in a young animal in full growth, and hence the great advantage of the whey or small quantity of skim milk which is so commonly added to the potato ration. It ought to be laid down as a general rule, that young creatures as well as adults ought to have a ration which contains the earthy elements of the bony system, the azotized elements of the flesh, and the fatty matter of the fat. From a series of experiments which I undertook, in concert with Messrs. Dumas and Payen, it appears that all the articles acknowledged the most powerful as fatteners, are those also that contain the largest proportions of fatty principles. The following substances contain the numerical quantities of matter soluble in ether in 100 parts:

Common maize	8.8	Dry lucern	3.5
Beaked Lombardy maize	7.8	Meadow hay	3.8
Large white Parisian maize	8.1	African wheat straw	3.2
Rice	0.8	Ditto Alsace	2.2
Oats	5.5	Ditto near Paris	2.4
Ditto	3.3	Oat straw	5.1
Rye	1.8	Bean meal	2.1
Rye flour	3.5	Beans	2.0
Hard Venezuela wheat	2.6	Haricots	3.0
Hard African wheat	2.1	Peas	2.0
Wheat flour	2.1	Lentils	2.5
Ditto	1.4	Potatoes	0.08
Fine bran	4.8	Mangel-wurzel	0.1
Coarse bran	5.2	Carrots	0.17
Dry clover	4.0	Oil-cake	9.0

M. Payen found that the oil was everywhere present in the seeds of gramineous plants. The embryo contains much, the husk less, the farinaceous portion still less. But maize and oil-cake contain about 9 per cent., whence the universally admitted superior fattening power of these two articles.

If we now discuss particularly one or more of the rations or allowances to oxen put up to fatten, or to milch-cows, we shall find that they uniformly contain a larger quantity of fatty matters than the secretions, excretions, and fat which have been produced under their influence. Thus a cow, in good condition, that produces 72 pints of milk, containing 3.3 lbs. of butter, while she eats 220 lbs. of meadow-hay, will have consumed more than 6.6 lbs. of matter soluble in ether—fatty substances. The quantity of fat contained in the food, over and above that which is discovered in the milk, has passed with the excrements, which are never entirely free from substances analogous to fatty matters.

With a view to comparing as accurately as is possible in inquiries of this kind, the quantity of fatty matter contained in the forage consumed by a cow, with that found in the milk and in the excrements, the following experiment was made. A cow, Esmeralda, No. 6 in the stable at Bechelbronn, calved on the 26th of September, and she was put to the bull again on the 4th of November. Up to the 22d of January (inclusive) Esmeralda received the usual allowance, viz:

After-math hay	11 lbs.
Oil-cake	22 "
Turnips	66 "
Wheat chaff	22 "

and the milk she gave in the course of the month of January, amounted on the—

	Pints.		Pints.
1st to	12.9	12th	12.3
2d	12.3	13th	11.4
3d	12.3	14th	11.4
4th	11.4	15th	12.3
5th	10.5	16th	10.5
6th	12.3	17th	11.4
7th	12.3	18th	10.5
8th	12.9	19th	11.4
9th	12.3	20th	11.4
10th	10.5	21st	11.4
11th	11.4	22d	11.4

The average quantity of milk, therefore, per day, in the course of the week that preceded the experiment, was 10.287 pints. From the 23d of January, when the ration was altered to—

Hay	16.5 lbs.
Chopped wheat straw	9.9 "
Beet-root	59.4 "

the quantity of milk yielded was:

January.	Pints.	January.	Pints.
23d	11.4	27th	11.6
24th	10.5	28th	11.4
25th	10.5	29th	11.4
26th	10.5	30th	11.4

on an average 11.8 pints per diem; a little more than with the former allowance.

The excrement passed by this cow was analyzed during four days, from the 24th to the 27th of January; the whole quantity being weighed moist every day, and well mixed, a mean sample of about **6 oz.** in weight was taken for analysis. This being stove-dried, the entire quantity of dry matter contained in the moist excrement was readily ascertained:

Dates.	Moist excrement.	Dry excrement.	Milk in pints.	Milk in lbs.
Jan. 24	40.7 lbs.	6.8 lbs.	10.5	13.5
25	41.8	7.3	10.5	13.5
26	62.1	9.0	10.5	13.5
27	43.4	7.1	10.5	13.5
	188.0	30.2	42.0	54.0

To ascertain the quantity of fatty or waxy substances contained in the food, the several samples were first treated with hot water, then with ether, and finally with a mixture of ether and alcohol boiling. The fatty element of the butter was determined by Peligot's method.

		Fatty Matters in the Food per cent.
Hay	1st Experiment	3.6
	2d ditto.	3.9
Straw	1st Experiment	2.4
	2d ditto.	2.0

		Fatty Matter in the Excrements and Milk per cent.
Excrements (dry)	1st Experiment	3.3
	2d ditto.	3.9
Milk		3.7

Let us say, that the proportion of fatty substance contained per cent. in the several articles consumed as food, was as follows—

Hay	3.7
Straw	2.2
Beet	0.1
Dry excrements	3.6
Milk	3.7

and we shall have the results of the experiment in this shape:

FOOD CONSUMED IN FOUR DAYS.		Fatty matter.
Beet	237.6 lbs.	1667.3 grs.
Hay	66.0	1776.1
Straw	39.6	6113.4
Fatty matter in food		24956.8
The excretions		21813.8
Fatty matter fixed or burned		3143.0

EXCREMENTS AND MILK IN FOUR DAYS.		Fatty matter.
Excrements	30.4 lbs.	7688.1 grs.
Milk	54.3	14125.7
Fatty matter in excrements and milk		21813.8

The natural conclusion from this experiment appears to be this: that the cow extracts from her food almost the whole of the fatty matter it contains, and that she converts this matter into butter.

It would perhaps be possible to make the proportion of butter contained in the milk to vary within certain limits. It is well known that the butter of cows in the same district varies notably according to the nature and abundance of the forage; the butter of the same country-side, for example, has been ascertained to contain 66 of margarine to 100 of oleine in summer, and 186 of margarine to 100 of oleine in winter. In the first case, the cows are grazing on the mountains, (the Vosges;) in the second they are eating dry fodder in the stall. I have besides had an opportunity to make a direct experiment upon this subject, which appears to me quite conclusive Having substituted for one half the allowance of hay an equivalent quantity of rape-cake, still very rich in oil, the cows were kept in excellent condition; but the butter was extremely soft, and had the taste of rape-seed oil to a degree that was perfectly intolerable.

I do not know a single instance in any of the systems followed at Bechelbronn, in which a milch-cow does not receive in her ration a quantity of matter analogous to fat, superior to that which is contained in the milk she yields. Having upon a certain occasion put a cow exclusively upon beet, I anticipated an unfavorable effect on her milk; and in fact a very sensible diminution in all its valuable elements occurred, and the animal began to suffer. By simply adding a few pounds of straw which had been taken away, the milk resumed its standard quality. The two rations thus composed may be contrasted under the two points of view of contents in fat and contents in organic matter.

In the ration of beet and straw (beet 119 lbs., straw 7½ lbs.) there were 140 dwts. 20 grs. of fatty matter; in that composed exclusively of beet (132 lbs.) there were but 38 dwts. 14 grs. of fat. The ill effects of the beet-root ration could not be ascribed to deficiency of inorganic elements, for the phosphate of lime it contained amounted to 37 dwts. 7 grs.—amply sufficient for all the purposes of the economy.

The information we have from M. Damoiseau, one of the most careful of the observers who have investigated the subject of the production of milk, confirms us in our views of the necessity of fatty matters in the daily ration of the milch-cow. The following are the elements of three of the rations for a cow in M. Damoiseau's establishment.

	No. 1.		No. 2.		No. 3.
Beet, or mangel-wurzel..	88 lbs.	Carrots	74 lbs.	Potatoes	55 lbs.
Bran....................	6.6	"	6.6	"	6.6
Pollard.................	5.5	"	5.5	"	5.5
Lucern..................	6.6	"	6.6	"	6.6
Oat-straw...............	13.2	"	13.2	"	13.2
Salt....................	0.11	"	0.11	"	0.11
	121.0		107.0		88.0
	Maximum.		Medium.		Minimum.
Quantity of milk yielded..	From 25 to 26 pts.		16 to 18 pts.		12¼ pts.

Let us now calculate the actual nutritive value of the different items in the above rations; or, selecting one, let us take that with the beet for particular consideration, as among the most usual.

6.6 lbs. of bran and 5.5 lbs. of pollard at....	5 per cent.	=0.60 of fatty matter.
6.6 lbs. lucern..............................	3 "	=0.33 "
13.2 lbs. oat-straw..........................	5 "	=0.60 "
		1.53

Whence it follows, that a cow here received upwards of 1½ lbs. of matter of a fatty nature with her food—a quantity more than sufficient to produce not only 16 or 18, but the maximum quantity of 25 or 26 pints of milk, very rich in cream. Did the cow receive an additional 40 lbs. of beet-root, she would find something like 12 lbs. more of solid matter in this article, composed especially of sugar, which she would burn to keep up her temperature, and nearly 25 dwts. of oily matter, which she would transfer to her milk,—to say nothing of new azotized principles which would be converted into caseine.

If we now, by an easy transition, pass to the phenomena of fattening, we still find that the principles which have been laid down can be most satisfactorily applied. Setting out from the numbers obtained from the experiments of Mr. Riedesel, which, in many points, agree with all I have seen myself, we arrive at the following conclusions.

An ox weighing 1320 lbs. avoird. will keep up his weight upon about 22 lbs. of good hay per diem. Put up to fatten, the same animal would require about twice this quantity, say 44 lbs., upon which he would gain at the rate of about 2 lbs. per day.

Now, if we even take Mr. Riedesel's conclusions as a little too favorable, as giving at least the maximum nutritive value to the hay and its equivalents, we may still admit, with him, that 22 lbs. of hay will produce about 17 pints of milk, or about 2 lbs. avoird. of flesh, containing 0.55 lbs., or rather more than ½ lb. of fat. Now, 22 lbs. of hay contain nearly 12 oz. 12 dwts. of principles soluble in ether, i. e. of fatty or waxy matter.

The fatting ox, consequently, fixes a certain proportion of these principles in the same way as the cow. There is only this difference, that the cow returns with the milk she yields a considerable quantity of the fat she finds in her food. There consequently exists an obvious relation between the formation of milk and fattening—a position which would gain support, did it require any, from a note which I owe to the politeness of M. Yvart, who, in summing up a long array of facts, concludes with these words: "The secretion of milk appears to alternate with that of fat. When a milch-cow fattens, she loses her milk;" and the converse of the proposition is no less true; when we would fatten a cow, we must let her go dry.

The breeds of kine admitted to be the best milkers remain long lean after the calving. In some of the short-horned English breeds, the quantity of milk is often very considerable shortly after the calving; but the animals are much disposed to get fat, and getting fat, the secretion of milk neither continues so long, nor is it so plentiful, as in some of the other less improved kinds. English hogs, which become much fatter than French hogs, appear not to be such good nurses. Now, if we admit that there is this intimate relation between the formation of milk and that of fat, we are obviously very near the admission, that articles of food containing fatty substances indispensable to the production of milk, are also indispensable to the production of animal fat. And, then, has it ever yet happened that animals have been fattened with food devoid of grease? I have not, for my own part, met with a single fact which countenances such a proposition. I have referred to the distinguished agriculturist, who attempted to fatten pigs upon potatoes, but who only succeeded by adding a certain quantity of graves to the food—an article which, as every one knows, contains a considerable proportion of fat in its composition.

M. Payen has, in fine, made some experiments which appear altogether conclusive, and from which it follows, that two Hampshire hogs which, having consumed 66 lbs. of gluten, and upwards of 30½ lbs. of starch, had only gained 17½ lbs.; while other two animals of the same breed, having been fed with 99 lbs. of the flesh of sheeps' heads, containing from 12 to 15 per cent. of fat, had gained 35 lbs. Yet, judging from elementary analysis, these two rations were almost identical; they contained the same quantity of dry nutritious matter. The first ration contained 26.4 lbs. of dry gluten, and 30.4 lbs. of starch; the second contained 20.9 lbs. of dry flesh, and 15.4 lbs. of fat. The quantities of carbon and azote were, therefore, a little higher in the vegetable than in the animal ration; but they differed notably in this, that the latter contained an equivalent of fat for the equivalent of starch contained in the former.

In a second experiment, four hogs, fed upon boiled potatoes, carrots, and a little rye, gained 117.7 lbs.; while other four animals, of the same age, and in the same conditions, but fed upon sheeps' heads, gained as many as 226.6 lbs.

In the course of these experiments, M. Payen was struck with this circumstance, that the increase in weight of an animal that is fattening being represented by 50 per cent. of water, 33.3 of fat, and 16.6 of azotized matter, the conviction is forced upon us that he actually fixes the greater proportion of the fat of his food in the cellular tissue of his body. The first hogs, for example, had eaten 14.74 lbs. of fat, and had gained 11.44 lbs. in weight; the four last referred to had had 18.48 of grease, and had increased 14.74 lbs. in weight.

It has now been the practice for several years, in various places, to maintain hogs in considerable numbers upon muscular flesh, horse-flesh; and it has been ascertained that the article, if extremely lean, though it keeps the animals in good heart and condition, though they grow and thrive on it, yet they will not fatten. When they are to be got ready for the butcher, they must, in addition, be put upon a course that is known to be proper to fatten them.

The scientific question of fattening having, of late years, attracted very general attention, the opinions which have now been announced have been very actively contested. Among other arguments, the general freedom from fat of the bodies of carnivorous animals, and the usual fat state of those of the herbivorous races, has been cited. Whales have even mistakenly been included in the list of fat vegetable feeders; but it is known to all naturalists, that the great majority of the whale tribes, the whole of those that inhabit the northern seas, are carnivorous. And, indeed, the mention of this fact leads me to revert to one of the most curious problems in the physics of the globe—that, to wit, presented by the vast amount of animal life amidst the waters of the ocean, and its support by a quantity of vegetables which to us appear altogether inadequate to such an end. The beautiful researches of M. Morren, however, seem calculated to throw some light on this interesting subject,—that inquirer having shown that certain animalcules possess the faculty of decomposing carbonic acid in the same way as vegetables; and it is probably in virtue of this power that the enigma is to be explained, of the source whence the myriads that people the deep derive their food.

But is it absolutely true that herbivorous animals only abound in fat? Who has not seen fat dogs and cats; and in the Cordilleras, where palm-trees abound, there is a particular species of bear, which lives in a great measure on the oily palm-nuts and young shoots of the palm-tree, which becomes remarkably fat, and proves a great attraction to the tigers of the country.*

Before coming to a close with this discussion, I think it right to refer to the experiments of M. Magendie, who has so well established the fact, that the chyle of animals fed on fat food contains a large quantity of fat; and that animals kept long on such food frequently become affected with what is called the fatty liver.†

To sum up, then, experiment demonstrates that hay contains a larger quantity of fatty matter than the milk and excretions which it forms; and that it is the same with all the other mixtures and varieties of food that are usually given to animals.

That oil-cake increases the production of butter, and that, like maize, it owes the fattening properties it possesses to the large quantity of oil it contains.

That there is the most perfect analogy between the production of milk and the fattening of animals; that potatoes, beet, carrot, and turnip, only fatten when they are conjoined with substances that contain fatty matters, such as straw, corn, bran, and oil-cake of various kinds.

That in equal weights, gluten mixed with starch, and flesh meat abounding in fat, have a fattening influence on the hog, which differs in the relation of 1 to 2.

Lastly, that fat food—food which will afford fat in the digestive canal—appears to be the indispensable condition of fattening. If it be necessary that the respiration be diminished or lessened in extent, this is only that the fatty substances taken into the stomach, and which have made their way into the blood, may not be oxidated, may not be burned; not that their formation may be favored.

All these facts are in such perfect harmony with the simple view of assumption and assimilation of fatty matters, that it is difficult to conceive on what foundation the opinion can repose which would have them composed out of their elements in the animal body. Nevertheless, I am myself the first to admit, that more extensive experience may lead to the modification or even entire change of the opinion which I advocate. The facts on which that opinion is based, despite their number, are not probably yet sufficient to constitute a perfectly satisfactory or conclusive theory. New researches

* These bears, evidently, cease to be carnivorous while they live on palm-nuts and leaves. For my own part, I do not think the point settled yet. The fatty matter of the generality of vegetables is wax rather than grease. And then some of the herbivorous tribes seem never to get fat.—ENG. ED.

† I may here state the contrary fact, as announced to me by a physiological friend, in whose report I place great reliance, that the chyle of animals fed with substances that give *mere traces* of waxy matter, contains fat or oil that can be collected in *large drops*—ENG. ED.

are, therefore, indispensable: it would be requisite to show, that a cow kept on a regimen abundant in point of quantity, but as poor as possible in matters analogous to fat, will continue to maintain her condition and yet yield milk abounding in cream; and that it is really possible, as some persons affirm, to fatten animals rapidly on roots and tubers alone.*

CHAPTER IX.

OF THE ECONOMY OF THE ANIMALS ATTACHED TO A FARM. OF STOCK IN GENERAL, AND ITS RELATIONS WITH THE PRODUCTION OF MANURE.

Agricultural industry generally extends to the breeding and fattening of cattle; the breeding, or at all events the maintenance, of horses; the breeding and feeding of sheep and swine. The circumstances, indeed, in which the tiller of the ground sees himself spared the necessity of attending to these matters, are rare exceptions to the general rule, and in fact only occur where it is easy to obtain abundant supplies of manure from without, or in those few favored spots where the fertility of the soil is such that it continues to yield its increase without addition in the shape of manure. In the vicinity of great centres of population, where dung can be bought cheap, or of guano islands, where a cargo costs a trifle, and in some tropical countries, large farming establishments may be found totally without stock in the shape of sheep and horned cattle. But in a general way the agriculturist is obliged to give himself up to the care of flocks and herds of one description or another; and, in fact, we now know that there is a certain and very indispensable relation to be maintained between the extent of surface under crop and the number of cattle to be provided for, variable as regards farms differently situated and circumstanced; but invariable when circumstances are the same, and the system of management pursued is similar in its principal features.

The question as to whether the cultivation of grain or other useful plants, or the rearing of cattle, is more profitable, which is often agitated, must receive a different solution in regard to each different locality. In one place it may be more advantageous to breed cattle or horses; in another to rear or fatten them: here, the production of milk, butter, and cheese, may be the best husbandry; there, the growth of hay, (as for miles round London on the north and west;) and again, wheat and the other cereal grasses may be the staples of

* Whoever would try experiments in this direction, must be careful to *mix his food;* one article alone never agrees. The Americans say, a pig will die upon pumpkins and upon apples alone; but he will live and fatten on a mixture of the two. I have myself seen scores of oxen fattened upon turnips, with a moderate allowance of straw or bog-hay; and have seen pigs get into admirable condition for the butcher on little more than potatoes.—Eng. Ed.

production. Even supposing that the growth of grain is that which is most advantageous on the whole, it by no means follows that the farmer shall give himself up to this exclusively; it is seldom that he can do so, indeed; he must have manure, and this entails the necessity of keeping cattle. If the latter, however, be the least profitable item in the economy of a particular domain, it will of course be kept within as narrow limits as possible.

In many places where the land is well adapted to the plough, and where the production of grain is unquestionably profitable, stock appears to offer few advantages; it sometimes happens, indeed, that the balance as regards the stall and cow-house is on the wrong side for the farmer, when the actual value of the forage that has been consumed is taken into the account. The loss is only made up for by the manure, which is in fact the return. This is the view that M. Crud obviously takes when he speaks of the stock upon a farm as *a necessary evil.** I am far from participating in his opinion; the cattle upon a farm are no evil, though they may be very necessary. To be satisfied of this, it is enough, in fact, to recollect the principle which has been established in treating of rotation courses, viz: *That in no case is it possible to export a larger quantity of organic matter, and particularly of organic azotized matter, from a farm, than is represented by the excess of the same description of matter contained in the manure consumed in the course of the rotation. By acting otherwise, the standard fertility of the soil would inevitably be diminished.*

This principle recognised, and I believe that it cannot be disputed, it is obvious that a portion of the produce of the fields must be returned to them to fecundate them anew, and it is precisely this portion of the forage crops destined to furnish manure that must be consumed in the stable and cow-house. Reasoning abstractly, the forage plants which it is not intended should quit the farm, might be buried directly as manure, without being made to pass through the bodies of animals; their fertilizing influence on the soil would come out sensibly the same; and this is what is done, in fact, so often as we manure by *smothering*. But we have scarcely made the first step in the rudiments of agriculture before we discover the immense advantages of following the usual custom, which first employs as forage for cattle the crops that are grown with a view to the production of manure. And we shall by and by find, in fact, that by adding to that portion of these crops a supplement of forage plants which it would be legitimate to export, without trenching upon the fundamental principle above laid down, we obtain the same quantity of manure, and turn the whole of this supplement into useful forces, or into animal products which possess a market value greatly superior to that of the forage before its assimilation. It is only the price of this portion of the forage, fixed or modified by the cattle on the farm, which can fairly be set down to the debit account of wool grown, of power created, and of flesh and dairy articles produced.

* Theoret. and Pract. Economy of Agricul. vol. ii. p. 235, (in French.)

As to the forage plants which are immediately turned into manure, it seems to me impossible to regard them as possessed of the proper market value; the farmer could not have sold them at this. In my mode of looking at the thing, the cost of producing the forage crop, and the value that it actually has, constitute a circulating capital, the annual interest of which, estimated at a certain rate, expresses the true cost-price or value of the manure employed in the course of a rotation. In a word, in my eyes, the value of the manure which gives fertility to the soil is represented by the price of the labor, the rent charge, &c.—by the general outlay entailed by the growth of the forage from which it is obtained.

I shall endeavor, by and by, to illustrate this topic by examples; but in order thoroughly to understand this mode of estimating the price of manure, there are several elements wanting, which I propose to assemble in this chapter. With this view, I shall first present the facts which I have been able to collect, or which I have myself had an opportunity of observing in reference to the economy of the domestic animals attached to a farm; and I shall then make an attempt to deduce the relation that exists between the consumption of forage and litter, animal reproduction and increase, and the formation of manure.

§ HORNED CATTLE.

It were foreign to the purpose of this work, did I enter into the natural history of the animals that are usually attached to farming establishments; neither will I pretend to discuss the relative merits of the different breeds of sheep and oxen, nor speak of the best methods of improving them. I confine myself to the varieties which I have on my own farm, or which I see on the farms of my neighbors, and upon which I have opportunity of making daily observations. It will be enough if I give a brief summary, in this place, of the general principles admitted by practical men of the highest name and authority upon these points.*

Between the external forms of animals and the internal organs essential to life, there is the most obvious and intimate connection. A broad and deep chest is the sure indication of ample lungs and a good general constitution. The *pelvis*, or bony cincture formed by the rump and haunches, ought to be spacious in the females. A small head is generally the indication of a good kind. Horns in our domestic animals must be regarded as objectionable rather than useful; and by adopting measures which tend to repress their growth, we undoubtedly favor both the production of flesh and wool. The strength of animals depends far more on the degree in which their muscular system is developed than on the mass of their bones; it is, besides, flesh, not bone, that has value in the butcher's eyes; so that the farmer's business is by all means to strive after a delicate but well-covered skeleton. Animals which have been indifferently

* Cline, in General Report of Scotland; Communication to the Board of Agriculture; Spencer on the choice of male animals for breeding from; Cully's Introduction, &c., on live stock, &c.

fed while young, have often the bony system very disproportionately developed.

Two modes are generally followed with a view to improving the external shape of domestic animals. One of these consists in only breeding from animals of the most faultless forms of the same race, and generally of close degrees of kindred; another in crossing females with the males of a neighboring race, each possessing in the greatest degree the qualities which it is held desirable to transmit to the future race. The former of these plans is spoken of as the method of *breeding in and in;* the second as the method *by crossing.*

Certain disadvantages have been stated as belonging to the system of breeding in, by the side of several unquestionable and more immediate advantages. While the race acquires small bones and shows a decided disposition to take on fat readily, it is said after several generations to lose in constitution, to become more subject to disease; the cows to give less milk, and the males, in losing their characteristic masculine forms, to show themselves less fit for propagation. The English breeders who take this view of the subject, are, therefore, in the habit of having recourse to males of the same race, but bred at a distance from themselves. I must for my own part say, that I have seen no reason to admit any ill effects from propagation continued in the same direct line. Our live-stock at Bechelbronn has not been otherwise renewed for a very long time, and without the race appearing to suffer in any way; our bulls, on the contrary, have very much improved.

Mr. Cline insists greatly on the selection of females not only of good shape, but so much above the mean height as to approach the standard of the males. When the bull is very much larger than the cow, the progeny is apt to fall off instead of improving; the reason for which Mr. Cline finds in the large size of the fœtus, the issue of a large male, which a small female can neither accommodate properly in her womb, send easily into the world, nor suckle duly when it is born. Whatever we think of this explanation, there can be no doubt of the propriety of giving the principle pointed at the most careful consideration in practice. Mr. Cline refers to the great improvement that has been effected in the breed of English horses mainly through crosses with small barbs and Arabian stallions; the introduction of Flemish mares would upon the same principle have been another means of still further improving the race. The neglect of this principle, Mr. Cline is of opinion, lies at the bottom of the numerous failures and disappointments that have been encountered in attempts to improve the breed of horses. A striking illustration of it occurred some years back, when bay horses of great height were in particular request; the Yorkshire breeders had their mares covered by the tallest stallions that could be found; but they immediately found that the progeny was merely long-legged, that it was narrow-chested, and without either weight or bottom.

Spencer acknowledges with breeders in general that the bodily and constitutional qualities are almost always those that preponderated in ancestors, and that the qualities of the father predominate in the posterity, particularly as regards oxen and sheep. This point settled, the choice of a good male is evidently the first point of consequence in attempting to improve a breed. As it is not possible, however, to find either a bull, or a tup, or a stallion, quite perfect, the one must be chosen that is most free from defect, particularly the defect or defects which we have it in view to correct in our breeding animals, our cows, ewes, and mares. Certainly no reasonable breeder would bring together animals that presented similar deficiencies; on the contrary, he will strive to have his female served by the male which shows all the qualities in the very highest degree that are most wanting in her. On the whole, the association of animals of the same race appears to me the best mode of continuing desirable qualities, especially when this is conjoined with ample supplies of good food to the young. The influence of feeding is immense; in my own neighborhood I see that the progeny of the Bechelbronn bulls are often inferior both in stature and shape to those that are brought up in our own stables.

Great size, however, is not always to be regarded as an improvement; height is by no means a constant indication of vigor of constitution. Improvement in those particulars of form and stature which are ascertained to be best suited to the circumstances of the locality, the climate, the pasture, &c., are the points to be especially attended to. It is above all indispensable to breed animals of vigorous constitution: over-refinement of original races has often led to indifferent conformation of body, and to undoubted delicacy of constitution, which has rendered the herd or the flock much more obnoxious to attacks of epizootic diseases.

The degree of refinement of an original stock is evidently connected with the quantity and quality of the forage of the district. In cold and mountainous districts, where the herbage is scanty, it is necessary to restrain the ambition of having highly-improved stock within considerably narrow limits; in such circumstances, the grand affair is to have a hardy race, not over nice in its food, which, through a considerable portion of the year, consists of but coarse grass.

The ox (*bos taurus*) has been reduced to domesticity from the remotest ages, and nothing but conjecture can be offered with regard to its original race. The animal accommodates himself with wonderful facility to the most opposite climatic circumstances; he multiplies with astonishing rapidity in the hottest regions of the tropics; unknown at the period of the conquest, he has now overrun the steppes of the vast basins of the Oronoco and the Amazons; and is met with in vast herds on the highest and coldest table-lands of the Andes, even up to the line of perpetual snow; wherever there is food, he appears to thrive; the extremes of temperature seem to have little or no influence upon him.

The buffalo (the *bos bubulus* of naturalists) is the only other member of this family that has been domesticated. He is fond of warmth, and is supposed to have been introduced into Italy towards the sixth century, from Eastern Asia. The buffalo is also found in Hungary and Greece; and wherever he is met with, he is made serviceable as a beast of draught and burden, and as food.

In breeding oxen, the great consideration is the bull. According to Thaer, the bull ought to have a short thick neck, the head short and small, the forehead broad and curled, the eyes black and sparkling, the ears long and well placed, the chest broad and deep, the body long, the legs short and columnar in shape.* A well-made bull would serve seventy or eighty cows were the season spread equally over the whole year; but as it is not so, Thaer thinks that twenty is as many as can properly be given to the same animal; and this, in fact, is the number which we adopt at Bechelbronn.

The cow gives more milk than any animal known. A great variety of external signs of a good milker have been particularized; but it may be said that there is none infallible. In a general way, I think that race has much to do with the point; the cow that is the offspring of a mother of a good kind, and a free milker, will herself be a good milker also. I will only add, that among the milch-kine which I have had an opportunity of observing, those that showed little tendency to take on fat, while they kept their appetite, have appeared to me to yield milk in largest quantity, and for the longest time.

The age at which it is advisable to put heifers to the bull, depends a good deal on the way in which they have been kept and brought up, and also on their growth. Young animals of a good kind, that have been well fed from the birth, and received all the care which contributes so powerfully to their development, will be ready to receive the bull when they are between a year and a half and two years old. At Bechelbronn, we bull the greater number of our heifers at the age of about eighteen months. Whenever they enter into heat with any thing like force, whatever their age, they ought to be put to the bull, or there is risk of the disposition to receive him dying away, and never returning; the heifer then begins to lay on fat, and ever after refuses the male. The rule, however, is not to allow the young female to be leaped until she is nearly at her full growth; this, in fact, is the season when the desire for the male usually first shows itself.

If there be no new indication of heat in the course of three or four weeks after the male has been admitted, there is reason to believe that the animal is pregnant. The cow goes with calf about forty weeks; the delivery generally takes place between the 277th and the 299th day after the access of the bull; but periods so short as 240 days, and others so long as 321 days, have been observed.†

The calf that is brought up with proper care is generally allowed to suck for five or six weeks; but it sometimes happens that even at three weeks old the quantity of milk supplied by the mother is insufficient; an additional quantity of food is therefore requisite. One of the best drinks for calves is made by mixing a proper quantity of oil-cake with tepid water; the large proportion of vegetable caseum, of oily matter, and of phosphates which the substance contains,

* Principles of Agriculture, vol. iv. p. 296.
† Teissier in Annals of French Agriculture, vol. ix. 2d series.

makes it peculiarly appropriate food for calves; diffused in water, it in fact bears a close resemblance to milk in point of chemical composition. It is now, too, that the calf begins to play with a little hay, so that it is always advisable to place some within its reach, the finest and softest portions being picked out.

But it is by no means necessary that the calf should ever be allowed to suck; it drinks without difficulty, or can be made to drink, as every dairy man and woman knows, by putting a finger or two into the animal's mouth under the surface of the drink. A little warm water is added to the milk during the first few days, in order to give it due warmth. Some begin from the very first to measure the milk; but those who are best informed upon the subject of breeding and rearing do nothing of the kind. Crud allows his calves to drink as much milk as they will take for the first week. After this time they have an allowance of about seven pints of new milk mixed with the same quantity of fresh whey. They are weaned at seven weeks. From the age of between nine and ten weeks to a year, a calf will consume about a fourth of the ration of a grown cow, say 6½ lbs. of hay per diem. During the second year, the allowance of hay may be estimated at about 13 lbs., or a little more; and in the third year it will amount to between 19 and 20 lbs. This is to be understood of cattle brought up carefully but frugally.

In some of the best dairies of Switzerland, the procedure is different. During the first six weeks the calves are allowed to drink as much milk as they will take without a surfeit. At a month old they are served with chopped hay and roots, or better still, if the season admits of it, with green clover or lucern, which they have at discretion till they are seventy days old. Treated in this way, a calf is nearly twice as large and twice as heavy as one that has been brought up economically. During the remaining 295 days that make up the first year, the animal is allowed from 8 to 9 lbs. of hay; and this quantity is doubled during the second year. By proceeding in this way, a heifer at two years old may herself be a mother and contributing to the produce of the dairy.

Our procedure at Bechelbronn is calculated on the Swiss plan. The calves suck till they are six or seven weeks old, being put to the cows night and morning. Any thing they leave is milked off. After numerous trials by gauging and weighing, I find that our calves take each during the forty-two days they are allowed to suck, from 528 to 600 pints of milk; in other words, from 14½ to 18½ pints per diem. The quantity of milk which a calf takes immediately after its birth, does not indeed amount to any thing like even the smaller of these quantities: still it is considerable.

A calf which weighed at its birth on the 18th of May 108.9 lbs., after having sucked, weighed 112.4 lbs.; so that it had taken 3.5 lbs. of milk to its meal; and as it had two of these in the day, 7.0 lbs. in all. The same calf, thirteen days afterwards, weighed 130.9 lbs.; and after having sucked, 139.0 lbs.; it had therefore taken 8.1 lbs. to its meal, or 16.2 lbs. per day.

About the third week after birth, our calves have hay of the best quality set before them ; they take very little at first, but they soon get accustomed to it, and at weaning time it commonly suffices for their support. It may happen, however, that at this period they fall off a little, but they soon recover again ; still, if any of them appear delicate, it will be prudent to allow about a couple of quarts of milk a day mixed with water, for some little time, which is gradually withdrawn as the animal becomes accustomed to its new food.

Calves grow with great rapidity during the suckling time. The only experimental data with which I am acquainted in regard to the increase of weight of calves during the first period of their lives, are those of M. Perrault de Jotemps. These observations I shall associate with those which I have myself made at Bechelbronn, where, by a happy coincidence, we have the same Swiss race of cattle as Messrs. Perrault at Feuillasse. The weight of three calves at birth was found by M. Perrault to be :

No. 1	70.4 lbs.
No. 2	83.8
No. 3	80.8
Average	78.0

At Bechelbronn the weight of six calves at birth was :

No. 1, born in May	108.9 lbs.
February	88.0
Ditto	90.2
April	100.1
June	88.8
May	101.2
Average	96.2

M. Ernest Perrault found that a calf, No. 1, during the first eighteen days of its life increased on an average 2.8 lbs. per diem ; No. 2 increased at the rate of 1.8 lbs. per day ; and No. 3 at the rate of 2.7 lbs. per day ; average increase, 2.4 lbs. per day. Another calf, born at Feuillasse, which weighed 101.2 lbs., when nineteen days old weighed 151.2 lbs ; so that it had gained 50 lbs., or at the rate of 2.6 lbs. per diem ; a rate which corresponded precisely with what was observed in the case of nine other calves fed for the butcher, the average increase of which per diem was 2.7 lbs., during which each has had about 19.3 pints of milk daily.

The conclusions come to at Bechelbronn bear a close resemblance to those of Feuillasse :

A calf which at birth weighed	108.9 lbs.	
Weighed 13 days afterwards	139.0	
Increase in 12 days	30.1	Increase per day, 2.5 lbs.
A calf born 12th of Feb. weighed	88.0 lbs.	
On the 30th of March it weighed	171.6	
Increase in 46 days	83.6	Increase per day, 1.8.
The same calf, weaned the 21st of April, weighed	193.6 lbs.	
Increase in 21 days	22.0	Increase per day, 1.03.

It is obvious, therefore, that from the time of weaning, the growth ceases to be so rapid; the transition from the milk diet to one of hard dry food, is often critical for young animals; and I have already said that it is one at which they frequently lose weight.

If we reckon the daily increase from birth, that is to say, for 69 days of mixed alimentation, we have 1.5 lb. for the quantity.

Crescent, born the 27th of June, weighed	88.8 lbs.	
Eleven days later	112.1	
Increase	23.3	per day, 2.1 lbs
At the age of 37 days he weighed	188.1	
Increase in 26 days	27.1	per day, 2.5
Six days afterwards he weighed	202.4	
Increase in 6 days	14.3	per day, 2.3
Another calf at birth weighed	101.2	
At weaning, aged 41 days	189.2	
Increase	88.0	per day, 2.1

These various observations give about 2.2 lbs. for the average daily increase of a calf in weight during the period it is sucking. The data of M. Perrault make it a little higher, 2.7 lbs. So that it may be assumed that a calf which is receiving from 15 to 19 pints of milk in the day, will be gaining 2.48, or very nearly $2\frac{1}{2}$ lbs. in weight per diem.

It will readily be understood that in places where milk is of considerable value, as in the neighborhood of cities, the farmer may find his profit in selling that article directly rather than in turning it into veal or beef, more especially if the usage of the district be to give the calves milk till they are three or even four months old. Nothing, in my eyes, can justify such a needless expenditure of milk; especially since I have had an opportunity of witnessing what I may call the *natural* course of rearing cattle in the steppes of South America. There the young animals only receive milk in any thing like quantity for two or three weeks; they soon get accustomed to live on grass. In the warmer countries of the earth, too, cows give much less milk than they do in temperate latitudes, and the secretion also dries up much sooner. The value of the milk, and the high price of butter and cheese, are unquestionably at the bottom of the immense slaughter that takes place in France among the calves, even at a very early age, when they are fat, but do not weigh more than from 110 to 112 lbs. This circumstance undoubtedly stands in the way of the production of meat in that country, and causes the notorious scarcity of meat of the best quality. Of the two millions of calves which it is calculated are slaughtered in France, $\frac{9}{10}$ths are killed before they are a month old, and when they do not weigh, one with another, more than from 90 to 110 lbs. But we have seen that at two months old the weight will have increased to from 154 to 176 lbs., more than half as much again; so that, by merely keeping the animals for one month more, the quantity of butcher-meat brought to market would be increased by about 120,000,000 lbs.*

It does not by any means follow, however, as the excellent authority I have quoted seems to think, that this increase of butcher-meat would add to the actual amount of food produced by the agri-

* Perrault de Jotemps, in Journal d'Agriculture, t. v.

cultural industry of the country. To produce 2 lbs. of veal, in fact, I have shown that something like 22 lbs. of milk must be consumed; but it is evident, that 1,200,000,000 of pounds of milk represent an amount of nutritive matter superior in value to 120,000,000 of pounds of veal. Could the production of the additional quantity of meat in the course of the second month be effected by means of any other food less costly than milk, which is itself a fluid of great value as food, with ordinary forage, for example, or linseed tea, or oil-cake, &c., the state of the question would be changed, and there would then be no doubt of the advantage to the community of the additional supply of butcher-meat. This indeed is so well understood, that all the efforts which have been made to improve upon the ordinary and simply natural mode of rearing young cattle have been directed with a view to economizing milk. The interesting work of M. Ernest Perrault, from which I am about to make several extracts, was not written with any other purpose.

M. Perrault set out with the view of ascertaining experimentally, 1st, whether the large quantity of milk generally allowed to sucking calves is really indispensable, and whether it is possible to diminish it without detriment to the animals; 2d, whether a portion of the milk can be replaced by hay-tea, an article prepared by pouring 14 or 15 pints of boiling water upon a pound of fine meadow-hay, and infusing for a few hours.

The observations were made upon three calves taken after weaning. A was kept for 94 days on the usual allowance to calves at Feuillasse; B had a smaller quantity of milk, and from the 42d day after birth had an increasing allowance of solid food; C in the course of the comparative experiment had 476 pints of hay-tea, and as it is impossible to regard the infusion as of higher nutritive value than the article from which it is prepared, I shall set down this drink as equal to 28½ lbs. of hay. The allowance of milk was stopped 48 days after the weaning.

A, B, and C were kept on their respective rations for 95 days. The three were kept for the first 18 days on milk entirely, during which it was calculated that each had had from the mother 337 pints. Here are the rations in a tabular and comparative manner.

A. On the usual allowance.			B. On a reduced allowance of milk.			C. On hay-tea.		
	Milk in pints.	Hay or an equivalent.		Milk in pints.	Hay or an equivalent.		Milk in pints.	Hay or an equivalent.
		lbs.			lbs.			lbs.
Food, 94 days ...	1460	374	Food, 95 days ..	1216	396	Food, 95 days ..	232	591
Suckling, 18 days	348	..	Suckling, 18 days	348	..	Suckling, 18 days	348	..
Days 112........	1808	374	Days 113........	1564	396	Days 113........	580	591

It would have been desirable to have had these three calves weighed immediately after the termination of the experiment; as this was not done, the results have not the whole of the precision that seems desirable. Nevertheless, M. Perrault from his observations concludes:

1st. That A, kept on milk alone, weighed at birth... 88 lbs
At the age of 452 days.......... 771.0

Total increase 683.0
Increase per day 1.5

2d. That B, on the reduced allowance of milk, weighed at birth .... 83.6 lbs
At the age of 224 days 404.2

Total increase 320.6
Increase per day 1.4

3d. That C, on hay-tea, weighed at birth .... 111.2 lbs.
At the age of 101 days 270.6

Total increase 159.4
Increase per day 1.67

M. Perrault's general inference is, that the calf which had the hay-tea ration grew more rapidly than either of the other two brought up either on pure or on dilute milk. The differences, however, are within the limits of the variations noted in animals that are reared on the same ration.

If we reduce the various articles consumed in these experiments to food of the same nutritive value—to hay, for example—we find, that—

A	consumed in 112 days,	1357 lbs.	of hay*
B	" 113 "	1137	"
C	" 113 "	906	"

The minimum ration for the maintenance of calves, to which M. Perrault comes from his experiments, differs little from that which we think amply sufficient at Bechelbronn; and our animals certainly are not inferior to those of Feuillasse. This fact may be judged of by the following particulars, which I have selected as affording the elements of contrast with M. Perrault's B and C:

Sophy weighed at birth .... 100.1 lbs.
At the age of 102 days 279.4

Total increase 179.3
Increase per day 1.76

Food consumed: Milk 525 parts, weighing 684 lbs. = hay 297 lbs.
Hay 539

In all 836

Rosa at birth weighed .... 96.8 lbs.
At the age of 239 days 473.0

Total increase 376.2
Increase per day 1.58

This calf consumed: Milk 528 pints, weighing 684 lbs. = hay 297 lbs.
Hay 1773

In all 2070

* Milk must be regarded in the light of forage, so that its equivalent should be stated. Assuming milk to consist of 12.61 dry matter and 87.39 water, I find by direct analysis

Without calling in the assistance of hay-tea, consequently, by bringing up on milk for seven weeks, and giving forage as soon as possible, it is obvious that we obtain results fully as good as those of M. Perrault.

I have said that it was during the period when calves are sucking, or receiving a regular allowance of milk, that the increase of weight was most rapid. As the animal approaches the term of its complete development, the weight, in an equal interval of time, increases at a progressively diminishing rate; but from the data which I have collected, but which are not very extensive, it appears that the increase is very regular until the full growth is attained. From this period the animal continues stationary, if he merely receives the ration of maintenance; any variation observed is purely accidental, and loss or gain one day is compensated by gain or loss on another. The adult animal, which does not lay on fat, thus acquires a standard weight, which is preserved for a term of years unchanged, until the period of decrepitude and decay arrives.

It is not unimportant to ascertain the progressive increase in weight of cattle; the balance is a means which the breeder and feeder ought not to neglect; it is a powerful check upon his servants, and a sure tell-tale in regard to the state of the stock at any moment. A conscientious herdsman is a most precious person on a farm; but the more I study breeding and feeding, the more I am satisfied that the most trustworthy agent of all is the balance. Frequent weighings are necessary, in order to keep a regular account of the state of the cow-houses. I here append such absolute observations as I have made on the increase of weight in horned cattle, with an expression of regret that I have not been able to present my reader with more numerous data.*

Names of Beasts.	Weight at birth.	Age when weighed.	Weight at this time.	Increase per diem.	REMARKS.
	lbs.	Days.	lbs.	lbs.	
Victoria	82	56	180	1.93	Under a year old.
Ditto	"	156	288	1.45	
Susan	80	168	176	1.55	
Ditto	"	168	270	1.25	
Gallop	88	82	178	1.21	
Ditto	"	164	254	1.43	
Ditto	"	264	388	1.59	
Schwartz	90	83	202	1.47	
Sophy	91	102	254	1.76	
Mignonne	90	103	216	1.34	
Ditto	"	203	312	1.56	
Margot	88	108	224	1.38	
Ditto	"	190	304	1.56	
Ditto	"	290	482	1.83	
James	88	119	216	1.25	
Ditto	"	201	284	1.07	
Ditto	"	301	452	1.32	

that 100 of this dry matter contains 4.0 of azote: so that 100 of milk contains 0.50 azote. This shows that 230 of milk are required to replace 100 of good meadow-hay containing 1.50 azote.

* As the weights were merely relative, I have neglected the fractions in turning the French kilogramme into avoird. pounds. I have, however, given the true increments of weights per diem —ENG ED.

Names of Beasts.	Weight at birth.	Age when weighed.	Weight at this time.	Increase per diem.	REMARKS.
	lbs.	Days.	lbs.	lbs.	
Petrel	88	147	270	1.36	
Ditto	"	229	378	1.40	
Ditto	"	329	538	1.49	
Rosa.	88	239	430	1.58	
Stern	90	275	570	1.91	
Ditto	"	357	732	1.93	
Ditto	"	436	880	2.00	From 1 to 3 yrs. old.
Chastel	89	465	972	2.09	
Ditto	"	547	1080	2.00	
Eichaas	90	730	976	1.34	
Ditto	"	811	1074	1.34	
Castor, 2d bull	100	796	1740	2.26	Become fat, killed.
Castor, 1st bull	98	1009	1602	1.65	Ditto.

By way of pendant to these results, I give two series of weighings, one of which has reference to the increase of weight of a heifer, on which I made a number of consecutive observations; the other was undertaken with a view to ascertain the variations which milch-kine, aged more than three years, may experience:

Dates heifer weighed.	Weight.	Gain between one weighing and another.	Time elapsed.	Increase in 24 hours.	Remarks.
	lbs.	lbs.	Days.		
Sept. 5	369.8	42	3	1.40	Fed with hay and roots.
" 9	374.0				
" 23	393.8	19.8	14	1.41	
Nov. 3	429.0	35.2	41	0.85	
" 28	469.4	40.4	25	1.21	
Jan. 29	550.0	80.6	62	1.32	
Apl. 21	678.4	128.4	82	1.50	Was fed on green clover at discretion.
July 9	884.4	206.0	79	2.48	
Aug. 3	950.4	66.0	25	2.64	

TABLE OF MILCH-KINE THREE YEARS OF AGE AND UPWARDS.

Names.	Age.		1st weighing.	2d weighing.	Difference.	Interval between the weighings.	Differences per day.
			lbs. avoir.	lbs. avoird.		Days.	lbs.
Esmeralda	3	1	1445	1555	110	82	+1.3
Orphan	3	2	1315	1434	119	"	+1.5
Galatea	6	"	1540	1394	146	"	—1.7
Gitana	6	"	1320	1386	66	"	+0.8
Hannchen	7	"	1100	1236	136	"	+0.2
Paysanne	7	"	1449	1540	91	"	+1.1
Raffalea	8	"	1672	1727	55	"	+0.6
Prima Donna	8	"	1784	1654	130	"	—1.5
Formosa	9	"	1573	1601	28	"	+0.4
Belle et Bonne	11	3	1346	1287	59	"	—0.6

Taking the preceding numbers as the authority, and until we have a larger number of weighings, I think we may conclude that the living weight of cattle of the Swiss breed increases by the following quantities per diem:

During the period of suckling at the rate of 2.4 lbs.
Under three years 1.5
Above three years 0.2

The increase of weight of growing animals depends much on the kind of food they have; and it is matter of great moment to know the precise amount of fodder which neat cattle require in order to thrive. Those who have treated of it specially, are as far from being agreed as to the proper ration; and then many who have specified the kind and quantity of the food, have neglected the ages, the absolute weights, the amount of labor required, and the milk obtained from the animals. It is subject of simple observation, that an animal of great size, all things else being equal, will require a larger quantity of forage than another of less bulk.

Once the allowance of food is well established, it is greatly to be desired that it be continued with the greatest regularity. Nothing is more injurious to cattle than stinting. Still, there is a term in every year when the live stock, or some portion of them, at least, are almost necessarily stinted in their food ; in the depth of winter the animals that are not put up to fatten, consume little or nothing but straw. At this season, consequently, the stock fall off considerably in flesh, in strength, and in the milk they give ; and when the loss has been very great, the animals are sometimes too far gone to recover when the spring has come round. This state of things is greatly to be deplored, and, indeed, ought to be viewed as most prejudicial ; it will be altogether impossible to advance the economy of neat cattle to the point of perfection which it is fitted to attain, until means are taken to secure every portion of the stock, at every period of the year, a sufficiency of properly nutritious food. Happily, with the progress of agriculture, this condition is becoming every year more and more easy ; the introduction of roots, (turnips and mangel-wurzel,) and of tubers, (potato,) into the routine of every farm that is respectably managed, supplies a fodder, through the whole of the winter, that is equivalent to the grass and other green meats of spring and summer.

Thaer fixes at 13 lbs. the quantity of hay per diem which a cow requires for her maintenance in perfect condition ; and if the animal be in milk, he allows as many as from 22 to 33 lbs. But the ration must vary, as I have said, with the weight of the animal. M. Perrault states 27 lbs. as the allowance for a milch-cow weighing about 880 lbs. ; he having, in his experience, found that an animal in milk required about 6¼ lbs. of hay for every 220 lbs. of living weight. Pabst, who paid great attention to the feeding of cattle, admits, that for the ordinary allowance of an ox doing nothing, or of a cow which is dry, 3.85, or upwards of 3¾ lbs. of hay, are required for each 220 lbs. of carcass weight ; 4.4, or about 4½ lbs., if the animal be a draught-ox ; and 6.6, or upwards of 6½ lbs., if it be a milch-cow.

The inquiries which I have made into this subject have led me to conclusions somewhat different ; from which I infer, that the relation between the weight of the living animal and the necessary fodder is not an invariable quantity. A very large ox or cow, relatively to its weight, requires less food than an animal of smaller dimensions. And this circumstance is a grand argument with those breeders who are in favor of very large cattle ; they say, that if a large ox consumes more food than a small one, still the increase of consumption is by no means in the ratio of the increase of weight.

The milch-cows at Bechelbronn have no more than 33 lbs. of hay per head per diem, or the equivalent of this quantity of forage. But the smallest creature on the farm, at the time my experiments were made did not weigh less than 1110 lbs. (79 stone, 4 lbs.) ; the relation of the living weight to the food being, therefore, as 100 is to 2.73, say $2\frac{3}{4}$.

The largest cow, again, weighed 1784 lbs., (127 stone, 4 lbs.,) so that the relation is here as 100 is to 1.85, or $1\frac{17}{20}$ths. The average relation, taking the whole of the cows in the stable, came out as 100 is to 2.25; in other words, for every 100 lbs. of carcass weight, $2\frac{1}{4}$ lbs. of meadow-hay per day had to be allowed.

It thus appears, from these inquiries, that growing animals require more food relatively to their weight than when they are adult. The young animals, upon which I made my observations, were from 5 to 20 months old; and for this age I found that for every 100 of living weight 3.08, or upwards of $3\frac{1}{2}$ lbs., of hay were required. The following table will give my conclusions at a glance:

Animals.	Ages in days.	Weight of the lots.	Hay consumed per lot in 24 hours.	Average age of the several animals.		Average weight of the several animals.	Hay consumed per head in 24 hours.	Hay per cent. of living weight consumed.	REMARKS.
		lbs.	lbs.	months.	days.	lbs.	lbs.	lbs.	
No. 1	107	418	14.3	3	26	209	7.15	3.43	In February.
2	119								
3	126	484	14.7	4	6	242	7.35	3.06	ditto.
4	130								
5	205	1267	36.0	5	9	311.8	18.0	2.89	July.
6	170								
7	158								
8	115								
9	163	..	11.0	5	10	297.0	5.5	3.70	February.
12	206	..	9.3	6	23	361.4	4.65	2.53	Septem.
9	328	2479	.66.0	9	5	495.8	33.0	2.66	July.
3	290								
4	289								
1	252								
10	239								
9	341	2536	62.7	9	18	507.3	31.35	2.47	ditto.
3	303								
4	302								
1	265								
10	252								
11	304	..	22.0	10	..	627.0	11.0	3.50	February.
12	371	..	19.1	13	5	550.0	9.55	3.48	ditto.
13	748	2063	62.7	20	5	1027.0	31.35	3.05	ditto.
14	483								
Average per cent. of living weight								3.08	

In the course of the experiments, the calves were kept on good meadow-hay, allowed them at will, according to our usual custom. The hay that was put into the crib once a day was weighed, and an account was kept and deducted of any that had been left of the previous day's allowance. The length of time during which each several experiment was continued, varied from 2 to 13 days; and I have thought it right to indicate the season of the year, lest that should have any influence. To sum up, then, it may be said, that for every 100 of living weight neat cattle require:

For simple sustenance (Pabst)...........	0.75 or ¾ lbs. meadow-hay.	
When laboring (Pabst).................	2.0	" "
When in milk (Pabst)...................	3.0	" "
" " (Perrault)................	3.12	" "
" " (Boussingault, large cows)	3.73	" "
Growing rapidly (Boussingault)..........	3.08	" "

The forage ought to be given to cattle with great regularity, and care should be taken that they do not eat too hastily. Generally speaking, they have their allowance three times a day, constituting so many meals, which, however, are well divided, the whole quantity for each meal not being placed before the animal at once. This precaution is particularly necessary when the allowance consists of green fodder. The watering should take place in the intervals between meals, the animals being driven to the trough night and morning; though, when the heat is excessive, it is better to water them three times a day. The water ought to be of good quality, though, if it have no deleterious substance dissolved in it, cattle seem to make no objection to that which is turbid, and which cannot, we should think, be very palatable. Our cattle are watered, during a part of the year, with water from a shaft pierced through a highly argillaceous soil. Cattle seem to dislike excessively cold water; they then drink as little as possible. The cattle in the great South American plains, drink water at a temperature of from 85° to 97° F. In Europe, the best water in point of temperature in winter is that of a deep well.

Every one is familiar with the taste which herbivorous animals show for salt, and this is one of the articles which is advantageously made to enter into the ration when its price is not too high. In France, it is absolutely necessary to use the article with extreme parsimony—a circumstance which I much regret, and which I cannot but view as prejudicial to rural economy:—[in England, where the odious salt-tax has been got rid of, salt, of the most beautiful quality, is one of the cheapest of all manufactured substances.] I know that many feeders do not think salt indispensable; but their authority is opposed by that of some of the highest names in Germany and England, and my own mind has long been made in regard to the value, to the excellent effects of this substance. I ascertained, for instance, that milch-kine, though they would not do upon potatoes alone, throve very well when they had from two or two and a quarter ounces of common salt added to the ration. A celebrated English breeder, Mr. Curwen, recommends about 3½ ounces of salt to be given daily to cows and heifers in calf, and to draught-oxen, and something less to fatting oxen, to young animals, and to calves.*

The high price of salt in France does not allow us to be so liberal at Bechelbronn; yet we make a distribution of the article three times a week, and in quantities which bring the allowance to something more than about an ounce and a half per day. By way of eking out the allowance of saline matter, we further supply, from time to time, a quantity of Glauber salt, which comes in all to rather

* Sinclair, Agriculture.

more than half an ounce per head per diem. The use of this salt, sulphate of soda, has long been common in Alsace, and also on the other side of the Rhine; and its effect on the health of horses and of sheep, as well as of horned cattle, has been recognised as highly advantageous. In Wurtemberg, the horses have, very commonly, 725 grains, neat cattle 463 grains, sheep 305 grains, and swine 250 grains of Glauber salt twice a week.*

Salt appears to be more especially useful in hot weather and in warm climates. In the steppes of South America, it is held by the llama keepers as an axiom that cattle cannot live without salt. Wherever a flock thrives particularly well, it may be averred, *a priori*, that there is a *salado* there, a salt-lick of the North Americans, or place where there is a salt-spring. In the savannas that are without saline springs, the herdsmen make a distribution of salt every day. On the plateau, or table-land, of Nueva Granada, common salt is replaced with Glauber salt, as in Alsace and Wurtemburg, and, I may say, that it was matter of much interest to me to find the same custom prevailing on the table-lands of the Andes as upon the banks of the Rhine.

§ II. MILCH-KINE.

I have already had occasion to say, that the signs by which the qualities of kine as milkers were sought to be appreciated, are somewhat deceitful. Still, I am far from denying that practice and experience do not enable many persons to pronounce with some certainty upon this particular. The power of doing so, however, is in some sort the peculiar privilege of him who possesses it; at least, I have seen all the general rules that have been laid down on the subject fail: I have seen cows of the most opposite conformations equally productive. I have also said, that race or descent had much to do with this quality; the heifer that comes of a mother, a good milker, will be very likely to turn out a good milker also. The legitimate way, therefore, of obtaining a good race of milch-kine, is to breed them from a stock that is already noted in this respect. At the time of my penning these lines, there are two animals on the farm that are remarkable as milch-kine: one is a tall, unseemly animal, the bones projecting, and altogether thin and miserable; the other is a small cow, with rounded outlines everywhere, the bony frame but little conspicuous; her skin soft, her hair sleek and fine. Nevertheless, these two animals have one character in common—the udder is of extraordinary size.

We ought not to be hasty in judging of the value of a milch-cow after the first calf; age has great influence on the secretion of milk. It is generally allowed that a cow does not attain to her maximum capacity of yielding milk until she has passed her sixth year.

With regard to the means we have of judging of the age of a cow, they are principally derived from the horns. The teeth do not afford us any indication, as in the horse and sheep. In the ox, about

* Communicated by M. Schattenmann.

the fifth year, there is a ring formed about the root of each horn; in the cow, this ring makes its appearance after the first calving, and from this epoch there is a new ring formed each year, which pushes on the former one. In aged animals these rings have become faint, and can scarcely be counted. It is also evident that the horns which, in early life, were thicker at the base, and tapered gradually towards the tips, about the ninth or tenth year of the animal's life present an opposite conformation; they exhibit a kind of constriction at the roots. The depression above the eye increases with age, and the false hooves become long and often bent.

Thaer reckons that, one with another, in well-regulated establishments, cows will continue in milk for about 280 days, and yield in all about 2265 pints, or 283 gallons. But it is certain, that the yielding of a cow varies greatly with circumstances, race, age, climate, and individual. The cows that graze at liberty in South America, do not give more than about three pints of milk per diem; which, as it is almost wholly used in bringing up the calf, the dairy is there of very little importance. In established farms, a cow is reckoned to yield about 40 lbs. of cheese per annum. Mr. Curwen estimates the quantity of milk at 6580 pints, or 822 gallons, per cow; M. Perrault states it at but 2992 pints, or 374 gallons; and Mr. Low gives the quantity at 5994 pints, or 749¼ gallons. The differences between these several quantities are obviously enormous, and can scarcely be reconciled with any conceivable diversity of circumstances. They are probably connected with the method taken to ascertain the quantities.

The following table comprises the whole of the statements with which I am acquainted.

Places.	Authorities.	Weight of the cows.	Hay consumed per day.	Milk per annum.	Milk per day.	Observations.
		lbs.	lbs.	pts.	pts.	
France: La Feuillasse (Ain)	Perrault de Jotemps	880	27.5	2992	8.2	Cows in the house.
Lompries (Ain) . .	D'Angeville.	605	14.3	1610	4.4	Do.
Roville (Meurthe . .	De Dombasle.	"	22.0	2492	5.9	Do.
Lyonnais (montagnes)	Grognier.	"	"	1284	3.5	Cows ill fed in winter.
Bechelbronn (Bas-Rhin)	Le Eel and Boussingault.	"	33.0	"	"	
England	Low.	"	"	5994	16.2	Do.
Do.	Curwen.	"	28.6	6580	17.8	Do.
Belgium: Antwerp . . .	Schwertz.	"	27.2	4495	12.3	Do.
Do.	Schwertz.	"	27.2	3967	10.9	At grass & in the house.
Holland: Low Countries .	Schwertz.	"	"	3400	9.2	In the house, winter.
Do.	Aiton.	686	"	7066	19.4	
Campine . . .	Schwertz.	"	"	9313	26.5	
Saxony: Meissen	Schweitzer.	567	18.6	2687	17.3	Kept in the house.
Altenburg . . .	Schmalz.	"	30.8	3412	9.2	
Austria: Carinthia . . .	Burger.	825	"	2752	7.5	Well fed.
Prussia: Moeglin	Thaer.	"	22.0	2648	7.2	Kept in the house.
Neighborhood of Berlin	Thaer.	"	27.5	3004	8.2	
Switzerland	D'Angeville.	1036	"	2992	8.2	Do.
Hoffwyll . .	D'Angeville.	1320	38.5	4685	12.8	Well fed.

At Bechelbronn we have seven cows whose allowance per head is 33lbs. of hay per diem. The milk is measured night and morning, and the quantity given by each cow is particularly noted. The herd consisted of Raffalea, 8 years old, whose milk failed the 21st of April, and reappeared the 18th of June without her having calved;—La Paysanne, 7 years old, whose milk ceased the 21st of February, and she calved the 29th of April;—Prima Donna, 8 years old: milk stopped February 19th, calved December 5th;—Formosa, 9 years old: ceased milking 1st April, calved 2d June;—La Gitana, 6 years old: ceased milking 30th September, calved 9th November;—Galatea, 6 years old: ceased milking 9th July, calved 2d October;—Belle et Bonne, 11¼ years old: ceased milking the 15th February, calved 3d April. These seven cows gave in the course of the year, neglecting fractions, 30576 pints, or 3822 gallons of milk. In the month of January, in round numbers, 1870 pints; in February, 1260 pints; March, 1260 pints; April, 1657 pints; May, 2527 pints; June, 3726 pints; July, 4180 pints; August, 3661 pints; September, 2913 pints; October, 2622 pints; November, 2540 pints; December, 2360 pints; having, one with another, given 546 gallons of milk, and milked on an average 302½ days each; the entire herd having milked during 2118 days, and the average quantity yielded by each cow having been 14.6, say 14½ pints for every day she was in milk; the quantity for each day of the year amounts to about 11.9, say 10 pints.

June, July, and August are obviously the months most productive of milk, during which the cows had scarcely any other food than clover. The average quantity for these months was undoubtedly raised from three of the cows having calved in March, April, and May, so that these were severally giving their largest measures during the three summer months.

It may be enough to state, that the largest quantity of milk is obtained in the course of the three first months after calving; the produce then will amount to 18, 20, and even 24 pints per day, while the mean quantity during the whole time of milking will very little exceed 12 pints.

The observations for the year 1842, which I referred to some short way back, showed a mean of 14.6, say 14½ pints of milk for each cow. But in the mode of reckoning pursued, there were sources of error, which have been avoided in the estimates just given. The only mode of securing accuracy of result is to take the quantity of milk yielded by each cow between the period of calving one year to the same event the following year. This mode of reckoning gives the quantity 13 pints per day for each cow, which I am disposed to adopt as the standard for the Swiss breed, fed with 33 lbs. of good meadow-hay, or an equivalent in wholesome roots, &c. I am also disposed to look upon 310 as the mean number of days during which a cow will give milk after calving.

We sometimes see quantities of milk mentioned as given by particular cows that are truly surprising, and that seem even calculated to excite suspicion of the veracity of the reporters. Some have spoken of cows that gave 44 and 52½ pints of milk a day for several months. M. Crud says that cows of great size indeed have even given as many as 70.4 pints in twenty-four hours; and Thaer goes still further when he states that persons worthy of every credit say they have seen cows in first-rate pastures, which, at the height of their milking time, produced as many as from 74 to 82½ pints of milk in the twenty-four hours. Such a flux of milk can only be very temporary, and indeed must occur but very rarely. The herdsmen at Bechelbronn have often diverted me with tales of such marvels; but since I have accurately gauged the dairy produce of the farm, I have met with nothing which would lead me to credit their reality. We have had cows indeed which have given 26½, and even 31½ pints a day for several weeks; but these are still very far from the quantities which have been mentioned to me.

Good feeding is undoubtedly required in order that cows may produce milk abundantly; but I believe that the influence of particular kinds of forage on the production of milk is often greatly exaggerated. Each breeder or feeder seems to have his own favorite article, however, so that there is nothing like uniformity among them; with one it is the carrot that is in the ascendant; with another it is the beet that is supreme; there is no root, in fact, which has not alternately had its apologists and detractors. The truth lies between the extremes here as it does in so many other instances; and I am satisfied that each and all the roots and other articles of forage that are generally introduced into the ration of milch-kine, are calculated to produce abundance of good milk; it is only necessary that the substances be allowed in ample quantity, that no mistake be committed in regard to the nutritive equivalents of the several articles. I do not hesitate to add, that the opinions of the generality of farmers and dairymen on the subject are based on observations which are always more or less imperfect.

It is but a few years ago that a series of experiments were undertaken at Bechelbronn, with a view to ascertain whether the particular nature of each of the several articles consumed by milch-kine influenced the quantity or chemical constitution of the milk in any appreciable manner. The purpose of these inquiries being purely practical,—having been undertaken with a special eye to the dairy and its produce, the inquiry was confined to the articles that are usually given to cows with us. These necessarily vary with the season, but I have already said that the dole to each head is equivalent to 33 lbs. of meadow-hay, which, indeed, always enters in considerable quantity into the ration, whatever else be given,—unless, indeed, the animals are exclusively upon green meat, when, of course, the use of every thing else is suspended. In winter the hay is mixed with beet, potatoes, turnips, or Jerusalems. In spring the hay is gradually replaced by green fodder, which in the first instance is rye cut green, and by and by clover. The experiments which I shall now detail were made upon a cow which had calved two hundred days, and was again pregnant.

1st EXPERIMENT.

200 DAYS AFTER CALVING.

The cow fed on hay alone gave 65.42 pints of milk in the course of seven days, or 9.34 pints per day. This milk consisted of:

Caseum	3.0	Solids 12.4
Butter	4.5	
Sugar of milk	4.7	
Ash of caseum	1.0	
Water	87.7	
	100.0	

2d EXPERIMENT.

207 DAYS AFTER CALVING.

Fed with turnips and cut straw, (the ration consisting of turnips equal to 29.7 lbs., and straw equal to 3.3 lbs. of hay,) the same cow gave in the course of eight days 84.4 pints of milk, or 10.5 pints per day. The composition of this milk was:

Caseum	3.0	Solids 12.4
Butter	4.2	
Sugar of milk	5.0	
Ash of caseum	0.2	
Water	87.6	
	100.0	

The animal discussed her provender with good appetite, but the ration was too large; about 11 lbs. of the turnips being left each day unconsumed.

3d EXPERIMENT.

215 DAYS AFTER CALVING.

The ration here consisted of:

Field-beet, an equivalent for 29.7 lbs. of hay.
Chopped straw " 3.6 "

In the course of fourteen days the quantity of milk obtained amounted to 137.6 pints, or 9.8 pints per diem, and was composed as below:

Caseum	3.4	Solids 12.9
Butter	4.0	
Milk sugar	5.3	
Ash of caseum	0.2	
Water	87.1	
	100.0	

4th EXPERIMENT.

229 DAYS AFTER CALVING.

The ration consisted of:

Raw potatoes equivalent to 29.7 lbs. of hay.
Chopped straw " 3.6 "

In the course of eleven days the cow gave 96.1 pints of milk, or at the rate of 8.7 pints per day, the fluid consisting of:

Caseum	3.4	Solids 13.5
Butter	4.0	
Milk sugar	5.9	
Ash of caseum	0.2	
Water	86.5	
	100.0	

The cow did not do well upon this regimen: she became heated, and refused one-half the straw. In a general way, we do not give tubers to a greater extent than is equivalent to one-half of the allowance of hay, in which proportion cows do very well upon raw potatoes.

5th EXPERIMENT.

240 DAYS AFTER CALVING.

The forage here consisted of the full allowance of hay, or 33 lbs. In the preceding experiment the milk, which had hitherto kept up to from about 9½ to 10½ pints a day, fell suddenly to little more than 8½ pints. To ascertain whether the fall was owing to the potato regimen or not, the cow was returned to the ration of hay under which in the 1st experiment the daily average of milk was 9.3 pints. In the course of thirty days 188 pints of milk were collected, at the rate of 6.2 pints per day. The declension in the quantity secreted consequently cannot be ascribed to the potatoes which were given in the 4th experiment.

6th EXPERIMENT.

270 DAYS AFTER CALVING.

The ration here was raw potatoes, with salt and straw—the ration of the fourth experiment, with the addition of about 2½ oz. of salt. The animal ate this salted ration with appetite ; she also made away with the whole of the chopped straw, and it agreed well with her ; nevertheless, the milk continued to decrease in quantity ; it had fallen off to 5.9, say 6 pints a-day.

7th EXPERIMENT.

290 DAYS AFTER CALVING.

In this trial the ration consisted of Jerusalem potatoes equivalent to 33 lbs. of hay, under which the milk may be said to have remained stationary, though it was above rather than under the 6 pints per diem, as in the 6th experiment. In composition it was as follows :

Caseum	3.3	Solids 12.5
Butter	3.5	
Sugar of milk	5.5	
Ash of caseum	0.2	
Water	87.5	
	100.0	

The quantity of the milk had obviously decreased from the first down to the two last experiments ; but its chemical constitution does not appear to have varied during the entire course of the trials ; the varied regimen has had no influence on the proportions in which its several ingredients are encountered. But there was still one point to be ascertained, viz : whether the milk secreted very shortly after the delivery differed from that which was formed at a period remote from that epoch.

8th EXPERIMENT.

A cow which had calved twenty-four days before, and, upon a mixed regimen of hay and green clover, was giving at the rate of 18.6 pints of milk a-day, was brought under observation. Analysis showed this milk to consist of :

Caseum	3.0	Solids 11.2
Butter	3.5	
Sugar of milk	4.5	
Ash of caseum	0.2	
Water	88.8	
	100.0	

9th EXPERIMENT

35 DAYS AFTER CALVING.

The same cow, upon green clover, was now producing 21.2 pints of milk a-day, and of the following composition:

Caseum	3.1	Solids 13.2
Butter	5.6	
Sugar of milk	4.2	
Ash of caseum	0.3	
Water	86.8	
	100.0	

This milk evidently presents a larger quantity of butter than appears in any of the preceding analyses. But no hasty conclusion must be drawn from this; for the succeeding experiments will exhibit a change equally sudden in the proportion of the fatty element, but in a different way.

In a second series of experiments I set myself the task of ascertaining whether green fodder had any such remarkable influence on the production of milk, and especially of its fatty element, or butter.

1st EXPERIMENT.

BEGUN 176 DAYS AFTER THE CALVING.

The ration here consisted of articles of winter fodder:

Potatoes equivalent to 16.5 lbs. of hay.
Hay " 16.5 "

Upon which the cow had long been kept, though the milk was only measured during the last six days. The quantity was 16.3 pints a-day, and consisted of:

Caseum	3.3	Solids 13.5
Butter	4.8	
Sugar of milk	5.1	
Ash of caseum	0.3	
Water	86.5	
	100.0	

2d EXPERIMENT.

182 DAYS AFTER THE CALVING.

Mixed regimen: Green clover equivalent to 16.5 lbs. of hay.
Hay " 16.5 "

Upon which the quantity of milk was at the rate of 17 pints a-day.

3d EXPERIMENT.

193 DAYS AFTER THE CALVING.

Green meat: Clover equivalent to 33 lbs. of hay

Quantity of milk, 17.2 pints a-day, composed of:

Caseum	4.0	Solids 11.2
Butter	2.2	
Sugar of milk	4.7	
Ash of caseum	0.3	
Water	89.7	
	100.0	

The small quantity of butter here induced me to repeat the analysis, but the result came out very nearly the same, the quantity being still but 2.35 per 100.

4th EXPERIMENT.

204 DAYS AFTER THE CALVING.

Green fodder: same quantity as before.

Milk per day 13.7 pints, composed of:

Caseum	3.7	Solids 12.6
Butter	3.5	
Sugar of milk	5.2	
Ash of caseum	0.2	
Water	87.4	
	100.0	

It would therefore appear that fresh-cut clover has no such virtue as that of increasing the quantity of milk given by cows. Under the winter fare, in fact, the milk produced in the course of the twenty-four hours amounted to 16.7 pints; under green clover it was but 14.9 pints. It would be a great mistake, however, as I conceive, to ascribe the diminution here to the use of the green forage; it is due, I apprehend, exclusively to the greater length of time that has elapsed since the period of calving.

The chemical composition of the milk varied little, as I have already incidentally remarked, in the course of these experiments. The differences in respect of the caseum, by which let me say I understand the whole of the azotized constituents, the whole *flesh* of the milk, rarely exceed one-hundredth part. The proportion of the fatty element varies suddenly, and, as it seems, independently of the various circumstances in which the cows are placed.

The general inference from these experiments, then, is that the nature of the food does not exert any marked influence on the quantity and chemical constitution of the milk (I do not now speak of the *quality* of the fluid) if the cows but receive the proper nutritive equivalents of the several sorts of provender. It is of great importance to insist on this point; for it is quite certain, that if the weight of the several rations be not calculated according to that of the equivalents, variations in the secretion of milk would be forthwith conspicuous; but then these variations would have the increase or diminution of the provender allowed as their cause.

When cows are kept through the winter upon straw alone, they cease to give milk; but on the return of green forage, in the spring, the secretion is restored. The re-appearance of the milk in this case, however, is not connected with the coming in of the fresh provender, but with the return of plenty; the animals are not only fed, from having been starved, but they are more than fed; they have something to spare, which their economy turns partly into milk.

In well-managed establishments, where a good system of husbandry secures an abundant supply of good nutritive provender to the cattle during winter, the produce of the dairy during this season differs much less from that of the summer than is generally supposed. I am besides persuaded that we estimate the nutritive powers of green forage at too low a rate, and that when cattle are upon wet clover or lucern, they are in fact much more effectually nourished than under ordinary circumstances.

If it be true, as it evidently is, that the quantity of milk produced depends especially upon the absolute quantity of nutritive food consumed, it is not so with the quality of the fluid. It is undeniable, that the milk of spring and summer, formed upon green and succulent food, is much more palatable than that of the winter season; the butter is also much finer and better flavored. The green herbs of our pastures undoubtedly contain volatile principles which are dissipated and lost in the processes of drying and fermentation which they undergo in their conversion into hay. If chemistry be powerless in seizing such principles, it still informs us of the possibility of introducing a variety of articles into the food of cows which have the property of communicating those qualities which we prize in milk. In all grazing countries certain vegetables are pointed out as giving, in the vulgar opinion, a particular aroma to the flavor of milk.

§ III. FATTENING OF CATTLE.

Under a parity of circumstances, feeding cattle for the butcher may occasionally be found more advantageous than the dairy to the farmer. In feeding for the market there is, in the first place, a quicker return for the outlay, than in keeping milch-kine through the whole of the year. In the first operation, the capital is realized at the end of four or five months; that which is employed in producing milk, and butter, and cheese, is always lying out, like a sum at interest.

The quantity of food requisite to bring cattle intended for the butcher into condition, does not vary less than that which is required to secure a plentiful production of milk. Thus the stature, the age, the race of the individual, and the relative proportions of flesh and fat which we would have laid on, all imply varied doles of various kinds of forage. The age in especial has to be considered; for in putting up a young animal to fatten, we have both flesh and fat to form. This is what always occurs in the fatting of oxen of two years old, and of pigs of ten or eleven months. The increase in living weight experienced at various ages is not equally owing to accumulation of fat; this indeed may be so in the case of beasts, the muscular system of which has already attained complete development, but it is otherwise with young and still growing animals.

Practice does much in enabling us to select the animals that will fatten readily. In a general way it is well to choose young animals that have a large chest, the body bulky and rounded, the ribs finely arched, the bones small, the limbs short, the neck thick for its length, the skin soft, pliant, velvety to the touch, and moveable over the body, particularly over the ribs, the tail should be scanty, the buttocks not deeply cleft, but fleshy—well *breeched*, as the phrase runs in some districts. The look of the animal should be sharp and bold; the horns slender, whitish, and rather transparent. The animal must have been cut while he was still at the teat.

The celebrated English breeder, Robert Bakewell, succeeded, after a long and troublesome course of experiments, in creating a race of neat-cattle and of sheep which show themselves particularly disposed to take on fat. The fundamental principles established by Bakewell, after all his experience, are these: that smallness of bone, fineness of skin, and cylindrical shape of body, are the surest indications in cattle of the disposition to lay on fat readily, and upon the smallest quantity of provender. The most striking features in the breed obtained by Bakewell, commonly known as the *Dishley breed*, may be summed up in the following terms:

1st. The animal low on his legs.

2d. The back-bone straight.

3d. The carcass rounded and almost cylindrical.

4th. The chest deep and large.

An ox is held to have grown rapidly and well, when at the age of three years he weighs from 1016 to 1051 lbs. avoirdupois, from 72 to 75 stone. The disposition to fatten young is also a precious quality in the beast which it is intended to bring up for the butcher; the feeder comes the sooner at his return. Sinclair thinks, that independently of good constitution, which is indispensable, this quality is derived especially from meekness of disposition, from good temper; and as docility is generally the result of good treatment in early life, young animals ought always to be treated with great gentleness and made perfectly familiar.

The different races do not all yield meat of the same quality, and this quite independently of age. The best meat has a very decided and characteristic flavor after it is dressed, which indifferent meat wants, or which is replaced by a savor that is disgusting rather than agreeable. The fat in the best meat, as well as being laid on superficially, is distributed through the substance of the muscles, so as to give the flesh a marbled appearance.

In fattening cattle, it is perhaps of more importance than in general feeding, that the provender should be distributed regularly; plenty of soft litter, and the greatest attention to cleanliness, aid materially in fattening. The cow-house ought to be dark and quiet; in a word, all the conditions ought to be combined which conduce to sleep, and secure freedom from disturbance of every description.

The age at which cattle fatten most readily is that of from 7 to 8 years.* Animals under this age, which have not yet come to their full growth, will nevertheless get into excellent condition; but they require both longer time and more food, for the reason, apparently, that they are still forming both flesh and fat.

In fattening during winter, which is done almost exclusively with hay in some countries, an ox weighing 748 lbs., upon 40 lbs. of hay per diem, will increase by about 2 lbs. daily. According to Mr.

* This is as in the original, and may be true, but in England and Scotland we have seldom an opportunity of proving it so.—ENG. ED.

Low, an ox weighing 770 lbs., and consuming about 2223 lbs. of turnips per week, if he thrive, will gain in the same space of time nearly a stone in weight. Admitting that the equivalent number for turnips is 676, I find that the ration of hay for this allowance comes out 47.8 lbs., having produced exactly 2 lbs. of increase.

In the information obtained in the Rhenish provinces by M. Moll, in regard to the fattening of cattle under the influence of a regimen which would give 11 lbs. of hay to every 100 lbs. of dead weight, the animal will increase one third in weight in the course of three or four months.

To these general results I add a few particular facts, which are, indeed, the only data in rural economy that can ever be received as having much value.

In a series of experiments which he undertook, Mr. Robert Stephenson proposed to compare the progress of the increase in weight of oxen upon different alimentary regimens. Starting with the principle which we have already established, that animals consume a quantity of food in proportion to their weight or size, when they are under the same conditions, he had of course to divide his stock into several lots, each made up of animals of as nearly as possible the same weight. Oxen of two years old, brought up on the same farm, and kept in the same manner, were the subjects of experiment. I shall select one experiment, in which the observations were made upon three lots of six beasts each. The weight of each lot was ascertained before and after the experiment, which was carried on for 119 days.

The first lot was put upon white turnips, linseed-oil cake, beans, and oats ; and for the last 24 days, each beast had 20 lbs. of potatoes every day in addition.

The second lot was fed like the first, with this difference, that it had no cake, and that during the last 24 days the quantity of potatoes allowed was but 10 lbs. per diem.

The third lot had no other provender than turnips.

Here are the weights and the nature of the provender consumed by the animals during the 119 days, with a column added containing the equivalent in hay corresponding with each of the articles consumed :

	LOT I.		LOT II.		LOT III.		
Provender.	Weight in lbs.	Equivalent in hay in lbs.	Weight in lbs.	Equivalent in hay in lbs.	Weight in lbs.	Equivalent in hay in lbs.	Equivalent assumed.
White turnips	1518	171.6	1628	184.8	1122	127	885
Swedes	13336	1973.4	13384.8	1980	12012	1777.6	676
Beans	358	1559.8	358	1559	"	"	23
Oil-cake	389	1768	"	"	"	"	22
Oats	173	279	173	279	"	"	62
Potatoes	479	151	239.8	77	"	"	315
Ration expressed in hay		5904		3971		1905	
Hay consumed per day per head		49.7		34.3		16.0	
Hay per 100 of the living weight		4.01		3.03		2.0	

It therefore plainly appears that the lot which had the largest allowance of provender, the food which contained the greatest quantity of azotized principles—of *flesh*, in fact—produced the largest amount of dead weight in a given time, and that the lot which had the shortest allowance increased in the smallest measure both in flesh and fat—results which might have been readily foreseen. It is also apparent, from the table, that in proportion to the nutritive value of the articles consumed by each lot, the increase in carcass weight was greatest in that which received its allowance in the least bulk. Thus reducing the different rations to a standard forage, we find that in the first lot, which was most plentifully supplied, 100 of hay gave 4.2 of increased weight; while the same allowance of hay produced 6 in the third lot, which was fed parsimoniously. This fact is most readily explained: over a certain limit, the more food an animal receives, the smaller is the fraction which is assimilated and turned to use in the body. Breeders have consequently discovered, that it is by no means generally advantageous to push animals beyond a certain point of fatness. The excess of weight which is obtained with the assistance of quantities of food, exaggerated as it were, no longer compensates for the additional expense incurred. This is a circumstance which Mr. Stephenson's experiments also illustrate, and indeed they led him to the conclusion which has just been stated. Judging by the market price of the several articles of provender employed by this distinguished breeder, the first lot appears to be that, the fattening of which turned out the least advantageously: while each pound weight of flesh produced here cost about 5*d*., the price of production in the second lot did not much exceed 4*d*. ($4\frac{1}{5}$th;) in the third it was a little more, ($4\frac{4}{5}$ths.)

With these observations of Mr. Stephenson, we find the following numbers to express the daily increase in weight of the cattle during the period of fattening:

	Average weight of the oxen before fattening.	Hay consumed per day and per head.	Increase per head in 119 days.	Increase per day and per head.
	lbs.	lbs.	lbs.	lbs.
1st lot	1115	49.7	247.5	2.
2d "	1016	34.3	231.6	1.9
3d "	794	16.	112.6	0.9

The weight of the several animals must also be taken into account, in seeking to estimate the increase realized upon every 100 lbs. of live weight during the fattening.

In the 1st lot—100 of live weight in 119 days gained,		22.2
2d "	"	22.8
3d "	"	14.2

It is seldom that cattle are fattened in the house upon clover or lucern in the green state; nevertheless, animals will fatten upon this forage with great rapidity. An ox will eat as much as 1 cwt. of clover cut in flower in the course of the day. In case the green food should relax the bowels too much, a fraction of the allowance may be given dried, and towards the end of the fattening a little cake may be given. But these additions do not appear to me indispensable; they are always attended with additional cost: and I have frequently seen cows, upon green clover at discretion, acquire a remarkable degree of fatness, although they had not ceased to be regularly milked.

In those countries, the nature of whose climate is favorable to pasturage, the rearing of cattle presents immense advantages; but the animals can only be fattened in those that are the most fertile. The meadow that suffices for the growth and keep of a bullock will not always bring the animal into condition for the butcher. Those countries where the climate is moist, but long droughts rarely felt, where neither the summer heats nor the winter colds are excessive—the conditions, in fact, which are met with in the beautiful pasture lands of England, in especial—are those that prove most favorable to the rearing and feeding of cattle. The pasture lands of Normandy and Brittany in France, of Switzerland and Holland, several of the provinces watered by the Rhine, &c., are also remarkable for their luxuriant herbage. In such situations and with such advantages, the grand object with the farmer is the production and fattening of cattle. Whenever it has been possible to lay down extensive and productive meadows, it is now beginning to be clearly understood that the introduction of even the best system of rotation were to make a false application of agricultural science. In my opinion, there is no system of rotation, however well conceived and carried out, which will stand comparison in point of productiveness with a natural meadow, favorably situated and properly attended to. The reason of this is obvious, and follows from the very principles which we have laid down in treating of rotations. The whole object in the best system of husbandry is to make the earth produce the largest possible quantity of organic matter in a given time. But in such a system we are limited by the climate, inasmuch as we are obliged so to arrange matters that our crops shall always attain to complete maturity; the consequence of which is, that with all our pains the soil remains unproductive during a certain number of weeks and months towards the end of autumn, in the early spring, and through the whole of the winter. But upon meadow lands, vegetation is incessant; the winter even does not interrupt it completely; it still revives and makes progress on the bright days; and in the spring it proceeds when the mean temperature is but a few degrees above the freezing point of water, and never ceases until it is checked again by the severer cold of winter. It is therefore easy to obtain conviction that a given surface of meadow land must necessarily produce a larger quantity of forage than land laid out in any other way. It is true that the forage thus obtained will not, like the cereal grasses, answer immediately for the support of man; but it nevertheless concurs powerfully in this by producing milk, and butter, and cheese, and in breeding and fattening cattle: let there be added to all these advantages of what may be called a permanent vegetation, that the cost of keeping it in order is infinitely less, and that there is no risk to be run from failures of crops, and the vast advantages of meadow or pasture land will meet us with all their force.

On the banks of the Elbe, in Holland, in the neighborhood of Arnheim, the meadows are depastured during one year, and cut, and their produce made into hay the following year, and so on alternately. The cattle are fed in the house with the hay during the winter. They are driven out into the pastures in May. In the Low Countries, it has been found that to fatten a large ox a surface of meadow land of about 9960 square yards, upon which he will pasture during five or six months, was necessary. In the bottoms of greatest fertility near Dusseldorf, it has been calculated that to keep a cow an extent of surface equal to about 1800 square yards was necessary.

In countries which possess rich pasture lands, oxen are put to fatten immediately upon the richest of them. In the valley of the Auge, in Normandy, these meadows are designated as herbages. A meadow of this kind requires a rich, damp soil, capable of retaining moisture. It is, therefore, to a considerable extent dependent upon its subsoil. In the district mentioned, the soil of the pastures consists of a thick layer of vegetable mould resting upon clay; it is therefore very rare that this meadow land feels the effect of drought; it is, indeed, only in the early spring that the pasture upon such lands sometimes fails, in which case the stock must of course be assisted with hay, the quantity being gradually diminished as the season advances.

M. Dubois finds that a lean ox weighing 473 lbs., after fattening in the valley of the Auge, will weigh 763 lbs., so that he will have gained 290 lbs.; the degree of fatness attained in this district is often prodigious. M. Dubois mentions oxen which weighed when fat 1760 lbs., upwards of 125 stone, and he speaks of one which attained the enormous weight of 2750 lbs., upwards of 196 stone.

The height of the oxen fattened in the *herbages* of the Auge varies from 4 ft. 7 in. to 5 ft. 3 in. measured at the haunch; when thoroughly fat, the four quarters will weigh from 550 lbs. to 990 lbs., the hide will weigh from 70 lbs. to 116 lbs., and they will yield from 100 lbs. to 150 lbs. of tallow.

It is calculated that on the meadows of the greatest fertility, a surface of 2760 square yards are required to fatten a large ox; on meadows of medium fertility, a surface of 4680 square yards are required to fatten an ox of medium size; on those of the third quality, a surface of 3720 square yards is deemed necessary to fatten a small ox.

M. Dubois calculates the quantity of green fodder consumed by an ox during the eight months when he is fattening, as equivalent to 6600 lbs. in dry hay; this, at least, is the quantity that the extent of meadow required to fatten one ox would produce. The average ration of green forage per diem is, therefore, equivalent to about 27 lbs. of hay, a quantity which appears small, and which would be so in effect, were not the oxen kept so long in the meadows. M. Dubois, indeed, observes that in the stall, with a ration composed of from 11 lbs. to 13 lbs. of linseed oil-cake and 26 lbs. of hay, an ox will become sufficiently fat for the butcher in seventy days, and will acquire nearly the same weight that he would have gained in the course of seven or eight months in the meadows. There is nothing surprising in this fact, inasmuch as the ration mentioned by M. Dubois, in our mode of viewing it, is equivalent in nutritive value to at least 81 lbs. weight of hay; the quantity of oil-cake alone is enough to supply a good pound weight of fat per diem.

In old Friesland, where the pastures are excellent, results are obtained which may be compared with those of the meadows in the valley of the Auge; an ox of from 770 lbs. to 990 lbs. weight will be pushed to a weight of from 1100 lbs. or 1650 lbs. on a surface of meadow land between 3000 and 3600 square yards in extent.

In the meadows of the Auge the fattening goes on even during the winter; the oxen are received into the pastures between the 15th of September and the 15th of November, and the animals pass the winter in the open field; but they receive from 12 lbs. to 26 lbs. of hay per diem until the month of April, when the grass has already grown sufficiently to suffice for their keep. These oxen are generally fat and ready for market in July.

In these observations of M. Dubois, the fattening has reference to the neat weight of the carcass, sinking the offal, as it is said, or estimating the weight by the quarter. The most esteemed quarters are the hind quarters, which are found to weigh rather less than the fore quarters, although the difference is less, the higher the condition of the animal.

It is long since various means have been devised for ascertaining the neat weight of a living animal, or in other words, the weight which the carcass will have when it has been embowelled, flayed, and the head and fat cut off. These various parts compose what is called the offal. It is readily to be conceived that one grand feature in the excellence of an ox must consist in the great relative weight of the carcass properly so called in comparison with the offal; but it may easily be imagined also that the relations in the weight of these two different portions of the living animal will vary according to the state of fatness, and also according to the breed and the age of the beast.

Mr. Anderdon has found that an ox which is not absolutely lean will give for every 100 lbs. of his absolute weight:

Of marketable meat......................53.5 lbs.
An ox somewhat fatter will yield..........55 "
And one completely fat as many as.........62.2 "

Mr. Layton Coke's estimate is:

For a lean ox ....60 per cent. of marketable meat.
For an ox in middling condition..65 "
And for a fat ox................73 "

These estimates appear to me exaggerated, and I much doubt from the sales of cattle which we make ourselves, whether they would readily be admitted by the buyers; they are in fact too high as regards the available meat.

From a great number of actual trials made with animals of about two years old, and which were all as nearly as possible in the same condition, Mr. Stephenson was enabled to determine with great accuracy the actual weight of the butcher's meat in contrast with the entire weight of the animal. Mr. Stephenson comes to the following conclusions:

Butcher's meat per cent	57.7
Tallow	8.0
The hide	5.5
The entrails and offal	28.8
	100.0

The precise quantities of marketable meat and of offal have also been determined by Mr. Mallo in an ox of the Durham breed which was slaughtered in his presence. The weight of the animal on its feet was 1496 lbs.

	lbs.	Per centage of live weight.
The two fore quarters weighed	405.9	55.4
The two hind "	423.5	
The skin	62.7	4.2
The tallow	112.0	7.5
The blood	110.0	7.4
The head, fat, and entrails	381.7	25.5
	1496.0	100.0

These relations as to meat, tallow, and skin agree in a very considerable measure with the estimates of Mr. Stephenson.

Sir John Sinclair gives the following numbers as the results obtained in connection with an ox of the Devonshire breed, slaughtered at the age of 3 years and 10 months.

Weight of the living animal, 1549.6 lbs.

	lbs.	Per centage of the live weight.
Butcher's meat, the four quarters	1083.5	70.0
The skin	84.9	5.5
Tallow	143.2	9.2
Entrails and blood	163.6	10.5
Head and tongue	36.7	2.4
Feet	17.1	1.4
Heart, liver, and lungs	20.4	1.3
	1549.4	100.0

The animal here was not in prime condition. On the whole, the relations as stated by Mr. Stephenson may be taken as those that will be found nearest the average truth, and as his numbers are deduced from numerous actual experiments, I feel disposed to adopt them. M. Dubois has found that an ox which will weigh 473 lbs., sinking the offal, will be brought by fattening to the weight of 763 lbs. We have, therefore, for the weight of an animal as it stands:

Before fattening	828 lbs.
After fattening	1336
Gain in weight	508

The fattening having been effected in eight months, the absolute increase in weight per diem will amount to 2 lbs.; the increase per cent. upon the weight is 61.4.

We have seen that during the fattening, the mean consumption, reckoning the provender in hay, amounts to 6600 lbs.; the increase obtained being 508 lbs. gives 16.9 lbs. of living solid for every 220 lbs. of hay consumed. Lastly, the mean ration being settled by M. Dubois at 26.4 lbs. of hay per head and per diem, and the weight of the animal on being taken into the meadow being 828.3 lbs., this ration corresponds to 7.1 lbs. of hay for every 220 lbs. weight of the living animal.

To sum up from the fact just stated on the subject of fattening, it appears that the increase per day is :

According to	Thaer	0.93	per cent. on the hay consumed.
"	Low	0.91	"
"	Stephenson, 1st lot	0.94	"
"	" 2d	20.99	"
"	" 3d	0.45	"
"	Dubois	0.95	"

§ IV. OF HORSES.

In what follows I shall limit myself to the consideration of the horse in his relation to agricultural industry, and shall give the result of certain experiments which I have made upon his growth with a view of solving the question, much disputed in various places at the present time, whether or not the general farmer can breed horses with advantage to himself.

The horse employed in farm labor ought to be spirited and strong ; attention to external form is only to be given in so far as it is an indication of the qualities that are required. He ought therefore to be broad in the chest and in the haunches, and his muscular system must in general be decidedly developed. A horse of considerable size, if he be otherwise exempt from defects. is generally preferable to a small animal ; he is stronger, takes longer steps, and does more for his keep than the other. We are not to require in the draught-horse the vivacity and amount of spirit which we look for in the saddle horse, yet he ought to have that liveliness which is almost always a sign of health in animals.

Thaer does not approve of the practice commonly followed at this time of mixing with good draught horses the blood of stallions of elegant shape, but little adapted to stand hard work. Although this remark is not without truth, it is still impossible to deny that in many cases the employment of stallions of some breeding has much improved the race of draught-horses in various districts. It is not, besides, unworthy of attention, that it is really important for the farmer to have a breed which he can readily dispose of to advantage, particularly in those countries where horses for cavalry and artillery service are in request. My own observations would lead me to say, that the breeds in France are frequently improved by crossing with stallions of the royal studs. The effect from this procedure has not perhaps been so great as might reasonably have been expected, still evident progress has been made.

The mare will take the stallion at about the age of three years ; but it is seldom that the animal is covered at so early an age ; on the farm she will be at least five or six years of age before this is allowed, especially if the animal is to be worked during the time she is with foal ; and the same consideration leads us to say, that a mare ought not to be covered oftener than once in two years, although it is very possible to have a foal from her every year, for she frequently comes into season towards the 11th day after foaling, and she goes with young for a term which varies between 333 and 346 days.

A brood mare may be employed in ordinary work during the first period of her pregnancy; but when the time is further advanced, when she is in the tenth month, for example, every possible precaution must be taken against accident. This is the period at which we withdraw our brood mares from the common stable, and put them into separate boxes. After she has foaled, the mare receives in small quantities and frequently repeated, warm drinks and bran mashes. While she is giving suck, her food ought to be of a more substantial or better kind than that which is generally allowed.

The mare may be put to light work twenty days after she has foaled; but it is requisite not to demand any thing like exertion from her within eight or ten weeks after this event; she then goes out accompanied by her foal, which is generally suckled for about one hundred days. Foals are frequently brought up in the stable or in the loose box; this is our practice in Alsace; but it is well, with a view to the growth and health of the young animal, that it be taken out every day. On quitting the teat, foals are fed upon choice hay; in the course of the second year a portion of the hay should be replaced by an allowance of oats, and in the season the use of green clover cannot be too highly recommended.

According to Thaer, the daily allowance to a horse of middling height, and doing ordinary work, may be regarded as good when it consists of:

Hay	8.2 lbs.	= Hay	8.2 lbs
Oats	9.2	= Ditto	14.2
Allowance reckoned in hay			22.4

In England the following allowance has been particularly mentioned as that of certain well-conducted stables.

Cut hay	11.0 lbs.	= Hay	11.0 lbs.
Cut straw	2.2	= Ditto	0.55
Oats	11.0	= Ditto	16.9
Beans	1.1	= Ditto	4.7
Allowance reckoned in hay			33.2

According to M. Tassey, veterinary surgeon in the Municipal Guard of Paris, the provender of the horses in this corps in 1840 consisted of:

Hay	11 lbs.	= Hay	11 lbs.
Oats	8	= Ditto	12
Straw for litter	11	= Ditto	2½
Total allowance			25

The same authority reckons that horses employed in severe draught receive or require:

Hay	16½ lbs.	= Hay	16½ lbs.
Oats	17	= Ditto	26
Total allowance			42½

Until very lately (previously to 1840) the allowance of troop horses in the French army consisted for the reserve cavalry of:

Hay	11 lbs. =	Hay	11 lbs.
Oats	8 =	Ditto	12
Straw	11 =	Ditto	2½
		Total allowance	25½

For the cavalry of the line:

Hay	8.8 lbs. =	Hay	8.8 lbs.
Oats	7.5 =	Ditto	11.5
Straw	11. =	Ditto	2.7
		Total allowance	23.0

For the light cavalry:

Hay	8.8 lbs. =	Hay	8.8 lbs.
Oats	6.6 =	Ditto	10.1
Straw	11. =	Ditto	2.7
		Total allowance	21.6

Influenced by the consideration of the frequent indifferent quality of hay, and its injurious effect upon the health of the horse, it was decided in 1841 to replace a portion of the hay ration by a larger quantity of oats, an article much less liable to be adulterated, or to be indifferent in quality. The allowance now consisted for the reserve cavalry of:

	lbs.		lbs.
Hay	8.8 =	Hay	8.8
Oats	9.2 =	Ditto	14.2
Straw	11 =	Ditto	2.7
		Total allowance	25.7

For the cavalry of the line:

	lbs.		lbs.
Hay	6.6 =	Hay	6.6
Oats	8.8 =	Ditto	13.5
Straw	11 =	Ditto	2.7
		Total allowance	22.8

For the light cavalry:

	lbs.		lbs.
Hay	6.6 =	Hay	6.6
Oats	8.3 =	Ditto	12.8
Straw	11 =	Ditto	2.7
		Total allowance	22.1

From what precedes, it appears that the substitution of oats for hay was made upon a calculation which squares well with the theoretical inferences in regard to the relative nutritive powers of these two articles.

The allowance to the horse ought to be distributed into three portions, constituting as many meals, and put before him in the morning before going to work, in the middle of the day, and in the evening; he is generally watered at meal times. It is also highly advantageous to the health of the horse that he be made to work with a certain regularity. Our horses at Bechelbronn, upon an allowance equivalent to 33 lbs. of hay, work from 8 to 10 hours a day, having an hour's rest at midday.

There is, of course, a certain relation between the height, or, if you will, the weight of the horse, and the quantity of provender he requires. Some attention, as we have seen, has been given to this point, in connection with horned cattle; but with reference to the horse I know of no data but such as I myself possess. Seventeen horses and mares, aged from 5 to 12 years, and having each provender equivalent to 33 lbs. of meadow hay, weighed together 18,190 lbs. The mean weight of each horse being represented by the number 1070 lbs., we perceive that for every 100 lbs. of live weight, 6.7 lbs. of meadow-hay are required for the daily ration, the horses working from 8 to 10 hours a day. This relation differs very little from that which we have obtained in reference to cattle.

I was anxious to ascertain the rate of growth of the horse; and in connection with our breed, which have a mean weight of about 1100 lbs., I found that the foals weighed as follows:

Names.	Date of birth.	Weight at birth.	Date of weighing.	Weight at weaning.	Number of days suckled.	Increase of weight during the suckling.	Increase of weight per diem.
		lbs.		lbs.	days.	lbs.	lbs.
Filly of Chevreuil	25 May, 1842	110	20 Aug. 1842	294.8	87	184.8	2.1
Filly of Hechler	12 June, 1842	113	7 Sept. 1842	286.	87	172.	1.9
Filly of Brunette	12 June, 1842	113	7 Sept. 1842	354.	87	241.	2.7

The mean increase per day during the period of suckling in the three cases quoted above, therefore, appears to have been rather more than 2 and $\frac{2}{10}$ths lbs. avoirdupois.

Immediately after weaning, young horses appear to experience an arrest of their growth for some short time, an event which indeed happens to animals generally. I found, for example, that Chevreuil's filly, which on the day of weaning weighed 294 lbs., nine days afterwards weighed but 288 lbs., and had consequently lost 6 lbs.

I shall add a few weighings of horses further advanced in age, although still young:

Alexander, a colt, weighed at birth 110 lbs.; at the age of 128 days, 337 lbs.; increase 227 lbs., or about 1.8 per diem: 51 days afterwards, 490 lbs.; increase 105 lbs., or per day 1.4 lb.

Finette, a filly, weighed, when weaned at the age of 86 days, 295 lbs.; 83 days afterwards, 396 lbs.: increase 101 lbs.; per day, 1.1 lb.

Hechler's filly weighed, when weaned at the age of 87 days, 286 lbs.; 65 days afterwards, 358 lbs.: increase 72 lbs., or per day 1.10 lb.

From what precedes we may conclude:

1st. That foals, the issue of mares weighing from 960 to 1100 lbs., weigh at birth about 112 lbs.

2d. That during suckling for three months, the weight increases in the relation of 278 to 100, and that the increase corresponds very nearly to 2 and $\frac{2}{10}$ lbs. avoirdupois for each individual per diem.

3d. That the increase of weight per diem of foals, from the end of the first to the end of the second year, is about $1\frac{3}{10}$ lbs. avoirdupois; and that towards the third year, the increase per day falls something under 1 lb. avoirdupois. After three years complete, the period at which the horse has very nearly attained his growth and development, any increase becomes less and less perceptible. These conclusions in regard to the horse, differ very little from those which I have had occasion to draw in connection with horned cattle.

I have also made a few experiments with reference to the quantity of provender consumed by foals in full growth, and have found that Alexander, Finette, and Hechler's filly, weighing together 1106 lbs., consume per day:

Hay	19.8 =	Hay	19.8
Oats	7 =	Ditto	11
		Total allowance	30.8
		Per head	10.22

The mean weight of these foals was 368.6 lbs., so that the hay consumed for every hundred pounds of live weight was 2.85 lbs., with which allowance the daily increase amounted to about 1.2 lb. Consequently, a mixed provender, equivalent to 100 lbs. of hay, had produced 12 lbs. of live weight. I must confess that this result appears to be somewhat too favorable, but I can only set down the numbers as they presented themselves to me.

The flesh of the horse is not generally used, or at least openly used, as food for man, though there are countries in which it is exposed for sale and commonly eaten. At Paris, indeed, in times of scarcity, horse-flesh has been consumed in quantity. During the Revolution, a knacker exposed publicly for sale, in the Place de Grève, joints from the horses which he had killed, and the sale continued for three years without any ill effect; in 1811, a scarcity obliged the Parisians to have recourse to the same kind of food, and it is said, indeed, that the traffic in horse-flesh as an article of human sustenance is still continued to a very considerable extent in the French metropolis; at the present moment, a distinguished writer on Medical Police, M. Parent-Duchatelet, has even proposed to legalize the sale of horse-flesh as food for man.

§ V. OF HOGS.

There is perhaps no farming establishment which does not keep a certain number of hogs, a measure by which offal of all kinds that would go directly to the dunghill, is turned to the very best account. The dairy, the kitchen-garden, and the kitchen, all yield their contingent of food to the pig-stye, which is moreover an excellent means of using up certain portions of the harvest. But the rearing and fattening of hogs, although frequently looked upon as matters of course, and requiring very little care, do in fact demand considerable attention and certain conveniences in situation. The rearing of hogs, in a general way, may be said to suit the small farmer better than the great agriculturist.

Our common domestic hog appears to derive his origin from the common wild hog of Europe. The breeds are extremely numerous. The black hog, covered with rather fine hair, and commonly found in Spain, is a native of Africa. This is the race which has been carried to South America, where it has multiplied in a truly surprising manner. It grows rapidly; and if it has little to recommend it with reference to fattening, it is nowise nice in the matter of food and general entertainment; the flesh is excellent when the animal has been kept upon the banana, and fattened off upon Indian corn.

The hogs of the east of Europe are remarkable for their size; they are of a deep gray color, and have very long ears; they are not very prolific, the brood swine having rarely more than four or five at a birth. The Westphalian breed, on the contrary, though they resemble the last, are highly prolific, the litter generally consisting of from ten to twelve. In Bavaria the hogs are remarkable for the smallness of their bones and the readiness with which they take on fat. Lastly, the Chinese race, which is common in England, and begins to extend on the continent, differs from those hitherto known, in having the back straight or even hollow, and the belly large. This breed is also remarkable for its quietness; the pork which it yields is of the very best quality.

One of the great advantages connected with the hog being its extreme fecundity, it is important to have a breed which is distinguished in this respect. There are some brood swine which have regularly borne ten to fifteen, and even eighteen pigs at a litter; a more general number is eight or nine.

According to Thaer, the hog that is disposed to take on fat is distinguished by length of body, long ears, and a pendulous belly. The hog attains his growth at the end of about a year, until which time the female ought not to be put to the boar. One boar generally suffices for about ten females.

The hog, as all the world knows, is an animal the least dainty in his food; he is omnivorous, nothing comes amiss to him; but his food is by no means matter of indifference when the quality of the flesh comes to be considered. Thaer seems to think that maize is of all articles that which is the best for feeding swine; and I have had occasion to verify the accuracy of his conclusion in South America, where I may add it is found that the oily fruit of the palm-tree contributes powerfully to the fattening.

Husbandry, in regard to the hog, comprises two distinct periods: the growth of the animal, and his fattening. It is generally admitted that it is most advantageous not to fatten swine for the butcher until they have completed or nearly completed their growth. A hog which has been well kept from the period of its birth, may be put up to fatten at the age of about a year. The female shows signs of heat at the age of about five or six months, and goes with young on an average 115 days, and will produce regularly two litters per annum; when particularly well kept, she may have three litters in the course of from thirteen to fourteen months.

The hogs which are destined to be fattened for the knife are generally cut at the age of six weeks, particularly if they are to be put up to fatten at the age of nine or ten months, as is often done. Almost all the varieties of roots and grain produced upon the farm are suitable for the maintenance of the hog; but in Alsace, and I believe generally, the staple is the steamed potato, with which are associated various articles in smaller quantity, such as peas, and barley and rye meal, &c.

The farrow sow ought to have food by so much the more abundant and nutritious as she is required to suckle a larger number of pigs. Our allowance at Bechelbronn to the hog with five young ones during the six weeks of suckling is as follows:

	lbs.		lbs.
Steamed potatoes	24.75	= hay	7.8
Rye meal	2.45	= "	4.0
Skim milk	13.2	= "	6.2
Total allowance			18.0

After the fifth week, when the animal is no longer giving suck, the ration consists of:

	lbs.		lbs.
Steamed potatoes	12.1	= hay	7.3
Rye meal	1.0	= "	1.6
Skim milk (sour)	6.5	= "	3.3
Total allowance			12.2

This allowance is gradually reduced to the end of the second month after the farrowing, when the animal is upon the maintenance ration of the farm, consisting of:

	lbs.		lbs.
Steamed potatoes	16.5	= hay	5.2

The potatoes are mixed with dish washings, which certainly contribute to improve their nutritive power, although I am altogether at a loss to estimate the value of the article.

The young pigs begin to taste the food given to the mother at the age of about a fortnight, but they never take to this kind of food freely until they are four or five weeks old and are weaned; up to this time they have an allowance of skim milk and whey. To five pigs at the time of weaning we allowed per day:

	lbs.				
Steamed potatoes	22.0	= hay	7.3	per head	1.4
Rye meal	1.0	= "	1.6	"	0.33
Skim milk	6.6	= "	2.8	"	0.57
	29.6		11.7		

This allowance was modified by degrees; the quantities of milk and rye meal were gradually abridged, and the proportion of potatoes increased, so that about the third month the allowance per head was from 11 to 13 lbs. of potatoes mixed with greasy water. This is the regimen, equivalent to about 5 lbs. of hay, upon which our store pigs are maintained until they are put up to fatten. During the three months which follow the weaning, therefore, we may reckon that each animal has consumed 3.8 lbs. of meadow-hay per day, and that from the third month the consumption may be represented by 5.2 lbs. of the same article.

We have attempted in vain to replace the potato by rape or madia oil-cake; the pigs refused it obstinately; but they showed no objection to poppy seed, walnut and linseed cake; during the season they will also eat clover, and are partly maintained upon this plant. In summer they are put entirely upon green meat, animals from five to six months old consuming about 19 lbs. of clover a-day, a quantity which represents very nearly 5 lbs. of clover hay.

The hog may be fattened at any age; but as we have already said, it is not generally advisable to fatten before he is ten months or a year, some say fifteen months or a year and a half old, at which period the animal is undoubtedly in flesh and at its full growth. The other extreme limit appears to be about five years; but it is only a brood sow that is ever kept to five years of age. It is generally allowed that twelve weeks are required to bring a hog into prime condition, when he ought to have a layer of fat under the skin upwards of an inch in thickness. Sixteen weeks may be required to obtain an animal really fat; and twenty weeks to have him at the highest point that is attainable. The hog requires to be fed regularly. After weaning, pigs should have five or six meals in the course of the day; the number of meals is diminished gradually, and towards the end of two months they amount to but three in all.

I was curious to ascertain the weight of pigs at the moment of their birth, so as to determine their rate of increase during the period of suckling. On the 5th of September a sow farrowed a litter of five.

		lbs.
No. 1	weighed	2.205
No. 2	"	3.025
No. 3	"	2.476
No. 4	"	2.750
No. 5	"	3.300
Weight of the litter......		13.756
Average weight per head..		2.751

On the 11th of October the weight of the litter was 86.9 lbs., or 17.3 lbs. per head: increase in thirty-six days, 73.2 lbs.; per head, 14.6 lbs.; per day, 0.409 lbs. On the 15th of November the weight was 177 lbs.: increase in thirty-five days, 90.2 lbs.; per head, 18 lbs.; per day, 0.506. During the thirty-six days of suckling, consequently, 100 of live weight at birth had become 632.

In another instance, I found that eight pigs which at the time of weaning weighed 114 lbs., or 14.3 lbs. per head, at a year old weighed 1320 lbs., or 165 lbs. per head: increase in eleven months 1206 lbs., or 150 lbs. per head.

The increase per diem since the weaning had been 0.4,—not quite half a pound; and as the food consumed may be represented by 5.2 lbs. of hay per day and per head, it will follow that 100 of forage had produced 8.58 of live weight. This ratio is too high, however; for these pigs besides the regular allowance had whey and various scraps of which no account was kept; and we know that whey alone contains a considerable quantity of the representatives both of flesh and fat.

Baxter came to some interesting conclusions on the growth and fattening of young hogs. Four animals each of the age of nine months weighed at the beginning of the experiment 458.2 lbs.; twenty-one days afterwards, 620.8 lbs.: increase of weight 162.6 lbs., to obtain which there were consumed:

	lbs.		lbs.
Barley	151	equivalent to hay,	250
Beans	140.8	"	611
Malt grains	440	"	257
			1229

So that a quantity of nutritive matter represented by 100 lbs. of hay produced 13.21 lbs. of live weight.

Assuming the weight of each pig of nine months old before the fatting to have been 29 lbs., the increase per head was 40.6 lbs. in the course of twenty-one days, or at the rate of 1.9 lbs. each. Baxter reckoned the carcass weight, sinking offal, at 7.4 per cent.

One of the pigs between nine and ten months old weighing 159.5 lbs., at the end of twenty days weighed 198.8 lbs.: increase in twenty days, 39.3 lbs.; increase per day, 1.9 lbs. During these twenty days, the animal had consumed 188 lbs. of barley, equivalent to 314 lbs. of hay. The increase would consequently give for every 100 lbs. of hay consumed an increase of live weight of 12.52, say 12½ lbs.

Arthur Young by keeping pigs of a year old on peas-meal obtained the following results:

No. 1	weighed 99.0;	35 days afterwards,	157.5;	gain, 58.5;	per day	1.672
No. 2	" 91.7;	42 "	145.4;	" 53.7	"	1.876
No. 3	" 86.6;	63 "	139.4;	" 52.8	"	0.836

I shall here give two series of observations made at Bechelbronn on the fattening of hogs. September 6th, 1841, seven hogs, aged fifteen months each, already in good condition, were put up to fatten. They had hitherto had the usual hog's food—sour milk and boiled potatoes after weaning; by and by from 11 to 15 lbs. of potatoes, whey, and dish washings. The seven porkers weighed 1691.8 lbs.; or 241.670 lbs. each. The increase had been at the rate of 0.528, rather better than half a pound per day and per head, supposing them to have weighed 13.7 lbs. each, at the time of weaning.

	lbs.
After fattening, 20th December, the 7 swine weighed	2101.0
Before " 6th September "	169.8
Increase in 104 days, 409.2 lbs.; or per head	58.3
Increase per day and per head	0.572

In the course of the 104 days, there were consumed :

	lbs.		lbs.
Barley	772	equivalent to hay	1144
Peas	1042.8	"	4171
Potatoes	9504	"	8296

Greasy water and whey—quantity not determined 8333.6

So that with the provender equivalent to 100 lbs. of hay 4.91 lbs. of live weight had been produced.

These seven porkers, slaughtered, yielded :

Hogs.	Weight alive.	Weight after bleeding.	Weight of the blood.	Weight of the porkers without heads or feet.	Weight of heads and offal.
	lbs.	lbs.	lbs.	lbs.	lbs.
1	323.0	312	11	268.0	44.0
2	259.0	248	11	208.4	39.6
3	283.0	272	11	208.2	63.8
4	316.0	305	11	263.2	41.8
5	264	257.4	6.6	220.0	37.4
6	259.6	250.8	8.8	213.4	37.4
7	393.8	376.2	17.6	321.2	55.0
	2098.6	"	"	1702.6	"

It must be admitted that these animals had increased both in flesh and in fat; but in spite of this, the experiment appears to be unfavorable to the opinion that fat in animals is the effect of direct assimilation of the substance. The whole increase of weight in the seven porkers had been 409.2 lbs. ; supposing 27 per cent. of fat in this increase, its amount must have been 110.4 lbs. in all ; but the food consumed did not contain more than 57.4 lbs. It would therefore be necessary to admit that the food which had not been taken into the account had contained as many as 53.0 lbs. of fatty matter, which I own does not appear to me probable. But no definitive conclusion can be drawn from the circumstance, owing to the actual state of fatness of the animals, when they were specially put up to fatten, not having been ascertained ; perhaps the absolute quantity of fat already accumulated is greater at this time than is generally supposed.

I find, for instance, that a young porker of 196.9 lbs., killed at the time when he might have been put up to fatten specially, yielded as many as 26.3 per cent. of fat.

The following are the data afforded by the fattening of the farm porkers for 1842 :

Nine porkers, from thirteen to fifteen months old, and already in good condition, were put upon the full fatting allowance on the 1st of October, on which day they weighed :

	lbs.
	1940
Nov. 28th, after having been bled, they weighed	2307.8
Increase in 58 days per head and per day	38.2

In the course of fifty-eight davs the hogs had consumed:

	lbs		lbs.
Rye.	776	equivalent to hay	1141.8
Peas	1302	"	5209
Potatoes........	4796	"	1861
Greasy water and whey undetermined....			8221.8

The nine animals gave 1746.8 lbs. of meat, fat and lean, or 75.7 per cent. of their weight as they stood alive; besides which, 141.9, say 142 lbs. of lard were obtained from the internal parts. Now supposing that in the increase of weight obtained in the course of fifty-eight days, the fat were to be represented by 29 per cent., the fat fixed would amount to 100.1 lbs.; while the whole of the fatty substances contained in the food consumed would not amount to more than 59.6 lbs. It would therefore be imperative on us, did we maintain that all the fat was obtained ready formed from without, to suppose that the whey and dish washings administered in indeterminate quantity, had introduced 40.5 lbs. of fat into the bodies of the animals. Some experiments which are going on at Bechelbronn at this moment will, I trust, settle the question definitively as to whether during the fatting of hogs and other animals there is any formation of fat at the cost of the starch and sugar of the food.

The observations which I have made on the fattening of hogs may be summed up in these terms:

Provender expressed in lbs. of hay.	Increase per day.	Live weight obtained from every 100 lbs. of provender.	Age of the animal at the beginning of experiment.	Duration of the experiment.	
lbs.	lbs.	lbs.	Months.	Months.	Days.
5.1	0.440	8.58	1	11	"
13.5	0.836	13.21	9	"	21
15.7	1.958	12.52	9½	"	20
"	1.254	"	12	"	"
11.6	0.572	4.91	15	"	104
15.4	0.660	4.21	14	"	58

The allowance to the hogs in the preceding observations was always abundant. To determine the quantity of potatoes consumed each day by a hog in full growth, and whose weight was known, I had him weighed at intervals, as well as the potato ration, placed before him at will, which he ate daily, and found that when he weighed:

lbs.		lbs.		lbs.	Per 100 of the live weight in hay
138	he ate	11	equivalent in hay to	3.4	2.52
145		13.2	"	4.2	2.89
160.5		15.4	"	4.8	30.1
184.8		17.6	"	5.5	3.02

To these observations on the keep and fatting of horned cattle, horses, and hogs, I would gladly have added remarks of like extent on the growth and fattening of sheep; unfortunately, I have only been able to obtain very imperfect information on this branch of rural economy. I have, however, sought to ascertain approximately the relations which exist between the weight of a young animal, the food consumed, and the increase in live weight, by means of the following experiment:

Two sheep six months old weighed together	134.2 lbs.
Sixteen days afterwards they weighed	151.8
Total increase	17.6
Increase per day, per head............	0.55

In the sixteen days the two sheep ate:

Hay............	22.0	=	Hay............	22.0
Potatoes........	53.3	=	Hay............	16.9
				38.9
Or per head in hay				19.45
Or per head per day....................				1.21

This would give us about 2.9 of hay provender per cent. of the live weight, so that a ration which should be represented by 100 of hay would be followed by an increase on the weight of a sheep of six months old of 27.7 per cent.

§ VI. OF THE PRODUCTION OF MANURE.

The forage consumed on the farm being the source of the manure produced there, it would seem that it must be easy to calculate the value of all that comes from the stables and cow-houses day after day. I do not mean the mass or weight of the dejections here, for it is certain that the more or less watery nature of the food materially influences the weight of the dung produced; and if a common mode of calculating the quantity of dung by merely multiplying the weight of food consumed by three be correct in some cases, it is very far from the truth in others. The dung produced on the farm must be calculated on different grounds from this; and without pretending to any degree of accuracy which is really unattainable, it is still very possible to get at the quantity of azote which is contained in the litter and in the dejections, so as to be able to refer to a standard the quantity of manure made.

Were not the azotized principles of the food partly exhaled by animals, the whole quantity not appearing in the excretions, it is obvious that it would suffice to have ascertained the quantity of azote contained in the food, to be in a condition to decide on that contained in the dung added to the litter. But this cannot be done; to be convinced of the fact, it is enough to take the least complex case, that of a full-grown horse, receiving as his allowance per day:

Hay	22 lbs.	containing	1775.3	grains of azote.
Oats......	11	"	1389.4	"
Straw	11	"	308.7	"
Litter	8.8	"	108.0	"
	Azote		3581.4	

Now assuming 2 per cent. as the contents in azote of dry farm-yard dung, we see that the food consumed by the horse, speaking theoretically, might or should form 25.5 lbs. of dry manure. But we have seen that a horse or cow will exhale from 355.0 to 416.8 grs. of azote, which is all derived from the food, and is consequently lost to the dung-heap. Now 385.9 grs. of azote represent 2.75 lbs. of dry manure; so that the dry dung produced by the horse kept in the stable, will be reduced from 25.5 lbs. to 23.1 lbs. In the course of a year, upon this calculation, the azote exhaled will diminish the weight of dry dung produced by one horse by a quantity equal to 1045 lbs.

The azote of the food of a cow is still more considerable in quantity, and the loss to the dunghill proportionally larger; inasmuch as to the amount she exhales, must be added all that goes to constitute the milk she gives. Practical men, without pretending to get at the cause of the thing, have long been aware of the fact, that a cow produces less dung than a horse; and the truth of this is readily demonstrated on scientific grounds. Suppose a cow, consuming the equivalent of 33 lbs. of hay, and giving about 17 pints of milk per day:

33 lbs. of hay contain............	2670 grs. of azote,	
44 " straw for litter contain..	123 "	
Azote........	2793	= 19.8 lbs. of dung supposed to be dry.

But in the 24 hours, there have been of

Azote exhaled..................................	385.9 grains, and of
Azote in 17 pints, or 22.7 lbs. of milk carried off,	802.7 grains,
	1188.6 = 8.8 of dry dung.

The 33 lbs. of hay digested by the cow, consequently, the litter added, have only produced 8.8 of dry dung. The azote of the food, of which we find no account in the dejections, amounts per annum to nearly 30 cwts., (3300 lbs.,) the deficiency in the case of the horse amounting to no more than 1045 lbs., (9 cwts. 1 qr. 9 lbs.)

The estimation of the dung produced by growing animals, presents several special difficulties, inasmuch, as besides the azote exhaled from the lungs, there is the quantity that is fixed in the living body.

In one of the experiments which I have related, it appears that a calf six months old, consuming:

Hay 9.6 lbs. containing	1069.8 azote,
Discharged by its dejections	838.3 "
Azote fixed or exhaled in 24 hours ..	231.5 "

The azote lost to the manure by the fixing of azote is therefore very considerable, in the case of young animals as well as of milch-kine. We find, for example, that for every 100 lbs. weight of hay consumed:

A horse supplies the equivalent of	51 lbs.	of dry standard dung,
A milch-cow	32 "	"
A calf of six months	40 "	"

To estimate with any rigor the quantity of azotized manure which ought to result from the forage consumed on the farm, it were necessary to know the proportion of azote contained in the bodies of all the animals entertained upon it. Having the increase of weight that occurred in the stable, cow-house, pig-stye, and poultry-yard, we should then be in a condition to know the precise quantity of dung which it would be necessary to retrench from that which the forage ought to have produced, had there been no production of animal matter, had the whole of the azote of the food passed through the live-stock to the dung-hill. Unfortunately, we have no very precise data by which we might calculate the quantity of azote contained in a living animal. I shall, nevertheless, endeavor to apply such as we possess.

From a few practical experiments, and the information at my command, I admit that the following substances in their usual state contain per cent.:

	Moisture.	Dry matter.	Salts.	Azote.
Beef-flesh	77	23	1.0	3.5
Veal	77	23	"	"
Blood	80	20	0.9	3.0
Skin	60	40	1.0	7.2
Hair	9	81	2.0	13.8
Horn	9	91	0.7	14.4
Beef bones (tibia)	30	70	"	"
An entire skeleton	36	64	35.0	5.2
Brain, intestines, &c.	81	10	1.0	2.9
Fat freed from skin	20	80	"	1.9

These data applied to the various parts which enter into the constitution of the animals which up to this point have engaged our attention, we should have for the quantity of azote per cent. contained:

In horned cattle	3.47
In the horse	3.64
In the hog	3.80
In the sheep	3.66
Average	3.64

For every 100 lbs. of live weight produced on the farm, consequently, we may, without probably being a great way from the truth, presume that there has been 3.6 of azote fixed, azote obtained from the forage, and which, consequently, cannot go to the dung-heap; in other words, every 100 lbs. of live weight produced, deprive the establishment of 180 lbs. of dry standard dung, or nearly 18 cwts. of moist farm-yard dung.*

We may be allowed, therefore, to entertain the hope that we shall one day be able, from the quantity of forage consumed upon a farm, to calculate the actual quantity of manure which we shall have at our disposal To arrive at this result, it would indeed only be necessary to subtract the manure represented by the azote exhaled from and fixed in the bodies of the stock, from the amount of azo-

* This discussion will undoubtedly extend by and by to phosphoric acid. I shall only say at this time, that from the results obtained in the case of a pig, the phosphoric acid appears to be in the proportion of from 2 to 3 per cent. of the live weight.

tized manure represented by the whole quantity of forage, were it to be used immediately. To obtain results of any accuracy, how ever, it were necessary to possess data both more numerous and more precise than any we have at present. This perfection of *co-efficients* must be viewed as an affair for the future; agricultural science has almost every thing to create.

In estimating the quantity of manure from the forage consumed, it has been supposed that there is no loss. With reference to the stall or cow-house, a careful husbandman may approach this perfection, by doing almost the contrary of all that is usually done now-a-days; *i. e.* by taking every precaution against waste; but it is obvious that in so far as the stable is concerned, there must always be a considerable and inevitable loss; all that falls upon highways and byways is irretrievably gone. It is, indeed, matter of ordinary calculation that in consequence of their work out of doors, the horses upon a farm do not afford more than about two-thirds of the dung which ought to be obtained from the provender consumed. Some experiments made in the stables at Bechelbronn show that the loss in this way may amount to one quarter of the whole amount of dejections; still, as the animals are for the major part engaged on the land of the farm, it is obvious that what falls there is by no means lost. To supply my reader with definite sums from a particular instance, upon which he may fix his mind, I shall state for his information that in the course of 1840–41,* my stock at Bechelbronn, consisting of sixteen head of cattle, eleven calves, twenty-seven horses, and (?) hogs, consumed 333,579 lbs. or 148 tons, 18 cwts. 1 qr. 15 lbs. of forage, containing 6925 lbs. of azote, and produced upon their original weight 20,821 lbs. of flesh, fat, and milk, containing with the addition of a calculated quantity for loss from out of door droppings, exhalation by the lungs, &c., 2631 lbs. of azote. The forage and the litter, from their contents in azote, ought to have produced about 15,356 cwt. of moist farm-yard dung; they, however, produced no more than 9522 cwt.; and, in fact, we see that there had been a consumption of azote by arrest within the bodies of the stock, by exhalation from their lungs, and by loss, amounting to 2631 lbs.; by an equivalent quantity of dung, therefore, had the absolute produce necessarily been diminished.

Thaer allows that articles of dry forage and litter double their weight in becoming converted into dung. The statement which I have just made agrees on the whole pretty well with this estimate. In our cow-house ration, one-half only is generally hay, the other half consists of roots and tubers. The dry forage and litter consequently amount to 4660 cwt., which according to Thaer ought to become changed into 9320 cwt. of dung, a number not very wide of that to which we have come. Sinclair reckons the dung of the cow-house at four times the weight of the litter, a view which neither accords with Thaer's estimate nor with our experience.

I think it altogether unnecessary to insist on the importance to

* Twelve months, I presume.—Eng. Ed.

the farmer of a foreknowledge of the quantity of manure which he may reasonably calculate on obtaining from a known weight of forage consumed upon his premises. Of the various methods proposed for arriving at this information, that which I have employed, and which is based on ascertaining the amount of azote, appears to me the best calculated to supply satisfactory results, particularly when experience shall have corrected or confirmed the numbers which I have adopted as the elements of my calculations.

I have already said that any supplementary forage, or forage added to that which is indispensable to the production of manure, generally acquires, by the fact of its conversion into power or into exportable substances, a value superior to that which it could have had of itself in the market-place. This additional forage is that fraction of the provender, the azote of which figures in the statements that have just been made as azote exhaled or assimilated and fixed. We find, in fact, in representing this forage which is lost to the dung-heap, but gained to power and exportable articles, that in the stall, 100 lbs. of hay yield 8.6 lbs. of live weight, and 40.8 lbs. of milk, and that in the hog-stye, 100 lbs. of hay yield 21 of living weight. In the stable, again, the azote fixed, exhaled, or lost amounts to nearly 1540 lbs., represented by about 1218 cwts. of hay, which have yielded 1504 lbs. of live weight, due in great part to the birth and growth of foals, in addition to the force represented by 8370 days' work.

CHAPTER IX.

METEOROLOGICAL CONSIDERATIONS.

§ 1. TEMPERATURE.

THE phenomena of vegetation are always accomplished under the influence of a certain temperature. If, in addition, the concurrence of light, air, moisture, and various inorganic substances, be required, it is still perfectly certain that all of these agents only contribute to the development of a plant when they are assisted by a due measure of heat, variable with reference to the different vegetable species, and comprised within limits that are rather far apart, but essential. Germination, for example, takes place at a temperature a few degrees above the freezing point of water, 38° or 39° F., and at one indicated by 100° or 120° of the same scale. The forests of tropical countries thrive in a hot, moist atmosphere, which often marks upwards of 100° F.; and I met with a saxifrage upon the Andes, at an elevation of 15,748 feet above the level of the sea, beyond the line of perpetual snow, and very near the line of perpetual congelation.

Some families of plants require a temperature not only high, but that never falls below a certain very limited degree; the majority of the intertropical plants are in this predicament. There are others which, imperatively requiring a high temperature for their growth and perfection, nevertheless suspend their powers during the winter, and bear without detriment degrees of cold of great intensity: among the number may be cited the larch-pine, which abounds in Siberia, and stands the utmost rigors of its climate, where the thermometer at mid-winter frequently falls to 30° and even 40° below zero, F.

The meteorological habitudes or dispositions of plants being extremely various, it follows, that the geographical distribution of plants is a consequence of the distribution of heat over the surface of the globe—of climate.

The earth we inhabit appears to have a heat proper to itself; it is a heated body in progress of cooling. It is found, in fact, that as the centre of the earth is approached, as mines penetrate more deeply below its surface, the temperature increases. Below a very limited distance from the surface, the temperature ceases to be affected by variations in the temperature of the general atmosphere; from the point of *invariable temperature* the subterranean heat increases uniformly at the rate of 1° cent. (1.8° Fahr.) for every 101 feet of descent.

The depth at which the point or stratum of *invariable temperature* is met with, varies in different places, and is mainly affected by the extent of the thermometrical variations in the superincumbent air in the course of the year. In the higher latitudes, consequently, the depth is very considerable; at Paris, for example, M. Arago has found that a thermometer, buried at 26¼ feet under the surface, does not remain absolutely stationary. In climates of greater constancy, as may be conceived, the layer of invariable temperature will be found much nearer the surface; were the temperature of the air invariable, the layer of invariable temperature would necessarily be found at the surface of the ground. In countries under and close to the equator, this, in fact, is found to be the case. From a series of observations which I made in South America, between the 2d parallel of southern and the 11th of northern latitude, I found that, near the line, the layer of invariable temperature is found nearly at the surface; the thermometer, placed in a hole about one foot deep, under the shade of an Indian cabin, or a shed, does not vary by more than from one-tenth to two-tenths of a degree cent.

It was probably under the influence of the internal or proper heat of the globe, according to M. de Humboldt, that the same species of animals which are now confined to the torrid zone, inhabited, in former and remote ages, the northern hemisphere, covered as it then was by arborescent ferns and stately palms. It is easy to imagine how, as the surface of the earth cooled, the distribution of climates became almost exclusively dependent on the action of the solar rays, and how also those tribes of plants and of animals, the organization of which required a higher temperature and more equable climate gradually died out and disappeared.*

* Humboldt's Central Asia, v. iii. p. 98.

In the state of stability to which the surface of the globe appears actually to have attained, the sun must be considered as the agent which most directly influences the temperature of our atmosphere. The length of the day, the number of hours during which the sun is above the horizon, coupled with the height to which he ascends, such is the cause with which the temperature of each particular latitude is primarily connected; and, in looking at the subject practically, it is found to be so precisely; not only is the mean temperature of the year dependent on the length of the days, and the meridian altitude of the sun, but the mean temperature of each month in the year is essentially connected with the same circumstances. In the northern hemisphere, the temperature rises from about the middle of January, slowly at first, more rapidly in April and May, to reach its maximum point in July and August, when it begins to fall again until mid-January, when it is at its minimum.

The highest mean annual temperature is, of course, observed in the neighborhood of the equator; between 0° and 10° or 12° of latitude on either side, at the level of the sea, where, besides the equality of day and night, the sun, always elevated, passes the zenith twice a year. The observations that have been made up to this time, lead us to conclude that this temperature oscillates between 26° and 29° cent.; 78.8° and 84.2° Fahr.

Did the earth present unvarying uniformity of surface, not only with reference to elevation but to constitution, so that the power of absorbing and of radiating heat should be everywhere alike, the climate of a place would depend almost entirely on its geographical position: the points of equal temperature would be found on the same parallels of latitude, or, to employ the happy expression introduced by M. de Humboldt, the *isothermal* lines would all be parallel with the equator. But the surface of our planet is covered with undulations and asperities, which cause its outline to vary to infinity; and then the soil is dry, or swampy; it is a moving desert of sand, or covered with umbrageous and impenetrable forests; and all this causes corresponding varieties in climate, for the surface becomes heated in different degrees as it is in one or other of these conditions. Another very important consideration is, that the surface is a continent, or an island in the ocean: the climate of a country, or a district, is vastly influenced by its proximity to or distance from the sea. The difficulty, the slowness, with which such a mass of liquid as the ocean becomes either heated or cooled, is the cause of the temperate character both of the summers and winters of the shores it bathes, and the islands of moderate dimensions it surrounds. As we penetrate great continents from the sea-board, we find that the temperature both of summer and winter becomes extreme, and the difference between the mean summer and mean winter temperature is great; and again we find, that places which have considerably different latitudes, have still very nearly the same mean annual temperature. The mean temperature of Paris, in latitude 48° 50′, is about 51.4° F.; that of London, in lat. 51° 31′, is 50.7° F.; that of Dublin, in lat. 53° 23′, is 49.1° F.; and that of Edinburgh, in lat 55° 57′, is 47.4° F.

An island, a peninsula, and the sea-shore, consequently, enjoy a more temperate and equable climate—the summers less sultry, the winters more mild. On the shores of Glenarm, in Ireland, in latitude 55°, the myrtle vegetates throughout the year as in Portugal; it rarely freezes in winter; but the heat of summer does not suffice to ripen the grape. Under the very same parallel, however, at Konigsberg, in Prussia, they experience a cold of 17° and 18° below zero of Fahrenheit's scale in the winter. The ponds and little lakes of the Feroe Islands, although situated in N. lat. 62°, never freeze, and the mean winter temperature is very nearly 40° F. On the coasts of Devonshire, in England, the winters are so mild, that the orange-tree, as a standard, will there carry fruit; and the agave has been seen to flourish, after having lived both winter and summer, for twenty-eight years, in the open air, uninjured.

One of the grand characteristics of what may be called a *maritime* climate, is the less difference which occurs between the temperature of summer and that of winter. At Edinburgh, for instance, the difference only amounts to 19° F.; at Moscow, which is nearly on the same parallel, the difference amounts to 50° F.; and at Kasan, (lat. 56°,) it is as much as 56.3° F.

The influence of extensive continents, or remoteness from the sea-board, does not seem merely to render a climate extreme, increasing at once the heat of summer and the cold of winter. The collective observations on temperature, made in Europe and in Asia, show that the mean annual temperature decreases as we penetrate more into the interior of continents towards the east. Humboldt ascribes this diminution of temperature partly to the refrigerating action of the prevailing winds. While the mean annual temperature of Amsterdam (N. lat. 52° 22′) is 49.6° F., that of Berlin (N. lat. 52° 31′) is 47.4° F.; that of Copenhagen (N. lat. 55° 41′) is 46.7° F.; and that of Kasan (N. lat. 55° 48′) is but 35.9° F.

The highest temperature which has yet been registered, as occurring in the open air, appears to have been observed by Burckhardt, in Upper Egypt; the thermometer indicated 47.5° cent., upwards of 117° F. The lowest was seen by Captain Back, in North America, when the thermometer fell to —.56° cent., 68.8° F. below zero.

§ II. DECREASE OF TEMPERATURE IN THE SUPERIOR STRATA OF THE ATMOSPHERE.

The temperature rises rapidly as we ascend in the atmosphere; places among the mountains always possess a climate more severe as they are higher above the level of the sea. Even under the equator, height of position modifies the seasons so much, that the hamlet of Antisana, which is within one degree of south latitude, but which is upwards of 13,000 feet above the sea level, has a mean temperature which does not differ much from that of St. Petersburgh. Near it, but at a still greater height, the summit of Cyambe, covered by an immense mass of everlasting snow, is cut by the equinoctial line itself.

The cold which prevails among lofty mountains, is ascribed to the dilatation which the air of lower regions experiences in its upward ascent, to a more rapid evaporation under diminished pressure, and to the intensity of nocturnal radiation.

Places which are situated upon the same mountain-chain, nearly in the same latitude, and at the same height, have often very different climates. The temperature which would be proper to a place perfectly isolated, is necessarily modified by a considerable number of circumstances. Thus the radiation of heated plains of considerable extent, the nature of the color of the rocks, the thickness of the forests, the moistness or dryness of the soil, the vicinity of glaciers, the prevalence of particular winds, hotter or colder, moister or drier, the accumulation of clouds, &c., are so many causes which tend to modify the meteorological conditions of a country, whatever its mere geographical position. The neighborhood of volcanoes in a state of activity does not appear to affect the temperature sensibly; thus Puracé, Pasto, Cumbal, which have flaming volcanoes towering over them, have not warmer climates than Bogota, Santa Rosa, De Osos, Le Param de Hervè, &c., situated on sand-stone or syenite.

From the whole series of observations which I had an opportunity of making on the Cordilleras, it appears that one degree of temperature, cent., 1.8° F., corresponds to 195 metres, or 649.4 feet of ascent among the equatorial Andes. In Europe, it has been ascertained that the decrease of temperature in ascending mountains, is more rapid during the day than during the night—during summer than during the winter; for example, between Geneva and Mount St. Bernard, to have the Fahrenheit thermometer fall one degree, it is necessary to ascend:

In spring	326.1 feet.
In summer	336.6
In autumn	382.2
In winter	422.2

It sometimes happens, however, that in winter, in a zone of no great elevation, the temperature increases with the elevation—a fact which Messrs. Bravais and Lottin observed in the 70° of N. lat., in calm weather; at an elevation between 1312 and 1640 feet, the rise amounted to as many as 6° centigrade, 10.8° Fahrenheit.

In no part of the globe is the diminution of temperature, occasioned by a rise above the level of the sea, more remarkable than among equatorial mountain ranges; and it is not without astonishment that the European, leaving the burning districts which produce the banana and cocoa-tree, frequently reaches, in the course of a few hours, the barren regions which are covered with everlasting snow. "Upon each particular rock of the rapid slope of the Cordillera," says M. de Humboldt, "in the series of climates superimposed in stages, we find inscribed the laws of the decrease of caloric, and of the geographical distribution of vegetable forms."*

* Humboldt's Central Asia, vol. iii. p. 236

In the hottest countries of the earth, the summits of very lofty mountains are constantly covered with snow; in the elevated and cold strata of the atmosphere, the watery vapor is condensed, and falls in the state of hail and snow. In the plain, hail melts almost immediately; the fusion is slower upon the mountains; and for each latitude there is a certain elevation where hail and snow no longer melt perceptibly. This elevation is the *inferior limit of perpetual snow.*

The accidental causes which tend to modify the temperature of a climate, also act in raising or lowering the snow-line. On the southern slope of the Himalaya, for example, the snow line does not descend so low as it does upon the northern slope; and in Peru, from 14° to 16° of S. latitude, Mr. Pentland found the perpetual snow-line, at an elevation of 1312 feet higher than it is under the equator.

Elevation above the level of the sea, consequently, has the same effect upon climate as increase in latitude. Upon mountain ranges, vegetation undergoes modification in its forms, becomes decrepit, and disappears towards the line of perpetual snow, precisely as it does within the polar circle, and for no other than the same reason, viz., depression of temperature.

The constancy and the small extent of variation which occurs in the temperature of the atmosphere under the equator, enables us to indicate with some precision the point of mean temperature below which there is no longer any vegetation. In ascending Chimborazo I met with this point at the height of 15774.5 feet, where the mean temperature approached 35° F., and where consequently the saxifrages, which root among the rocks, must still receive a temperature of from 41° to 43° F. during the day, inasmuch as far beyond the inferior snow-line, at an elevation of 19,685 feet above the sea-line, I saw a thermometer suspended in the air, and in the shade mark 44.6° F.

In considering the extension of vegetation towards the polar regions, we discover plants growing in very high latitudes in places which have a mean temperature much below that which I believe to be the limit of vegetable life on the mountains of the equatorial region. In these rigorous climates vegetation is suspended by the severity of the cold during the greater portion of the year; it is only during the brief and passing heat of summer that the vegetable world wakes from its long winter sleep. Nova Zembla, lat. 73° N., the mean temperature of whose summer is between 34° and 35° F., is, perhaps, like the perpetual snow-line of the equator, the term of vegetable existence. It is also to the very remarkable heat of the summer in countries situated at the northern extremity of the continent of Asia, remarkable if it be contrasted with the intensity of the winter cold, that man succeeds in rearing a few culinary vegetables in those dreadful climates. At Jakoustk, in 62° of N. lat., and where mercury is frozen during two months of the year, the mean temperature of summer is very nearly 64° F. We have here, as M. de Humboldt observes, "a well-characterized continental climate," examples of which indeed are frequent in the north of America. At

Jakoustk wheat and rye sometimes yield a return of 15 for 1, although at the depth of a yard the soil which grows them is constantly frozen.*

The limit of perpetual snow being much lower upon the mountains of Europe than in tropical countries, agriculture ceases at a much less elevation. At a height of 6560 feet above the level of the sea, the vegetables of the plain have almost entirely disappeared. In Northern Switzerland the vine does not grow at an elevation of more than 1800 feet above the sea-line; maize scarcely ripens at an elevation of 2850 feet, while in the Andes it still affords abundant harvests at an elevation of 8260 feet. On the plateau or table land of Los Pastos, fields of barley are seen at upwards of 10,000 feet above the level of the sea; but on the northern slope of Monte Rosa, in Switzerland, barley fails at an elevation of about 4260 feet; on the southern slope, indeed, it reaches a height of about 6560 feet; and this great variation in the ultimate limit of barley is frequently observed with reference to the same plant grown upon opposite aspects of a mountain range. The difference is ascribed to local influences; thus, it is a well-ascertained fact, that on the mountains of the northern hemisphere vegetation reaches a much higher latitude upon southern than upon northern exposures; but a general law, and one applicable to every latitude, is, that the higher we rise above the level of the sea, the scantier does vegetation become, the later do harvests reach maturity; but as the heat of the atmosphere increases with the elevation, it follows that there is an obvious relation between the time a crop is upon the ground and the mean temperature of the place or season where it grows. We have still to examine this relationship.

§ III. METEOROLOGICAL CIRCUMSTANCES UNDER WHICH CERTAIN PLANTS GROW IN DIFFERENT CLIMATES.

In discussing the conditions of temperature under which the various plants that are common in our European agriculture come to maturity, we are led to conclusions which are not without interest. A knowledge of the mean temperature of a place situated between the tropics suffices of itself to give us an idea of the nature of its agriculture; in fact, the temperature of each day differs little from that of the entire year, during which vegetable life proceeds without interruption. It is altogether different with regard to countries situated beyond the limits of the torrid zone. The mean annual temperature is not then a datum sufficient to enable us to appreciate the agricultural importance of a country. In order to know what the earth will produce, the temperature proper to the different seasons of the year must be known; in a word, it is the mean temperature of the cycle in which vegetation begins and ends that it imports us to ascertain, in order to learn what the useful plants are which may be required of the soil.

In examining the question which now engages us, we first inquire what time elapses between the sprouting of a plant and its maturity,

* Humboldt's Central Asia, vol. iii. p. 49.

and then we determine the temperature of the interval which seperates these two extreme epochs in vegetable life. In comparing these data with reference to the same species of plant grown in Europe and America, we arrive at the following curious result, that the number of days that elapse between the commencement of vegetation and the period of ripeness, is by so much the greater as the mean temperature is lower. The duration of the life of the vegetable would be the same, however different the climate, were this temperature identical; it will be longer or it will be shorter as the mean temperature of the cycle itself is lower or higher. In other words, the duration of the vegetation appears to be in the inverse ratio of the mean temperature; so that if we multiply the number of days during which a given plant grows in different climates, by the mean temperature of each, we obtain numbers that are very nearly equal. This result is not only remarkable in so far as it seems to indicate that upon every parallel of latitude, at all elevations above the level of the sea, the same plant receives in the course of its existence an equal quantity of heat, but it may find its direct application by enabling us to foresee the possibility of acclimating a vegetable in a country, the mean temperature of the several months of which is known.

CULTIVATION OF WHEAT, ALSACE.

In 1835 we sowed our wheat on the 1st of November; the cold set in shortly after the plant had sprung, and the harvest took place the 16th of July, 1836. The vegetation during the last days of autumn is so slow and irregular, that it may be assumed without sensible error, that it really begins in spring, when the frosts are no longer felt; from this period only does it proceed without interruption. For Alsace I regard this period as beginning with the 1st of March.

The period of the growth was, therefore, 137 days, the mean temperature was 59° F., (3083° F.)

Tremois wheat, this same year, required 131 days to ripen under a mean temperature of between 60° and 61° F., (7925° F.)

At Paris, setting out from the 31st of March, wheat generally requires 160 days to attain maturity, the mean temperature being about 56° F., (8960° F.)

At Alais the month of February having generally but few days of heat, it may be regarded as the epoch when the continued vegetation of autumn-sown wheat commences. The harvest taking place on the 27th of June, the number of days which it requires to ripen is 146, the mean temperature being between 57° and 58° F. (8322° F.)

CULTIVATION OF WHEAT IN AMERICA.

At Kingston, New York, the wheat is sown in autumn, vegetation suspended through the winter resumes its activity in the beginning of April, and the harvest takes place about the 1st of August. The crop is therefore growing during about 122 days under the influence of a mean temperature of 63° F. (7680° F.)

In the same place Tremois wheat is sown in the beginning of May, and the harvest takes place towards the 15th of August, so that it is 106 days on the ground under a mean temperature of 68° F., (7208° F.)

At Cincinnati the wheat sown in the end of February is harvested in the 2d week in July, say the 15th day, the crop is therefore 137 days on the ground under a mean temperature of between 60° and 61° F. (8288° F.)

INTERTROPICAL REGION.

Wheat sown at the end of February was reaped on the 25th of July at Zimijaca, plain of Bogota, having been 147 days on the ground, the mean temperature being between 58° and 59° F., (8526° F.)

At Quinchuqui the vegetation of wheat begins in February and ends in the month of July, say, 181 days; and I found the mean temperature to be between 57° and 58° F.

At Venezuela, according to M. Codazzi, wheat to ripen requires 92 days at Turmero, mean temperature between 75.2° and 76° F., (6918° F.;) 100 days at Truxillo, mean temperature 72.1° F., (7210° F.)

CULTIVATION OF BARLEY.

Of the cereals, barley is that which succeeds in the most diversified climates. It comes to maturity under the burning heats of the tropics; and in regions where the mean and constant temperature is scarcely 52° F., fields of barley of great beauty are still encountered.

At Alsace (Bechelbronn) barley sown at the end of April was harvested on the 1st of August. It had remained 92 days on the ground, the mean temperature having been between 66° and 67° F., (6118° F.)

Winter barley sown on the 1st of November was cut on the 1st of July. Reckoning the period of active vegetation from the 1st of March, it was 122 days in coming to maturity, the mean temperature having been between 58° and 59° F., (7076° F.)

At Alais winter barley is harvested on the 18th of June. Assuming that, as in the case of wheat, the 1st of February is the date of commencing vegetation, it must have taken 137 days to come to maturity under a mean temperature between 55° and 56° F.

In Egypt upon the banks of the Nile barley is sown in the end of November, and the harvest takes place at the end of February, at an interval therefore of 90 days, and the mean temperature of the winter at Cairo is nearly 70° F., (6300° F.)

At Kingston, North America, the barley is sown in the beginning of May, and the crop is cut towards the beginning of August, in about 92 days, therefore, the mean temperature being between 66° to 67° F.

At Cumbal under the line there is no fixed period for sowing barley. It is generally put into the ground on the approach of the rainy season about the 1st of June, and it is then reaped about the middle of November; it therefore stands on the ground for about 168 days, and the mean temperature is between 51° and 52° F.

At Santa Fé de Bogota they reckon about four months between the barley seed-time and harvest, or about 122 days, the mean temperature being between 58° and 59° F.

CULTIVATION OF MAIZE, OR INDIAN CORN.

In the neighborhood of Bechelbronn the maize which sprouted on the first of June yielded an abundant harvest on the 1st of October, the mean temperature having been 68° F.

In South America maize comes to maturity in the course of three months, say 92 days, the mean temperature being between 81° and 82° F.; but on the elevated plains, as that of Santa Fé, maize will require six months to come to maturity, say 183 days, and there the mean temperature is 59° F.

CULTIVATION OF THE POTATO.

In 1836 our potatoes at Bechelbronn were put into the ground on the 1st of May, and the crop was gathered on the 15th of October, after 157 days, therefore, the mean temperature having been about 65° F.; but in ordinary years, when the temperature is less elevated than that of 1836, the potato crop is generally gathered at the end of October, after 183 days, the mean temperature having been as before nearly 59° F.

In the neighborhood of Alais potatoes are planted at the end of March and taken up about the 1st of September, after five months or 153 days, the mean temperature of which has been 70° F.

According to M. Codazzi potatoes are grown near the lake of Valencia, (Venezuela,) in 120 days, and the mean temperature of Maracaibo near the lake is 78° F.

According to the same observer, the potato still yields good crops at Merida in the Cordilleras, where the mean temperature is between 71° and 72° F., and the growth lasts about 4½ months.

On the temperate levels of New Granada at Santa Fé I saw potatoes set in the middle of December immediately after the rainy season, and the harvest was gathered in the course of the first week in June, the crop therefore was at least 200 days in the ground, the mean temperature having been between 58° and 59° F.

On the occasion of my ascent of the volcanic mountain, Antisana, I ate on the 4th of August some potatoes which had just been gathered, and which had been planted in the beginning of November, so that the crop had been 276 days in the ground, the mean temperature of the country being 52° Fahr.

But this is not yet the superior limit to the cultivation of potatoes under the equator. They are still grown at Cambugan, the mean temperature of which scarcely exceeds 49° Fahr., the plant remaining nearly eleven months in the ground, and the crop being frequently lost from frosts that occur at this great elevation in the course of the months of November and January.

CULTIVATION OF THE INDIGO PLANT.

In Venezuela, in plantations very near the level of the sea, the first crop is cut about eighty days after sowing. The mean temperature is there between 81° and 82° Fahr. In other countries where the mean temperature ranges between 72° and 74° Fahr., which must be regarded as the limits to the growth of indigo, the first cutting takes place 3½ months or 106 days after the sowing. In India the first cutting seems generally to occur about ninety days after the sowing, and the mean temperature of the two winter months and of the summer months when the crop is on the ground, at Bombay is about 76° Fahr.

I shall terminate this section by calling the attention of vegetable physiologists to a fact which appears to have escaped them. It is this: that plants in general, those of tropical countries very obviously so, spring up, live, and flourish in temperatures that are nearly the same. In Europe and in North America, an annual plant is subjected to climatic influences of the greatest diversity. The cereals, for example, germinate at from 43° to 47° or 48°; they get through the winter alive, making no progress; but in the spring they shoot up, and the ear attains maturity at a season when the temperature, which has risen gradually, is somewhat steady at from 74° to 78° Fahr.

In equinoctial countries things pass differently: the germination, growth, and ripening of grain take place under degrees of heat which are nearly invariable. At Santa Fé the thermometer indicates 79° Fahr. at seed as at harvest time. In Europe the potato is planted with the thermometer at from 50° to 54° Fahr., and it does not ripen until it has had the heats of July and August. But we have just seen that this plant grows, slowly indeed, but regularly, in places where the temperature, nearly invariable, does not rise above 48.2° or 50° Fahr.

Germination, and the evolution of those organs by which vegetables perform their functions in the soil and in the air, take place at temperatures that vary between 32° and 112° Fahr.; but the most important epoch in their life, ripening, generally happens within much smaller limits, and which indicate the climate best adapted to their cultivation, if not always to their growth; for the vine grows lustily in many places where its fruit never ripens. To produce drinkable wine, a vineyard must have not only a summer and an autumn sufficiently hot; it is indispensable in addition that at a given period—that, namely, which follows the appearance of the seeds—there be a month, the mean temperature of which does not fall below 19° cent. or about 66½° Fahr., a fact of which conviction may be obtained from the following table which I borrow from M. de Humboldt:

	Temperature of summer.	Temperature of autumn.	Temperature of the hottest month.
Bordeaux........	70° Fahr.	58°	73.3° F. very favorable
Frankfort, A. M..	65	50	66.0
Lausanne	65.2	49.7	65.8
Paris............	65.8	52.2	66.2
Berlin..	63.2	48.0	64.4 Wine scarcely drinkable.
London	62.9	51.2	64.1 Vine not cultivated.
Cherbourg.......	61.9	54.4	63.2 "

In high latitudes the disappearance of vigorous vegetation in plants may depend quite as much on intensity of winter colds as on insufficiency of summer heat. The equable climate of the equatorial regions is therefore much better adapted than that of Europe to determine the extreme limits of temperature between which vegetable species of different kinds will attain to maturity. Thus it has been found that the vine between the tropics is productive in temperatures that vary from 69° F. to 79° or 80°. I shall terminate with a list of the temperatures favorable to the particular vegetables in the success of which man is more especially interested.

	Maximum.	Minimum.		Maximum.	Minimum.
The cocoa, or chocolate bean	82° F.	73° F.	Pine-apple	"	68
Banana....................	"	64	Melon	"	67
Indigo	"	71	Vanilla	"	68
Sugar-cane	"	71	Guaduas..................	"	77
Cocoa-nut	"	78	The vine	79	74
Palm.......................	"	78	Coffee	79	74
Tobacco..................	"	65	Anise.......................	77	66
Manihot..................	"	72	Wheat	74 (?)	74
Cotton-tree	"	67	Barley......................	74	59
Maize	"	59	Potatoes	75 (?)	52
Haricots..................	"	59	Arachaca..................	75	49
Orchil	"	72	Flax	74	54
Rice.......................	"	75	Apple	72	59
Calabash	"	72	Oak	67	61
Carica papaya.............	"	66			

§ IV. COOLING THROUGH THE NIGHT; DEW, RAIN.

When the sky is clear and calm during the night, vegetables cool down and very soon show a temperature inferior to that of the air which surrounds them. This property of cooling in such circumstances belongs to all bodies; but all do not possess it to the same degree. Organic substances, for instance, such as wool or cotton, feathers, &c., radiate powerfully and sink low; polished metals, on the contrary, have a very weak emissive or radiating power; and air and the gases in general radiate still more feebly.

Inasmuch as a body is continually emitting heat, its temperature can only remain stationary so long as it receives from surrounding objects at every instant a quantity of caloric precisely equal in quantity to that which it loses from its surface.

From the moment that these incessant exchanges cease to be in a state of perfect equality, the temperature of a body varies; it may even experience a considerable degree of cooling if it is exposed during a clear night in an open spot. In such circumstances, a body gives off towards all the visible parts of the heavens more heat than it receives; for the higher regions of the atmosphere are excessively cold, a fact which is proved by the rapid diminution of temperature experienced on ascending mountains, or by rising into the air in balloons. The internal temperature of the globe, the tendency of which would be to compensate the loss experienced by the body which radiates, has scarcely any effect in lessening the cooling, because it is propagated with extreme slowness, in consequence of the indifferent conducting powers of the earthy substances of which its crust is composed. The air, lastly, which surrounds the radiating body, does not warm it save in the most minute, inappreciable degree, and rather by its contact than by transmitting to it rays of heat, for the gases have only very limited emissive powers. It is even in consequence of the small amount of this power that the stratum of air in contact with the surface of the ground, does not by any means sink in the same proportion as the surface upon which it lies. Thus, in the circumstances which I have indicated, a thermometer laid upon the ground always indicates a temperature lower than that which is proclaimed by one suspended in the air; and the difference is by so much the greater as the radiating power of the bodies exposed is more decided, or as it may take place into a greater extent of the heavens. Every cause which agitates the air, which disturbs its transparency, which contracts the extent of the visible sphere, interferes with nocturnal radiation, and therefore with cooling. A cloud, like a screen, compensates either in whole or in part according to its proper temperature, for the loss of heat which a body upon the surface of the earth experiences in radiating into space. Wind, by continually renewing the air which is in contact with the surface of bodies tending to cool by radiation, always diminishes its effect to a certain extent. It is for this reason that a cloudless sky and a calm atmosphere, when nocturnal radiation attains its maximum, are most dangerous or injurious to our harvests.

In a night which combines all the conditions favorable to radiation, a thermometer of small size laid upon the grass will be found to mark from 10° to 14° or 15° Fahr. below the temperature of the surrounding atmosphere. Thus in the temperate zone in Europe, as Mr. Daniell has observed, the temperature of meadows and heaths is liable to fall during ten months of the year by the mere effect of nocturnal radiation to a temperature below the freezing point of water; this is particularly apt to happen both in spring and autumn, when the destructive effects of radiation are most to be apprehended, the nocturnal radiations of those seasons frequently lowering the temperature several degrees below the freezing point.

A few observations which I made upon nocturnal radiation at different heights in the Cordilleras, seem to indicate that its effects there are less decided than in Europe, in consequence perhaps of the greater quantity of heat acquired by the ground in the course of the day. It appears that in this mountain range it rarely freezes at a height less than 6560 feet above the level of the sea; although there are certain circumstances there which favor nocturnal radiation so much, that it is really impossible to indicate any very precise limits. In a general way it may be said that the crops of those plains which are sufficiently elevated to have a mean temperature of from 50° to 58° Fahr. are exposed to suffer from frost; it frequently happens that a crop of wheat, barley, maize, or potatoes, of the richest appearance, is destroyed in a single night by the effect of radiation. In Europe during the fine nights of April and May, when the air is calm and the sky clear, buds, leaves, and young shoots are frequently cut off, are frozen; in a word, although a thermometer in the air indicates several degrees above the point of congelation. Market gardeners and others who are much exposed to loss from this cause, ascribe the effect to the light of the moon of the months of April and May; and they ground their opinion upon the fact that when the sky is clouded, the destructive effects of frost are not apparent, although the same temperature of the atmosphere be indicated by the thermometer.

In the lower ranges of the Cordilleras, farmers also ascribe the same injurious consequences to the light of the moon, with this difference, that according to them the destructive influence continues throughout the year; and it is not unworthy of remark that, in the neighborhoods of Paris and of London, the mean temperature of the months of April and May (from 50° to 57°, or 58° F.) represents exactly the invariable climate of those places among the Andes, where the effects of frost upon vegetation are particularly to be apprehended. M. Arago has shown, that the cold ascribed to the light of the moon is nothing but a consequence of the nocturnal radiation, at a season when the thermometer in the air is frequently at from 40° to 43° F. and the sky is clear and calm. At this temperature a plant, radiating into space, readily falls below the point of congelation, and then the hopes of the gardener and farmer are destroyed. The phenomenon takes place particularly in a bright night: and if the moon happen to be up when it occurs, the influence is ascribed by the vulgar to her light. Were the sky clouded, the principal condition to radiation would be wanting; the temperature of objects on the surface of the ground would not fall below that of the surrounding medium, and plants would not freeze unless the air itself fell to 32° F.

The observation of gardeners, therefore, as M. Arago remarks, was not in itself false, it was only incomplete. If the freezing of the soft and delicate parts of vegetables in circumstances when the air is several degrees above the freezing point, be really due to the escape of caloric into planetary space, it must happen that a screen placed above a radiating body, so as to mask a portion of the heavens, will either prevent or at least diminish the amount of the cooling. And that this takes place in fact, appears from the beautiful experiments of Dr. Wells. A thermometer, placed upon a plank of a certain thickness, and raised about a yard above the ground, occasionally indicates in clear and calm weather from 6° to 7° or 8° F. less than a second thermometer attached to the lower surface of the plank. It is in this way that we explain the use of mats, of layers of straw, in a word, of all those slight coverings which gardeners are so careful to supply during the night to delicate plants at certain seasons of the year. Before men were aware that bodies on the surface of the ground became colder than the air which surrounds them during a clear night, the rationale of this practice was not apparent; for it was altogether impossible to conceive that coverings so slight could protect vegetables from a low temperature of the air.

The means indicated, as simple as they are effectual in protecting plants in the garden, are rarely applicable in farming, where the surface to be preserved is always very extensive. Nevertheless, in severe winters, the frost by penetrating the ground would frequently destroy the fields sown in autumn, were it not that in high latitudes the snow which covers the surface becomes a powerful obstacle to excessive cooling, by acting at one and the same time as a covering, and as a screen preventing radiation. As a covering, because snow is one of the worst of conductors, one of those substances which for a given thickness opposes the passage of heat most effectually; it is, therefore, an obstacle almost insurmountable to the earth beneath it getting into equilibrium in point of temperature with the atmosphere. As a screen, because in sheltering the ground it prevents it from undergoing the cooling which it would not fail to experience in clear nights by radiation into the open firmament. It is familiarly known in many parts of Europe, that the accidental want of the usual covering of snow will cause the loss of the autumn-sown crops of grain It is on the surface of the snow that the great depression of temperature takes place; and the substance being a very bad conductor, the earth cools in a much less degree. In the month of February, 1841, I made some experiments, which show that the snow which covers the ground acts in the manner of a screen. I had first a thermometer upon the snow, the bulb of the instrument being covered by from 0.078 to 0.117 of an inch of snow in powder; second, a thermometer, the bulb of which was situated completely under the layer of snow in contact with the ground; third, a thermometer in the open air, at about 37 or 38 feet above the surface, on the north of a building. The layer of snow was about four inches in thickness, and had covered a field sown with wheat for a month. The sun shone brightly upon the field on those days when my experiments were made.

Feb. 11. Five o'clock in the evening; the sun has been hidden by the mountains for half an hour; the sky is unclouded, the air very calm: thermometer under the snow, 32° F.; thermometer upon the snow, 29° F.; thermometer in the air, 36.3° F.

Feb. 12. The night very fine, no clouds, the air calm. At seven o'clock in the morning, the sun is not yet upon the field: thermometer under the snow, 26.2° F.; thermometer upon the snow, 10° F.; thermometer in the air, 26.3° F.

At half-past five in the evening, the sun behind the mountains: thermometer under the snow, 32° F.; thermometer upon the snow, 29° F.; thermometer in the air, 37.5° F.

Feb. 13. At seven in the morning; the sky gray, the air slightly in motion: thermometer under the snow, 28° F.; thermometer upon the snow, 17° F.; thermometer in the air, 25° F.

At half-past five in the evening; the air calm, the sky cloudless, the sun already concealed for some time: thermometer under the snow, 32° F.; thermometer upon the snow, 30° F.; thermometer in the air, 40° F.

Feb. 14. Seven in the morning, wind W., a fine rain falling: thermometer under the snow, 32° F.; thermometer upon the snow, 32° F.; thermometer in the air, 35.7° F.

When we reflect upon the losses occasioned to farmers and market gardeners by frosts that are entirely due to nocturnal radiation at seasons of the year when vegetation has already made considerable progress, we ask eagerly if there be no possible means of guarding against them. I shall here make known a method imagined and successfully followed by South American agriculturists with this view. The natives of the upper country in Peru who inhabit the elevated plains of Cusco are perhaps more than any other people accustomed to see their harvest destroyed by the effects of nocturnal radiation. The Incas appear to have ascertained the conditions under which frost during the night was most to be apprehended. They had observed that it only froze when the night was clear and the air calm: knowing consequently that the presence of clouds prevented frost, they contrived to make as it were artificial clouds to preserve their fields against the cold. When the evening led them to apprehend a frost—that is to say, when the stars shone with brilliancy, and the air was still—the Indians set fire to a heap of wet straw or dung, and by this means raised a cloud of smoke, and so destroyed the transparency of the atmosphere from which they had so much to apprehend. It is easy in fact to conceive that the transparency of the air can readily be destroyed by raising a smoke in calm weather; it would be otherwise were there any wind stirring; but then the precaution itself becomes unnecessary, for with air in motion, with a breeze blowing, there is no reason to apprehend frost from nocturnal radiation.

The practice followed by the Indians just described is mentioned by the Inca Garcillaso de la Vega in his Royal Commentaries of Peru. Garcillaso in the imperial city of Cusca, and in his youth, had frequently seen the Indians raise a smoke to preserve the fields of maize from the frost.*

The cooling of bodies occasioned by nocturnal radiation is always accompanied by a deposite of moisture upon their surface under the form of minute globules: this is dew. The ingenious experiments of Wells having demonstrated that the appearance of dew always follows, never precedes the fall in temperature of the bodies on which it is deposited, the phenomenon cannot be attributed to any thing more than a simple condensation of the watery vapor contained in the air, comparable in all respects to that which takes place upon the surface of a vessel containing a fluid that is colder than the air.† The quantity of moisture dissolved in the atmosphere

* The good effects of smoke in preventing nocturnal congelation are also signalized by Pliny the naturalist.

† Arago, Annuaire des Longitudes, Année 1837, p. 160.

is by so much the greater as the temperature is higher. In very warm climates the dew is so copious as to assist vegetation essentially, supplying the place of rain during a great part of the year.

According to some meteorologists dew is most copious near the sea-board of a country; very little falls in the interior of great continents, and indeed is said only to be apparent in the vicinity of lakes and rivers.* I cannot agree in any statement of this kind made so absolutely. I have never had occasion to see more copious dews than those which occasionally fall in the steppes of San Martin to the east of the eastern Cordilleras, and at a very great distance from the sea; the dew was so copious that for several nights I found it impossible to employ an artificial horizon of black glass in order to take the meridian altitude of the stars; the moment the apparatus was exposed there was such a quantity of water deposited on the surface that it soon gathered into drops and trickled off. I found it necessary to have recourse to mercury to reflect the star under observation. During the clear calm nights the turf of these immense plains receives a considerable quantity of moisture in the form of dew, which materially assists vegetation, and by its evaporation tempers the excessive heat of the ensuing day. In tropical countries the forests contribute to keep down the temperature. In very hot countries it is rare to be out in a cleared spot, when the night is favorable to radiation, without hearing drops of water, produced by the copiousness of the dew, falling continually from the surrounding trees, so that forests contribute further to produce and to maintain springs by acting as condensers of the watery vapor dissolved in the air. I might cite a number of observations upon this point which I made in the forest of Cauca. In the bivouac between the 4th and 5th of July, 1827, the night was magnificent; nevertheless, in the forest which began at the distance of a few yards from our encampment, it *rained abundantly;* by the light of the unclouded moon we could see the water running from the branches.

It is possible that the transpiration from the green parts of the trees might have been added to the dew condensed, and so increased the intensity of the phenomenon which I have described; but I rather incline to believe that the cooling of the leaves by way of radiation had by far the largest share in the production of this dew-rain. It is true that of all the leaves which form the crown of a tree, those whose surface is exposed and radiate freely into space intercept, as would a screen, the radiation of the leaves and branches which are not so exposed; but, as M. de Humboldt has observed, if the leaves and branches which crown a tree cool directly by emission, those which are situated immediately beneath them by radiating towards the lower parts of the leaves which are already cooled a greater quantity of heat than they receive, their temperature will also necessarily fall, and the cooling will thus be propagated from above downward until the whole mass of the tree feels its effects It is thus that the ambient air circulating through the leaves becomes

* Kaemtz, Meteorology, translated by W. Walker, London, 1844.

cooled during bright nights, and to judge from the influence which forests exert in lowering the temperature of a country, it is enough to recollect with M. de Humboldt that by reason of the vast multiplicity of leaves, a tree, the crown of which does not present a horizontal section of more than about 120 or 130 square feet, actually influences the cooling of the atmosphere by an extent of surface several thousand times more extensive than this section.

The proportion of watery vapor which a gas will retain in solution is by so much the greater as the temperature of the air is higher. All the causes which cool air saturated with watery vapor occasion, as we have already seen, the precipitation of a certain quantity of moisture.

When this condensation takes place in the midst of a gaseous mass, the precipitated water collects into small floating vesicles, which trouble the transparency of the medium that momentarily holds them in suspension. Mists, fogs, and clouds are collections of these vesicles; a fog, as a celebrated naturalist said, is a cloud in which one is, and a cloud is a fog in which one is not.

The vesicles of clouds tend towards the earth, like all heavy bodies, but by reason of their specific lightness the resistance of the air which they displace lessens the rapidity of their descent. When they are of larger size they coalesce and form drops of water which fall with greater celerity. When these drops pass through strata of very dry air they undergo partial evaporation, and this is the reason wherefore there is sometimes less rain upon plains than upon mountains. In opposite circumstances it is the inverse phenomenon that is observed; the drops increase in size in passing through the inferior strata of an atmosphore saturated with moisture, condensing vapor in their course. This is what happens most generally.

In taking a survey of a large amount of observations, meteorologists have inferred that the annual quantity of rain varies with the latitude; that, greatest at the equator, it gradually lessens as higher northern and southern latitudes are attained; this is as much as saying that the quantity of rain is greater as the temperature of the climate is higher. But to this rule there are numerous exceptions; for instance, under the line at Payta on the sea-coast it only rains very rarely; a shower of rain is an event, and when I visited the country eighteen years had elapsed since they had had any thing of a fall of rain. Local causes have the greatest influence upon the fall of rain, so that countries on the same parallel of latitude are far from being equally distinguished by dryness or humidity.

It is believed that in Europe it rains more heavily and more frequently in the day than in the night. In the equinoctial regions, at least in those parts that I have visited, it would seem that the opposite rule held good. Every one in South America allows that it rains principally during the night, and the observations which I made in the neighborhood of Marmato enable me to state that of 7.874 inches of rain which fell in the month of October, 1.336 inches fell in the day, 5.638 inches in the night; of 8.881 inches which fell in the month of November, 0.707 inches came down in the day,

8.174 inches in the night; of 5.934 inches which fell in December, 0.786 inches fell in the day, 5.148 inches in the night.

Two series of observations taken in the same country at two stations not far from one another, but situated at very different elevations, seem to confirm, in reference to the equatorial regions, the conclusions of European meteorologists as to the fact that the annual quantity of rain which falls diminishes as the height above the level of the sea increases. They also show that in latitudes which do not differ materially, more rain falls where the mean temperature is 68° F. than where it is 58° F.

Marmato lies in N. lat. 5° 27″, and 75° 11″ (?) W. long., at a height of 4676 feet above the level of the sea; Santa Fé in N. lat. 4° 36″, W. long. 75° 6″, at a height of 8692 feet above the level of the sea. And while the quantity of rain at the former place amounted, according to my own observations for 1833, to 60 inches, according to Caldas, in 1807, at the latter there fell but 39.4 inches.

In temperate climates the quantity of rain that falls varies with the seasons. Near the equator, where the temperature remains constant throughout the year, the rainy season commences precisely at the period when the sun approaches the zenith; and whenever the latitude of a place in the torrid zone where it rains is of the same denomination and equal to the declination of the sun, storms occur. In such circumstances the sky in the morning is of remarkable purity, the air is calm, the heat of the sun insupportable. Towards noon clouds begin to show themselves upon the horizon, the hygrometer does not advance towards dryness as it usually does, it remains stationary, or even falls towards extreme humidity. It is always after the sun has passed the meridian that the thunder is heard, which being preceded by a light wind is soon followed by a deluge of rain. In my opinion the permanence or incessant renovation of storms in the bosom of the atmosphere is a capital fact, and is connected with one of the most important questions in the physics of our globe, that of the fixation of the azote of the air by organized beings.

The most recent inquiries show dry atmospherical air to consist in volume of:

Oxygen	20.8
Azote	79.2

The air contains in addition from 2 to 5 10,000ths of carbonic acid gas, and quantities perhaps still smaller of carbureted combustible gas. The experiments of M. Theodore de Saussure, as well as those of Professor Liebig, have further demonstrated in it traces of ammoniacal vapor.

I have already shown that animals do not directly assimilate the azote of the atmosphere. Azote is nevertheless an element essential to the constitution of every living being, and is met with indifferently in either kingdom of nature. If we inquire into the source of this principle in connection with the herbivorous tribes of animals, we find it as an element in the food which sustains them. If we next inquire into the immediate origin of the azote which enters into the constitution of vegetables, it is discovered in the manure which proceeds more especially from animal remains; for vegetables, to thrive, must receive azotized aliment by their roots. We thus come to apprehend that plants supply animals with their azote, and that these restore it to plants when the term of their existence is accomplished; we are led to discover, in a word, that living organic matter derives its azote from dead organic matter.

This view leads us to conclude that the amount of living matter on the surface of the globe is restricted; that its limits are in some sort determined by the quantity of azote in circulation among organized beings; but the question must be viewed from a loftier eminence, and we must ask what is the origin of the azote which enters into the constitution of organic matter considered as a whole?

If we now turn to the possible sources or magazines of azote, we shall find, setting aside organized beings and their remains, that there is in truth but one, the atmosphere. It is therefore extremely probable that all living beings have previously obtained their azote from the atmosphere, just as it seems very certain that they have thence derived their carbon.*

The most reasonable supposition in the actual state of science, is to consider the ammoniacal vapors diffused through the atmosphere as the prime source of the azotized principles of vegetables, and then through them of animals; a consequence of which hypothesis would be to assume with Liebig, that carbonate of ammonia existed in the atmosphere before the appearance of living things upon the face of the earth.

The phenomena and effects of thunder-storms appear to me calculated to support this opinion. It is known, in fact, that so often as a succession of electrical sparks passes through moist air, there is formation and combination of nitric acid and ammonia. Now nitrate of ammonia is one of the constant ingredients in the rain of thunder-storms. But nitrate of ammonia, being a fixed salt, cannot exist in the atmosphere in the state of gas or vapor; and then it is not the nitrate, but the carbonate of ammonia that has been signalized in the air. In bringing to mind the series of reactions of which I have spoken, it is not difficult to perceive how the nitrate of ammonia, precipitated in thunder-showers, and thus brought into contact with calcareous rocks, should suffer decomposition, pass into the state of carbonate on the return of fair weather, and become fitted to undergo diffusion in the state of vapor through the atmosphere. We should in this way be led to regard the electrical agency, the flash of lightning, as the means by which the azote of the atmosphere is made fit for assimilation by organized beings. In Europe, where thunder-storms are rare, an office of so much importance will perhaps be accorded reluctantly to the electricity of the clouds; but in tropical countries no difficulty would probably be felt on the matter. In the torrid zone, thunder-storms happen in one place or another not only every day, but every hour, and even every minute of every

* Boussingault, Annales de Chimie, t lxxi. 1839.

hour throughout the year; so that an observer, placed at the equator, were he endowed with organs of sufficient delicacy, would never lose the roll of the thunder.

As the equator is quitted, the times at which rain falls become less specific or periodical. Under the tropics, the rains of thunder-storms, which are always the most copious, fall while the sun is in the neighborhood of the zenith. In the northern hemisphere, the greatest quantity of rain falls during winter; and at places somewhat far south on the temperate zone, the summer rain is altogether insignificant. In assuming the number 100 to express the whole annual quantity of rain, we should have in

	Madeira.	Lisbon.
Winter	51	40
Spring	16	34
Summer	3	3
Autumn	30	23

Less rain falls in the eastern parts of Europe than in the western. The annual rain, too, is distributed very unequally over the different seasons, as has been shown by M. Gasparin in a remarkable paper. If we express by 100 the quantity of rain gauged in a year, we should have for each season:

	In the west of England.	West of France.	East of France.	Germany.	St. Petersburg.
Winter	26	23	20	18	14
Spring	20	13	23	22	18
Summer	23	25	29	37	37
Autumn	31	34	28	23	30

The quantity of rain which falls in the course of a year varies considerably according to the climate; to form an idea of the extent of these variations, it is enough to notice the results obtained at different observatories; but it is less the annual quantity of rain that falls, than the way or quantities in which it is distributed over the different months of the year, which interests the farmer; upon this distribution, in fact, in many districts, depend the productiveness and fertility of the soil. I add a table of the mean quantities of rain in inches and 10ths, that fall at London in the different months of the year:

Jan.	Feb.	March.	April.	May.	June.	July.	Aug.	Sept.	Oct.	Nov.	Dec.
in.	in.	in.	in.	in.	in.	in.	in.	in.	in.	in.	in.
1.45	1.25	1.17	1.29	1.61	1.72	2.39	1.80	1.84	2.08	2.20	1.72

§ V. ON THE INFLUENCE OF AGRICULTURAL LABORS ON THE CLIMATE OF A COUNTRY IN LESSENING STREAMS, ETC.

A question of great importance, and that is frequently agitated at this time, is, as to whether the agricultural labors of man are influential in modifying the climate of a country or not? Do extensive clearings of woods, the draining and drying up of great swamps, which certainly influence the distribution of heat during the different seasons of the year, also exert an influence on the quantity of running water of a country, whether by lessening the quantity of rain which falls, or by promoting the more speedy evaporation of that which has fallen?

In some districts it has been held, that the streams which had been used as moving powers, have very sensibly diminished. In other places, the rivers are said to have shrunk visibly; and in others, springs that were formerly abundant, have almost dried up. Observations to this effect appear to have been principally made in valleys, surmounted by mountains; and it is generally asserted, that the falling off in the springs and streams, had followed close upon the period at which the woods, scattered over the surface of the country, were cleared away without any kind of reserve.

These statements, which may be assumed as facts, seem to indicate that where the woods have been felled, it rains less than it did formerly; this, indeed, is the general opinion entertained on the subject; and were it admitted, without further examination, the natural inference from it would be, that the extension of agriculture diminishes the annual quantity of rain which falls in a country. But at the same time that the facts as stated have been observed, it has further been noticed that since the clearing of the surface from forests, the torrents and rivers which seemed to have lost in amount of regular supply of water, had become subject to sudden and extraordinary risings which had proved the cause of numerous and grave calamities. In the same way, springs that are generally all but dry, have been seen to burst forth again abundantly after violent storms. These latter observations, as may readily be imagined, are of a kind that should lead us not lightly to embrace the vulgar opinion, which maintains that the cutting down of the woods has had the effect of lessening the mean annual quantity of rain: it is not only not impossible that this quantity has not varied, but it may even happen that the mass of water which passes over the bed of a stream, supposed shrunken, is actually the same as ever it was; the only difference may be, that now the flow is much less regular than it used to be: in former times the bed was always and more moderately full; at present it is excessively full at intervals only. It is very possible, therefore, that here as elsewhere, occasionally, the appearance of the fact has been taken for the reality. It were very important to discover some natural index to a solution of the question at issue: whether or not the destruction of the forests that once covered the face of a district of country, had had the effect of lessening the mean annual fall of rain?

The lakes which are met with in plains, and at different levels in mountain ranges, seem to me peculiarly well calculated to throw light on this subject. Lakes may, in fact, be received as natural gauges of the running waters of a country. If the mass of the water contained in the lakes undergo change in one direction or another, it is obvious that this change, and the direction in which it has occurred, will be proclaimed by the state or mean level of the lake or lakes, which will differ for the same reason that it does at different seasons of the year, viz. as drought or rain prevails. The mean level of the lake or lakes of a district will, therefore, fall, if the quantity of water which flows through that district diminishes; the level, on the contrary, will rise, if its streams increase; and it will remain stationary if the afflux and efflux of the lake continue unchanged. In the following remarks, I shall attach myself particularly to observations upon lakes which have no outlet, by reason of the facility with which any even slight change in the level of these must be discovered. I shall not, however, neglect those lakes which have an exit by a stream or canal, because I believe that the study of these may also lead to accurate enough results; the only point requiring preliminary remark, is the sense in which the words, change of level, are to be taken.

Geologists admit, that the level of the waters upon the surface of the globe has everywhere undergone great changes, whether attention be directed to the shores of the sea or to those of great inland lakes. This fact is universal, and is questioned by none, but great diversity of opinion prevails in regard to the cause of the phenomenon. Some pretend, that in many cases the change of level is only apparent,—that the body of water has not sunk, but that the shores have risen; others, again, maintain that there has been a true diminution in the mass of fluid, a true drying up, and that its level has actually sunk. I shall not, in this place, enter upon the great geological question; the variations which are there signalized are often of vast extent, and involve the supposition of violent catastrophes, which, in a general way, were long anterior to the historical epoch. I shall only refer to changes of level observed in lakes by our ancestors or contemporaries; in a word, to facts which have taken place under the eyes of men, inasmuch as it is the influence of their agricultural labors upon the meteorological state of the atmosphere, which I am seeking to appreciate. The facts upon which I shall more particularly dwell, were observed in South America; but I shall show that what is true with regard to this continent, is true also with reference to any other continent.

One of the most interesting portions of Venezuela is, undoubtedly, the valley d'Aragua. Situated at a short distance from the seaboard, possessed of a warm climate, and of a soil fertile beyond example, it combines within itself all the varieties of agriculture that belong in peculiar to tropical regions; on the hillocks which rise in the bottom of the valley, are seen fields which bring to mind the agriculture of Europe. Wheat succeeds pretty well upon the heights which surround La Vittoria. Bounded on the north by a chain of hills which run parallel with the sea-board, and to the south by the range which separates it from Llanos, the Aragua Valley is limited on the east and west by a series of lesser elevations, which shut it in completely. In consequence of this peculiar configuration of country, the rivers which rise in its interior have no outlet to the ocean; their waters accumulate in the lowest part of the valley, and form the beautiful lake Valentia. This lake, which M. de Humboldt says exceeds the lake Neufchâtel in size, is raised about 1300 feet above the level of the sea; it is about ten leagues in length, and about two leagues and a half where it is widest.

At the time when M. de Humboldt visited the Aragua Valley, the inhabitants were struck with the gradual diminution which had been going on in the waters of the lake during the last thirty years. It was enough to compare the statements of older writers with its condition at this time, to obtain conviction that the waters had, in fact, very much diminished. Oviedo, for instance, who visited the valley frequently towards the end of the sixteenth century, says, that the town of New Valencia was founded in 1555, at the distance of half a league from the lake; in 1800, M. de Humboldt ascertained that the lake was upwards of 549 yards, or upwards of 3¼ miles, instead of about 1¼ mile from its banks.

The appearance of the surface also gives new proof of the fact of the recession of the water; certain hillocks which rise in the plain still preserve the title of islands, which, undoubtedly, they formerly received with propriety, when they were surrounded by water. The land which had been left by the retreat of the lake, soon became transformed into beautiful plantations of cotton-trees, bananas, and sugar-canes. Buildings which had been erected on the banks were left, year after year, further and further from them. In 1796, new islets made their appearance. An important military position, a fortress built in 1740, in the Isle de la Cabrera, was then upon a peninsula. Finally, in two islets of granite, M. de Humboldt discovered, several yards above the level of the lake, a bed of fine sand, mixed with fresh-water shells. These facts, so certain, so unquestionable, did not pass without numerous explanations from the wise men of the country, who, as if by common consent, fixed upon a subterranean exit for the waters of the lake. M. de Humboldt, after the most careful examination of all the circumstances, did not hesitate to ascribe the diminution of the waters of the lake Valencia to the extensive clearings which had been effected in the course of half a century in the Aragua valley. "In felling the trees which covered the crowns and slopes of the mountains," says this celebrated traveller, "men in all climates seem to be bringing upon future generations two calamities at once—a want of fuel and a scarcity of water."*

In the year 1800, the population of this favored valley, where the cultivation of indigo, of cotton, of cocoa, and the cane had made immense progress, was as dense as it was in the most thickly populated districts of England or France, and every one was delighted with the appearance of comfort that prevailed in the numerous villages of this industrious country.

Twenty-five years after M. de Humboldt, I explored in my turn the Valley d'Aragua, having fixed my residence in the little town of Maracaibo. The inhabitants had now remarked that for several years, not only had the lake ceased to diminish, but that it had even risen very perceptibly. Some fields that were formerly covered with cotton plantations were now submerged. The Isles de las Nuevas Aparacidas, which had risen from the waters in 1796, had again become shoals dangerous to navigation; the tongue of earth, De la Cabrera, on the north side of the valley, had become so narrow that the slightest rise in the water of the lake covered it completely; a

* Humboldt, vol. v. p. 173.

continuous N.E. wind was sufficient to flood the road which led from Maracaibo to New Valencia; in short, the fears which the inhabitants of the lake had entertained for so long a period had entirely changed their nature; they were now no longer afraid of the lake drying up; they saw with dismay that if the water continued to rise as it had done lately, it would in no long space of time have drowned some of the most valuable estates, &c. Those who had explained the diminution of the lake by supposing subterraneous canals, now hastened to close them up in order to find a cause for the rise in the level of the water.

In the course of the last twenty-two years important political events had transpired. Venezuela no longer belonged to Spain; the peaceful valley d'Aragua had been the theatre of many a bloody contest; war to the knife had desolated this beautiful country and decimated its inhabitants. On the first cry of independence raised, a great number of slaves found freedom by enlisting under the banners of the new republic; agricultural operations of any extent were abandoned, and the forest, which makes such rapid progress in the tropics, had soon regained possession of the surface which man had won from it by something like a century of sustained and painful toil. With the increasing prosperity of the valley many of the principal tributaries to the lake had been turned aside to serve as means of irrigation, so that the beds of some of the rivers were absolutely dry for more than six months in the year. At the period which I now refer to, the water was no longer used in this way, and the beds of the rivers were full. Thus with the growth of agricultural industry in the Valley d'Aragua, when the extent of cleared surface was continually on the increase, and when great farming establishments were multiplied, the level of the water sunk; but by and by, during a period of disasters, happily passing in their nature, the process of clearing is arrested, the lands formerly won from the forest are in part restored to it, and then the waters first cease to fall in their level, and by and by show an unequivocal disposition to rise.

I shall now, without, however, quitting America, carry my readers into a district where the climate is analogous to that of Europe, where the surface is occupied by immense fields, covered with the cereals as with us. I speak of the table-lands of New Granada, of those valleys raised from 10,000 to 13,000 and 14,000 feet above the level of the sea, in which the mean temperature throughout the year ranges from 58° to about 62° Fahr. Lakes are frequent in the Cordilleras; and it would be easy for me to describe a great number; I shall, however, confine myself to those which became subjects of observation in former times.

The village of Ubaté is now situated in the neighborhood of two lakes. Some seventy years ago these two lakes formed but one; the old inhabitants saw the water shrinking and new fields presenting themselves year after year. At this present time fields of wheat of extraordinary luxuriance occupy levels that were completely inundated 30 years ago.

It is enough indeed to perambulate the neighborhood of Ubaté, to consult the old sportsmen of the country, and to refer to the annals of the various parishes, to be satisfied that extensive forests have been cut down in the whole of the surrounding country: the clearing, in fact, still continues; and it is certain that the recession of the waters, although much slower than it was in former times, has not yet entirely ceased.

A lake, situated in the same valley, to the east of Ubaté, deserves our particular attention. By means of barometric measurements, made with extreme care, I found that this lake had precisely the same level as that of Ubaté. Nearly two centuries ago, it was visited by Piedrahita, Bishop of Panama, an author of great accuracy, to whom we owe the history of the conquest of New Granada. He states this lake to be ten leagues in length, by three leagues in breadth; but Dr. Roulin having had occasion, a few years ago, to make a plan of the lake, he found it a league and a half in length, by one league in breadth; my own impression is, that at the time Piedrahita wrote, there was but a single lake, extending all the way from Ubaté to Zimijaca, not two lakes as at present, a supposition which would take away every thing like exaggeration from the statement of Piedrahita. But the fact of the retreat of the waters of these lakes is unquestioned; the inhabitants of Zimijaca all know that the village was founded close to the lake, whereas, at the present time, it is nearly a league from its banks. Formerly, there was no difficulty in obtaining all the building timber that was wanted; the mountains which rose from the valley on either hand were covered up to a certain height with the trees proper to these cold regions; the oak of the Andes abounded; numerous myrcias were also in existence, from which abundance of wax was obtained: at the present time these mountains are almost stripped of their trees, an event mainly brought about by the eagerness to procure fuel in manufacturing salt from the springs of Taosa and Enemocon.

To these authentic facts, which I could multiply and support by many others of a similar kind, it may be replied, that the diminution of the water, incontestable as it is, might perhaps have taken place without the clearing away of the forests. It may indeed be maintained, that the drying up of the waters is owing to a totally different cause, altogether unknown to us; that it must be ranked among the numerous phenomena, the reality of which we perceive, but without being able to render any account of their cause.

I cannot, in the instance last quoted, as in that of the lake of Valencia, refer to any increase of the lake connected with the suspension of agriculture, or the reappearance of the forest. I might, however, adduce in favor of the opinion which I defend, the slowness with which the decrease in the lakes of the valley of Ubaté has lately gone on, and since the felling of trees in the neighborhood has almost entirely ceased. Extensive plots of fertile ground are now no longer left dry and available to the husbandman by the retreat of the lake; he already begins to think of means for obtaining by artifice that which nature, assisted by the clearing of the country presented him with in former times. In the year 1826 there was a speculation on foot for draining the valley completely by cutting a canal and letting off the water. Further proof of the fact which I am urging is obtained in another way. It is not difficult to show, that lakes in the immediate vicinity of those that have shrunk most remarkably, but around which no destruction of the forest has taken place, have undergone no change in their level. The lake of Tota, situated at no great distance from the valley of Ubaté, at an elevation that must approach 13,000 feet above the level of the sea, in a region where vegetation has almost entirely disappeared, has preserved its pristine level unaltered. The lake is nearly circular; and Piedrahita, in 1542, gives it two leagues in breadth. It is subject to violent storms, which render its navigation dangerous; and even travelling along its banks, from the particular circumstances in which the road is situated, with the lake on one hand and a perpendicular cliff upon the other, is not without risk. In 1652, the road passed as it does at present, the water laving the foot of the same rocks, and its level having suffered no change, any more than the sterile country which surrounds it.

I shall conclude what I have to say on the lakes of South America by speaking of that of Quilatoa, because it has been made the subject of accurate observations sufficiently remote from one another—1740 and 1831.

Living at Latacunga, a town situated at no great distance from Cotopaxi, the traveller is sure to hear of the wonders of the Laguna da Quilatoa. From time to time this lake, it is said, casts forth flames which set fire to the shrubs and withered grass that grow upon its banks, and frequent detonations are heard, the sound of which extends to a great distance. M. de la Condamine, in 1738, visited this lake, which he found of a circular form, and about 1278 feet in diameter; on the 28th November, 1831, I also visited the Lake of Quilatoa. It cannot be better compared to any thing than to the crater of a volcano filled with water; I found it nearly 13,000 feet above the level of the sea, in the cold region, therefore; and indeed it is surrounded with immense pastures; but the information which I obtained from the shepherds in the neighborhood of the Lake of Quilatoa, dissipated all that was marvellous in its history; they had never seen any flames issue from its bosom, they had never heard any detonations; in short, I found the lake as M. de la Condamine appears to have found it, every thing having remained for nearly a century without change.

The study of the lakes which are so common in Asia, would probably supply conclusions similar to those deduced from observations made in South America, viz., that the waters which irrigate a country diminish as the forests are cleared away, and as agriculture extends. The recent labors of M. de Humboldt, which have thrown so much new light upon this quarter of the world, appear to leave no doubt upon the subject. After having shown that the system of the Altai is about to lose itself by a succession of slopes in the steppes of Kirgiz, and that consequently the Ural chain is not connected with that of the Altai, as was generally believed, this celebrated geographer shows, that precisely in the situation where the Alghinic mountains are usually set down, a remarkable region of lakes commences, which extend into the plains that are traversed by the Ichim, the Omsk, and the Obi.* It would appear that these numerous lakes are remainders as it were of an immense sheet of water, which formerly covered the whole of the country, and which had become divided into so many particular lakes by the configuration of the surface. In crossing the steppe of Baraba, in his way from Tobolsk to Baraoul, M. de Humboldt perceived everywhere that the drying up of waters increases rapidly under the influence of the cultivation of the soil.

Europe also possesses its lakes; and we have still to examine them from the particular point of view which engages us. M. de Saussure, in his first inquiries in regard to the temperature of the lakes of Switzerland, examined those which are situated at the foot of the first line of the Jura. The Lake of Neufchâtel is eight leagues in length, and its greatest breadth does not exceed two leagues. On visiting it, Saussure was struck with the extent which this lake must formerly have possessed; for, as he says, the extensive level and marshy meadows which terminate it on the south-west, had unquestionably been covered with water at a former period.

The Lake of Bienne is three leagues long and one broad; it is separated from the Lake of Neufchâtel by a succession of plains that were probably inundated.

Lake Morat is also separated from the Lake of Neufchâtel by low and level marshes, which beyond all question were formerly submerged. Unquestionably, adds Saussure, the three great lakes of Neufchâtel, Bienne, and Morat, were formerly connected, and formed one great sheet of water.†

In Switzerland, as in America and Asia, the old lakes, those that may be spoken of under the title of the primitive lakes, and which occupied the bottoms of the valleys when the country was uncultivated and wild, have become divided, and now form a variable number of smaller and independent lakes. I shall wind up the present subject by referring to the observations of Saussure upon the Lake of Geneva, which may be looked upon as the starting point of the admirable works of this distinguished philosopher.

Saussure admits, that at an epoch long anterior to the times of history, the mountains which surround this lake were themselves submerged; a great catastrophe let off this immense collection of water, and by and by the current possessed no more than the bottom of the valley; the Lake of Geneva was formed.

In merely considering the monuments left by man, it is impossible to doubt that within 1200 or 1300 years the waters of the Lake of Geneva have gradually fallen in their level. It is evidently upon the levels which have thus been left that the quarter de Rive, and the lower streets of the city of Geneva have been built. This de

* Humboldt, Fragmens Asiatiques, t. i. p. 40–50.
† Saussure, Voyage dans les Alpes, t. ii. chap. 6.

pression of the surface, continues Saussure, is not merely the effect of any deepening of the bed of the Rhone, by which the lake is discharged; it has also been produced by a diminution in the quantity of water which flows into it.

The conclusions which it seems legitimate to draw from the observations of Saussure are, that in the course of from 1200 to 1300 years the quantity of running water has sensibly diminished in the districts around the Lake of Geneva. No one will, I apprehend, deny that in this long period there have not been extensive clearings of forest lands in Switzerland, and a continual increase in the extent of cultivated land in this beautiful country. Here, consequently, as elsewhere, an attentive examination of the levels of the lakes leads us to conclude, that where extensive clearings from forest have been effected, where agriculture has extended, that there has in all probability been diminution of the running waters which irrigate the surface; while in those districts where no change has been effected, the amount of running stream does not appear to have undergone any variation.

The effect of forests considered in this point of view would therefore be to keep up the amount of the waters which are destined for mills and canals; and next to prevent the rain-water from collecting and flowing away with too great rapidity. That a soil covered with trees is further less favorable to evaporation than ground that has been cleared, is a truth that all will probably admit without discussion. To be aware that it is so, it is enough to have travelled, a short time after the rainy season, upon a road which traverses in succession a country that is free from forests, and one that is thickly wooded. Those parts of the road that pass through the unencumbered country are found hard and dry, while those that traverse the wooded districts are wet, muddy, and often scarcely passable. In South America, more perhaps than anywhere else, does the obstacle to evaporation from a soil thickly shaded with forests, strike the traveller. In the forests the humidity is constant, it exists long after the rainy season has passed; and the roads that are opened through them remain through the whole year deeply covered with mire: the only means known of keeping forest ways dry, is to give them a width of from 260 to 330 feet, that is to say, to clear the country in their course.

If once the fact is admitted that running streams are diminished in size by the effect of felling the forests and the extension of agriculture, it imports us to examine whether this diminution proceeds from a less quantity of rain, or from a greater amount of evaporation, or whether perchance it may be owing to the practice of irrigation.

I set out with the principle that it must be next to impossible to specify the precise share which each of these different causes has in the general result; I shall, nevertheless, endeavor to appreciate them in a summary way. The discussion will have gained something if it be proved that there may be diminution of running streams in consequence of clearing off the forests alone, without the whole of the causes being presumed to act simultaneously.

With regard to irrigation, it is necessary to distinguish between that case in which an extensive farm has been substituted for an impenetrable forest, and that in which an arid soil, which never supported wood, has been rendered productive by the industry of man. In the first case it is very probable that irrigation will have contributed but little to the diminution in the mass of running water; it may readily be imagined that the quantity of water used up by a dense forest will equal, at all events, if not exceed, that which will be required by any of the vegetables which human industry substitutes for it. In the second case, that is to say, where a great extent of waste country has been brought under cultivation, there will evidently be consumption of water by the vegetation which has been fostered upon the surface; agricultural industry will thus tend to diminish the quantity of water which irrigates a country. It is extremely probable that it is to a circumstance of this kind that we must ascribe the diminution of the lakes which receive so large a proportion of the running streams of the north of Asia. It is almost unnecessary to add, that in circumstances of this kind the effect which is due to the simple evaporation of rain-water is not increased; the loss by this means must be rather less, because from a surface covered with plants evaporation takes place more slowly than from one that is devoid of vegetation.

In the considerations which I have presented upon the lakes of Venezuela, of New Granada, and of Switzerland, the diminution may be directly ascribed to a less mean annual quantity of rain; but it may with equal reason be maintained to be a simple consequence of more rapid evaporation.

There are, in fact, a variety of circumstances under the influence of which the diminution of running streams can be shown to be connected with more active evaporation. I shall confine myself to the mention of two particular instances, one noticed by M. Desbassyns de Richemond, in the Island of Ascension; the other is from observations by myself, and is among the number of facts which I registered during my residence for several years at the mines of Marmato.

In the Island of Ascension there was an excellent spring situated at the foot of a mountain originally covered with wood; this spring became scanty and dried up after the trees which covered the mountain had been felled. The loss of the spring was rightly ascribed to the cutting down of the timber. The mountain was therefore planted anew, and a few years afterwards the spring reappeared by degrees, and by and by flowed with its former abundance.

The metalliferous mountain of Marmato is situated in the province of Popayan, in the midst of immense forests. The stream along which the mining works are established is formed by the junction of several small rivulets which take their rise in the table-land of San Jorge. The country which overlooks the establishment is thickly wooded.

In 1826, when I visited the mines for the first time, Marmato consisted of a few miserable cabins, inhabited by negro slaves. In

1830, when I quitted the country, Marmato had the most flourishing appearance; it was covered with workshops, it had a foundry of gold, machinery for grinding and amalgamating the ores, &c., and a free population of nearly three thousand inhabitants. It may be readily imagined, that in the course of these four years an immense quantity of timber had been cut down, not only for the construction of machinery and of houses, but as fuel, and for the manufacture of charcoal. For facility of transport, the felling had principally gone on upon the table-land of San Jorge. But the clearing had scarcely been effected two years before it was perceived that the quantity of water for the supply of the machinery had notably diminished. The volume of water had been measured by the work done by the machinery, and actual gauging at different times showed the progressive diminution of the water. The question assumed a serious aspect, because at Marmato any diminution in the quantity of the water, which is the moving power, would be of course attended with a proportional diminution in the quantity of gold produced. Now, in the Island of Ascension, and at Marmato, it is highly improbable that any merely local and limited clearing away of the forest should have had such an influence upon the constitution of the atmosphere as to cause a variation in the mean annual quantity of rain which falls in the country. More than this, as soon as the diminution of the stream at Marmato was ascertained, a pluviometer, or rain-gauge, was set up, and in the course of the second year of observation a larger quantity of rain was gauged than in the first year, although the clearing had been continued; still there was no appreciable increase in the size of the running stream.

A couple of years of observation are unquestionably insufficient to show any definitive variation in the annual quantity of rain that falls. But the observations made at Marmato still establish the fact that the mass of running water had diminished in spite of the larger quantity of rain which fell. It is therefore probable that local clearings of forest land, even of very moderate extent, cause springs and rivulets to shrink, and even to disappear, without the effect being ascribable to any diminution in the amount of rain that falls.

We have still to inquire, whether extensive clearings of the forest—clearings which embrace a whole country—cause any diminution in the quantity of rain that falls. Unfortunately, the observations which we have upon the quantity of rain which falls in particular districts, are only of sufficient antiquity and accuracy in Europe to be worthy of any confidence, and there the soil was cleared before observation, in the generality of instances, began.

The United States of America, where the forests are disappearing with such rapidity, will probably one day afford elements for the complete and satisfactory solution of the question, whether or not the cutting down of forests causes any diminution in the quantity of rain which falls in the course of the year.

In studying the phenomena accompanying the fall of rain in the tropics, I have come to a conclusion which I have already made known to many observers. My own opinion is, that the felling of forests over a large extent of country has always the effect of lessening the mean annual quantity of rain.

It has long been said, that in equinoctial countries the rainy season returns each year with astonishing regularity. There can be no doubt of the general accuracy of this observation, but the meteorological fact must not be announced as universal and admitting of no exception; the regular alternation of the dry and rainy season is as perfect as possible in countries which present an extreme variety of territory. Thus, in a country whose surface is covered with forests and rivers and lakes, with mountains and plains, and table-lands, the periodical seasons are quite distinct. But it is by no means so where the surface is more uniform in its character. The return of the rainy season will be much less regular if the soil be in general dry and naked; or if extensive agricultural operations take the place of the primeval forest; if rivers are less common, and lakes less frequent. The rains will then be less abundant; and such countries will be exposed, from time to time, to droughts of long continuance. If, on the contrary, thick forests cover almost the whole of the territory, if its rivulets and rivers be numerous, and agriculture be limited in extent, irregularity in the seasons will then take place, but in a different way; the rains will prevail, and in some seasons they will become as it were incessant.

The continent of America presents us, on the largest scale, with two regions placed in the same conditions as to temperature, but in which we successively encounter the circumstances which are most favorable to the formation and fall of rain in one case, and to its absence in the other.

Setting out from Panama, and proceeding towards the south, we encounter the Bay of Cupica, the provinces of San Bonaventura, Choco, and Esmeraldas; in this country, covered with thick forests and intersected with a multitude of streams, the rains are almost incessant; in the interior of Choco, scarcely a day passes without rain. Beyond Tumbez, towards Payta, an order of things entirely different commences: the forests have entirely disappeared, the soil is sandy, agriculture scarcely exists, and here rain is almost unknown. When I was at Payta, the inhabitants informed me that it had not rained for seventeen years! The same want of rain is common in the whole of the country which surrounds the desert of Sechura, and extends to Lima; in these countries rain is as rare as trees are.

In Choco, where the soil is thickly covered with trees, it rains almost continually; and on the coasts of Peru, where the soil is sandy, without trees, and devoid of verdure, it never rains; and this, as I have said, under a climate which enjoys the same temperature, and whose general features and distance from the mountains are nearly the same. Piura is not more remote from the Andes of Assuay than are the moist plains of Choco from the Western Cordillera.

The facts which have now been laid before the reader seem to authorize me to infer—

1st. That extensive destruction of forests lessens the quantity of running water in a country.

2d. That it is impossible to say precisely whether this diminution is due to a less mean annual quantity of rain, or to more active evaporation, or to these two effects combined.

3d. That the quantity of running water does not appear to have suffered any diminution or change in countries which have known nothing of agricultural improvement.

4th. That independently of preserving running streams, by opposing an obstacle to evaporation, forests economize and regulate their flow.

5th. That agriculture established in a dry country, not covered with forests, dissipates an additional portion of its running water.

6th. That clearings of forest land of limited extent may cause the disappearance of particular springs, without our being therefore authorized to conclude that the mean annual quantity of rain has been diminished.

7th, and lastly. That in assuming the meteorological data collected in intertropical countries, it may be presumed that clearing off the forests does actually diminish the mean annual quantity of rain which falls.*

* These meteorological observations are highly interesting, and worthy of every consideration. That unforesting a country makes it absolutely drier, seems unquestionable; but whether that be in consequence of less rain falling, or of that which falls going further, making more show, cannot be easily determined. It does not seem very legitimate to decide, that because a country is covered with wood, therefore it is wet: the converse of that proposition appears much more probable—viz., that because a country is wet, therefore it is covered with trees. There is one part of the ocean which is called by mariners "The Rains;" because it rains there almost ceaselessly, as it does in the province of Choco: but "The Rains" has no forests to account for its dripping sky. Did that region consist of dry land instead of salt-water, then doubtless its surface would be covered, as that of Choco is, with an impenetrable forest. The subject is adverted to here, however, not to discuss the general question, but to throw out the suggestion that under the hand of man, the soil and even the climate of our immense Australian possessions might possibly be improved. Drought is the grand enemy of Australian settlers; and the country is generally barren of wood.

Governors, district governments, and farmers, and all who are interested in the prosperity of the colony, surely ought to encourage, by every possible means, the growth of the taller trees and shrubs that are indigenous to the country.

Expeditions might be made once or twice a year, at the proper season, for scattering or planting the seeds of these trees or shrubs. Could every knoll within a hundred miles of Sidney be seen crowned with a thick screen of leafy trees, there can be little doubt but that the rain which falls would be economized; and that the beds of the rivers, instead of being dry for eight or nine months, would be occupied all the year round by at least a moderate stream of water.—Eng. Ed.

THE END.

HISTORY AND BIOGRAPHY.

Arnold.—The History of Rome,

From the Earliest Period. By THOMAS ARNOLD, D. D. Reprinted entir from the last English edition. Two vols. 8vo, $5,00.

Arnold.—The Later Roman Commonwealth.

The History of the Later Roman Commonwealth. By THOMAS ARNOLD, D. D. Two vols. of the English edition, reprinted entire in 1 vol. 8vo, $2,50.

Arnold.—The Life and Correspondence of Thomas Arnold, D. D. By ARTHUR P. STANLEY, A. M. 2d American from the 5th London edition. One handsome 8vo. volume, $2,00.

Arnold.—Lectures on Modern History,

Delivered in Lent Term, 1842, with the Inaugural Lecture delivered in 1841. By THOMAS ARNOLD, D. D. Edited, with a Preface and Notes, by Henry Reed, M. A., Prof. of Eng. Lit. in the University of Pa. 12mo, $1,25.

Burnet.—Notes on the Early Settlement of the North-Western Territory.—By JACOB BURNET. One vol. 8vo, $2,50.

Coit.—The History of Puritanism.

Puritanism; or, a Churchman's Defence against its Aspersions, by an Appeal to its own History. By THOMAS W. COIT, D. D., Rector of Trinity Church, New Rochelle. 12mo, 528 closely-printed pages, $1,50. Reduced to $1,00.

Carlyle.—The Life of Schiller:

Comprehending an Examination of his Works. By THOMAS CARLYLE, Author of "The French Revolution," etc. 12mo, paper cover, 50 cts.; cloth, 75 cts.

Evelyn.—Life of Mrs. Godolphin.

By JOHN EVELYN, Esq. Now first published. Edited by Samuel Wilberforce, Bishop of Oxford. 12mo, paper cover, 38 cts.; cloth, 50 cts.

Frost.—The Life of Gen. Zachary Taylor,

With Notes of the War in Northern and Southern Mexico; with Biographical Sketches of the Officers who have distinguished themselves in the Mexican War. By JOHN FROST, LL. D., author of "The Book of the Army," etc., etc. One vol. 12mo, illustrated with Portraits and Plates. $1,00.

Guizot.—History of Civilization in Europe,

From the Fall of the Roman Empire to the French Revolution. By F. GUIZOT, late Professor of History, and Prime Minister of France. Translated by Wm Hazlitt. Four volumes, 12mo, cloth, $3,50.

Guizot.—History of the English Revolution

Of 1640, from the Accession of Charles I. to his Death. By F. GUIZOT, the Prime Minister of France; Author of "History of Civilization in Europe," etc. etc. Translated by William Hazlitt. In two volumes, 12mo. Paper cover, $1,00; or two volumes bound in one, cloth, $1,25.

Hull.—Revolutionary Services and Civil Life of Gen. William Hull, from 1775 to 1805. Prepared from his Manuscripts by his Daughter, Mrs. MARIA CAMPBELL: together with the History of the Campaign of 1812, and Surrender of the Post at Detroit, by his Grandson, JAMES FREEMAN CLARKE. One vol. 8vo, $2,00.

HISTORY AND BIOGRAPHY.

ohlrausch.—History of Germany,

From the Earliest Period to the Present Time. By FREDERICK KOHLRAUSCH Chief of the Board of Education for the Kingdom of Hanover, and late Professor of History in the Polytechnic School. Translated from the last German edition, by James D. Haas. One volume, 8vo, of 500 pages, with complete Index, $1,50.

ing.—The Argentine Republic.

Twenty-four Years in the Argentine Republic; embracing its Civil and Military History, and an Account of its Political Condition before and during the Administration of Gov. Rosas; his course of Policy, the Causes and Character of his interference with the Government of Montevideo, and the Circumstances which led to the Interposition of England and France. By Col. J. ANTHONY KING, an Officer in the Army of the Republic. One volume, 12mo, $1,00.

ahon.—History of England,

Embracing from the Peace of Utrecht to the Peace of Paris, 1763. By LORD MAHON. Edited, with the consent and revision of the author, by Henry Reed, LL. D. of the University of Pa. 2 vols. 8vo, $5,00.

ichelet.—The History of France,

From the Earliest Period. By M. MICHELET, Professor of History in the College of France. Two vols. 8vo, $3,50.

ichelet.—The History of the Roman Republic.

By M. MICHELET. Translated from the French, by Wm. Hazlitt. One vol., 12mo, 1,00. Paper cover, 75 cts.

ichelet—The Life of Martin Luther,

Gathered from his own Writings. By M. MICHELET. Translated by G. H. Smith, F.G.S. 12mo, paper cover, 50 cts.; cloth, 75 cts.

ichelet.—The People.

By M. MICHELET. Translated by G. H. Smith, F.G.S. 12mo, paper cover, 37 cts.; cloth, 62 cts.

apoleon.—Pictorial History

Of Napoleon Bonaparte, translated from the French of M. LAURENT DE L'ARDECHE, with Five Hundred spirited Illustrations, after designs by Horace Vernet, and twenty Original Portraits. Complete in two handsome volumes, 8vo, about 500 pages each, $3,50; or in one vol., $3,00.

'Callaghan.—History of New Netherland;

Or, *New-York under the Dutch.* By E. B. O'CALLAGHAN, Corresponding Member of the New-York Historical Society. Two 8vo. volumes, accompanied with a fac-simile of the Original Map of New Netherland, etc. $5,00.

owell.—Life of Major-General Zachary Taylor,

With an Account of his Early Victories, and Brilliant Achievements in Mexico; including the Siege of Monterey, and Battle of Buena Vista. By C. F. POWELL. 8vo, with Portrait. Paper cover, 25 cts.

owan.—History of the French Revolution;

Its Causes and Consequences. By F. MACLEAN ROWAN. Two volumes 18mo 75 cts.; or two vols. in one, 63 cts.

HISTORY AND BIOGRAPHY.

Stevens.—A History of Georgia,

From its First Discovery by Europeans to the Adoption of the Present Constitution in 1798. By Rev. William Bacon Stevens, M.D. Vol. I. 8vo, $2,50. *** To be completed in two volumes.

Taylor.—A Manual of History.

A Manual of Ancient and Modern History, comprising:—1. Ancient History, containing the Political History, Geographical Position, and Social State of the Principal Nations of Antiquity, carefully digested from the Ancient Writers and illustrated by the discoveries of Modern Scholars and Travellers. 2. Modern History, containing the Rise and Progress of the Principal European Nations, their Political History, and the Changes in their Social Condition, with a History of the Colonies founded by Europeans. By W. Cooke Taylor, LL.D., of Trinity College, Dublin. Revised, with Additions on American History, by C. S. Henry, D.D., Professor of History in the University of New-York. One handsome volume, 8vo, of 800 pages, $2,50.

☞ For convenience as a Class-Book, the Ancient or Modern portion can be had in separate volumes.

Twiss.—The Oregon Territory;

Its History and Discovery, including an account of the Convention of the Escurial; also, the Treaties and Negotiations between the United States and Great Britain—held at various times for the Settlement of a Boundary Line—and an examination of the whole question in respect to Facts and the Law of Nations. By Travers Twiss, D.C.L. 12mo, paper cover, 50 cts.; cloth, 75 cts.

POETRY.

American Poets.—Gems from American Poets.

Contains selections from nearly one hundred writers; among whom are Bryant, Halleck, Longfellow, Percival, Whittier, Sprague, Brainerd, Dana, Willis, Pinckney, Alston, Hillhouse, Mrs. Sigourney, L. M. Davidson, Lucy Hooper, Mrs. Embury, Mrs. Hale, etc., etc. One vol. 32mo, frontispiece, 37½ cts.

Amelia.—Poems.

By Amelia (Mrs. Welby), of Louisville, Ky. Sixth edition. One volume, 12mo, $1,25; gilt leaves, $1,50; morocco, $2,50.

The same on large and fine paper, with illustrations on steel from paintings by Wier. One vol. 8vo, $2,50; gilt leaves, $3,00; morocco, $4,00.

Brownell.—Poems.

By H. H. Brownell. One vol. 12mo, price 75 cents.

Burns.—The Complete Poetical Works

Of Robert Burns, with Explanatory and Glossarial Notes, and a Life of the Author. By James Currie, M.D. Illustrated with six Steel Engravings. 16mo, $1,25; gilt edges, $2,00; morocco, $2,50.

Butler.—Hudibras.

By Samuel Butler. With Notes and a Literary Notice, by the Rev. T. R. Nash, D.D.; illustrated with Portraits, and containing a new and complete Index. 16mo, $1,50; gilt edges, $2,25; morocco, $3,00.

Byron.—Childe Harold's Pilgrimage.

A Romance. By Lord Byron. Illustrated, 16mo, $1,25; gilt edges, $2,00; morocco, $2,50; cheap edition, 18mo, 50 cts.

POETRY

Byron.—The Complete Poetical Works of Lord Byron.

Collected and arranged, with illustrative Notes, by Thos. Moore, Lord Jeffrey, Sir Walter Scott, Prof. Wilson, J. G. Lockhart, Thomas Campbell, etc., etc. Illustrated with a Portrait, and ten elegant Steel Engravings. One vol. 8vo, cloth, $4,00, cloth, extra gilt leaves, $5,00, morocco extra, $6,50; or a cheaper paper edition, with portrait only, $2,50.

Cowper.—The Complete Poetical Works

Of William Cowper, Esq., including the Hymns, and Translations from Mad. Guion, Milton, etc., and Adam, a Sacred Drama, from the Italian of Battista Andreini, with a Memoir of the Author. By the Rev. Henry Stebbings, A.M. One vol. 16mo, 800 pages, $1,50; gilt edges, $2,25; morocco, $3,00.

Campbell.—The Complete Poetical Works

Of Thomas Campbell. Illustrated with a fine Portrait and several handsome Steel Engravings. One vol. 16mo, $1,50; gilt edges, $2,25; morocco, $3,00.

Dante.—The Vision of Hell, Purgatory, and Para-

dise, of Dante Alighieri. Translated by the Rev. Henry Cary, A.M. With a Life of Dante, Chronological View of his Age, Additional Notes and Index. Illustrated with Twelve Steel Engravings, from Designs by John Flaxman, R.A., and a finely engraved Portrait. One elegantly printed volume, 16mo, $1,50; gilt edges, $2,25; morocco, $3,00.

Griswold.—The Sacred Poets of England and Ame-

rica, for Three Centuries. Edited by Rufus W. Griswold. Illustrated with Steel Engravings. One handsome 8vo vol., 550 pages, $2,00; gilt edges, $2,50.

List of Authors.—George Gascoine, Edmund Spenser, Michael Drayton, Sir Henry Wotton, Barnabe Barnes, Sir John Davis, Francis Davison, Joseph Bryan, John Donne, Ben Jonson, Thomas Carew, George Sandys, Sir J. Beaumont, Phineas Fletcher, William Drummond, Giles Fletcher, Henry King, James Shirley, George Wither, Robert Herrick, Frances Quarles, Thomas Heywood, Richard Crashaw, Patrick Carey, W. Habington, Edmund Waller, John Milton, Jeremy Taylor, Sir E. Sherburne, Henry More, Abraham Cowley, Andrew Marvell, Henry Vaughan, Geo. Herbert, Thomas Randolph, Richard Baxter, John Quarles, Sir R. Blackmore, Thomas Flatman, Rev. John Norris, Isaac Watts, Thomas Parnell, Edward Young, Robert Blair, James Thomson, Charles Wesley, James Merrick, Chris. Smart, William Cowper, John Logan, N. Cotton, J. Grahame, Anne Steele, A. M. Toplady, J. Scott, Hannah More, Anna L. Barbauld, T. Dwight, J. Q. Adams, W. Wordsworth, James Montgomery, James Hogg, S. T. Coleridge, Robert Southey, W. Herbert, C. C. Colton, R. Heber, H. Barton, H. K. White, J. Pierpoint, George Croly, A. Norton, R. H. Dana, W. Knox, J. A Hillhouse, H. H. Milman, Bp. Mant, Felicia Hemans, [illegible] Sigourney, C. Wilcox, J. N. Eastburn, W. B. Peabody, H. Knowles, Bp. Doane, J. Keble, Robert Pollok, J. Moultrie, G. W. Bethune, Wm. Croswell, J. G. Whittier, Sir Robert Grant, W. C. Bryant, A. C. Cox, Isaac Williams.

Halleck.—The Complete Poetical Works

Of Fitz-Greene Halleck. Now first collected. Illustrated with elegant Steel Engravings from Paintings by American Artists. One vol. 8vo, $3,50; gilt edges, $4,00; morocco, $6,00.

Hemans.—The Complete Poetical Works

Of Felicia Hemans, printed from the last English edition, edited by her Sister. Illustrated with Six Steel Engravings. Two beautifully printed and portable volumes, 16mo, $2,50; gilt edges, $4,00; morocco, $5,00.

Hemans.—Songs of the Affections,

By Felicia Hemans. One volume, 32mo, gilt, 31 cts.

Lord.—Poems,

By William W. Lord. 12mo, illuminated cover, 75 cts.

POETRY.

Moore.—The Complete Poetical Works

Of THOMAS MOORE, beautifully printed in clear legible type, in exact imitation of the recent corrected London edition. Illustrated with numerous fine Steel Engravings and an elegantly-engraved Portrait of the Author. One volume, 8vo, $4,00 ; gilt edges, $5,00 ; morocco, $7,00. Cheap edition, Portrait and Vignette, only $2,50.

Moore.—Irish Melodies,

By THOMAS MOORE, with the original prefatory Letter on Music. 32mo, 38 cts.

Moore.—Lalla Rookh,

An Oriental Romance. By THOMAS MOORE. 32mo, cloth gilt, 38 cts.

Milton.—The Complete Poetical Works

Of JOHN MILTON, with Explanatory Notes and a Life of the Author, by the Rev. Henry Stebbing, A. M. Illustrated with Six Steel Engravings. One volume, 16mo, $1,25 ; gilt edges, $2,00 ; morocco, $2,50.

Milton.—Paradise Lost,

By JOHN MILTON. With Notes, by Rev. H. Stebbing. One volume, 18mo, cloth, 38 cts. ; gilt leaves, 50 cts.

Pollok.—The Course of Time,

By ROBERT POLLOK. With a Life of the Author, and complete Analytical Index, prepared expressly for this edition. 32mo, frontispiece, 38 cts.

Poetic Lacon (The), or Aphorisms from the Poets.

Edited by BEN. CASSEDAY. Miniature size, 38 cts.

Pope.—The Works

Of ALEXANDER POPE. Complete, with Notes by Dr. Warburton, and illustrations on Steel. 16mo, $1,50 ; gilt edges, $2,25 ; morocco, $3,00.

Southey.—The Complete Poetical Works

Of ROBERT SOUTHEY, Esq., LL. D. The ten volume London edition in one elegant volume, royal 8vo. Illustrated with a Portrait and several fine Steel Engravings, $3,50 ; gilt edges, $4,50 ; morocco, $6,50.

CONTENTS.—Joan of Arc, Juvenile and Minor Poems, Thalaba the Destroyer, Madoc, Ballads and Metrical Tales, The Curse of Kehama, Roderick, the last of the Goths, The Poet's Pilgrimage to Waterloo, Lay of the Laureate, Vision of Judgment, Oliver Newman, &c.

Scott.—Lady of the Lake,

A Poem. By Sir WALTER SCOTT. 18mo, cloth, 38 cts. ; gilt edges, 50 cts.

Scott.—Marmion,

A Tale of Flodden Field. By Sir WALT. SCOTT. 18mo, 38 cts. ; gilt edg., 50 cts.

Scott.—Lay of the Last Minstrel,

A Poem. By Sir WALTER SCOTT. 18mo, 25 cts. ; gilt edges, 38 cts.

Scott.—The Poetical Works

Of Sir WALTER SCOTT, Bart. Containing Lay of the Last Minstrel, Marmion, Lady of the Lake, Don Roderick, Rokeby, Ballads, Lyrics, and Songs, with a Life of the Author. Illustrated with six Steel Engravings. One vol. 16mo, $1,25 ; gilt edges, $2,00 ; morocco $2,50.

POETRY.

Thomson.—The Seasons,

A Poem. By JAMES THOMSON. One volume, 32mo, cloth, gilt leaves, 38 cts.

Tasso.—The Jerusalem Delivered

Of TORQUATO TASSO. Translated into English Spenserian verse, with a Life of the Author, by J. H. Wiffen. Two volumes of the last London edition, reprinted in one elegant 16mo volume, illustrated with a finely engraved Portrait and several beautiful Steel Engravings, $1,50 ; gilt edges, $2,25 ; morocco, $3,00.

Token of Affection, Friendship, Love, Remembrance,

The Heart. 31 cents each volume.

Young.—Night Thoughts.

The Complaint, or Night Thoughts. By EDWARD YOUNG, D. D. Miniature size volume, elegantly printed, 38 cts.

RELIGIOUS.

Arnold.—Rugby School Sermons.

Sermons preached in the Chapel of Rugby School, with an address before Confirmation. By THOMAS ARNOLD, D.D. One volume, 16mo, 50 cts.

Anthon.—An Easy Catechism for Children;

Or, The Church Catechism with Scripture Proofs. By HENRY ANTHON, D.D., Rector of St. Mark's Church, New-York. Part 1, price 6¼ cts.

Anthon.—Catechisms on the Homilies of the Church.

By HENRY ANTHON, D.D. 18mo., paper cover, 6¼ cts.

A Kempis.—Of the Imitation of Christ:

Four Books by THOMAS à KEMPIS. 16mo, $1,00. Reduced to 75 cts.

Burnet.—An Exposition of the XXXIX. Articles

of the Church of England. By GILBERT BURNET, D.D., late Bishop of Salisbury. With an Appendix, containing the Augsburg Confession, Creed of Pope Pius IV, &c. Revised and corrected with copious Notes and Additional References, by the Rev. James R. Page, A.M. One handsome 8vo volume, $2,00

Burnet.—The History of the Reformation of the

Church of England By GILBERT BURNET, D.D., late Lord Bishop of Salisbury; with the Collection of Records and a copious Index, revised and corrected, with Additional Notes and a Preface, by the Rev. E. Nares, D.D. Illustrated with twenty-three engraved Portraits. 4 vols., $8,00. Reduced to $6,00.

A cheap edition is printed, containing the History in three volumes, without the Records, which form the fourth volume of the above. Price, $2,50.

Beaven.—A Help to Catechising.

For the use of Clergymen, Schools, and Private Families. By JAMES BEAVEN D.D., Professor of Theology at King's College, Toronto. Revised and adapted to the use of the Protestant Episcopal Church in the United States. By Henry Anthon, D.D. 18mo, paper, 6¼ cents.

Bradley.—Family and Parish Sermons:

Preached at Clapham and Glasbury. By the Rev. CHARLES BRADLEY. From the seventh London edition. Two volumes in one, 8vo, $1,25.

SCIENCE AND THE ARTS.

I. AGRICULTURE.

Bouissangault.—Agricultural Chemistry.

Rural Economy, in its Relations with Chemistry, Physics, and Meteorolog or, Chemistry applied to Agriculture. By J. B. BOUISSANGAULT. Translate with Notes, etc., by Geo. Law, Agriculturist. 12mo, over 500 pages, $1,5C

Falkner.—The Farmer's Manual:

A Practical Treatise on the Nature and Value of Manures, founded from E periments on various crops; with a brief account of the most Recent Disc veries in Agricultural Chemistry. By F. FALKNER and the Author of "Briti Husbandry." 12mo, cloth, 50 cents.

Marshall.—The Farmer's Hand-Book:

Being a Full and Complete Guide for the Farmer and Emigrant: comprising– The Clearing of Forest and Prairie Lands; Gardening; Farming Generally Farriery; The Management and Treatment of Cattle; Cookery; The Co struction of Dwellings; Prevention and Cure of Disease; with copious Table Recipes, Hints, &c., &c. By JOSIAH T. MARSHALL. 12mo, illustrated, $1,0(

Thomson.—Experimental Researches

On the Food of Animals and the Fattening of Cattle, with Remarks on the Fo of Man: founded on Experiments made by order of the British Governme by ROBERT DUNDAS THOMSON, M. D., Prof. University of Glasgow. 12m cloth, 50 cts. Paper cover, 38 cts.

II. ARCHITECTURE AND MECHANICS.

Arnot.—Gothic Architecture,

Applied to Modern Residences. Containing designs for Entrances, Hal Stairs and Parlors, Window Frames and Door Panelling, the jamb and lal Mouldings on a large scale; the decoration of Chimney Breasts and Mantel Panelling and Graining of Ceilings, with the appropriate furniture. The wh illustrated with Working and perspective Drawings, and forming all the n cessary parts of a modern dwelling. By D. H. ARNOT, Architect. Now publishi in Nos., to be completed in one vol., 4to.

Byrne.—Encyclopædia of Machines, Mechanics, an

Engineering. Comprising Working Drawings, and Description of every in portant Machine in practical Use in the United States, Great Britain, etc., i cluding a complete Treatise on Mechanics, Machinery, and Engine Wor By OLIVER BYRNE, Civil Engineer.

*** This work (now in preparation) will be published in Nos., at 25 cents each. It will co prise nearly 2000 pages, and 1500 Engravings.

Bourne.—A Catechism of the Steam Engine.

Illustrative of the Scientific principles upon which its operation depends, a the practical details of its structure, in its application to Mines, Mills, Stea Navigation and Railways. With various suggestions of Improvement. E JOHN BOURNE, C. E., 16mo, 75 cts.

Hodge.—The Steam Engine:

Its Origin and gradual improvement, from the time of Hero to the present da as adapted to Manufactures, Locomotion and Navigation. Illustrated with Plates in full detail, numerous wood-cuts, &c. By PAUL R. HODGE, C. One volume folio of plates, and letterpress in 8vo, $10.

SCIENCE AND THE ARTS.

Lafever.—Beauties of Modern Architecture;

Consisting of forty-eight Plates of Original Designs, with Plans, Elevations, and Sections—also a Dictionary of Technical Terms; the whole forming a complete Manual for the Practical Builder. By M. LAFEVER, Architect. Large 8vo, $5.

III. MEDICINE AND CHEMISTRY.

De Leuze.—Practical Instruction in Animal Magnetism.

By J. P. F. DE LEUZE. Translated by Thomas C. Hartshorn. Revised edition, with an Appendix of Notes by the Translator, and Letters from eminent Physicians and others, descriptive of cases in the U. S. 1 vol. 12mo, $1.

Fresenius.—Chemical Analysis.

Elementary Instruction in Chemical Analysis. By Dr. C. RHEMIGIUS FRESENIUS. With a Preface by Prof. Liebig. Edited by I. Lloyd Bullock. 12mo, cloth, $1.

Hall.—The Principles of Diagnosis.

By MARSHALL HALL, M. D. F. R. S., &c. Second edition, with many improvements, by John A. Sweet. One volume, 8vo, $2.

Liebig.—Familiar Letters on Chemistry,

And its relation to Commerce; Physiology, and Agriculture. By JUSTUS LIEBIG, M. D. Edited by John Gardner, M. D. One vol., 25 cts., bound.

Leger.—Animal Magnetism,

Or Psychodunamy. By THEODORE LEGER, M. D., late Prof. of Anatomy at the Practical School, Paris. One vol. 12mo, $1,25.

Wilson.—On Healthy Skin:

A Popular and Practical Treatise on Healthy Skin; with Rules for the Medical and Domestic Treatment of Cutaneous Diseases. By ERASMUS WILSON, F. R. S. 12mo, illustrated, $1.

IV. USEFUL ARTS.

Cooley.—The Book of Useful Knowledge:

A Cyclopædia of Six Thousand Practical Receipts, and Collateral Information in the Arts, Manufactures and Trades; including Medicine, Pharmacy, and Domestic Economy: designed as a compendious Book of Reference for the Manufacturer, Tradesman, Amateur, and Heads of Families. By ARNOLD JAMES COOLEY, Practical Chemist. Illustrated with numerous Wood Engravings. One volume, 8vo, of 650 pages. Price $2,00, bound.

Parnell.—Applied Chemistry,

In Manufactures, Arts, and Domestic Economy. Edited by E. A. PARNELL. Illustrated with numerous Wood Engravings, and Specimens of Dyed and Printed Cottons. Paper cover, 75 cts.; cloth, $1.

Ure.—Dictionary of Arts, Manufactures, and Mines;

Containing a clear Exposition of their Principles and Practice. By ANDREW URE, M. D., F. R. S., &c. Illustrated with 1450 Engravings on wood. One thick volume, with Supplement complete, $5.

Ure.—A Supplement to Dr. Ure's Dictionary.

8vo, 200 cuts, $1,50.

V. MILITARY.

Halleck.—Elements of Military Science and Art;

Or, a Course of Instruction in Strategy, Fortification, Tactics of Battles, &c.; embracing the Duties of Staff, Infantry, Cavalry, Artillery and Engineers; adapted to the Use of Volunteers and Militia. By H. WAGER HALLECK, A. M., Lieut. of Engineers of U. S. Army. One vol. 12mo, illustrated, $1,50.

MISCELLANEOUS.

Acton; or, The Circle of Life.

A Collection of Thoughts and Observations, designed to delineate Life, Man, and the World. 12mo, $1,25; gilt edges, $1,50; extra gilt, $1,75.

Ansted.—The Gold Seeker's Manual.

Being a practical and Instructive Guide to all Persons emigrating to the newly discovered Gold Regions of California. By Prof. D. Ansted. 12mo, 25 cts.

Arthur.—Tired of Housekeeping.

By T. S. Arthur, author of "Insubordination," etc., etc. 18mo, frontisp., 38 cts. Forming one of the series of "Tales for the People and their Children."

Adler.—A Dictionary of the German and English

Languages. Compiled from Hilpert, Flugel, Greib, &c. By G. J. Adler, Prof. German in the University of New-York. 1 large vol., 8vo, 1400 pp. $5.

Agnel.—Chess for Winter Evenings;

Or, Useful and Entertaining Lessons on the Game of Chess. Compiled from the best English sources and translations from the French. By H. R. Agnel. Illustrated with fine Steel Plates, from paintings by R. W. Weir. 12mo, $1,75.

Appletons' Steamboat and Railroad Companion:

Being a Traveller's Guide through New England and the Middle States, with Routes in the Southern and Western States and also in Canada; forming likewise a complete Guide to the White Mountains, Catskill Mountains, &c., Niagara Falls, Trenton Falls, &c., Saratoga Springs, and other watering places; and containing full and accurate Descriptions of all the Principal Towns, Villages, the Natural and Artificial Curiosities in the vicinity of the routes; with Distances, Fares, &c. Illustrated with thirty Maps and numerous Engravings. By W. Williams. One very neat volume, $1,25.

Appletons' Southern and Western Guide Book.

Accompanied with numerous Maps and Plans of cities. By W. Williams. One volume 16mo.

Appletons' New-York City Guide.

Accompanied with a Map. 18mo. 38 cts.

Appletons' New City Maps,

Of New-York, Boston, and Philadelphia. Price, 12½ cts. each, in a case.

Appletons' Library Manual:

Containing a Catalogue Raisonne of upwards of 12,000 of the most important Works in every department of Knowledge, and in all Modern Languages. Part I. Subjects alphabetically arranged. II. Biography, Classics, Miscellaneous, and Index to Part I. 1 vol. 8vo, 450 pages, $1,00; half bound, $1,25.

Arnold.—The Miscellaneous Works

Of Thomas Arnold, D.D., with nine additional Essays, not included in the English collection. One vol. 8vo, $2.

Blanchard.—Tourist and Travellers.

By Laman Blanchard. One volume 24mo. Illustrated, 25 cts.

Brooks.—Four Months among the Gold Finders in

California; being the Diary of an Expedition from San Francisco to the Gold Districts. By J. Tyrrwhitt Brooks, M. D., 8vo. paper, 25 cts.

MISCELLANEOUS.

Bond.—Golden Maxims;

Or, a Thought for every Day in the Year. Devotional and Practical. Selected by the Rev. ROBERT BOND. 32mo, 31 cts.

Bryant.—What I saw in California.

Being the Journal of a Tour, by the Emigrant Route and South Pass of the Rocky Mountains, across the Continent of North America, the Great Desert Basin and through California, in the years 1846, 1847. By EDWIN BRYANT, late Alcalde of San Francisco. Sixth edition, with an Appendix containing Accounts of the Gold Mines, various Routes, Outfit, etc., etc., with maps of California and the Gold Region. 12mo, $1,25.

Californian Guide Book;

Comprising Col. Fremont's Geographical Account of Upper California; Major Emory's Overland Journey, and Captain Fremont's Narrative of the Exploring Expedition to the Rocky Mountains and to Oregon and California, accompanied with a Map of the various Routes, and a Map of the Gold Regions. 8vo, 50 cts.

Chapman.—Instruction to Young Marksmen,

In all that relates to the general construction, Practical Manipulation, etc., etc. as exhibited in the Improved American Rifle. By JOHN RATCLIFFE CHAPMAN Civil Engineer. Illustrated with plates. 12mo, $1,25.

Cooley.—The American in Egypt;

With Rambles through Arabia Petræa and the Holy Land, during the years 1839–40. By JAMES EWING COOLEY. Illustrated with numerous Steel Engravings; also Etchings and Designs by Johnson. 8vo, of 610 pages, $2.

Corbould.—The History and Adventures of Margaret Catchpole, a Suffolk Girl.

By the Rev. RICHARD CORBOULD. 8vo, paper cover, 2 Steel plates, 25 cts.

Don Quixote de La Mancha.

Translated from the Spanish of Miguel Cervantes Sauvedra. Embellished with eighteen Steel Engravings. 16mo, $1,50.

Drury.—Friends and Fortune;

A Moral Tale, by ANNE HARRIET DRURY. 12mo, 75 cts.; paper, 50 cts.

Dumas.—Marguerite De Valois;

An Historical Romance. By ALEXANDER DUMAS. 8vo, paper cover, 25 cts.

Edwards.—Voyage up the River Amazon;

Including a Residence at Para. By W. H. EDWARDS. 12mo, cloth, $1,00; paper cover, 75 cts.

Ellis.—Prevention better than Cure;

Or, the Moral Wants of the World we live in. By Mrs. ELLIS. 12mo, cloth, 75 cts.

Ellis.—The Women of England;

Their Social Duties and Domestic Habits. By Mrs. ELLIS. 1 vol. 12mo, 50 cts.

Ellis.—The Mothers of England;

Their Influence and Responsibility. By Mrs. ELLIS. 1 vol. 12mo, 50 cts.

Ellis.—The Minister's Family;

Or, Hints to those who would make Home happy. By Mrs. ELLIS. 18mo, 38 cts

MISCELLANEOUS.

Ellis.—First Impressions;
Or, Hints to those who would make Home happy. By Mrs. Ellis. 18mo, 38 cts.

Ellis.—Somerville Hall;
Or, Hints to those who would make Home happy. By Mrs. Ellis. 16mo, 38 cts.

Ellis.—Family Secrets; or, The Dangers of Dining Out. By Mrs. Ellis. 16mo, 38 cts.

Ellis.—Social Distinctions; or, Hearts and Homes.
By Mrs. Ellis. 8vo.

*** It is written in the spirit of Modern Progressive Philosophy, and seems to be illustrative of great social truths, while aiming a well directed blow against cruel and proscriptive usages of society."—*Nat. Era.*

Ellis.—A Voice from the Vintage; or, The Force of Example. Addressed to those who Think and Feel. By Mrs. Ellis. 8vo, paper, 6 cts.

Embury.—Nature's Gems; or, American Flowers in their Native Haunts. By Emma C. Embury. With twenty Plates of Plants carefully colored after Nature, and Landscape Views of their Localities, from drawings taken on the spot, by E. W. Whitefield. One imperial octavo volume, printed on the finest paper and elegantly bound. Price $6.

Emory and Fremont.—Notes of Travel in California.
Comprising the prominent Geographical, Agricultural, Geological and Mineralogical features of the country; also the Route from Fort Leavenworth, in Missouri to San Diego, in California, from the official Reports of Col. Fremont and Major Emory. 8vo, paper, 25.

Everett.—A System of English Versification.
Containing Rules for the Structure of the different kinds of Verse. Illustrated by numerous examples from the best Poets. By Erastus Everett, A. M. 12mo, 75.

Frost.—Travels in Africa.
The Book of Travels in Africa, from the Earliest Ages to the present time. Compiled from the best authorities, by John Frost, L.L. D. 12mo, illustrated with over 100 plates, $1.

Frost.—The Book of the Indians of North America.
Their Manners, Customs, and Present State. Compiled from the most recent authorities. By John Frost, L.L. D. 12mo, illustrated, $1.

Frost.—The Book of the Army:
Comprising a General Military History of the United States, from the period of the Revolution to the present time, with particular Accounts of all the most celebrated Battles; compiled from the best authorities. By John Frost, L.L. D. Illustrated with numerous Engravings and Portraits. 12mo, $1,25.

Frost.—The Book of the Navy:
Comprising a General History of the American Marine, and particular Accounts of all the most celebrated Naval Battles, from the Declaration of Independence to the present time; compiled from the best authorities. By John Frost, L.L. D. With an Appendix, containing Naval Songs, Anecdotes, &c. Embellished with numerous original Engravings and Portraits. 12mo, $1.

MISCELLANEOUS.

Frost.—The Book of the Colonies:

Comprising a History of the Colonies composing the United States; from the Discovery in the 10th Century, to the Commencement of the Revolutionary War. Compiled from the best authorities. By JOHN FROST, L.L. D. 12mo, illustrated, $1.

Frost.—The Book of Good Examples:

Drawn from Authentic History and Biography. Designed to illustrate the beneficial Effects of Virtuous Conduct. By JOHN FROST, L.L. D. 12mo, illustrated, $1.

Frost.—The Book of Illustrious Mechanics of Europe and America.

Translated from the French of Edward Foncaud. Edited by JOHN FROST, L.L. D. 12mo, illustrated, $1.

Frost.—The Book of Anecdotes;

Or, the Moral of History, taught by Real Examples. By JOHN FROST, L. L. D. With illustrations. 12mo, $1,00.

Foster.—Biog., Lit., and Philo. Essays,

Contributed to the Eclectic Review. By JOHN FOSTER, author of "Essays on Decision of Human Character," etc. One vol. 12mo, $1,25.

Fremont.—Oregon and California Expedition.

Narrative of the Exploring Expedition to the Rocky Mountains in the year 1842, and to Oregon and North California in the years 1843–4. By Brevet Capt. J. C FREMONT. Reprinted from the official report ordered to be published by the United States Senate. One vol. 8vo, paper cover, 25 cts ; or printed on thick paper, bound, 63 cts.

Fullerton.—Ellen Middleton.

A Tale, by Lady GEORGIANA FULLERTON. 12mo, 75 cts.; paper cover, 50 cts.

Fullerton.—Grantly Manor.

A Tale, by Lady GEORGIANA FULLERTON. 12mo, 75 cts. ; paper cover, 50 cts.

Goldsmith.—Pictorial Vicar of Wakefield.

By OLIVER GOLDSMITH. Illustrated with upwards of 100 Engravings on Wood. 1 beautiful vol. 8vo, of 300 pages, $1,25; 12mo, 75 cts.; miniature size, 38 cts.

Grace Leslie; or, The History of a Month.

A Tale. 12mo, 75 cts.; paper cover, 50 cts.

Grant.—Memoirs of an American Lady:

With Sketches of the Scenery and Manners in America, as they existed previous to the Revolution. By Mrs. GRANT, (of Laggaw.) 12mo, paper cover, 50 cts. ; cloth, 75 cts.

Gil Blas. The Adventures of Gil Blas of Santillane.

Translated from the French of Le Sage, by T. SMOLLET. With an account of the Author's Life. Splendid Illustrations, 16mo, $1,50.

Guizot.—Democracy in France.

By Monsieur GUIZOT. 12mo, paper cover, 25 cts.

Jones.—My Uncle Hobson and I:

Or, Slashes at Life with a Free Broad-Axe. By PASCAL JONES. 12mo, paper cover, 50 cts.; cloth, 75 cts.

MISCELLANEOUS.

Kenny.—The Manual of Chess:

Containing the Elementary Principles of the Game. Illustrated with Diagrams, Recent Games, and Original Problems. By CHARLES KENNY. 1 vol. 32mo, cloth.

Kip.—The Christmas Holydays in Rome.

By the Rev. WILLIAM INGRAHAM KIP, M.A. 12mo, $1,00.

Lamb.—Literary Sketches and Letters;

Being the Final Memorials of Charles Lamb, never before published. By THOMAS NOON TALFOURD, one of his executors. 12mo, 75 cents.

Lamartine.—Les Confidences.

Confidential Disclosures. By ALPHONSE DE LAMARTINE. Translated from the French, by Eugène Plunkett. 12mo, paper cover, 25 cents; cloth, 50 cents.

Lamartine.—Les Confidences et Raphael.

Par M. DE LAMARTINE. Two volumes in one. 8vo, $1,00.

Lanman.—Summer in the Wilderness:

Embracing a Canoe Voyage up the Mississippi and around Lake Superior. By CHARLES LANMAN. 12mo, paper cover, 50 cents; cloth, 63 cents.

Letter-Writer.

The Useful Letter-writer, comprising a succinct Treatise on the Epistolary Art, and Forms of Letters for all ordinary Occasions of Life. Compiled from the best authorities. Frontispiece, 32mo, gilt leaves, 38 cents.

Lover.—Handy Andy:

A Tale of Irish Life. By SAMUEL LOVER. Illustrated with twenty-three characteristic Steel Engravings. One volume 8vo, cloth, $1 25, boards, $1; cheap edition, two Plates, paper, 50 cts.

Lover.—L. S. D.; Treasure Trove:

A Tale. BY SAMUEL LOVER. 8vo, two Steel Engravings, paper cover, 25 cts.

Manzoni.—The Betrothed.

I Promissi Sposi; The Betrothed. By ALESSANDRO MANZONI. A new translation. Two volumes, 12mo, paper cover, $1,00; cloth, $1,50.

Maxwell.—Fortunes of Hector O'Halloran

And his man Mark Antony O'Toole. By W. H. MAXWELL. One vol. 8vo, two plates, paper, 50 cents; twenty-four plates, boards, $1,00; cloth, $1,25.

Maxwell.—Hill-side and Border Sketches;

With Legends of the Cheviots and the Summer Muir. By W. H. MAXWELL. 8vo, paper cover, 25 cents.

McIntosh.—Two Lives;

Or, To Seem and To Be. A Tale, by MARIA J. McINTOSH, author of "Praise and Principle," etc. One vol. 12mo, paper cover, 50 cents; cloth, 75 cents.

McIntosh.—Aunt Kitty's Tales.

By MARIA J. McINTOSH. One vol. 12mo, paper cover, 50 cts; cloth, 75 cts.

McIntosh.—Charms and Counter-charms.

By MARIA J. McINTOSH. 12mo, paper, 75 cts.; cloth, $1,00.

MISCELLANEOUS.

Miles.—The Horse's Foot,

And how to keep it sound; with illustrations. By WILLIAM MILES. From the third English edition. 12mo, paper cover, 25 cents.

Reid.—A Dictionary of the English Language.

Containing the Pronunciation, Etymology, and Explanation of all words authorized by eminent writers: to which are added a Vocabulary of the Roots of English Words, and an Accented List of Greek, Latin, and Scripture Proper Names. By ALEXANDER REID, A.M., Rector of the Circus School, Edinburgh. With a Critical Preface by Henry Reed, Prof. of Eng. Lit. in the University of Pennsylvania. One vol. 12mo, of near 600 pages, well bound in leather, $1,00.

Republic, The, of the United States of America:

Its Duties to Itself, and its responsible relations to other Countries. Embracing also a Review of the late war between the United States and Mexico; its causes and results; and of those measures of Government which have characterized the democracy of the Union. 12mo, paper, 75 cents; cloth, $1,00.

Richardson.—Dogs, their Origin and Varieties:

Directions as to their General Management; with numerous Original Anecdotes. Also, Simple Instructions as to their Treatment under Disease. By H. D. RICHARDSON. Illustrated. 12mo, paper cover, 25 cents.

Roget.—Economic Chess-Board Companion;

By which the Game of Chess may be played in Railroad Cars and in Steamboats without any inconvenience. Invented by P. M Roget, M.D. 50 cents.

Rough and Ready Annual;

Or Military Souvenir. Illustrated with twenty Portraits and Plates. 12mo, morocco, $1,00.

Pure Gold from the Rivers of Wisdom:

A Collection of Short Extracts from the most Eminent Writers—Bishop Hall, Jeremy Taylor, Barrow, Hooker, Bacon, Leighton, Addison, Wilberforce, Johnson, Young, Southey, Lady Montague, Hannah More, etc. 32mo, 31 cents.

Sawyer.—A Plea for Amusements.

By FREDERICK W. SAWYER. 12mo, 50 cents.

Sewell.—Amy Herbert:

A Tale. By Miss SEWELL. Edited by the Rev. W. Sewell, B.D. One vol. 12mo, paper cover, 50 cents; cloth, 75 cents.

Sewell.—Laneton Parsonage:

A Tale. By Miss SEWELL. Edited by the Rev. W. Sewell, B.D. Three vols. 12mo, paper cover, $1,50; cloth, $2,25.

Sewell.—Margaret Percival:

A Tale. By Miss SEWELL. Edited by the Rev. W. Sewell, B.D. Two vols. 12mo, paper cover, $1,00; cloth, $1,50.

Sewell.—Gertrude:

A Tale. By Miss SEWELL. Edited by the Rev. W. Sewell, B.A. 12mo, cloth, 75 cents; paper cover, 50 cents.

Sewell.—Walter Lorrimer:

And other Tales. By Miss SEWELL. Illustrated with six colored plates 12mo, 75 cents.

MISCELLANEOUS.

Smith.—The Natural History of the Gent.

By ALBERT SMITH. Illustrated with numerous Wood Engravings. 18mo, 25 cts.

Smith.—The Natural History of the Ballet Girl.

By ALBERT SMITH. Illustrated with numerous Wood Cuts. 18mo, 25 cts.

Southgate.—Visit to the Syrian Church.

Narrative of a Visit to the Syrian [Jacobite] Church of Mesopotamia; with Statements and Reflections upon the Present State of Christianity in Turkey, and the Character and Prospects of the Eastern Churches. By the Rt. Rev. HORATIO SOUTHGATE, D.D. One vol. 12mo, with a Map, $1,00.

Southgate.—Narrative of a Tour through Armenia,

Kurdistan, Persia, and Mesopotamia: with an Introduction, and Occasional Observations upon the Condition of Mohammedanism and Christianity in those Countries. By the Rev. HORATIO SOUTHGATE. Two vols. 12mo, $1,50.

Southey.—Life of Oliver Cromwell.

By ROBERT SOUTHEY. One vol. 18mo, 38 cents.

Something for Every Body:

Gleaned in the Old Purchase, from fields often reaped. By ROBERT CARLTON, Esq. 12mo, paper cover, 50 cents; cloth, 75 cents.

Sprague.—History of the Florida War.

The Origin, Progress, and Conclusion of the Florida War: to which is appended a record of Officers, Non-commissioned Officers, Musicians, and Privates of the U. S. Army, Navy, and Marine Corps, who were killed in battle, and others who died from disease; as also the Names of Officers who were distinguished by Brevets, and the Names of others recommended; together with the Orders for Collecting the Remains of the Dead in Florida, and the Ceremony of Interment at St. Augustine, on the 14th Aug., 1842. By JNO. T. SPRAGUE, Brevet Capt. 8th Reg't U. S. Infantry. 1 vol. 8vo, $2,50.

Stewart.—On the Management of Horses.

Stable Economy: a Treatise on the Management of Horses, in relation to Stabling, Grooming, Feeding, Watering, and Working. By JOHN STEWART, Veterinary Surgeon. With Notes and Additions, adapting it to American Food and Climate, by A. B. Allen. 12mo, illustrated with 23 Engravings, $1,00.

Surenne.—Pronouncing French Dictionary.

The Standard Pronouncing Dictionary of the French and English Languages. Part I. French and English. Part II. English and French. By GABRIEL SURENNE, F.A.S.E. One vol. 12mo, near 900 pages, strongly bound, $1,50.

Taylor.—Anecdote Book.

Anecdotes of Zachary Taylor and the Mexican War. By TOM OWEN, the Bee-Hunter. Illustrated with Engravings. 8vo, 25 cents.

Tuckerman.—Artist Life;

Or Sketches of American Artists. By H. T. TUCKERMAN. One vol. 12mo, 75 cts.

Warner.—Rudimental Lessons in Music:

Containing the Primary Instruction requisite for all beginners in the Art, whether Vocal or Instrumental. By JAMES F. WARNER. 18mo, 50 cents.

Warner.—The Primary Note Reader;

Or First Steps in Singing at Sight. By JAMES F. WARNER. 12mo, 25 cts.

MISCELLANEOUS.

Wayland.—Real Life in England.

By Mrs. Wayland; with a Preface by President Wayland. 1 vol. 16mo, 38 cts

Whipple.—Essays and Reviews.

By Edwin P. Whipple. Two vols. 12mo, $2,25.

Heads of Contents.—Vol. I. Macaulay; Poets and Poetry of America; Talfourd; Words; James' Novels; Sidney Smith; Daniel Webster; Neal's History of the Puritans; Wordsworth; Byron; English Poets of the Nineteenth Century; Vagaries of Volition.—Vol. II. Old English Dramatists; South's Sermons; Romance of Rascality; The Croakers of Society and Literature; British Critics; Rufus Choate; Coleridge as a Philosophical Critic; Prescott's Histories; Prescott's Conquest of Peru; Shakspeare's Critics; Richard B. Sheridan; Appendix; Thomas Hood; Leigh Hunt; Thomas Carlyle; Novels of the Season.

Woman's Worth:

Or, Hints to Raise the Female Character. First American from the last English edition. With a Recommendatory Notice, by Emily Marshall. 18mo, 38 cts.

Zschokke.—Incidents of Social Life Amid the European Alps.

Translated from the German of J. H. Zschokke. 12mo, $1.

LAW BOOKS.

Anthon.—Contributions to Legal Science.

By John Anthon. 1 vol. 8vo. (In press.)

Holcombe.—A Selection of Leading Cases upon Commercial Law,

decided by the Supreme Court of the United States; with Notes and Illustrations. By James P. Holcombe. 1 vol. law sheep, $4.

Holcombe.—A Digest of the Decisions of the Supreme Court of the U. S.,

from its organization to the present time, by James P. Holcombe. A new edition, with an Appendix of Latest Cases. 8vo, $6.

Holcombe.—The Law of Debtor and Creditor,

In the United States and Canada. By James P. Holcombe. 8vo, $4.

*** These volumes are highly commended by Chief Justices Taney and Woodbury, Daniel Webster, Rufus Choate, Chancellor Kent, &c.

Smith.—A Compendium of Mercantile Law.

By the late John W. Smith. Greatly enlarged from the third and last English edition, by James P. Holcombe and Wm. T. Gholson. 1 vol. 8vo, law sh'p, $4.

Warren.—Law Studies:

A Popular and Practical Introduction to Law Studies, and to every Department of the Legal Profession—Civil, Criminal, and Ecclesiastical; with an Account of the State of Law in Ireland and Scotland, and occasional Illustrations from American Law. By Samuel Warren, F. R. S., author of "Ten Thousand a Year," etc. Revised, with an American Introduction and Appendix, by Thos. W. Clerke, Counsellor at Law. One vol. 8vo, 675 pages, $3,50.

JUVENILE.

Aunt Fanny's Story Book.

Illustrated with many Plates. 16mo, 50 cts.

Aunt Kitty's Tales.

Containing Florence Arnot, Blind Alice, and other Tales. By Maria J. McIntosh. 12mo, 75 cts.

www.ingramcontent.com/pod-product-compliance
Lightning Source LLC
LaVergne TN
LVHW021307110826
845150LV00003B/511

* 9 7 8 1 4 2 5 5 5 8 7 4 1 *